Methods in Enzymology

Volume 300

OXIDANTS AND ANTIOXIDANTS

Part B

METHODS IN ENZYMOLOGY

Methods in Enzymology

Volume 300

Oxidants and Antioxidants Part B

EDITED BY

Lester Packer

UNIVERSITY OF CALIFORNIA
BERKELEY, CALIFORNIA

ACADEMIC PRESS
San Diego New York Boston London Sydney Tokyo Toronto

This book is printed on acid-free paper. 

Academic Press
a division of Harcourt Brace & Company
525 B Street, Suite 1900, San Diego, California 92101-4495, USA
http://www.academicpress.com

Academic Press Limited
24-28 Oval Road, London NW1 7DX, UK
http://www.hbuk.co.uk/ap/

International Standard Book Number: 0-12-182201-X

PRINTED IN THE UNITED STATES OF AMERICA
98 99 00 01 02 03 MM 9 8 7 6 5 4 3 2 1

Table of Contents

Section I. Oxidative Damage to Lipids, Proteins, and Nucleic Acids

A. Lipids

B. Proteins and Nucleic Acids

Section II. Assays in Cells, Body Fluids, and Tissues

Section III. Oxidant and Redox-Sensitive Steps in Signal Transduction and Gene Expression

Section IV. Noninvasive Methods

Contributors to Volume 300

Article numbers are in parentheses following the names of contributors.
Affiliations listed are current.

IAN N. ACWORTH (31), *Department of Pharmacology, Massachusetts College of Pharmacy, Boston, Massachusetts 02115*

BHARAT B. AGGARWAL (34), *Cytokine Research Section, University of Texas M. D. Anderson Cancer Center, Houston, Texas 77030*

BRUCE N. AMES (10, 18), *Department of Molecular and Cell Biology, University of California, Berkeley, Berkeley, California 94720*

ELIAS S. J. ARNÉR (24), *Medical Nobel Institute for Biochemistry, Department of Medical Biochemistry and Biophysics, Karolinska Institute, S-17177 Stockholm, Sweden*

M. FLINT BEAL (31), *Department of Neurology, Massachusetts General Hospital, Boston, Massachusetts 02114*

KENNETH B. BECKMAN (18), *Department of Molecular and Cell Biology, University of California, Berkeley, Berkeley, California 94720*

ELIE BEIT-YANNAI (30), *Department of Pharmacology, School of Pharmacy, The Hebrew University of Jerusalem, Jerusalem, Israel 91120*

BARBARA S. BERLETT (25), *Laboratory of Biochemistry, National Heart, Lung, and Blood Institute, National Institutes of Health, Bethesda, Maryland 20892-0320*

ELLIOT M. BERRY (30), *Department of Human Nutrition and Metabolism, The Hebrew University of Jerusalem, Jerusalem, Israel 91120*

NEIL V. BLOUGH (22), *Department of Chemistry and Biochemistry, University of Maryland, College Park, Maryland 20742*

MIKHAIL B. BOGDANOV (31), *Department of Neurology, Massachusetts General Hospital, Boston, Massachusetts 02114*

STEPHEN BRAND (35), *ProScript, Inc., Cambridge, Massachusetts 02139*

ANA MARIA OLIVEIRA BRETT (32), *Departamento de Química, Universidade De Coimbra, Coimbra 3049, Portugal*

I. HENDRIKJE BUSS (13), *Department of Pathology, Christchurch School of Medicine, Christchurch, New Zealand*

STEPHEN E. BUXSER (28), *Discovery Technologies, Pharmacia and Upjohn, Kalamazoo, Michigan 49001*

HO ZOON CHAE (23), *Department of Biology, College of Sciences, Chonnam National University, Kwangju, Korea*

DANIEL F. CHURCH (9), *Department of Chemistry, Louisiana State University, Baton Rouge, Louisiana 70803*

ELAINE M. CONNER (35), *Department of Molecular and Cellular Physiology, Louisiana State University Medical Center, Shreveport, Louisiana 71130*

JAN R. CROWLEY (12, 16), *Department of Medicine, Washington University School of Medicine, St. Louis, Missouri 63110*

MARIA L. CRUZ (32), *Departamento de Farmácia, Faculdade de Ciências Médicas, Pontificia Universidade Católica, Campinas 13020-904, Brasil*

T. DE LEO (26), *Dipartimento di Fisiologia Generale ed Ambientale, Università di Napoli "Federico II," I-80134 Napoli, Italy*

ANTONIO DESTREE (35), *ProScript, Inc., Cambridge, Massachusetts 02139*

RUI M. B. DIAS (21), *Departamento de Química, Instituto Tecnológico e Nuclear, P-2685 Sacavém, Portugal*

S. DI MEO (26), *Dipartimento di Fisiologia Generale ed Ambientale, Università di Napoli "Federico II," I-80134 Napoli, Italy*

TAMMY R. DUGAS (9), *Department of Chemistry, Louisiana State University, Baton Rouge, Louisiana 70803*

PETER ELLIOTT (35), *ProScript, Inc., Cambridge, Massachusetts 02139*

PATRICIA EVANS (17), *Neurodegenerative Disease Research Centre, King's College, University of London, London SW3 6LX, United Kingdom*

DAVID FANBERSTEIN (43), *Department of Pharmaceutics, School of Pharmacy, The Hebrew University of Jerusalem, Jerusalem, Israel 91120*

JOHN L. FARBER (11), *Department of Pathology, Thomas Jefferson University, Philadelphia, Pennsylvania 19107*

SHARON FARFOURI (43), *Department of Pharmaceutics, School of Pharmacy, The Hebrew University of Jerusalem, Jerusalem, Israel 91120*

BALZ FREI (3, 7), *Linus Pauling Institute, Oregon State University, Corvallis, Oregon 97331*

JOSEPH P. GAUT (12), *Department of Medicine, Washington University School of Medicine, St. Louis, Missouri 63110*

PETER G. GEIGER (4), *Department of Biochemistry, Medical College of Wisconsin, Milwaukee, Wisconsin 53226*

ALBERT W. GIROTTI (4), *Department of Biochemistry, Medical College of Wisconsin, Milwaukee, Wisconsin 53226*

PETER A. GLASCOTT, JR. (11), *Rhône-Poulenc Rorer Research and Development, Drug Safety, Collegeville, Pennsylvania 19426-0994*

KISHORCHANDRA GOHIL (40), *Membrane Bioenergetics Group, Department of Molecular and Cell Biology, University of California, Berkeley, Berkeley, California 94720-3200*

MATTHEW B. GRISHAM (35), *Department of Molecular and Cellular Physiology, Louisiana State University Medical Center, Shreveport, Louisiana 71130-3932*

PETER L. GUTIERREZ (22), *Greenbaum Cancer Center, University of Maryland School of Medicine, Baltimore, Maryland 21201*

IVANO G. R. GUTZ (32), *Instituto de Quimica, Universidade De Sao Paulo, Sao Paulo 05599-970, Brasil*

BARRY HALLIWELL (17), *Pharmacology Group, King's College, University of London, London SW3 6LX, United Kingdom*

GARRY J. HANDELMAN (6), *Antioxidant Research Laboratory, U.S. Department of Agriculture–Human Nutrition Research Center on Aging, Tufts University, Boston, Massachusetts 02111*

DAVID R. HATHAWAY (45), *Bristol-Myers Squibb Company, Princeton, New Jersey 08543-4000*

ALFRED HAUSLADEN (38), *Department of Medicine, Duke University Medical Center, Durham, North Carolina 27710*

STANLEY L. HAZEN (12, 16), *Department of Cell Biology, Cleveland Clinic Foundation, Cleveland, Ohio 44195*

JAY W. HEINECKE (12, 16), *Departments of Medicine, Molecular Biology and Pharmacology, Washington University School of Medicine, St. Louis, Missouri 63110*

HAROLD J. HELBOCK (10, 18), *Department of Molecular and Cell Biology, University of California, Berkeley, Berkeley, California 94720*

MIDORI HIRAMATSU (27), *Institute for Life Support Technology, Yamagata Technopolis Foundation, Yamagata 990-2473, Japan*

ARNE HOLMGREN (24), *Medical Nobel Institute for Biochemistry, Department of Medical Biochemistry and Biophysics, Karolinska Institute, S-17177 Stockholm, Sweden*

UDO HOPPE (42), *Paul Gerson Unna–Skin Research Center, D-20245 Hamburg, Germany*

FONG F. HSU (12, 16), *Department of Medicine, Washington University School of Medicine, St. Louis, Missouri 63110*

YVONNE M. W. JANSSEN (36), *Department of Pathology, University of Vermont, Burlington, Vermont 05405*

SANG WON KANG (23), *Laboratory of Cell Signaling, National Heart, Lung, and Blood Institute, National Institutes of Health, Bethesda, Maryland 20892*

JOHN F. KEANEY, JR. (3, 7), *Evans Memorial Department of Medicine and Whitaker Cardiovascular Institute, Boston University School of Medicine, Boston, Massachusetts 02118*

ANTHONY J. KETTLE (14), *Free Radical Research Group, Department of Pathology, Christchurch School of Medicine, Christchurch, New Zealand*

RON KOHEN (30, 43), *Department of Pharmaceutics, School of Pharmacy, The Hebrew University of Jerusalem, Jerusalem, Israel 91120*

MAKIKO KOMATSU (27), *Institute for Life Support Technology, Yamagata Technopolis Foundation, Yamagata 990-2473, Japan*

WITOLD KORYTOWSKI (4), *Department of Biochemistry, Medical College of Wisconsin, Milwaukee, Wisconsin 53226*

ASHOK KUMAR (34), *Cytokine Research Section, University of Texas, M. D. Anderson Cancer Center, Houston, Texas 77030*

EGIL KVAM (33), *Department of Pharmacy and Pharmacology, University of Bath, Bath, Ba2 7AY, United Kingdom*

MAURO A. LA-SCALEA (32), *Instituto de Quimica, Universidade De Sao Paulo, Sao Paulo 05599-970, Brasil*

CHRISTIAAN LEEUWENBURGH (16), *Department of Medicine, Washington University School of Medicine, St. Louis, Missouri 63110*

HANS-ANTON LEHR (46), *Institute of Pathology, Johannes Gutenberg University, D-55101 Mainz, Germany*

RODNEY L. LEVINE (15), *Laboratory of Biochemistry, National Heart, Lung, and Blood Institute, National Institutes of Health, Bethesda, Maryland 20892*

GUDRUN LEWIN (44), *Research Institute for Antioxidant Therapy, 10559 Berlin, Germany*

SANDRA A. LEWISCH (15), *AtheroGenics, Inc., Norcross, Georgia 20071*

BEIBEI LI (22), *Department of Chemistry and Biochemistry, University of Maryland, College Park, Maryland 20742*

JAMES K. LIAO (37), *Cardiovascular Division, Department of Medicine, Brigham and Women's Hospital, Boston, Massachusetts 02115*

JIANKANG LIU (10), *Department of Molecular and Cell Biology, University of California, Berkeley, Berkeley, California 94720*

STEFFEN LOFT (19), *Institute of Public Health, University of Copenhagen, DK-2200 Copenhagen, Denmark*

LEONIDAS LYRAS (17), *Neurodegenerative Disease Research Centre, King's College, University of London, London SW3 6LX, United Kingdom*

MATILDE MAIORINO (5), *Department of Biological Chemistry, University of Padova, I-35121 Padova, Italy*

DOUGLAS R. MCCABE (31), *ESA, Inc., Chelmsford, Massachusetts 01824*

WEI PING MEI (42), *Paul Gerson Unna–Skin Research Center, D-20245 Hamburg, Germany*

MICHAEL D. MENGER (46), *Institute for Clinical and Experimental Surgery, University of Saarland, D-66421 Homburg/Saar, Germany*

EMILE R. MOHLER III (45), *University of Pennsylvania School of Medicine, Philadelphia, Pennsylvania 19104*

JASON D. MORROW (1), *Departments of Medicine and Pharmacology, Vanderbilt University School of Medicine, Nashville, Tennessee 37232*

JACKOB MOSKOVITZ (25), *Laboratory of Biochemistry, National Heart, Lung, and Blood Institute, National Institutes of Health, Bethesda, Maryland 20892-0320*

HIRO-OMI MOWRI (3), *Evans Memorial Department of Medicine and Whitaker Cardiovascular Institute, Boston University School of Medicine, Boston, Massachusetts 02118*

DIANNE M. MUELLER (16), *Department of Medicine, Washington University School of Medicine, St. Louis, Missouri 63110*

JAFFAR NOUROOZ-ZADEH (2, 8), *Division of Clinical Pharmacology and Toxicology, Department of Medicine, University College London, London W1N 8AA, England*

LESTER PACKER (39–41), *Membrane Bioenergetics Group, Department of Molecular and Cell Biology, University of California, Berkeley, Berkeley, California 94720-3200*

VITO J. PALOMBELLA (35), *ProScript, Inc., Cambridge, Massachusetts 02139*

ANTONIO M. PASTORINO (5), *Glaxo-Wellcome S.p.A. Medicine's Research Centre, Verona, Italy*

CHRISTINE PIEN (35), *ProScript, Inc., Cambridge, Massachusetts 02139*

STYLIANOS M. PIPERAKIS (20), *DNA Repair Laboratory, Institute of Biology, NCSR "Demokritos," Athens, Greece*

IGOR POPOV (44), *Research Institute for Antioxidant Therapy, 10559 Berlin, Germany*

J. MICHAEL POSTON (25), *Laboratory of Biochemistry, National Heart, Lung, and Blood Institute, National Institutes of Health, Bethesda, Maryland 20892-0320*

HENRIK ENGHUSEN POULSEN (19), *Department of Clinical Pharmacology, Rigshospitalet, University of Copenhagen, DK-2200 Copenhagen, Denmark*

JANE E. RASMUSSEN (16), *Department of Medicine, Washington University School of Medicine, St. Louis, Missouri 63110*

THOMAS J. RAUB (28), *Pharmaceutical Development I, Pharmacia and Upjohn, Kalamazoo, Michigan 49001*

SUE GOO RHEE (23), *Laboratory of Cell Signaling, National Heart, Lung, and Blood Institute, National Institutes of Health, Bethesda, Maryland 20892*

L. JACKSON ROBERTS II (1), *Departments of Pharmacology and Medicine, Vanderbilt University School of Medicine, Nashville, Tennessee 37232*

SASHWATI ROY (39, 40), *Membrane Bioenergetics Group, Department of Molecular and Cell Biology, University of California, Berkeley, Berkeley, California 94720-3200*

STEFAN RYTER (33), *Department of Internal Medicine, Southern Illinois School of Medicine, Springfield, Illinois 62702*

JASON M. SAMII (7), *Evans Memorial Department of Medicine and Whitaker Cardiovascular Institute, Boston University School of Medicine, Boston, Massachusetts 02118*

GERHARD SAUERMANN (42), *Paul Gerson Unna–Skin Research Center, D-20245 Hamburg, Germany*

GERI SAWADA (28), *Pharmaceutical Development I, Pharmacia and Upjohn, Kalamazoo, Michigan 49001*

CHANDAN K. SEN (36, 39, 40), *Membrane Bioenergetics Group, Department of Molecular and Cell Biology, University of California, Berkeley, Berkeley, California 94720-3200, and Department of Physiology, University of Kuopio, Finland*

SILVIA H. P. SERRANO (32), *Instituto de Quimica, Universidade De Sao Paulo, Sao Paulo 05599-970, Brasil*

GLENN T. SHWAERY (3, 7), *Evans Memorial Department of Medicine and Whitaker Cardiovascular Institute, Boston University School of Medicine, Boston, Massachusetts 02118*

MARTIN SPIECKER (37), *Cardiovascular Division, Department of Medicine, Brigham and Women's Hospital, Boston, Massachusetts 02115*

FRANZ STÄB (42), *Paul Gerson Unna–Skin Research Center, D-20245 Hamburg, Germany*

EARL R. STADTMAN (25), *Laboratory of Biochemistry, National Heart, Lung, and Blood Institute, National Institutes of Health, Bethesda, Maryland 20892-0320*

JONATHAN S. STAMLER (38), *Howard Hughes Medical Institute, Duke University Medical Center, Durham, North Carolina 27710*

ASPASIA M. TASSIOU (20), *Institute of Biology, NCSR "Demokritos," Athens, Greece*

JOÃO P. TELO (21), *Secção de Química Orgânica, Instituto Superior Técnico, P-1096 Lisboa codex, Portugal*

JENS J. THIELE (41), *Institut für Physiologische Chemie I, Heinrich-Heine-Universität, D-40225 Düsseldorf, Germany*

OREN TIROSH (30, 43), *Department of Pharmaceutics, School of Pharmacy, The Hebrew University of Jerusalem, Jerusalem, Israel 91120*

JOHN TURK (16), *Department of Medicine, Washington University School of Medicine, St. Louis, Missouri 63110*

REX M. TYRRELL (33), *Department of Pharmacy and Pharmacology, University of Bath, Bath, BA2 7AY, United Kingdom*

FULVIO URSINI (5), *Department of Biological Chemistry, University of Padova, I-35121 Padova, Italy*

PETER VAJKOCZY (46), *Neurosurgical Department, Klinikum Mannheim, University of Heidelberg, D-68167 Mannheim, Germany*

P. VENDITTI (26), *Dipartimento di Fisiologia Generale ed Ambientale, Università di Napoli "Federico II," I-80134 Napoli, Italy*

ABEL J. S. C. VIEIRA (21), *Secção de Química Orgânica, Instituto Superior Técnico, P-1096 Lisboa codex, Portugal*

EVANGELOS-E. VISVARDIS (20), *Institute of Biology, NCSR "Demokritos," Athens, Greece*

BRIGITTE VOLLMAR (46), *Institute for Clinical and Experimental Surgery, University of Saarland, D-66421 Homburg/Saar, Germany*

CHRISTINE C. WINTERBOURN (13), *Department of Pathology, Christchurch School of Medicine, Christchurch, New Zealand*

HENRY WONG (35), *ProScript, Inc., Cambridge, Massachusetts 02139*

HELEN C. YEO (10), *Roche Bioscience, Palo Alto, California 94304-1397*

HIDEKATSU YOKOYAMA (29), *Institute for Life Support Technology, Yamagata Technopolis Foundation, Yamagata 990-2473, Japan*

ADRIANA ZAMBURLINI (5), *Department of Biological Chemistry, University of Padova, I-35121 Padova, Italy*

ABRAHAM ZELKOWICZ (43), *Department of Pharmaceutics, School of Pharmacy, The Hebrew University of Jerusalem, Jerusalem, Israel 91120*

LUCIO ZENNARO (5), *Department of Biological Chemistry, University of Padova, I-35121 Padova, Italy*

LIANGWEI ZHONG (24), *Medical Nobel Institute for Biochemistry, Department of Medical Biochemistry and Biophysics, Karolinska Institute, S-17177 Stockholm, Sweden*

Preface

The importance of reactive oxygen and nitrogen species (ROS and RNS) and antioxidants in health and disease has now been recognized in all of the biological sciences and has assumed special importance in the biomedical sciences. Overwhelming evidence indicates that ROS play a role in most major health problems, that antioxidants play a critical role in wellness and health maintenance, and that by inhibiting oxidative damage to molecules, cells, and tissues prevent chronic and degenerative diseases.

We now know that ROS are essential for many enzyme-catalyzed reactions. Low levels of reactive oxygen and reactive nitrogen species are signaling molecules. At high concentration, these ROS are essential in the antitumor, antimicrobial, antiparasitic action, etc., of neutrophils and macrophages and contribute to oxidative damage to molecules, cells, and tissues.

In this volume we have updated the assays and methods applicable to assaying oxidative damage to biomolecules, particularly macromolecules and assays that are useful for evaluating activities in cells and tissues. New methods are swiftly evolving for assaying the oxidant- and antioxidant-redox-sensitive steps in cell signaling and gene expression. And, finally, the development of new technologies is allowing for the noninvasive evaluation of oxidant and antioxidant action. These developments and methodologies can be expected in the future to be applicable to clinical situations in which the early detection of disease remains a problem of paramount importance.

In bringing this volume to fruition, credit must be given to experts in various specialized fields of oxidant and antioxidant research. Our appreciation is to the contributors who, with those who helped select them, have produced this state-of-the-art volume on oxidant and antioxidant methodology. The topics included were chosen on the excellent advice of Bruce N. Ames, Enrique Cadenas, Balz Frei, Matthew Grisham, Barry Halliwell, William Pryor, Catherine Rice-Evans, and Helmut Sies. To these colleagues, I extend my sincere thanks and most grateful appreciation.

Lester Packer

METHODS IN ENZYMOLOGY

VOLUME I. Preparation and Assay of Enzymes
Edited by SIDNEY P. COLOWICK AND NATHAN O. KAPLAN

VOLUME II. Preparation and Assay of Enzymes
Edited by SIDNEY P. COLOWICK AND NATHAN O. KAPLAN

VOLUME III. Preparation and Assay of Substrates
Edited by SIDNEY P. COLOWICK AND NATHAN O. KAPLAN

VOLUME IV. Special Techniques for the Enzymologist
Edited by SIDNEY P. COLOWICK AND NATHAN O. KAPLAN

VOLUME V. Preparation and Assay of Enzymes
Edited by SIDNEY P. COLOWICK AND NATHAN O. KAPLAN

VOLUME VI. Preparation and Assay of Enzymes (*Continued*)
Preparation and Assay of Substrates
Special Techniques
Edited by SIDNEY P. COLOWICK AND NATHAN O. KAPLAN

VOLUME VII. Cumulative Subject Index
Edited by SIDNEY P. COLOWICK AND NATHAN O. KAPLAN

VOLUME VIII. Complex Carbohydrates
Edited by ELIZABETH F. NEUFELD AND VICTOR GINSBURG

VOLUME IX. Carbohydrate Metabolism
Edited by WILLIS A. WOOD

VOLUME X. Oxidation and Phosphorylation
Edited by RONALD W. ESTABROOK AND MAYNARD E. PULLMAN

VOLUME XI. Enzyme Structure
Edited by C. H. W. HIRS

VOLUME XII. Nucleic Acids (Parts A and B)
Edited by LAWRENCE GROSSMAN AND KIVIE MOLDAVE

VOLUME XIII. Citric Acid Cycle
Edited by J. M. LOWENSTEIN

VOLUME XIV. Lipids
Edited by J. M. LOWENSTEIN

VOLUME XV. Steroids and Terpenoids
Edited by RAYMOND B. CLAYTON

VOLUME XVI. Fast Reactions
Edited by KENNETH KUSTIN

VOLUME XVII. Metabolism of Amino Acids and Amines (Parts A and B)
Edited by HERBERT TABOR AND CELIA WHITE TABOR

VOLUME XVIII. Vitamins and Coenzymes (Parts A, B, and C)
Edited by DONALD B. MCCORMICK AND LEMUEL D. WRIGHT

VOLUME XIX. Proteolytic Enzymes
Edited by GERTRUDE E. PERLMANN AND LASZLO LORAND

VOLUME XX. Nucleic Acids and Protein Synthesis (Part C)
Edited by KIVIE MOLDAVE AND LAWRENCE GROSSMAN

VOLUME XXI. Nucleic Acids (Part D)
Edited by LAWRENCE GROSSMAN AND KIVIE MOLDAVE

VOLUME XXII. Enzyme Purification and Related Techniques
Edited by WILLIAM B. JAKOBY

VOLUME XXIII. Photosynthesis (Part A)
Edited by ANTHONY SAN PIETRO

VOLUME XXIV. Photosynthesis and Nitrogen Fixation (Part B)
Edited by ANTHONY SAN PIETRO

VOLUME XXV. Enzyme Structure (Part B)
Edited by C. H. W. HIRS AND SERGE N. TIMASHEFF

VOLUME XXVI. Enzyme Structure (Part C)
Edited by C. H. W. HIRS AND SERGE N. TIMASHEFF

VOLUME XXVII. Enzyme Structure (Part D)
Edited by C. H. W. HIRS AND SERGE N. TIMASHEFF

VOLUME XXVIII. Complex Carbohydrates (Part B)
Edited by VICTOR GINSBURG

VOLUME XXIX. Nucleic Acids and Protein Synthesis (Part E)
Edited by LAWRENCE GROSSMAN AND KIVIE MOLDAVE

VOLUME XXX. Nucleic Acids and Protein Synthesis (Part F)
Edited by KIVIE MOLDAVE AND LAWRENCE GROSSMAN

VOLUME XXXI. Biomembranes (Part A)
Edited by SIDNEY FLEISCHER AND LESTER PACKER

VOLUME XXXII. Biomembranes (Part B)
Edited by SIDNEY FLEISCHER AND LESTER PACKER

VOLUME XXXIII. Cumulative Subject Index Volumes I–XXX
Edited by MARTHA G. DENNIS AND EDWARD A. DENNIS

VOLUME XXXIV. Affinity Techniques (Enzyme Purification: Part B)
Edited by WILLIAM B. JAKOBY AND MEIR WILCHEK

VOLUME XXXV. Lipids (Part B)
Edited by JOHN M. LOWENSTEIN

Volume XXXVI. Hormone Action (Part A: Steroid Hormones)
Edited by Bert W. O'Malley and Joel G. Hardman

Volume XXXVII. Hormone Action (Part B: Peptide Hormones)
Edited by Bert W. O'Malley and Joel G. Hardman

Volume XXXVIII. Hormone Action (Part C: Cyclic Nucleotides)
Edited by Joel G. Hardman and Bert W. O'Malley

Volume XXXIX. Hormone Action (Part D: Isolated Cells, Tissues, and Organ Systems)
Edited by Joel G. Hardman and Bert W. O'Malley

Volume XL. Hormone Action (Part E: Nuclear Structure and Function)
Edited by Bert W. O'Malley and Joel G. Hardman

Volume XLI. Carbohydrate Metabolism (Part B)
Edited by W. A. Wood

Volume XLII. Carbohydrate Metabolism (Part C)
Edited by W. A. Wood

Volume XLIII. Antibiotics
Edited by John H. Hash

Volume XLIV. Immobilized Enzymes
Edited by Klaus Mosbach

Volume XLV. Proteolytic Enzymes (Part B)
Edited by Laszlo Lorand

Volume XLVI. Affinity Labeling
Edited by William B. Jakoby and Meir Wilchek

Volume XLVII. Enzyme Structure (Part E)
Edited by C. H. W. Hirs and Serge N. Timasheff

Volume XLVIII. Enzyme Structure (Part F)
Edited by C. H. W. Hirs and Serge N. Timasheff

Volume XLIX. Enzyme Structure (Part G)
Edited by C. H. W. Hirs and Serge N. Timasheff

Volume L. Complex Carbohydrates (Part C)
Edited by Victor Ginsburg

Volume LI. Purine and Pyrimidine Nucleotide Metabolism
Edited by Patricia A. Hoffee and Mary Ellen Jones

Volume LII. Biomembranes (Part C: Biological Oxidations)
Edited by Sidney Fleischer and Lester Packer

Volume LIII. Biomembranes (Part D: Biological Oxidations)
Edited by Sidney Fleischer and Lester Packer

Volume LIV. Biomembranes (Part E: Biological Oxidations)
Edited by Sidney Fleischer and Lester Packer

VOLUME 125. Biomembranes (Part M: Transport in Bacteria, Mitochondria, and Chloroplasts: General Approaches and Transport Systems)
Edited by SIDNEY FLEISCHER AND BECCA FLEISCHER

VOLUME 126. Biomembranes (Part N: Transport in Bacteria, Mitochondria, and Chloroplasts: Protonmotive Force)
Edited by SIDNEY FLEISCHER AND BECCA FLEISCHER

VOLUME 127. Biomembranes (Part O: Protons and Water: Structure and Translocation)
Edited by LESTER PACKER

VOLUME 128. Plasma Lipoproteins (Part A: Preparation, Structure, and Molecular Biology)
Edited by JERE P. SEGREST AND JOHN J. ALBERS

VOLUME 129. Plasma Lipoproteins (Part B: Characterization, Cell Biology, and Metabolism)
Edited by JOHN J. ALBERS AND JERE P. SEGREST

VOLUME 130. Enzyme Structure (Part K)
Edited by C. H. W. HIRS AND SERGE N. TIMASHEFF

VOLUME 131. Enzyme Structure (Part L)
Edited by C. H. W. HIRS AND SERGE N. TIMASHEFF

VOLUME 132. Immunochemical Techniques (Part J: Phagocytosis and Cell-Mediated Cytotoxicity)
Edited by GIOVANNI DI SABATO AND JOHANNES EVERSE

VOLUME 133. Bioluminescence and Chemiluminescence (Part B)
Edited by MARLENE DELUCA AND WILLIAM D. MCELROY

VOLUME 134. Structural and Contractile Proteins (Part C: The Contractile Apparatus and the Cytoskeleton)
Edited by RICHARD B. VALLEE

VOLUME 135. Immobilized Enzymes and Cells (Part B)
Edited by KLAUS MOSBACH

VOLUME 136. Immobilized Enzymes and Cells (Part C)
Edited by KLAUS MOSBACH

VOLUME 137. Immobilized Enzymes and Cells (Part D)
Edited by KLAUS MOSBACH

VOLUME 138. Complex Carbohydrates (Part E)
Edited by VICTOR GINSBURG

VOLUME 139. Cellular Regulators (Part A: Calcium- and Calmodulin-Binding Proteins)
Edited by ANTHONY R. MEANS AND P. MICHAEL CONN

VOLUME 140. Cumulative Subject Index Volumes 102–119, 121–134

VOLUME 141. Cellular Regulators (Part B: Calcium and Lipids)
Edited by P. MICHAEL CONN AND ANTHONY R. MEANS

VOLUME 142. Metabolism of Aromatic Amino Acids and Amines
Edited by SEYMOUR KAUFMAN

VOLUME 143. Sulfur and Sulfur Amino Acids
Edited by WILLIAM B. JAKOBY AND OWEN GRIFFITH

VOLUME 144. Structural and Contractile Proteins (Part D: Extracellular Matrix)
Edited by LEON W. CUNNINGHAM

VOLUME 145. Structural and Contractile Proteins (Part E: Extracellular Matrix)
Edited by LEON W. CUNNINGHAM

VOLUME 146. Peptide Growth Factors (Part A)
Edited by DAVID BARNES AND DAVID A. SIRBASKU

VOLUME 147. Peptide Growth Factors (Part B)
Edited by DAVID BARNES AND DAVID A. SIRBASKU

VOLUME 148. Plant Cell Membranes
Edited by LESTER PACKER AND ROLAND DOUCE

VOLUME 149. Drug and Enzyme Targeting (Part B)
Edited by RALPH GREEN AND KENNETH J. WIDDER

VOLUME 150. Immunochemical Techniques (Part K: *In Vitro* Models of B and T Cell Functions and Lymphoid Cell Receptors)
Edited by GIOVANNI DI SABATO

VOLUME 151. Molecular Genetics of Mammalian Cells
Edited by MICHAEL M. GOTTESMAN

VOLUME 152. Guide to Molecular Cloning Techniques
Edited by SHELBY L. BERGER AND ALAN R. KIMMEL

VOLUME 153. Recombinant DNA (Part D)
Edited by RAY WU AND LAWRENCE GROSSMAN

VOLUME 154. Recombinant DNA (Part E)
Edited by RAY WU AND LAWRENCE GROSSMAN

VOLUME 155. Recombinant DNA (Part F)
Edited by RAY WU

VOLUME 156. Biomembranes (Part P: ATP-Driven Pumps and Related Transport: The Na,K-Pump)
Edited by SIDNEY FLEISCHER AND BECCA FLEISCHER

VOLUME 157. Biomembranes (Part Q: ATP-Driven Pumps and Related Transport: Calcium, Proton, and Potassium Pumps)
Edited by SIDNEY FLEISCHER AND BECCA FLEISCHER

VOLUME 158. Metalloproteins (Part A)
Edited by JAMES F. RIORDAN AND BERT L. VALLEE

VOLUME 193. Mass Spectrometry
Edited by JAMES A. MCCLOSKEY

VOLUME 194. Guide to Yeast Genetics and Molecular Biology
Edited by CHRISTINE GUTHRIE AND GERALD R. FINK

VOLUME 195. Adenylyl Cyclase, G Proteins, and Guanylyl Cyclase
Edited by ROGER A. JOHNSON AND JACKIE D. CORBIN

VOLUME 196. Molecular Motors and the Cytoskeleton
Edited by RICHARD B. VALLEE

VOLUME 197. Phospholipases
Edited by EDWARD A. DENNIS

VOLUME 198. Peptide Growth Factors (Part C)
Edited by DAVID BARNES, J. P. MATHER, AND GORDON H. SATO

VOLUME 199. Cumulative Subject Index Volumes 168–174, 176–194

VOLUME 200. Protein Phosphorylation (Part A: Protein Kinases: Assays, Purification, Antibodies, Functional Analysis, Cloning, and Expression)
Edited by TONY HUNTER AND BARTHOLOMEW M. SEFTON

VOLUME 201. Protein Phosphorylation (Part B: Analysis of Protein Phosphorylation, Protein Kinase Inhibitors, and Protein Phosphatases)
Edited by TONY HUNTER AND BARTHOLOMEW M. SEFTON

VOLUME 202. Molecular Design and Modeling: Concepts and Applications (Part A: Proteins, Peptides, and Enzymes)
Edited by JOHN J. LANGONE

VOLUME 203. Molecular Design and Modeling: Concepts and Applications (Part B: Antibodies and Antigens, Nucleic Acids, Polysaccharides, and Drugs)
Edited by JOHN J. LANGONE

VOLUME 204. Bacterial Genetic Systems
Edited by JEFFREY H. MILLER

VOLUME 205. Metallobiochemistry (Part B: Metallothionein and Related Molecules)
Edited by JAMES F. RIORDAN AND BERT L. VALLEE

VOLUME 206. Cytochrome P450
Edited by MICHAEL R. WATERMAN AND ERIC F. JOHNSON

VOLUME 207. Ion Channels
Edited by BERNARDO RUDY AND LINDA E. IVERSON

VOLUME 208. Protein–DNA Interactions
Edited by ROBERT T. SAUER

VOLUME 209. Phospholipid Biosynthesis
Edited by EDWARD A. DENNIS AND DENNIS E. VANCE

VOLUME 210. Numerical Computer Methods
Edited by LUDWIG BRAND AND MICHAEL L. JOHNSON

Volume 227. Metallobiochemistry (Part D: Physical and Spectroscopic Methods for Probing Metal Ion Environments in Metalloproteins)
Edited by James F. Riordan and Bert L. Vallee

Volume 228. Aqueous Two-Phase Systems
Edited by Harry Walter and Göte Johansson

Volume 229. Cumulative Subject Index Volumes 195–198, 200–227

Volume 230. Guide to Techniques in Glycobiology
Edited by William J. Lennarz and Gerald W. Hart

Volume 231. Hemoglobins (Part B: Biochemical and Analytical Methods)
Edited by Johannes Everse, Kim D. Vandegriff, and Robert M. Winslow

Volume 232. Hemoglobins (Part C: Biophysical Methods)
Edited by Johannes Everse, Kim D. Vandegriff, and Robert M. Winslow

Volume 233. Oxygen Radicals in Biological Systems (Part C)
Edited by Lester Packer

Volume 234. Oxygen Radicals in Biological Systems (Part D)
Edited by Lester Packer

Volume 235. Bacterial Pathogenesis (Part A: Identification and Regulation of Virulence Factors)
Edited by Virginia L. Clark and Patrik M. Bavoil

Volume 236. Bacterial Pathogenesis (Part B: Integration of Pathogenic Bacteria with Host Cells)
Edited by Virginia L. Clark and Patrik M. Bavoil

Volume 237. Heterotrimeric G Proteins
Edited by Ravi Iyengar

Volume 238. Heterotrimeric G-Protein Effectors
Edited by Ravi Iyengar

Volume 239. Nuclear Magnetic Resonance (Part C)
Edited by Thomas L. James and Norman J. Oppenheimer

Volume 240. Numerical Computer Methods (Part B)
Edited by Michael L. Johnson and Ludwig Brand

Volume 241. Retroviral Proteases
Edited by Lawrence C. Kuo and Jules A. Shafer

Volume 242. Neoglycoconjugates (Part A)
Edited by Y. C. Lee and Reiko T. Lee

Volume 243. Inorganic Microbial Sulfur Metabolism
Edited by Harry D. Peck, Jr., and Jean LeGall

Volume 244. Proteolytic Enzymes: Serine and Cysteine Peptidases
Edited by Alan J. Barrett

Volume 245. Extracellular Matrix Components
Edited by E. Ruoslahti and E. Engvall

Volume 246. Biochemical Spectroscopy
Edited by Kenneth Sauer

Volume 247. Neoglycoconjugates (Part B: Biomedical Applications)
Edited by Y. C. Lee and Reiko T. Lee

Volume 248. Proteolytic Enzymes: Aspartic and Metallo Peptidases
Edited by Alan J. Barrett

Volume 249. Enzyme Kinetics and Mechanism (Part D: Developments in Enzyme Dynamics)
Edited by Daniel L. Purich

Volume 250. Lipid Modifications of Proteins
Edited by Patrick J. Casey and Janice E. Buss

Volume 251. Biothiols (Part A: Monothiols and Dithiols, Protein Thiols, and Thiyl Radicals)
Edited by Lester Packer

Volume 252. Biothiols (Part B: Glutathione and Thioredoxin; Thiols in Signal Transduction and Gene Regulation)
Edited by Lester Packer

Volume 253. Adhesion of Microbial Pathogens
Edited by Ron J. Doyle and Itzhak Ofek

Volume 254. Oncogene Techniques
Edited by Peter K. Vogt and Inder M. Verma

Volume 255. Small GTPases and Their Regulators (Part A: Ras Family)
Edited by W. E. Balch, Channing J. Der, and Alan Hall

Volume 256. Small GTPases and Their Regulators (Part B: Rho Family)
Edited by W. E. Balch, Channing J. Der, and Alan Hall

Volume 257. Small GTPases and Their Regulators (Part C: Proteins Involved in Transport)
Edited by W. E. Balch, Channing J. Der, and Alan Hall

Volume 258. Redox-Active Amino Acids in Biology
Edited by Judith P. Klinman

Volume 259. Energetics of Biological Macromolecules
Edited by Michael L. Johnson and Gary K. Ackers

Volume 260. Mitochondrial Biogenesis and Genetics (Part A)
Edited by Giuseppe M. Attardi and Anne Chomyn

Volume 261. Nuclear Magnetic Resonance and Nucleic Acids
Edited by Thomas L. James

Volume 262. DNA Replication
Edited by Judith L. Campbell

Volume 281. Vitamins and Coenzymes (Part K)
Edited by Donald B. McCormick, John W. Suttie, and Conrad Wagner

Volume 282. Vitamins and Coenzymes (Part L)
Edited by Donald B. McCormick, John W. Suttie, and Conrad Wagner

Volume 283. Cell Cycle Control
Edited by William G. Dunphy

Volume 284. Lipases (Part A: Biotechnology)
Edited by Byron Rubin and Edward A. Dennis

Volume 285. Cumulative Subject Index Volumes 263, 264, 266–284, 286–289

Volume 286. Lipases (Part B: Enzyme Characterization and Utilization)
Edited by Byron Rubin and Edward A. Dennis

Volume 287. Chemokines
Edited by Richard Horuk

Volume 288. Chemokine Receptors
Edited by Richard Horuk

Volume 289. Solid Phase Peptide Synthesis
Edited by Gregg B. Fields

Volume 290. Molecular Chaperones
Edited by George H. Lorimer and Thomas Baldwin

Volume 291. Caged Compounds
Edited by Gerard Marriott

Volume 292. ABC Transporters: Biochemical, Cellular, and Molecular Aspects
Edited by Suresh V. Ambudkar and Michael M. Gottesman

Volume 293. Ion Channels (Part B)
Edited by P. Michael Conn

Volume 294. Ion Channels (Part C)
Edited by P. Michael Conn

Volume 295. Energetics of Biological Macromolecules (Part B)
Edited by Gary K. Ackers and Michael L. Johnson

Volume 296. Neurotransmitter Transporters
Edited by Susan G. Amara

Volume 297. Photosynthesis: Molecular Biology of Energy Capture
Edited by Lee McIntosh

Volume 298. Molecular Motors and the Cytoskeleton (Part B)
Edited by Richard B. Vallee

Volume 299. Oxidants and Antioxidants (Part A)
Edited by Lester Packer

VOLUME 300. Oxidants and Antioxidants (Part B)
Edited by LESTER PACKER

VOLUME 301. Nitric Oxide: Biological and Antioxidant Activities (Part C) (in preparation)
Edited by LESTER PACKER

VOLUME 302. Green Fluorescent Protein (in preparation)
Edited by P. MICHAEL CONN

VOLUME 303. cDNA Preparation and Display (in preparation)
Edited by SHERMAN M. WEISSMAN

VOLUME 304. Chromatin (in preparation)
Edited by PAUL M. WASSERMAN AND ALAN P. WOLFFE

Section I

Oxidative Damage to Lipids, Proteins, and Nucleic Acids

A. Lipids

Articles 1 through 11

B. Proteins and Nucleic Acids

Articles 12 through 22

[1] Mass Spectrometric Quantification of F_2-Isoprostanes in Biological Fluids and Tissues as Measure of Oxidant Stress

By JASON D. MORROW and L. JACKSON ROBERTS II

Introduction

Free radicals derived primarily from oxygen have been implicated in the pathophysiology of a wide variety of human disease including atherosclerosis, cancer, neurodegenerative disorders, and even the normal aging process.[1] Much of the evidence for this, however, is indirect or circumstantial, largely because of limitations of methods that are available to quantify free radicals or the products they produce in biological systems. This is a particular problem when noninvasive approaches are used to assess oxidant injury in animals or humans.[2]

Measures of lipid peroxidation are frequently utilized to implicate free radicals in pathophysiological processes. These measurements include quantification of short chain alkanes, malondialdehyde, or conjugated dienes. Each of these, however, suffers from problems related to specificity and sensitivity, especially when utilized to quantify oxidant stress *in vivo*. Further, artifactual generation of these lipid peroxidation products can occur *ex vivo* and factors such as endogenous metabolism can affect levels of compounds measured.[2]

In 1990, we reported that a series of prostaglandin(PG) F_2-like compounds, termed F_2-isoprostanes (F_2-IsoPs), are produced *in vivo* in humans by a non-cyclooxygenase free radical-catalyzed mechanism involving the peroxidation of arachidonic acid.[3] Formation of these compounds initially involves the generation of four positional peroxyl radical isomers of arachidonic acid which undergo endocyclization to PGG_2-like compounds that are subsequently reduced to PGF_2-like compounds. Four F_2-IsoP regioisomers are formed, each of which can theoretically comprise eight racemic diastereomers. In support of the proposed mechanism of formation, we

[1] B. Halliwell and J. M. C. Gutteridge, *Methods Enzymol.* **186,** 1 (1990).

[2] B. Halliwell and M. Grootveld, *FEBS Lett.* **213,** 9 (1987).

[3] J. D. Morrow, K. E. Hill, R. F. Burk, T. M. Nammour, K. F. Badr, and L. J. Roberts, *Proc. Natl. Acad. Sci. USA* **87,** 10721 (1990).

0076-6879/99 $30.00

have obtained direct evidence both *in vitro* and *in vivo* that each of the four classes of regioisomers are formed.[4]

We have accumulated a large body of evidence that suggests that quantification of F_2-IsoPs represents a reliable and useful approach to assess lipid peroxidation and oxidant stress *in vivo*. First, the formation of F_2-IsoPs has been shown to increase dramatically in various animal models of free radical injury, and these increases correlate with measures of tissue damage. For example, circulating concentrations of F_2-IsoPs increase up to 100-fold following the administration of diquat to selenium-deficient rats or CCl_4 to normal rats.[3] These studies have also illustrated that quantification of F_2-IsoPs provides a much more sensitive and accurate method to assess lipid peroxidation *in vivo* compared with other markers. As an example, following administration of CCl_4 to rats, levels of F_2-IsoPs esterifed to hepatic lipids increased greater than 80-fold, whereas levels of malondialdehyde (MDA) increased only 2.7-fold.[5] In another study, we found that measuring F_2-IsoP afforded a significantly more sensitive indicator of CCl_4-induced lipid peroxidation compared with measurement of lipid hydroperoxides by mass spectrometry.[6] Further, concentrations of F_2-IsoPs are present that are easily detected in normal human biological fluids such as plasma and urine.[3] This allows the definition of a normal range which permits an assessment of small increases in the formation of F_2-IsoPs in settings of mild oxidant stress. In addition, F_2-IsoPs can be detected in virtually every type of biological fluid analyzed thus far, including plasma, urine, cerebrospinal fluid, bile, lymph, bronchoalveolar lavage fluid, and synovial fluid.[7] Levels of these compounds are also detectable in all types of tissues examined to date. These include liver, kidney, stomach, brain, lung, vascular tissue, muscle, and heart. Thus, the fact that F_2-IsoPs are detectable in various tissues and fluids provides the opportunity to assess the formation of these compounds at local sites of oxidant injury. Finally, we have shown that F_2-IsoPs are increased in a number of human disorders associated with enhanced oxidant stress, including chronic heavy cigarette smoking, atherosclerosis, hepatorenal syndrome, and systemic sclerosis.[7]

The precursor of the F_2-IsoPs is arachidonic acid. The vast majority of arachidonic acid present *in vivo* exists esterified to phospholipids. Previously, we reported the novel finding that F_2-IsoPs are initially formed *in*

[4] R. J. Waugh, J. D. Morrow, L. J. Roberts, and R. C. Murphy, *Free Rad. Biol. Med.* **23,** 943 (1997).

[5] A. W. Longmire, L. L. Swift, L. J. Roberts, J. A. Awad, R. F. Burk, and J. D. Morrow, *Biochem. Pharm.* **47,** 1173 (1994).

[6] W. R. Matthews, R. McKenna, D. M. Guido, T. W. Petry, R. A. Jolly, J. D. Morrow, and L. J. Roberts, *Proc. 41st A.S.M.S. Conf. Mass Spectrometry,* 865 (1993).

[7] J. D. Morrow and L. J. Roberts, *Prog. Lipid Res.* **36,** 1 (1997).

situ from arachidonic acid esterified to phospholipids and then subsequently released in the free form by phospholipases.[8] This observation provides the basis for an important concept regarding the assessment of isoprostane formation *in vivo* in that when isoprostane formation is quantified, total production of F_2-IsoPs may be more accurately assessed by measuring levels of both free and esterified compounds. The relative ratio of esterified and free compounds presumably would depend on the degree of activation of phospholipases in settings of oxidant injury.[5] For example, following administration of CCl_4 to rats, F_2-IsoP esterified to liver phospholipids increases rapidly with levels maximal at 2 hours, while increases in circulating levels of free F_2-IsoPs are delayed and maximal at 8 hours.[7] On the other hand, when quantified in tissues, the majority of F_2-IsoPs are esterified in phospholipids, and thus, analyzing levels of esterified compounds, as opposed to both free and esterified F_2-IsoPs, appears to provide an accurate assessment of oxidant injury in specific organs.[7]

The purpose of this chapter is to detail methods employed for the analysis of F_2-IsoPs from biological sources utilizing mass spectrometric techniques. Specific examples are given demonstrating both the utility and limitations of the assay. Procedures will be outlined for the analysis of both free F_2-IsoPs and F_2-IsoPs esterified to phospholipids. F_2-IsoPs from biological sources can only be quantified as free compounds using gas chromatography (GC)/mass spectrometry (MS). Thus, to measure levels of these compounds esterified to phospholipids, the phospholipids are first extracted from the tissue sample and subjected to alkaline hydrolysis to release free F_2-IsoPs. Free F_2-IsoPs are then quantified using the same procedure for the measurement of free compounds in biological fluids. The following methods, therefore, first outline the extraction and hydrolysis of F_2-IsoPs from tissue lipids. Subsequently, the method of analysis for free compounds is discussed.

Handling and Storage of Biological Fluids and Tissues for Quantification of F_2-Isoprostanes

As discussed, F_2-IsoPs have been detected in all biological fluids and tissues examined thus far. A potential drawback to measuring F_2-IsoPs as an index of endogenous lipid peroxidation is that they can be readily generated *ex vivo* in biological fluids such as plasma in which arachidonyl-containing lipids are present. This occurs not only when biological fluids or tissues are left at room temperature, but also when they are stored at

[8] J. D. Morrow, J. A. Awad, H. J. Boss, I. A. Blair, and L. J. Roberts, *Proc. Natl. Acad. Sci. USA* **89,** 10721 (1992).

$-20°C$.[7] However, we have found that the formation of F_2-IsoPs does not occur if biological fluids or tissues are processed immediately after procurement and if agents including butylated hydroxytoluene (BHT, a free radical scavenger) and/or triphenylphosphine (a reducing agent) are added to the organic solvents during extraction and hydrolysis of phospholipids.[7,8] Thus, samples to be analyzed for F_2-IsoPs should either be processed immediately or stored at $-70°C$. Ideally, samples should be rapidly frozen in liquid nitrogen prior to placement at $-70°C$. This is especially important for tissue samples, and we routinely snap freeze them with a clamp prior to storage at $-70°C$ since it is known that inner areas of tissue samples that are not snap frozen may remain in a liquid state for a period of time, even when placed at $-70°C$.

Extraction and Hydrolysis of F_2-Isoprostane-Containing Phospholipids in Tissues and Biological Fluids

To 0.05–1 g of tissue is added 20 ml of ice-cold Folch solution, $CHCl_3$/methanol (2:1, v/v), containing 0.005% BHT in a 40 ml glass centrifuge tube. As discussed, the presence of BHT during extraction and hydrolysis of lipids is important since it completely inhibits *ex vivo* formation of F_2-IsoPs during this procedure.[7] The tissue is then homogenized with a blade homogenizer (PTA 10S generator, Brinkmann Instruments, Westbury, NY) for 30 seconds and mixture allowed to stand sealed under nitrogen at room temperature for 1 hour. Four milliliters of aqueous NaCl (0.9%) is then added and the solution is vortexed vigorously and centrifuged at 800*g* for 10 minutes. After centrifugation, the upper aqueous layer is discarded and the lower organic layer carefully separated from the intermediate semisolid proteinaceous layer.

The organic phase containing the extracted lipids is then transferred to a 100 ml conical bottom flask and evaporated under vacuum to dryness. Four milliliters of methanol containing BHT (0.005%) and 4 ml of aqueous KOH (15%) is then added to the residue and incubated at 37°C for 30 minutes to effect hydrolysis and release of the F_2-IsoPs. The mixture is then acidified to pH 3 with 1 *M* HCl and diluted to a final volume of 80 ml with pH 3 water in preparation for extraction of free F_2-IsoPs as discussed below. Dilution of the methanol in the solution with water to 5% or less is necessary to ensure proper column extraction of F_2-IsoPs in the subsequent purification procedure.

For the extraction of lipids containing esterified F_2-IsoPs in biological fluids such as plasma, as opposed to tissue, a different method is used.[7,9]

[9] J. D. Morrow, B. Frei, A. W. Longmire, J. M. Gaziano, S. M. Lynch, Y. Shyr, W. E. Strauss, J. A. Oates, and L. J. Roberts, *New Engl. J. Med.* **332,** 1198 (1995).

We have found that the addition of BHT alone to the Folch solution used to extract plasma lipids does not suppress *ex vivo* lipid peroxidation entirely. On the other hand, the addition of both triphenylphosphine and BHT to the organic solvents used to extract plasma lipids entirely prevents autoxidation. Thus, to 1 ml of a biological fluid such as plasma in a 40 ml conical glass centrifuge tube is added 20 ml of ice-cold Folch solution, $CHCl_3$/methanol (2:1, v/v), containing 0.005% BHT and triphenylphosphine (5 mg). The mixture is shaken for 2 minutes and then 10 ml of 0.043% $MgCl_2$ in water is added. The mixture is shaken again for 2 minutes and then centrifuged at 800*g* for 10 minutes. The organic layer is removed and dried under nitrogen and the residue is then subjected to hydrolysis using the same procedure outlined above for tissue lipids.

Purification, Derivatization, and Quantification of F_2-Isoprostanes

Quantification of F_2-IsoPs by GC/negative-ion chemical ionization (NICI) MS is extremely sensitive with a lower limit of detection in the range of 1–5 pg using a deuterated internal standard with a $^2H_0$ blank of less than 5 parts per thousand. Thus, it is usually not necessary to assay more than 1–3 ml of a biological fluid or a lipid extract from more than 50–100 mg of tissue. Further, because urinary levels of F_2-IsoPs are high (typically greater than 1 ng/ml), we have found 0.2 ml of urine is more than adequate to quantify urinary F_2-IsoPs. The following assay procedure that is described and summarized in Fig. 1 is the method used for analysis of free F_2-IsoPs in plasma, but is equally adaptable to other biological fluids and hydrolyzed lipid extracts of tissues. Previously, we had reported an alternative method for the quantification of F_2-IsoPs in urine involving two thin-layer chromatographic steps,[10] but we have found that urinary F_2-IsoPs are accurately measured by the same assay utilized to quantify these compounds in plasma or other biological fluids.

Following acidification of 3 ml of plasma to pH 3 with 1 *M* HCl, 200–1000 pg of a deuterated standard is added. Previously, we had employed $[^2H_7]9\alpha,11\beta$-PGF_2 or $[^2H_4]PGF_{2\alpha}$ as an internal standard. We had synthesized the former compound from deuterium-labeled arachidonic acid, and the latter compound is available commercially. We have, however, begun to use a deuterium-labeled isoprostane, $[^2H_4]$8-iso-$PGF_{2\alpha}$, as an internal standard. This compound is also available from commercial sources (Cayman Chemical, Ann Arbor, MI) and we have found it to be of equal accuracy compared to the other standards. After addition of the internal standard, the mixture is vortexed and applied to a C_{18} Sep-Pak column

[10] J. D. Morrow and L. J. Roberts, *Methods Enzymol.* **233,** 163 (1994).

Biological fluid (1-3 ml) or
hydrolyzed lipid extract (50-1000 mg);
acidify to pH 3; add deuterated standard ([$^2H_4$] 8-iso-$PGF_{2\alpha}$)

↓

C_{18} and silica Sep-Pak extraction

↓

Formation of PFB esters

↓

TLC of F_2-IsoPs as PFB esters

↓

Formation of trimethylsilyl ether derivatives

↓

Quantification by selected ion monitoring GC/NICI MS

FIG. 1. Outline of the procedures used for the extraction, purification, derivatization, and mass spectrometric analysis of F_2-IsoPs from biological sources.

(Waters Associates, Milford, MA) preconditioned with 5 ml methanol and 5 ml of water (pH 3). The sample and subsequent solvents are eluted through the Sep-Pak using a 10 ml plastic syringe. The column is then washed sequentially with 10 ml of water (pH 3) and 10 ml heptane. The F_2-IsoPs are eluted with 10 ml ethyl acetate/heptane (50:50, v/v).

The ethyl acetate/heptane eluate from the C_{18} Sep-Pak is then dried over anhydrous Na_2SO_4 and applied to a silica Sep-Pak (Waters Associates). The cartridge is then washed with 5 ml of ethyl acetate followed by elution of the F_2-IsoPs with 5 ml of ethyl acetate/methanol (50:50, v/v). The ethyl acetate/methanol eluate is evaporated under a stream of nitrogen.

The F_2-IsoPs are then converted to pentafluorobenzyl (PFB) esters by treatment with a mixture of 40 μl of 10% pentafluorobenzyl bromide in acetonitrile and 20 μl of 10% *N,N*-diisopropylethylamine in acetonitrile at room temperature for 30 minutes. The reagents are then dried under nitrogen and this procedure is repeated to ensure quantitative esterification. After the second esterification, the reagents are dried under nitrogen and the residue subjected to TLC using the solvent chloroform/ethanol (93:7, v/v). Approximately 2–5 μg of the methyl ester of $PGF_{2\alpha}$ is chromatographed on a separate lane and visualized by spraying with a 10% solution of phosphomolybdic acid in ethanol followed by heating. The methyl ester of $PGF_{2\alpha}$ is chromatographed rather than the PFB ester because any contamination of the sample being quantified with the methyl ester of $PGF_{2\alpha}$ will not interfere with the analysis owing to the fact that the F_2-IsoPs are analzyed as PFB esters. Compounds migrating in the region of the methyl ester of $PGF_{2\alpha}$ (R_f 0.15) and the adjacent areas 1 cm above and below are scraped and extracted from the silica gel with ethyl acetate.

The ethyl acetate is dried under nitrogen and the F_2-IsoPs are then converted to trimethylsilyl ether derivatives by adding 20 μl *N,O*-bis(trimethylsilyl)trifluoroacetamide (BSTFA) and 10 μl dimethylformamide and incubating at 40°C for 20 minutes. The reagents are dried under nitrogen and the F_2-IsoPs are redissolved in 10 μl of undecane, which has been dried over calcium hydride, for analysis by GC/MS.

For analysis of F_2-IsoPs, we routinely use a Nermag R10-10C or Hewlett-Packard 5982A mass spectrometer interfaced with an IBM Pentium computer system. The F_2-IsoPs are chromatographed on a 15 m DB1701 fused silica capillary column (J and W Scientific, Folsom, CA). We have found this GC phase gives superior separation and resolution of the individual F_2-IsoPs compared to other columns. The column temperature is programmed from 190 to 300°C at 20°C/minute. Methane is used as the carrier gas for negative ion chemical ionization at a flow rate of 1 ml/minute. Ion source temperature is 250°C, electron energy is 70 eV, and the filament current is 0.25 mA. The ion monitored for endogenous F_2-IsoPs is the carboxylate anion *m/z* 569 (M − 181, loss of $\cdot CH_2C_6F_5$). The corresponding M − 181 ion for the $[^2H_4]$8-iso-$PGF_{2\alpha}$ internal standard is *m/z* 573.

Application of the Assay for Analysis of F_2-Isoprostanes in Biological Tissues and Fluids

As stated, we have successfully employed this assay to quantify F_2-IsoPs in a number of diverse biological fluids and tissues. Shown in Fig. 2 is a selected ion current chromatogram obtained from the analysis of F_2-IsoPs in plasma of a rat following treatment with CCl_4 to induce endogenous lipid peroxidation. The series of peaks in the upper *m/z* 569 selected ion current chromatogram represents different endogenous F_2-IsoPs. This patten of peaks is virtually identical to that obtained from all other biological fluids and tissues that we have examined to date. In the lower *m/z* 573 chromatogram, the single peak represents the $[^2H_4]$8-iso-$PGF_{2\alpha}$ internal standard that was added to the plasma sample. For quantification purposes, the peak denoted by an asterisk (*), which coelutes with the 8-iso-$PGF_{2\alpha}$ internal standard, is routinely measured. Using the ratio of the intensity of this peak to that of the internal standard, the concentration of F_2-IsoPs was calculated to 831 pg/ml, approximately 28-fold above normal. Normal

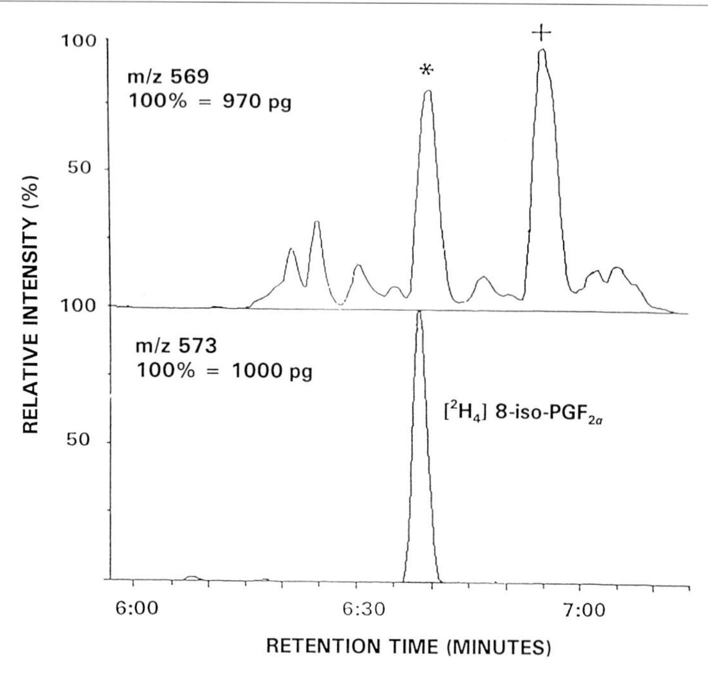

FIG. 2. Analysis of F_2-IsoPs in plasma obtained from a rat 4 hours after treatment with CCl_4 (2 ml/kg orogastrically) to induce endogenous lipid peroxidation. The *m/z* 573 ion current chromatogram represents the $[^2H_4]$8-iso-PGF$_{2\alpha}$ internal standard. The *m/z* 569 ion current chromatogram represents endogenous F_2-IsoPs. The peak in the upper chromatogram marked by an asterisk (*) is the one routinely used for quantification of the F_2-IsoPs. The peak represented by the plus (+) can comprise both F_2-IsoPs and cyclooxygenase-derived PGF$_{2\alpha}$. The concentration of F_2-IsoPs in the plasma using the asterisked (*) peak for quantification was calculated to be 831 pg/ml.

plasma concentrations of F_2-IsoPs in rats are 30 ± 11 pg/ml (mean ± 1 S.D.) and in normal human plasma are 35 ± 6 pg/ml (mean ± 1 S.D.) The reader is referred to previously published data for normal levels of F_2-IsoPs in other biological fluids and tissues.[7]

Quantification of the F_2-IsoPs based on the intensity of the asterisked (*) peak shown in Fig. 2 is highly precise and accurate. The precision is ±6% and the accuracy is 96%. Accuracy was determined by quantification of an added known amount of 8-iso-PGF$_{2\alpha}$, which cochromatographs with the asterisked (*) peak shown in Fig. 2. However, we have previously reported that quantifying the F_2-IsoPs based on the intensity of other peaks in the *m/z* 569 ion current chromatogram is less precise. The reason for this is that despite the wide region scraped on the TLC plate, there are additional F_2-IsoPs that migrate ahead of the region routinely included.

These compounds have similar GC retention times and coelute with other *m/z* 569 peaks representing F_2-IsoPs that chromatograph in the region of the TLC plate that is routinely scraped. We have found that minor differences in either preparation of the TLC solvents or scraping techniques of the TLC plate can result in partial inclusion of these compounds, which alters the intensity of some of the F_2-IsoP peaks. Fortuitously, however, none of these compounds cochromatographs on GC with the asterisked (*) peak that we routinely measure. In addition, $PGF_{2\alpha}$ derived from the cyclooxygenase cochromatographs with the peak noted by a plus (+). Thus, the intensity of this peak can also vary as a result of the presence of the enzymatic product $PGF_{2\alpha}$, which is not an F_2-IsoP. Therefore, if these additional F_2-IsoPs that chromatograph in adjacent areas of the TLC plate are partially included in the analysis, or if cyclooxygenase derived $PGF_{2\alpha}$ is present in the biological fluid, the precision and accuracy of the assay is not affected when the asterisked (*) peak is used for quantification.

As mentioned, F_2-IsoPs can be quantified in a wide variety of biological fluids and tissues. Quantification of these compounds in urine has been utilized as an assessment of systemic oxidant stress status in humans.[7] It should be mentioned that it is not known at the present time whether F_2-IsoPs that are excreted into the urine derive primarily from local production in the kidney, like that of cyclooxygenase-derived PGs, or whether they represent F_2-IsoPs that are filtered from the circulation. It is likely, however, that unmetabolized F_2-IsoPs present in the urine derive, at least in part, from the kidney based on several lines of evidence.[7] Thus, caution should be used in the interpretation of whether increased urinary F_2-IsoPs reflects increased systemic lipid peroxidation or enhanced lipid peroxidation in the kidney.

An attractive alternative approach for assessing systemic endogenous F_2-IsoP production is the measurement of urinary F_2-IsoP metabolites. Measurement of metabolites of these compounds in urine has the advantage not only of providing an integrated assessment of F_2-IsoP production over time, but also of circumventing the problem of artifactual generation of F_2-IsoPs *ex vivo* by autoxidation of arachidonic acid. In this respect, we have recently identified the major urinary metabolite of 8-iso-$PGF_{2\alpha}$ in humans as 2,3-dinor-5,6-dihydro-8-iso-$PGF_{2\alpha}$.[7] Identification of this compound provides the basis for development of an assay to quantify this metabolite as a means to obtain an integrated assessment of total endogenous F_2-IsoP production in humans. Work is currently underway in our laboratory toward this goal.

Although this chapter has focused on the quantification of F_2-IsoPs, we have identified other classes of IsoPs derived from the free radical-catalyzed peroxidation of arachidonic acid. Specifically, we have found that IsoP endoperoxide intermediates not only undergo reduction *in vivo* to form F_2-IsoPs, but can isomerize to form D/E-ring IsoPs (D_2/E_2-IsoPs) and can rearrange to form thromboxane-like compounds termed isothromboxanes. Formation of both classes of these compounds increase dramatically in animal models of oxidant stress and parallel the formation of F_2-IsoPs. Whether quantification of these compounds to assess oxidant stress status offers any advantage over measurement of F_2-IsoPs or would provide additional insights into the role of lipid peroxidation in various pathophysiological processes is an area of active investigation in our laboratory.

Summary

This chapter has outlined methods to assess lipid peroxidation associated with oxidant injury *in vivo* by quantifying concentrations of free F_2-IsoPs in biological fluids and levels of F_2-IsoPs esterified in tissue lipids. The mass spectrometric assay described herein is highly precise and accurate. A potential shortcoming with this approach is that it requires expensive instrumentation, i.e., a mass spectrometer. However, several immunoassays for an F_2-IsoP, 8-iso-$PGF_{2\alpha}$, have become available from commercial sources. At this time, the accuracy and reliability of these assay for quantifying F_2-IsoPs in biological fluids has not been fully validated by mass spectrometry. If these immunoassays prove to be a reliable measure of F_2-IsoPs, however, this should greatly expand the use of F_2-IsoPs to assess oxidant stress.

In conclusion, studies carried out over the past several years have shown that measurement of F_2-IsoPs has overcome many of the limitations associated with other methods to assess oxidant status, especially when applied to the measurement of oxidant stress *in vivo* in humans. Therefore, the quantification of F_2-IsoPs represents an important advance in our ability to assess the role of oxidant stress and lipid peroxidation in human disease.

Acknowledgments

Supported by NIH grants DK48831, GM42056, and GM15431.

[2] Gas Chromatography–Mass Spectrometry Assay for Measurement of Plasma Isoprostanes

By JAFFAR NOUROOZ-ZADEH

Introduction

Lipid peroxidation is thought to play an important role in the pathogenesis of many human diseases. However, increases in oxidative stress have been difficult to assess because methods for measuring lipid peroxidation products derived from specific fatty acids have been unavailable.

A major breakthrough was the discovery of F_2-isoprostanes.[1] The F_2-isoprostanes are made up of four unique families of prostaglandin F_2 (PGF_2)-like compounds produced by a free radical-mediated process from arachidonic acid (AA) and independent of the cyclooxygenase pathway (Fig. 1). Measurement of one of the major F_2-isoprostanes, 8-epi-$PGF_{2\alpha}$, has been shown to represent a sensitive and reliable marker of oxidative stress because (1) it can be specifically and accurately measured in biological samples; (2) the level is increased in response to prooxidants; and (3) the level is suppressed by dietary supplementation with antioxidants.[2]

Classical procedure for the isolation and quantitative determination of plasma prostanoids involves chromatography on a C_{18} and Si cartridge, followed by thin-layer chromatography (TLC) prior to final determination by gas chromatography–mass spectrometry/negative-ion chemical ionization (GC-MS/NICI).[3,4] Problems with the existing assay include a low recovery of PGF_2-like compounds due to losses during the TLC step[5,6] and the fact that the assay is time consuming.

This chapter describes an improved procedure for the isolation of isoprostanes from plasma or isolated lipoprotein fractions.[6,7] The modification involves the following changes: (1) Recovery of PGF_2-like compounds dur-

[1] J. D. Morrow, K. E. Hill, R. F. Burk, T. M. Nammour, K. F. Badr, and L. J. Roberts II, *Proc. Natl. Acad. Sci. USA* **87,** 9383 (1990).

[2] L. J. Roberts II and J. D. Morrow, *Biochim. Biophysc. Acta* **1345,** 121 (1997).

[3] J. D. Morrow, T. M. Harris, and L. J. Roberts II, *Anal. Biochem.* **184,** 1 (1990).

[4] D. F. Wendelborn, J. D. Morrow, and L. J. Roberts II, *Methods Enzymol.* **187,** 51 (1990).

[5] B. Sjöquist, E. Oliw, I. Lunden, and E. E. Änggård, *J. Chromatogr. B* **163,** 1 (1979).

[6] J. Nourooz-Zadeh, N. K. Gopaul, S. Barrow, A. I. Mallet, and E. E. Änggård, *J. Chromatogr. B* **667,** 199 (1995).

[7] N. K. Gopaul, E. E. Änggård, A. I. Mallet, D. J. Betteridge, S. P. Wolff, and J. Nourooz-Zadeh, *FEBS Letters* **368,** 225 (1995).

0076-6879/99 $30.00

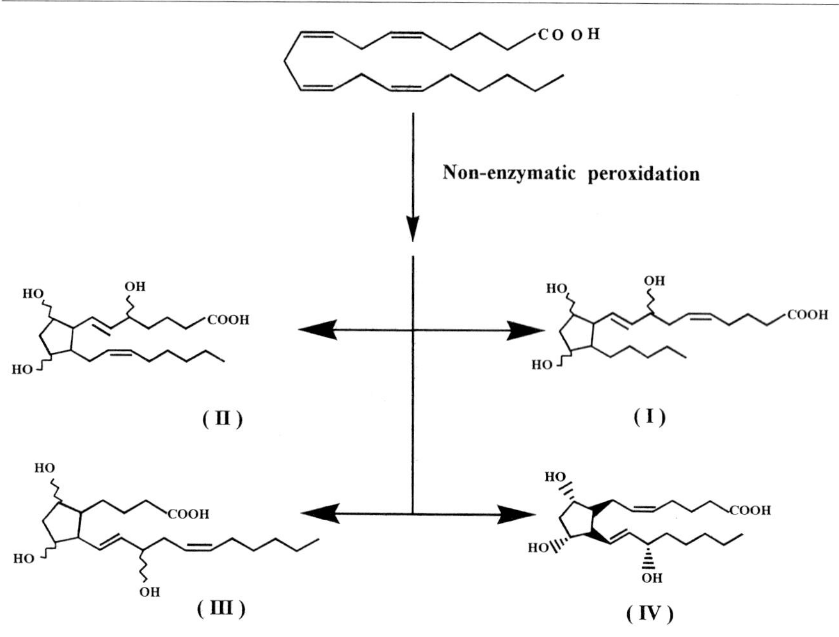

FIG. 1. Proposed structures for the F_2-isoprostane regioisomers formed during non-cyclooxygenase peroxidation of arachidonic acid (AA). Regioisomer **IV** represents the 8-epi-$PGF_{2\alpha}$.

ing the C_{18} chromatography step is improved. (2) The Si cartridge and the TLC steps are replaced with an aminopropyl (NH_2) cartridge. The NH_2 sorbent functions as an ion exchanger selectively binding organic compounds containing a carboxylate anion. Overall recovery of PGF_2-like compounds following chromatography on C_{18} and NH_2 cartridges is 75%. (3) Gas chromatographic separation of PGF_2-like compounds is improved.

Enrichment and Final Determination of Isoprostanes

Solid-Phase Extraction (SPE) Procedure

A. Free (Unesterified). Plasma samples (1 ml) are transferred into glass vials and subsequently 4 ml of 0.1 *M* HCl (pH 2) is added. The samples are centrifuged at 300*g* for 10 min at room temperature and $PGF_{2\alpha}$-d_4 (5 ng in 100 μl ethanol) is added as the internal standard. The supernatants are loaded onto C_{18} cartridges (500 mg, Waters, Millipore Inc., Milford, MA) conditioned with methanol (5 ml) and water (5 ml; pH 2). The cartridges are washed with water (10 ml; pH 2) followed by acetonitrile/water (15/85, v/v) to remove nonlipid materials. The lipids are eluted by washing the cartridges with 5 ml of hexane/ethyl acetate/2-propanol (85/10/5, v/v). The final eluates from the C_{18} cartridges are applied onto NH_2 cartridges (500 mg, Waters, Millipore), preconditioned with hexane (5 ml). The NH_2 cartridges are sequentially washed with 10 ml of hexane/ethyl acetate (15/85, v/v), acetonitrile/water (90/10, v/v), and acetonitrile. Isoprostanes are eluted by washing the NH_2 cartridges with 5 ml of ethyl acetate/methanol/acetic acid (85/10/5, v/v).

B. Total (Sum of Free and Esterified). Plasma samples (1 ml) are transferred into glass vials and 1 *M* aqueous potassium hydroxide (1 ml) is added. The samples are incubated for 30 min at 45°C to release bound (esterified) lipids. The pH is adjusted to 2 by sequentially adding 1 *M* HCl (1 ml) and 0.1 *M* HCl (3 ml) while vortexing. The samples are centrifuged at 3000*g* for 10 min at room temperature with $PGF_{2\alpha}$-d_4 (5 ng in 100 μl ethanol). The remaining procedure for the C_{18} and NH_2 chromatography steps is as described under (A).

Derivatization

Pentafluorobenzyl (PFB) Ester. The final eluate from the NH_2 chromatography step is dried under a stream of nitrogen at 40°. PFB-Br (40 μl; 10% in acetonitrile) and diisopropylethylamine (20 μl; 10% in acetonitrile) are added. The samples are kept at 40° for 45 min.

Trimethylsilyl (TMS) Ether. After the PFB-ester derivatization step, the samples are dried under nitrogen at 40°C. *N,O*-Bis(trimethyl)trifluoroacetamide (50μl) and diisopropylethylamine (5 μl) are added. After incubation at 45°C for 1 h or at 4°C for 12 h, the samples are dried under nitrogen at 40°C. The samples are redissolved in isooctane containing 0.1% *N,O*-bis(trimethyl)trifluoroacetamide (40 μl) and are stored at −20°C until GC-MS analysis.

Gas Chromatographic–Mass Spectrometric/Negative Ion Chemical Ionization (GC-MS/NICI)

Samples are analyzed with a Hewlett-Packard 5890 GC (Bracknell, UK) linked to a VG70SEQ (Fisons Instruments, Manchester, UK) using the NICI with ammonia as reagent gas. Prostanoids are separated on a SPB-1701 column (30 m × 0.25 mm ID; 0.25 μm film thickness, Supelco, Bellefonte, PA). Samples (2 μl) are injected into a temperature-programmed Gerstel injector (Thames Chromatography, Maidenhead, UK). The GC is programmed from 175 to 280°C at a rate of 30°C/min.

The GC-MS/NICI profile of a mixture of authentic PGF_2-like compounds following solid extraction on C_{18} and NH_2 cartridges is illustrated

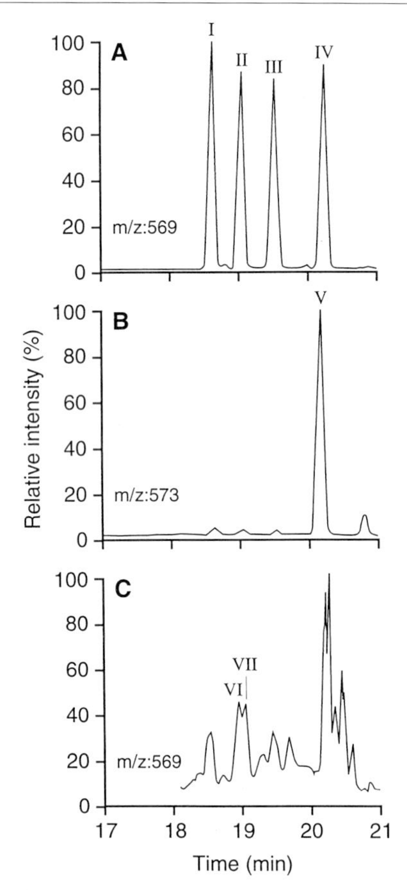

FIG. 2. GC-MS/NICI chromatograms of PGF_2-like compounds as PFB ester/TMS ether derivatives. (A) Authentic PGF_2 standards; (B) PGF_2-d_4; and (C) plasma. Peaks: (I) 9β,11α-PGF_2; (II) 9α,11α-PGF_2 (8-epi); (III) 9α,11β-PGF_2; (IV) 9α,11α-PGF_2; (V) 9α,11α-PGF_2-d_4; (VI) 15(R)-9α,11α-PGF_2(8-epi); and 15(S)-9α,11α-PGF_2(8-epi).

in Figs. 2A and B. The signals monitored at m/z 569 and 573 represent the carboxylated anion $[M\text{-}180]^-$ of PGF_2-like compounds and PGF_2-d_4, respectively. A typical chromatogram of total (sum of free and esterified) F_2-isoprostanes in human plasma is displayed in Fig. 2C. The peaks assigned as VI and VII in the plasma sample represent the 8-epi-15(R)-$PGF_{2\alpha}$ and 8-epi-15(S)-$PGF_{2\alpha}$, respectively. Inter- and intraassay coefficients of variation of the assay were 5% and 10%, respectively.

Applications

Using the combined SPE/GC-MS assay, plasma samples were analyzed for the content of free as well as total (sum of free and esterified) 8-epi-$PGF_{2\alpha}$. Plasma from healthy subjects or type II diabetics had no quantifiable levels of free 8-epi-$PGF_{2\alpha}$ at the detection limit, 0.02 nM.[6,7] On the other hand, the samples had subtantial amounts of total 8-epi-$PGF_{2\alpha}$. For 39 diabetic subjects, the levels of esterified 8-epi-$PGF_{2\alpha}$ varied from 0.32 to 2.19 nM with a mean of 0.93 ± 0.46 nM. The mean level of esterified 8-epi-$PGF_{2\alpha}$ in subjects with chronic renal failure (n = 38) was 0.68 ± 0.41 nM, in the range 0.21–2.47 nM. In subjects with primary hyperlipidemia (n = 15), the levels varied from 0.13 to 0.8 nM with a mean of 0.43 ± 0.22 nM. The mean level of esterified 8-epi-$PGF_{2\alpha}$ corresponding levels in healthy controls (n = 7 and 20, respectively) were 0.34 ± 0.06 nM and 0.28 ± 0.04 nM.

Acknowledgment

The British Heart Foundation is acknowledged for financial support.

[3] Preparation of Lipid Hydroperoxide-Free Low-Density Lipoproteins

By GLENN T. SHWAERY, HIRO-OMI MOWRI, JOHN F. KEANEY, JR., and BALZ FREI

Introduction

The oxidative modification of low density lipoproteins (LDL) has been implicated in the initiation and progression of atherosclerosis.[1] Considerable effort has been devoted to understanding the mechanisms of LDL oxidation both *in vivo* and *in vitro*. One assumption inherent in *ex vivo* studies of LDL oxidation is that the LDL isolated from plasma is similar to circulating or "native" LDL. In particular, it is often assumed that the extent of oxidation present in isolated LDL is similar to that in native LDL.

Lipid hydroperoxides formed during LDL isolation are important determinants of subsequent LDL resistance to metal ion-dependent oxidation

[1] D. Steinberg, S. Parthasarathy, T. E. Carew, J. C. Khoo, and J. L. Witztum, *N. Engl. J. Med.* **320**(14), 915 (1989).

0076-6879/99 $30.00

and can mask or alter effects of LDL antioxidant manipulation.[2] Thus, of paramount importance in studying the oxidation of LDL *ex vivo* is the prevention of adventitious oxidation during its isolation and handling in preparation for such assays. The purpose of this chapter is to describe a method for isolating LDL in a manner that prevents artifactual oxidation, yielding an LDL preparation containing the same amounts of lipid hydroperoxides, if any, as LDL in plasma.

Materials and Methods

We take meticulous care to exclude traces of adventitious metal ions in all aqueous buffers used for isolation of LDL. Buffers are prepared with double distilled water and treated overnight with approximately 100 g/liter Chelex 100 resin (Bio-Rad, Richmond, CA) to remove adventitious metals. Chelex is filtered out of the solution prior to use.

LDL Isolation

Blood is collected from healthy, normolipidemic, male donors, after an overnight fast, into Vacutainer Tubes (Becton Dickinson, Rutherford, NJ) containing 186 USP units sodium heparin/10 ml of whole blood. Plasma is prepared by centrifugation (1200*g*) at 4°C for 15 min. LDL is isolated from heparinized plasma by single vertical spin discontinuous density gradient ultracentrifugation as described.[3] Briefly, plasma density is adjusted to 1.21 g/ml with 0.3265g KBr/ml. A simple two-step gradient is formed by underlaying 3.5 ml of Chelex-treated saline (0.15 *M* NaCl) with 1.5 ml of plasma in a Beckman OptiSeal polyallomar ultracentrifuge tube. The samples are placed in a Beckman NVT-90 rotor and spun at 80,000 rpm (443,000*g*) for 45 min at 7°C in a Beckman L8-80 ultracentrifuge with slow acceleration and deceleration programs applied to minimize the gradient disturbance. The LDL appears as a dense yellow-orange (carotenoids) band approximately one-half the distance from the bottom of the tube. LDL is collected by piercing the tube with an 18-gauge needle and syringe and slowly withdrawing the LDL fraction ($\approx$0.8 ml) after puncturing the top of the tube to relieve any pressure from tube deformation.

Any remaining low molecular weight antioxidants and salts are removed by size-exclusion chromatography using a Sephadex G-25 (coarse grade, Pharmacia Biotech, Inc., Piscataway, NJ) column as previously described.[4]

[2] B. Frei and J. M. Gaziano, *J. Lipid Res.* **34,** 2135 (1993).

[3] B. H. Chung, J. P. Segrest, M. J. Ray, J. D. Brunzell, J. E. Hokanson, R. M. Krauss, K. Beaudrie, and J. T. Cone, *Methods Enzymol.* **128,** 181 (1986).

[4] K. L. Retsky, M. W. Freeman, and B. Frei, *J. Biol. Chem.* **268**(2), 1304 (1993).

Briefly, 6.25 g of fully hydrated Sephadex G-25 that has been treated for at least 30 min at 4°C with double-distilled water or Chelex-treated phosphate-buffered saline (PBS; 0.15 *M* NaCl and 0.01 *M* NaH_2PO_4) is filtered through a Büchner funnel under gentle vacuum and packed in an EconoPak column (Bio-Rad, Richmond, CA). The column is centrifuged at 1200*g* for 10 min at 4° and equilibrated with 2.5 ml of Chelex-treated PBS, and the centrifugation and equilibration steps are repeated. The LDL sample is then applied to the column and centrifuged at 1200*g* for 10 min. The eluate (LDL) is collected into a clean microfuge tube placed over the receiving end of the column with an approximate yield of 0.8 ml of LDL at approximately 1 mg/ml protein. Contaminating metal ions are removed by brief treatment (≈5 min on ice in the dark) of the sample with ≈40 mg Chelex 100 resin followed by filtration through a 0.2 μm filter (Acrodisc, Gellman Sciences, Ann Arbor, MI) to remove the resin. An aliquot of LDL is removed for measurement of protein by a modified procedure of Lowry.[5] The LDL is then assayed immediately for antioxidant and cholesteryl ester hydroperoxide (CEOOH) content. We typically prepare LDL fresh the morning of an experiment and use it immediately.

HPLC Analysis of LDL Antioxidant and Cholesteryl Ester Hydroperoxide Content

In order to measure lipid peroxidation in isolated LDL, we assay LDL CEOOH content using a sensitive and specific HPLC method with postcolumn chemiluminescence detection.[6] For the assay of CEOOH, isolated LDL (100 μg protein) is diluted to 250 μl with Chelex-treated PBS and precipitated with an equal volume of HPLC grade methanol (Sigma Chemical Co., St. Louis, MO), extracted with 10 volumes of HPLC-grade hexane (Aldrich Chemical Co., Milwaukee, WI), vortexed, and centrifuged at 500*g* for 10 min at 4°C. The hexane layer is dried under N_2 and reconstituted in ethanol (200 μl). LDL lipids in 50 μl ethanol (equivalent to 25 μg LDL protein) are separated using a 25 cm LC-18 column (5.0 μm particle size, Supelco, Bellefonte, PA) using 1:1 methanol:*tert*-butanol (v:v) as mobile phase at a flow rate of 1.0 ml/min and UV detection at 205 nm on a Hewlett Packard Series 1100 HPLC system essentially as described.[6] The CEOOH content of the samples is then detected by postcolumn chemiluminescence using a mixing tee and 7:3 (v:v) methanol:100 m*M* sodium borate buffer (pH 10) containing 1 m*M* isoluminol (Sigma) and 5 mg/liter microperoxidase (Sigma) as the postcolumn detection solution at a flow rate of 1.5 ml/

[5] G. L. Peterson, *Anal. Biochem.* **83,** 346 (1977).

[6] R. Stocker and B. Frei, *in* "Oxidative Stress: Oxidants and Antioxidants" (H. Sies, Ed.), p. 213. Academic Press, London, 1991.

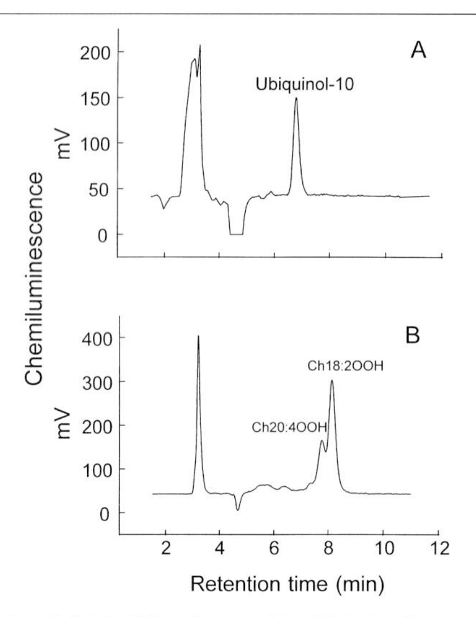

FIG. 1. LDL ubiquinol and CEOOH as detected by HPLC using postcolumn chemiluminescence. (A) Chromatogram of LDL lipids from samples isolated by vertical rotor ultracentrifugation and gel filtration. Ubiquinol-10 is detected with a retention time of 6.9 min in the absence of CEOOH (retention time, 8.0 min). (B) LDL isolated by density gradient ultracentrifugation for 24 h followed by dialysis for 24 h as described in Table I. The ubiquinol-10 has been completely consumed concomitant with the appearance of at least two peaks of cholesteryl ester oxidation products (Ch20:4OOH and Ch18:2OOH) with retention times of approximately 8 min.

min.[7] This method is extremely sensitive with a detection limit of 10 pmol/mg LDL protein or 1 CEOOH for every 200 LDL particles.

Comparison with Other Methods of LDL Isolation

LDL isolated by two-step gradient vertical rotor ultracentrifugation[3] for 45 min and size-exclusion chromatography[4] was found to be free of CEOOH as detected by HPLC with postcolumn chemiluminescence (Fig. 1, Table I). The ubiquinol-10 peak (retention time, 6.9 min) from this freshly prepared, unoxidized LDL was subsequently assigned a value of 100% for comparison. On careful storage of this clean preparation at 4°C in the dark, we observed autoxidation of the sample manifested by ubiquinol-10 consumption and the accumulation of CEOOH in LDL (Table I). This loss of ubiquinol-10 is indicative of the earliest events in LDL

[7] B. Frei, Y. Yamamoto, D. Niclas, and B. N. Ames, *Anal. Biochem.* **175**(1), 120 (1988).

TABLE I
LDL OXIDATION DURING ISOLATION[a]

LDL isolation method	LDL sampling	CEOOH (pmol/mg LDL protein)	Ubiquinol-10 (% control)	Vitamin E (% control)
Vertical rotor	After isolation	≤10	100	100
	Stored 48 h	41	45	91
Density gradient	After isolation	42	0	100
	Dialyzed 24 h	5612	0	83
Differential density gradient	After isolation	67	6	106
	Dialyzed 24 h	1971	4	106

[a] LDL samples were subjected to lipid extraction either immediately after isolation by ultracentrifugation or after subsequent storage or dialysis as indicated. Vertical rotor LDL isolation involved a single 45 minute ultracentrifugation as described in the text.[3] Density gradient isolation involved a four-step gradient with centrifugation for 20 h as described.[10] Differential density gradient isolation involved a 20-h centrifugation followed by a 20-h three-step gradient centrifugation as described.[11] Values are the means of 2 samples from different subjects that varied by less than 15%.

oxidation and is in agreement with data from several studies showing the rapid loss of ubiquinol-10 with little or no changes in vitamin E content.[8,9]

Alternatively, methods involving long periods of centrifugation and extensive dialysis[10,11] yielded LDL samples with significant accumulation of CEOOH (Ch20:4OOH and Ch18:2OOH) and depleted antioxidant content as a result of adventitious oxidation during preparation (Table I). We find that LDL prepared in this manner is devoid of ubiquinol-10. Although the CEOOH content was relatively small on completion of centrifugation using these methods, we found that removal of any remaining small molecular weight plasma antioxidants and EDTA by dialysis leads to rapid accumulation of CEOOH even with precautions to minimize such oxidation, such as protection from light and performing dialysis at 4°C with deoxygenated solutions (Table I). Vitamin E levels in LDL do not appear compromised by small quantities of CEOOH detected after centrifugation. However, significant oxidation occurs during dialysis in the density gradient procedure[10] and 17% of the LDL vitamin E is consumed (Table I). Further oxidation of cholesteryl esters upon storage of LDL preparations results

[8] R. Stocker, V. W. Bowry, and B. Frei, *Proc. Natl. Acad. Sci. USA* **88**(5), 1646 (1991).

[9] D. L. Tribble, J. J. M. van den Berg, P. A. Motchnik, B. N. Ames, D. M. Lewis, A. Chait, and R. M. Krauss, *Proc. Natl. Acad. Sci. USA* **91,** 1183 (1994).

[10] A. H. Terpstra, C. J. Woodward, and F. J. Sanchez-Muniz, *Anal. Biochem.* **111,** 149 (1981).

[11] A. H. M. Terpstra and A. E. Pels III, *Fresenius Z. Anal. Chem.* **330,** 149 (1988).

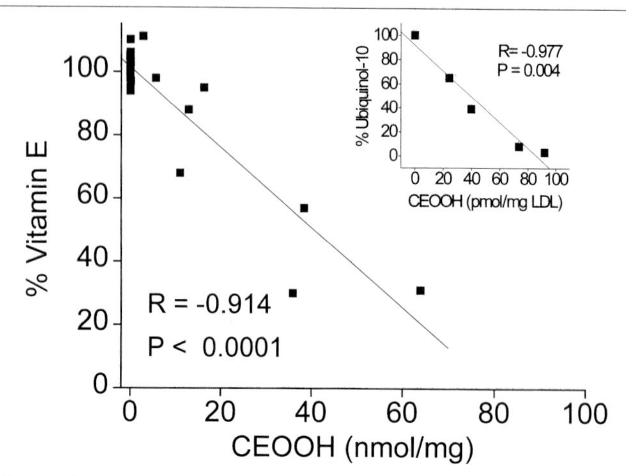

FIG. 2. Concomitant disappearance of ubiquinol-10 and vitamin E and formation of CEOOH during LDL preparation and storage. LDL was prepared by the three methods described in Table I and ubiquinol-10 (*inset*), vitamin E, and CEOOH were measured using HPLC. Some samples were stored for up to 48 h at 4°C in the absence of EDTA. There was no loss of vitamin E during the early stages of oxidation with CEOOH levels at pmol/mg LDL protein. On significant accumulation of nmol CEOOH /mg LDL protein, LDL vitamin E content decreased linearly with as much as a 70% loss in some samples. This pattern of antioxidant consumption with concomitant accumulation of CEOOH is indicative of the earliest stages of LDL oxidation.

in further decreases in LDL vitamin E (Fig. 2). In contrast, small accumulations of CEOOH are associated with complete consumption of ubiquinol-10 (Fig. 2).

Conclusions

Lipoproteins have been routinely isolated from plasma based on their flotation characteristics during ultracentrifugation in differential and density gradients. Modifications of the method described by Redgrave[12] have been used combining standard fixed angle or swinging bucket rotors with one or two centrifugations of 20 h or more.[11] More recently, with the advent of near-vertical and vertical rotors, simple two-step gradients have been employed with durations of ultracentrifugation as brief as 25 min for isolation of LDL. These methods described by Chung and colleagues[3] have several advantages, including significantly shorter centrifugation times and ease of preparation with underlaying of density adjusted plasma in a simple

[12] T. G. Redgrave, *Anal. Biochem.* **65,** 42 (1975).

two-step gradient; the methods also minimize apoprotein redistribution and oxidation of lipoproteins that may occur during longer periods of ultracentrifugation. In support of this third advantage, we have found that preparation of LDL by vertical rotor ultracentrifugation and removal of low molecular weight contaminants by Sephadex G-25 column chromatography yield a preparation that is free of detectable cholesteryl ester oxidation products and possesses a full complement of antioxidants. However, more conventional methods of LDL isolation compromise the antioxidant status of LDL particles and yield LDL which is "seeded" with lipid hydroperoxides which have been shown to be directly associated with decreased resistance to Cu^{2+}-mediated oxidation.[2]

Acknowledgments

This work was supported by NIH grants HL55834, HL03195, and HL59634 (to J.F.K.) and HL49954 (to B.F.). J. F. Keaney, Jr., is the recipient of a Clinical Investigator Development award from the NIH.

[4] Lipid Hydroperoxide Analysis by High-Performance Liquid Chromatography with Mercury Cathode Electrochemical Detection

By WITOLD KORYTOWSKI, PETER G. GEIGER, and ALBERT W. GIROTTI

Introduction

Peroxidative modification of unsaturated lipids in cell membranes, lipoproteins, and other organized systems is attracting widespread biomedical interest because of its possible role in pathological conditions such as ischemia–reperfusion injury, neurodegeneration, atherogenesis, and carcinogenesis.[1,2] This interest has stimulated the development of highly sensitive techniques for assessing lipid peroxidation based on measurement of intermediates/products such as hydroperoxides, aldehydes, and alkanes.[3,4] Though long popular because of their relative simplicity, "bulk"-type assays based on determination of thiobarbituric acid-reactive by-products, conju-

[1] A. W. Girotti, *J. Free Rad. Biol. Med.* **1,** 87 (1985).
[2] B. Halliwell and J. M. C. Gutteridge, "Free Radical Biology and Medicine," Clarendon Press, New York, 1989.
[3] T. F. Slater, Methods in Enzymology, Vol. 105, p. 283 (1984).
[4] W. A. Pryor and L. Castle, Methods in Enzymology, Vol. 105, p. 294 (1984).

0076-6879/99 $30.00

gated dienes, or iodide-reducible hydroperoxides often lack specificity and may (depending on the assay) detect only part of the peroxide population or peroxidation process. These limitations have prompted the relatively recent introduction of highly sensitive approaches for separating and determining relatively long-lived intermediates such as lipid hydroperoxides (LOOHs). Most of these involve LOOH separation by high-performance liquid chromatography (HPLC) combined with some type of high sensitivity detection. Detection based on chemiluminescence (CL) of oxidized luminol[5,6] or isoluminol[7,8] has been used quite extensively. We recently developed a new approach based on reversed-phase HPLC with reductive-mode electrochemical detection on a renewable mercury drop cathode.[9–11] Adapted from earlier prototypes,[12,13] this approach, designated HPLC-EC(Hg), is vastly superior to EC detection with static thin film electrodes[14,15] because the latter inevitably lose sensitivity over time due to analyte deposition. When compared with HPLC-CL, HPLC-EC(Hg) has the advantage of (i) not requiring any postcolumn chemistry, (ii) being less susceptible to interference, and (iii) attaining lower detection limits, which approach 100 fmol for cholesterol-derived hydroperoxides.[9,10] Two applications of HPLC-EC(Hg) are described in this chapter: (i) determination of LOOHs in mammalian cells exposed to minimal oxidative stress, and (ii) an improved assay for determining activity of the selenoenzyme phospholipid hydroperoxide glutathione peroxidase (PHGPX).

General Procedures

Chemicals, Biochemicals, and Standards. HPLC grade methanol, 2-propanol, and acetonitrile are obtained from Burdick and Jackson (Muskegon, MI). Cholesterol, reduced glutathione (GSH), NADPH, and glutathione peroxidase (GPX) are from Sigma (St. Louis, MO); desferrioxamine (DFO)

[5] T. Miyazawa, *Free Rad. Biol. Med.* **7,** 209 (1989).

[6] T. Miyazawa, K. Fujimoto, T. Suzuki, and K. Yasuda, Methods in Enzymology, Vol. 233, p. 324 (1994).

[7] B. Frei, Y. Yamamoto, D. Niclas, and B. N. Ames, *Anal. Biochem.* **175,** 120 (1988).

[8] Y. Yamamoto, Methods in Enzymology, Vol. 233, p. 319 (1994).

[9] W. Korytowski, G. J. Bachowski, and A. W. Girotti, *Anal. Biochem.* **213,** 11 (1993).

[10] W. Korytowski, P. G. Geiger, and A. W. Girotti, *J. Chromatogr. B* **670,** 189 (1995).

[11] W. Korytowski and A. W. Girotti, *in* "Analysis of Free Radicals in Biological Systems" (A. E. Favier, J. Cadet, B. Kalyanaraman, M. Fontecave, and J. L. Pierre, Eds.), p. 165, Birkhauser Verlag, Basel, 1995.

[12] M. O. Funk, M. B. Keller, and B. Lewison, *Anal. Chem.* **52,** 773 (1980).

[13] M. O. Funk, P. Walker, and J. C. Andre, *Bioelectrochem. Bioenerget.* **18,** 127 (1987).

[14] K. Yamada, J. Terao, and S. Matsushita, *Lipids* **22,** 125 (1987).

[15] W. Korytowski, G. J. Bachowski, and A. W. Girotti, *Anal. Biochem.* **197,** 149 (1991).

from Ciba-Geigy (Suffern, NY); and 1-palmitoyl-2-oleoyl-*sn*-glycero-3-phosphocholine (POPC) from Avanti Polar Lipids (Birmingham, AL). The photosensitizing dyes merocyanine 540 (MC540) and chloroaluminum phthalocyanine disulfonate ($A1PcS_2$) are supplied respectively by Eastman Kodak (Rochester, NY) and Dr. J. van Lier (University of Sherbrooke). PHGPX is isolated from rat testis as described.[16] Specific activity of the purified enzyme, as determined by coupled assay,[17] is typically 2400–2600 U/mg protein. The following hydroperoxide standards are prepared by dye-sensitized photooxidation of parent lipids,[15] followed by HPLC separation of products: 3β-hydroxy-5α-cholest-6-ene 5-hydroperoxide(5α-OOH); 3β-hydroxycholest-4-ene 6α-hydroperoxide(6α-OOH); 3β-hydroxycholest-4-ene 6β-hydroperoxide (6β-OOH); 3β-hydroxycholest-5-ene 7α-hydroperoxide (7α-OOH); 3β-hydroxycholest-5-ene 7β-hydroperoxide (7β-OOH); POPC-OOH, 1-palmitoyl-2-oleoyl-*sn*-glycero-3-phosphocholine hydroperoxide.

Sample Extraction. In 1.5 ml polypropylene microfuge tubes, samples for LOOH analysis are typically brought to a volume of 0.5 ml with Chelex-treated PBS. EDTA (0.1 m*M*) is included in order to chelate and retain any redox metal ions in the aqueous phase. Samples are then mixed with 0.8 ml of chloroform/methanol (2 : 1, v/v) and extracted by vortexing for ~1 min. After centrifugation, each upper layer is discarded and 0.4 ml of the lower (organic) layer is recovered and dried under nitrogen. Dried samples are stored at −20° C. LOOHs are quantitatively extracted using this procedure.[18]

Instrumentation. Chromatography is accomplished on an Ultrasphere XL-ODS column (4.6 × 70 mm; 3 μm particles) from Beckman Instruments (San Ramon, CA), using an Isco integrated HPLC system (Lincoln, NE) interfaced with an EG&G-Princeton Model 420 electrochemical detector (Princeton, NJ). The detector is equipped with a hanging mercury drop electrode; operating potential is set at −150 mV versus Ag|AgCl. Separations are carried out at ambient temperature (25–27°C). A typical mobile phase consisting of methanol/acetonitrile/2-propanol/aqueous 1 m*M* sodium perchlorate plus 10 m*M* ammonium acetate (72 : 11 : 8 : 9 v/v) is sparged continuously with high purity argon (>99.998%; BOC Gases, Chicago, IL) that has been passed first through an OMI-1 oxygen scrubber (Supelco, Bellefonte, PA) to reduce O_2 concentration to <10 ppb, and then through a presaturating mobile phase scrubber. An overnight purging

[16] A. Roveri, A. Casasco, M. Maiorino, P. Dalan, A. Calligaro, and F. Ursini, *J. Biol. Chem.* **267,** 6142 (1992).

[17] J. P. Thomas, M. Maiorino, F. Ursini, and A. W. Girotti, *J. Biol. Chem.* **265,** 454 (1990).

[18] G. J. Bachowski, W. Korytowski, and A. W. Girotti, *Lipids* **29,** 449 (1994).

with argon or 1 h purging with high purity helium, followed by 30 min column washing, is usually sufficient to achieve a background current <0.5 nA, which allows operation at the highest detector sensitivity, 0.1 nA full scale. The mobile phase is delivered at a flow rate of 1.7 ml/min against a back pressure of approximately 2000 psi. Samples in 2-propanol are injected manually through a 10 μl loop in the Rheodyne injection valve. A fresh mercury drop is dispensed and equilibrated 3–4 min before each injection. Stainless steel tubing is used throughout to prevent O_2 contamination of the mobile phase.[9,10]

Other Techniques. Total LOOH content of standards and highly peroxidized samples is determined by iodometric assay.[17,18] Measurement of PHGPX activity by conventional coupled enzymatic assay is carried out as described[17]; this is used for purified enzyme and enzyme in cell extracts when relatively large amounts of sample material are available. Murine leukemia L1210 cells are grown in RPMI 1640 medium supplemented with 1% fetal calf serum and other factors,[19] and either lacking or containing sodium selenite (60 n*M*). Reseeding into fresh medium is carried out on alternate days. Cells given selenite are Se-replete and designated Se(+), while those not given selenite are Se-deficient and designated Se(−). Ten million cells of either type in exponential growth contain 0.14 ± 0.01 mg of total lipid and 0.64 ± 0.04 mg of total protein.[19]

Detection of Lipid Hydroperoxides in Oxidatively Stressed Cells

Cells may accumulate LOOHs when their ability to scavenge activated oxygen species (e.g., superoxide, hydrogen peroxide) or to reduce/detoxify LOOHs themselves is overwhelmed. LOOHs can be deleterious for at least two reasons: (i) relatively high polarity, which perturbs membrane structure/function; (ii) susceptibility to iron-catalyzed degradation, which triggers damaging chain peroxidation reactions. For studying factors that influence LOOH formation on the one hand vs one-electron or two-electron turnover on the other,[1,2] one can generate LOOH species in many different ways in model membranes and cells, e.g., exposure to Fe(III)/ascorbate, azo initiators, or dye-sensitized photodynamic action. When mediated by singlet oxygen ($^1O_2$), the latter approach, unlike the first two which generate free radicals,[2] produces LOOHs in a steady, linear fashion as $^1O_2$ adds to unsaturated lipids.[20] This makes LOOH yield more predictable and reproducible. Another advantage is that reactions can be stopped by simply removing light, which is more reliable than adding a metal chelator or

[19] F. Lin, P. G. Geiger, and A. W. Girotti, *Cancer Res.* **52,** 5282 (1992).

[20] A. W. Girotti, *Photochem. Photobiol.* **51,** 497 (1990).

lowering temperature, for example, in the case of free radical-mediated reactions.

Photooxidation Procedure. A typical protocol for photoperoxidizing suspended cells (illustrated for the L1210 line) is as follows. Exponentially growing Se(+) cells are centrifuged (1000*g*, 5 min) and resuspended to a titer of $\sim 1.0 \times 10^6$/ml in 1% serum/Se(+) RPMI medium (total lipid ~14 μg/ml). The cell suspension (typically 25 ml in a culture flash) is dark-incubated for 30 min with 5–10 μM MC540, a $^1O_2$-generating sensitizer.[21] The cells are then transferred to 30-mm culture dishes (2.5 ml per dish), which are illuminated at room temperature on a translucent plastic platform over a twin bank of 40-W cool-white fluorescent tubes. Light intensity (fluence rate) at the platform surface is measured with a Yellow Springs radiometer (Model 65A) and maintained at 6–7 W/m^2. After exposure to a given light fluence, the cells from each dish are transferred to a microfuge tube, resuspended in 0.5 ml of PBS/EDTA, and extracted (see *General Procedures*). Recovered lipid fractions are dried, redissolved in a minimal volume of isopropanol (25–50 μl), and then analyzed by HPLC-EC(Hg). A mixture of cholesterol hydroperoxide (ChOOH) and phospholipid hydroperoxide (PLOOH) standards is typically analyzed alongside the cell samples in order to establish analyte retention times and responsiveness.

Lipid Hydroperoxide Profiles. Figure 1 shows the HPLC-EC(Hg) profile of an artificial mixture of photochemically generated LOOH standards, including 7α,7β-OOH, 5α-OOH, 6β-OOH, and POPC-OOH. The three ChOOHs are well separated not only from one another, but also from a PLOOH, which elutes as a partially resolved doublet reflecting structural microheterogeneity.[10] Figure 2 shows chromatograms of lipid extracts from Se(+) L1210 cells that had been photodynamically stressed. At a light fluence of ~0.08 J/cm^2, one sees approximately 8 well-defined EC peaks over a 12 min range (Fig. 2, pattern a). A fivefold greater fluence produces more intense signals (Fig. 2, pattern c; note the 2.5-fold lower sensitivity setting than in Fig. 2b). Only the peak at 2.8 min is observed in the light control (Fig. 2, pattern a), indicating that MC540 sensitization is necessary for generation of the other signals. That all peaks except the one at 2.8 min represent hydroperoxides is inferred from their >90% diminution after GSH/PHGPX treatment (compare Fig. 2, patterns d and c). Peaks 1, 2, 3, and 4 (Fig. 2) are assigned as 7α,7β-OOH, 5α-OOH, 6α-OOH, and 6β-OOH, respectively, based on retention times and increased intensity on spiking samples with the respective standards (see Fig. 1). These ChOOHs

[21] B. Kalyanaraman, J. B. Feix, F. Seiber, J. P. Thomas, and A. W. Girotti, *Proc. Natl. Acad. Sci. USA* **84,** 2999 (1987).

[22] M. Maiorino, C. Gregolin, and F. Ursini, Methods in Enzymology, Vol. 186, p. 448 (1990).

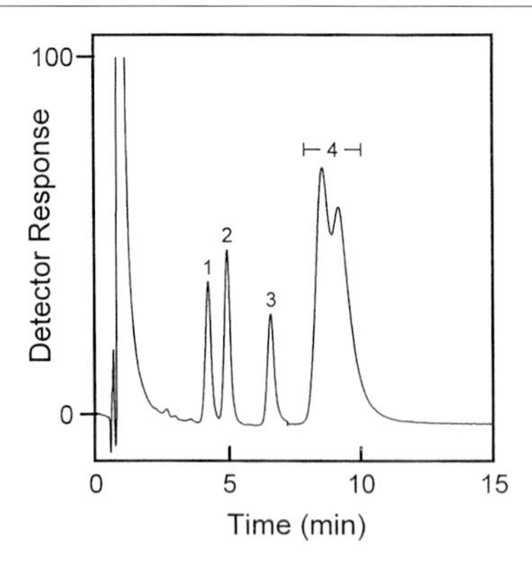

FIG. 1. HPLC-EC(Hg) chromatogram of lipid hydroperoxide standards. Hydroperoxide assignments, retention times, and amounts injected are as follows: (1) unresolved $7\alpha,7\beta$-OOH (4.24 min, 20 pmol); (2) 5α-OOH (4.95 min, 20 pmol); (3) 6β-OOH (6.61 min, 20 pmol); (4) POPC-OOH (8.58 min and 9.19 min, 200 pmol). Full scale detector sensitivity was 2.0 nA.

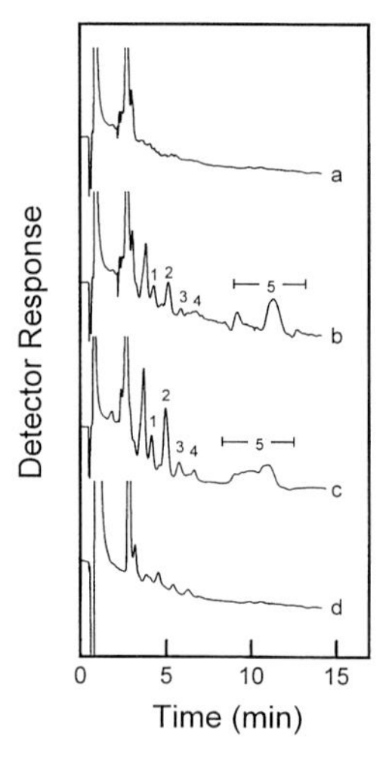

FIG. 2. HPLC-EC(Hg) chromatogram of lipid hydroperoxides from photooxidized L1210 cells. Se(+) cells (1.0×10^6/ml in 1% serum/RPMI medium) were exposed to the following doses of broad-band visible light: (a) 385 mJ/cm^2 in the absence of sensitizer; (b) 77 mJ/cm^2 and (c) 385 mJ/cm^2, both in the presence of 5 μM MC540. (77 mJ/cm^2 and 385 mJ/cm^2 correspond to 2 min and 10 min of irradiation, respectively.) Samples were recovered and lipid extracts representing 5.5×10^6 cells (~80 μg of total lipid) were analyzed. Sample (d) represents photooxidized cells (385 mJ/cm^2) after solubilization in 0.1% Triton X-100 and 1 h of dark incubation in the presence of 5 mM GSH and 19 μg/ml PHGPX at 37°C. Peak assignments are as follows: (1) $7\alpha,7\beta$-OOH; (2) 5α-OOH; (3) 6α-OOH; (4) 6β-OOH, (5) PLOOH family. Full-scale detector sensitivity was 0.2 nA (a, b, d) and 0.5 nA (c).

are presumed to derive mainly from plasma membrane cholesterol because MC540 localizes in the plasma membrane,[21] where most of the cellular cholesterol is found. Peaks 5 represent a family of *sn*-2 peroxidized phospholipids because they disappear upon treating solubilized cell samples with Ca^{2+} and phospholipase A_2 (not shown). Detection of 5α-OOH and 6-OOHs confirms $^1O_2$ involvement in the photoreaction.[23] How the 7-OOHs originate is not entirely clear, although 5α-OOH rearrangement may be at least partially responsible.[23,24] Quantitative analysis based on EC responses of peroxide standards indicates that the ChOOHs shown in Fig. 2b collectively account for ~0.05 mol % of cellular cholesterol and the PLOOHs ~0.03 mol % of cellular phospholipid. This photodynamic insult is found to be minimally toxic, resulting in <5% loss of clonogenicity.[19] Clearly, therefore, HPLC-EC(Hg) permits LOOH identification and quantitation at very low levels of oxidative stress lethality.

HPLC-EC(Hg)-Based Assay for Measuring Activity of Phospholipid-Hydroperoxide Glutathione Peroxidase

PHGPX is a monomeric selenoenzyme (~20 kDa) which can catalyze the reduction of a wide variety of peroxides, including LOOHs, and may play an important role in cellular detoxification of these species.[22,25] In the classical approach for determining activity,[17,22] PHGPX-catalyzed reduction of a PLOOH substrate is coupled to the glutathione reductase-catalyzed reduction of glutathione disulfide, using GSH and NADPH as coreductants, and correcting for any nonspecific oxidation of NADPH in the absence of PLOOH or GSH. The sensitivity of this method is compromised by (i) the upper limit of cellular material in the assay mixture (~10^7 cells/ml) before the sample becomes too turbid for spectrophotometric measurement at 340 nm, and (ii) substantial nonspecific decay of NADPH (~0.005 AU/min) compared with PHGPX-dependent decay for low-activity samples. As a result, 0.3–0.6 U/10^7 cells is estimated to be the practical limit for measuring PHGPX activity under coupled assay conditions as specified.[17,22] A better approach for assessing activity would be based on direct measurement of disappearance of the hydroperoxide substrate and/or appearance of the alcohol product. One such approach[26] involves reversed phase HPLC separation of a phosphatidylcholine hydroperoxide from its corresponding alcohol and monitoring of conjugated diene absorbance at 232 nm. Here

[23] M. Kulig, and L. L. Smith, *J. Org. Chem.* **38,** 3639 (1973).

[24] L. L. Smith, "Cholesterol Autoxidation," Plenum Press, New York, 1981.

[25] F. Antunes, A. Salvador, and R. E. Pinto, *Free Rad. Biol. Med.* **19,** 669 (1995).

[26] Y. Bao, S. J. Chambers, and G. Williamson, *Anal. Biochem.* **224,** 395 (1995).

we describe a method based on much more sensitive HPLC-EC(Hg) determination of PHGPX-catalyzed reduction of 7β-hydroperoxycholesterol (7β-OOH).[27] 7β-OOH has the following advantages over other LOOHs as a PHGPX substrate: (i) relatively short retention time and low detection limit (~0.1 pmol) on HPLC-EC(Hg); (ii) stability with respect to epimerization or rearrangement (unlike 7α-OOH or 5α-OOH).[24]

Preparation of Peroxide Substrate. 7β-OOH is prepared by photooxidation of cholesterol (10 mM) in chloroform, using $A1PcS_2$ (40 μM) as the sensitizer. The stirred solution is irradiated for 12–14 h at 5° C, using a quartz–halogen source (fluence rate ~0.35 W/m^2). After irradiation, chloroform is evaporated under a stream of nitrogen; the sample is dissolved in methanol and subjected to two HPLC purification steps. First, four major ChOOHs are separated from one another and parent cholesterol, using a semipreparative ODS column (Ultrasphere, 10 × 250 mm; 5 μm particles) and methanol/acetonitrile/2-propanol/water (76:11:7.5:5.5 (v/v)) as the mobile phase, which is pumped at 3 ml/min. Analyte identities and retention times are as follows: 7α- and 7β-OOH mixture (single peak at 17 min); 5α-OOH (20 min); 6α-OOH (24 min); 6β-OOH (27 min); cholesterol (64 min). The fraction containing $7\alpha,7\beta$-OOH is collected, evaporated to dryness under nitrogen, dissolved in hexane, and rechromatographed, using a silica column (Ultrasphere, 4.6 × 75 mm; 3 μm particles), a mobile phase consisting of hexane/2-propanol (97.5:2.5 (v/v)) pumped at 1.5 ml/min, and UV detection at 212 nm. 7β-OOH and 7α-OOH are resolved under these conditions, the retention times being 10.5 min and 13.2 min, respectively. The 7β-OOH fraction is recovered, dried down, dissolved in 2-propanol, and stored in a −20° freezer, where it is stable for several months.

Activity Measurement. The PHGPX assay to be described applies for purified enzyme as well as enzyme in cell extracts. In the latter case, a 0.2 ml aliquot of cells (~10^8/ml) is solubilized in 1% Triton X-100, centrifuged, and the supernatant solution recovered for analysis. The complete composition of a typical reaction mixture (1.0 ml) is as follows: sample (a specified amount of purified PHGPX or solubilized cell supernatant), 5 mM GSH, 50 μM DFO, 0.1% Triton X-100, Chelexed PBS, and 50 μM 7β-OOH (added last). At various times, 50 μl aliquots are removed, mixed with 0.2 ml of Chelexed PBS, and extracted with 0.4 ml of chloroform/methanol (see *General Procedures*). After centrifugation, 0.2 ml of the lower phase is recovered and dried under nitrogen; the residue is dissolved in 50 μl of 2-propanol and analyzed by HPLC-EC(Hg). A representative decay profile for 7β-OOH using purified PHGPX is shown in Fig. 3. Pseudo first-order

[27] W. Korytowski, P. G. Geiger, and A. W. Girotti, *Biochemistry* **35,** 8670 (1996).

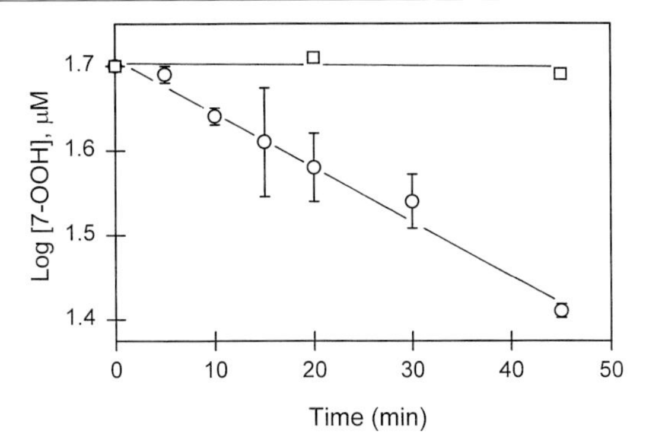

FIG. 3. Time course for the PHGPX-catalyzed reduction of 7β-OOH. The reaction was carried out at 37°C in Chelex-treated PBS containing 5 m*M* GSH, 50 μ*M* DFO, 0.1% Triton X-100, 3.9 μg/ml PHGPX, and 50 μ*M* 7β-OOH; starting volume was 1.0 ml. At the indicated times, 100 μl aliquots were extracted and recovered lipid fractions were analyzed by HPLC-EC(Hg) (○). A control system containing everything except PHGPX was also analyzed (□).

kinetics are observed over at least a 45-min period. No reaction occurs when (i) the sample is heated at 80° for 15 min before assaying (not shown); (ii) GSH or PHGPX is omitted (Fig. 3); or (iii) "classical" glutathione peroxidase (GPX) is substituted for PHGPX.[17] This indicates that catalytic activity specific to PHGPX is being observed. Figure 4 shows that the initial rate of 7β-OOH reduction (calculated from pseudo first-order

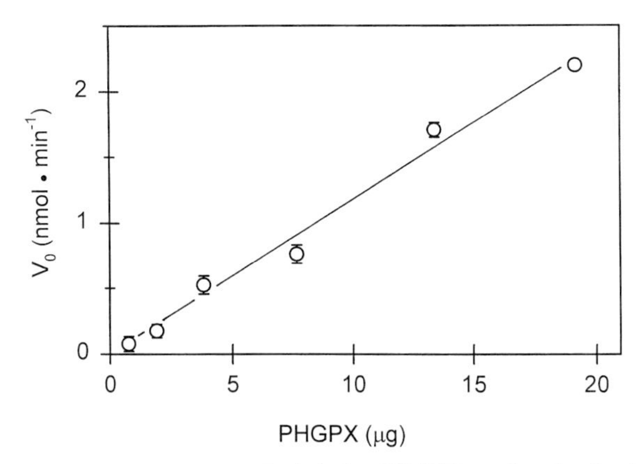

FIG. 4. Initial rate of 7β-OOH decay (V_o) during PHGPX-catalyzed reduction of 7β-OOH as a function of increasing amount of enzyme in micrograms. Additional details are provided in Fig. 3. Specific enzymatic activity calculated from the slope of the curve is 119 ± 15 U/mg protein.

TABLE I
EFFECT OF SELENIUM DEFICIENCY ON PHGPX ACTIVITY OF L1210 CELLS[a]

Time (days)	Enzyme activity (U/10^7 cells)	
	Se(+)	Se(−)
6	0.42 ± 0.03	0.12 ± 0.04
42	0.42 ± 0.04	0.04 ± 0.03
90	0.45 ± 0.01	0.02 ± 0.02

[a] Cells growing in 1% serum/RPMI medium supplemented with Na_2SeO_3 (60 n*M*) were either maintained on Na_2SeO_2 [Se(+)] or deprived of it [Se(−)]. After the indicated growth periods, PHGPX activity was measured, using 7β-OOH as peroxide substrate and HPLC-EC(Hg) for analysis. Values are means ± SD of three measurements.

data as in Fig. 3) is linearly dependent on the amount of PHGPX protein in the assay mixture up to at least 20 μg.

The HPLC-EC(Hg)-based assay for PHGPX has the following advantages over the conventional coupled assay: (i) very high sensitivity, reliability, and reproducibility; (ii) an essentially negligible correction for background reactions; and (iii) no limit imposed by sample turbidity. This permits measurements on cells expressing very low enzymatic activity. We estimate that with 10^8 cells/ml and a 1 h reaction period, the limit of detection is ~10 mU/10^7 cells (based on 1 U = 1 nmol/min). A practical aspect of this is illustrated by measurement of PHGPX activity in Se-deficient L1210 cells. When cells growing in 1% serum with Se supplementation are taken off the latter, PHGPX activity drops,[19] but after 5–6 days reaches a level close to the detection limit of the coupled assay, i.e., ~0.5 U/10^7 cells, disallowing any accurate measurement of residual activity beyond this point. However, using the more sensitive HPLC-EC(Hg)-based assay, we find that PHGPX activity of Se(−) cells continues to decline after 6 days, reaching ~4% of the Se(+) level after about 3 months (Table I).

Summary

In addition to the applications described, HPLC-EC(Hg) can be used for determining LOOHs in lipoproteins[28] and for monitoring LOOH detoxification in cells.[27] As it continues to be developed and refined, this approach

[28] J. P. Thomas, B. Kalyanaraman, and A. W. Girotti, *Arch. Biochem. Biophys.* **315,** 244 (1994).

should prove to be valuable not only for ultrasensitive determination of lipid-derived peroxides, but protein- and nucleic acid-derived peroxided as well.

Acknowledgments

This work was supported by USPHS Grants CA70823 and CA72630 from the National Cancer Institute, and by FIRCA Grant TW00424 from the Fogarty International Center.

[5] Measurement of Lipid Hydroperoxides in Human Plasma and Lipoproteins by Kinetic Analysis of Photon Emission

By ANTONIO M. PASTORINO, ADRIANA ZAMBURLINI, LUCIO ZENNARO, MATILDE MAIORINO, and FULVIO URSINI

Introduction

Lipid hydroperoxides participate in reactions leading to oxidative damage of biomolecules. Besides this detrimental effect, it has been suggested that lipid hydroperoxides could play a physiological role as well, by accepting reducing equivalents from substrates undergoing a physiological redox transition. The chemical (lipid peroxidation) vs the enzymatic (lipoxidase-catalyzed) nature of hydroperoxidation of lipids fits this dual aspect.[1]

The effects of lipid hydroperoxides in biological systems rely on free radical and non-free-radical reactions and, reasonably, on biophysical effects on membranes.

The molecule-assisted homolytic decomposition of hydroperoxides and, more likely, the one-electron transfer reactions with metal ion, both producing free radicals, account for a major pathway of initiation of lipid peroxidation,[1,2] whereas the heterolytic decomposition via a nonradical nucleophilic displacement leads to oxidation of nucleophiles such as thiols and methionine residues.[3] Finally, by decreasing the packing of lipids and increasing the water concentration in the lipid phase of monolayers, trace amounts

[1] F. Ursini, M. Maiorino, and A. Sevanian, *in* "Oxidative stress: Oxidants and Antioxidants" (H. Sies, Ed.), p. 319. Academic Press, London, 1991.

[2] J. Terao, *in* "Membrane Lipid Oxidation," Vol. I (C. Vigo-Pelfrey, Ed.), p. 219. CRC Press, Boca Raton, 1990.

[3] P. J. O'Bien, *Can. J. Biochem.* **47,** 485 (1969).

0076-6879/99 $30.00

of lipid hydroperoxides perturb the interface, as shown by polarity-sensitive fluorescent probes.[4]

The accurate quantitative analysis of lipid hydroperoxides in biological samples is a difficult task because of their low concentration and instability. This is one of the reasons why, most frequently, lipid peroxidation is measured by analyzing the content of products of decomposition or rearrangement of lipid hydroperoxides, which are more easily detected by colorimetric or immunological reactions.

The direct analysis of lipid hydroperoxides by enzymatic reduction[5] is specific, but the sensitivity of the detection is limited by the absorbance coefficient or fluorescence of NADPH. Titration with redox compounds such as iodine[6] or ferrous ion complexes[7] is not—at least in our hands—as specific as desirable for analyzing biological samples, such as plasma or native liproproteins.

The recently introduced chromatographic–cheluminescent procedures,[8,9] in which lipid hydroperoxides are measured by chemiluminescence emission in the presence of luminol and an iron catalyst, are, so far, the most sensitive.

We have implemented a new procedure based on chemiluminescence emission, addressing some methodological aspects in order to improve sensitivity, specificity, and the accuracy of the analysis.[10,11] (1) An innovative, very sensitive "single-photon counting" instrument was used, allowing the direct analysis of lipid hydroperoxides in plasma or intact lipoproteins, even in the presence of inhibitors of the chemiluminescent reaction, thus escaping from extraction and chromatography, which could bias the analysis. (2) The reaction takes place in the presence of a detergent which provides a homogeneous environment for different substrates and minimizes the occurrence of side reactions. (3) The specificity of the reaction for lipid hydroperoxides was tested during development of the standardized analytical procedure by the complete disappearance of the photon emission,

[4] T. Parasassi, A. M. Giusti, E. Gratton, E. Monaco, M. Raimondi, G. Ravavagnan, and O. Sapora, *Int. J. Radiat. Biol.* **65,** 329 (1994).

[5] M. Maiorino, A. Roveri, F. Ursini, and C. Gregolin, *J. Free Rad. Biol. Med.* **1,** 203 (1985).

[6] S. M. Thomas, W. Jessup, J. M. Gebicki, and R. T. Dean, *Anal. Biochem.* **176,** 353 (1989).

[7] J. Nourooz-Zadeh, J. Tajaddini-Sarmadi, and S. P. Wolff, *Anal. Biochem.* **220,** 403 (1994).

[8] Y. Yamamoto, Methods in Enzymology, Vol. 233, p. 319 (1994).

[9] T. Miyazawa, K. Fujimoto, T. Sizuki, and K. Yasuda, K., Methods in Enzymology, Vol. 233, p. 324 (1994).

[10] A. Zamburlini, M. Maiorino, P. Barbera, A. M. Pastorino, A. Roveri, L. Cominacini, and F. Ursini, *Biochim. Biophys. Acta* **1256,** 233 (1995).

[11] A. Zamburlini, M. Maiorino, P. Barbera, A. Roveri, and F. Ursini, *Anal. Biochem.* **232,** 107 (1995).

following enzymatic reduction of lipid hydroperoxides by phospholipid-hydroperoxide glutathione peroxidase (PHGPx, EC 1.11.1.12)[12] and glutathione. (4) The photon emission was analyzed using the fitting of the exponential decay, which permits the detection of possible oxidative side reactions and thus contributes to the definition of optimized analytical condition.

We report here the most updated kinetic procedure for analyzing lipid hydroperoxides by chemiluminescence, where mathematical fitting of the integrated equation describing the whole kinetics of photon emission is used for calculating the hydroperoxide concentration in the sample.

This approach provides: (i) a better understanding and a continuous monitoring of the reaction chemistry underlying the photon emission; (ii) a better analytical precision, since the effect of noise and background is minimized; and, finally, (iii) an indirect evaluation of the effect of antioxidants possibly present in the sample.

Mechanism of Photon Emission by Luminol

Lipid hydroperoxides, reacting with hemin (or a suitable hemoprotein) in the presence of luminol, produce a transient chemiluminescence. In the adopted experimental conditions, the photon emission rate reaches a maximum in about 3–6 sec, then slowly decreases to background values in 40–60 sec.[11]

This photon emission is a complex multistep reaction, where reagents evolve in a complex pathway toward an unstable intermediate, which decomposes leading to an excited state of aminophthalate anion (AP*).

Different mechanisms have been suggested for the generation of oxidizing free radicals from lipid hydroperoxides and heme compounds.

Peterson *et al.*[13] and Rosen and Rauckman[14] suggested a Fenton-like reaction, where hydroperoxides reduce ferric hematin to ferrous, thus producing lipid peroxy radicals (LOO·). In a second, concerted step, ferrous hematin reduces a second LOOH originating an alkoxy radical (LO·) and restoring ferric hematin. Alternatively, Tappel[15] proposed that hematin-catalyzed homolytic scission of LOOH generates a LO· and a hematin hydroxy radical. This in turn reacts with a second molecule of LOOH, originating a LOO· and restoring ferric hematin.

[12] A. Roveri, M. Maiorino, and F. Ursini, Methods in Enzymology, Vol. 233, p. 202 (1994).

[13] D. A. Peterson, J. M. Gerrard, G. H. R. Rao, and J. G. White, *Prostagland. Med.* **4,** 73 (1980).

[14] G. M. Rosen and E. J. Rauckman, *Mol. Pharmacol.* **17,** 233 (1989).

[15] A. L. Tappel, *in* "Autoxidation and Antioxidants," Vol. I (W. O. Lundberg, Ed.), p. 325. Wiley, New York, 1961.

Dix and Marnett[16] have shown that LOOH in the presence of hematin gives rise to an epoxyallylic radical, which, on oxygen addition, generates LOO·.

The spectroscopic and chromatographic behavior of phospholipid hydroperoxides, decomposing in the presence of myoglobin,[17] is in agreement with the latter mechanism.

All the above mechanisms argue for the production of LOO·, and this is indirectly confirmed by the detection of chemiluminescence when luminol is omitted from the reaction mixture. This photon emission is produced during termination reaction by disproportionation of LOO· generating singlet oxygen (Russel mechanism).[18] Unfortunately, the yield of photons in this reaction is too low for analytical purposes, the sensitivity threshold of detection being 1 nmol of hydroperoxide, whereas in the presence of luminol 1 pmol can be easily detected.

Notably, the chemiluminescence emission detected in the presence of luminol and hemin is two orders of magnitude lower when hydrogen peroxide substitutes for lipid hydroperoxides (unpublished observation). This is practically useful since a possible bias in the measurement is virtually eliminated and indicates that the luminol oxidizing species is different in the case of lipid hydroperoxides or hydrogen peroxide: most likely LOO· and hypervalent state of iron, respectively.

The oxidation of luminol by LOO· competes with termination reactions via radical–radical interaction and reduction by compounds different from luminol. This aspect has been addressed when the analytical procedure has been optimized by adopting a relatively high concentration of luminol (30 μM).

The initiating radical reacts with luminol giving rise to a luminol radical which reduces oxygen to superoxide anion radical ($O_2^{\cdot -}$), although this could also be generated by reoxidation of ferrous hematin, if present.[19]

In the second step of the concerted reaction, luminol radical adds superoxide, giving rise to a hydroperoxide intermediate. The hydroperoxide intermediate decomposes, through cyclization and nitrogen loss, to the excited state of the aminophthalic acid monoanion. This species decays to the fundamental state, emitting photons.[19,20]

This mechanism accounts for: (a) the oxygen requirement of the reaction; (b) the inhibition by superoxide dismutase (SOD)[21]; and (c) the sec-

[16] T. A. Dix and L. J. Marnett, *J. Biol. Chem.* **260,** 5351 (1985).

[17] M. Maiorino, F. Ursini, and E. Cadenas, *Free Rad. Biol. Med.* **16,** 661 (1994).

[18] G. A. Russel, *J. Am. Chem. Soc.* **90,** 1056 (1957).

[19] M. Lind, *J. Am. Chem. Soc.* **105,** 7655 (1983).

[20] S. Ljiunggren and J. Lind, *J. Am. Chem. Soc.* **105,** 7662 (1983).

[21] S. Koga, M. Nakano, and K. Uehara, *Arch. Biochem. Biophys.* **289,** 223 (1991).

ond-order kinetics of the relationship between lipid hydroperoxides and integrated photon emission rate.[11]

Detergents (in our conditions, peroxide-free Triton X-100) produce more rapid and reproducible kinetics of photon emission. This is apparently due to several reasons: (a) the physical state of the peroxidic substrate is homogeneous in the sample; (b) peroxidative side reactions are prevented by substrate dilution; and (c) hemin dimers, which exhibit a reactivity with hydroperoxides much lower than that of monomers, dissociate to monomers.[22,23]

Mathematical Model of Kinetics of Chemiluminescence Emission

To describe the kinetics of photon emission, a minimal model was worked out from the above reaction mechanism. This is consistent with the concept that the overall kinetics of a complex series of reactions is mainly dependent on that of the slowest, rate-limiting step.

Pulse radiolysis experiments clearly showed that the initial free radical oxidation of luminol is the rate-limiting step of the process and that all following reactions have rate constants approaching diffusion limit.[24]

It has been also calculated that the half-life of AP*, which decays to the ground state emitting photons, is very short ($t_{1/2} < 5$ ns).[25] Thus, the photon emission rate accounts for the rate of formation of the excited emitting species.

The minimal model applied is

$$\mathrm{A} \xrightarrow{k_1} \mathrm{B} \xrightarrow{k_2} \mathrm{C} \rightarrow h\nu$$

where A is related to lipid hydroperoxide concentration; B is an intermediate, possibly the luminol hydroperoxide; C is the excited state of aminophthalate, and k_1, k_2 are complex constants actually accounting for more than one reaction.

The integration of the system of differential equations for two consecutive first-order reactions gives for B the following equation:

$$\mathrm{B} = \mathrm{A}\frac{k_1}{k_2 - k_1}(e^{-k_1 t} - e^{-k_2 t}) \tag{1}$$

which describes the time course of the transient intermediate B.[26]

[22] T. A. Dix, F. Fontana, A. Panthani, and L. J. Marnett, *J. Biol. Chem.* **260,** 5358 (1985).
[23] S. B. Brown, T. C. Dean, and P. Jones, *Biochem. J.* **117,** 741 (1985).
[24] J. H. Baxendale, *J. Chem. Soc. Faraday Trans.* **69,** 1665 (1973).
[25] J. H. Baxendale, *Chem. Comm.* **22,** 1489 (1971).
[26] A. A. Frost and R. G. Pearson, *in* "Kinetics and Mechanism," p. 166. Wiley, New York, 1961.

The photon emission rate (c.p.s.) is proportional to the rate of formation of C from B; therefore,

$$\text{c.p.s.} = dC/dt = k_2\text{B} \tag{2}$$

Combining Eqs. (1) and (2), we obtain

$$\text{c.p.s.} = k_2\,\text{A}\,\frac{k_1}{k_2 - k_1}\,(e^{-k_1 t} - e^{-k_2 t}) \tag{3}$$

Equation (3) describes the time course of c.p.s. as a function of A expressed as c.p.s. This value is related to the amount of hydroperoxides in the sample.

Fitting Kinetics of Photon Emission Rate and Calibration

For fitting the data to the model, the initial time of the chemiluminescent reaction was identified as that of the last point giving background readings before the sudden c.p.s. increase. The data were processed with a nonlinear regression analysis program operating by successive iterations. Starting from initial approximate estimates, the program finds the best parameter fitting the equation, when the chi-square (χ^2) difference between two successive iterations approaches a minimum value (usually less than 1%).

The algorithm also fits the best average background and the end of the reaction and corrects for it the best fit consistent with the model.

Thus, the instrumental data (c.p.s. vs time) were fitted with Eq. (3), as modified by introducing a background offset value:

$$\text{c.p.s.} = k_2\,\text{A}\,\frac{k_1}{k_2 - k_1}\,(e^{-k_1 t} - e^{-k_2 t}) + \text{offset} \tag{4}$$

An example of fitting of the photon emission is reported in Fig. 1.

From the fitting, the A value can be determined with good precision.

The validity of the applied model is confirmed by the quality of the fit (Snederor's F-stat > 1000) but also by the evidence that the k_1 and k_2 values are relatively constant while widely ranging the hydroperoxide content and thus the A value.

In a series of calibration curves where 2 to 10 pmol of PLPC-OOH were analyzed, k_1 was 0.71 ± 0.28 and $k_2 = 9.47 \times 10^{-2} \pm 7.40 \times 10^{-3}$ ($\text{sec}^{-1} \pm$ S.D., $n = 10$).

These values have been used as provisional estimates to start the iterative analysis for the best fit of data produced with unknown samples.

As previously observed also for integrated photon emission,[11] the A value is a nonlinear function of the hydroperoxide concentration (Fig. 2),

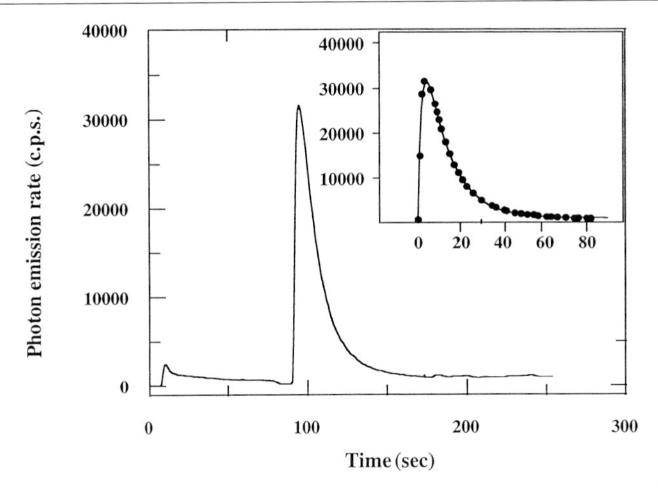

FIG. 1. Photon emission from 10 pmol of PLPC-OOH. *Inset:* The best fit of kinetic equation [Eq. (4)].

while data of the calibration curve fit the exponential equation

$$A = q\,x^{n} \quad (5)$$

where x is the actual amount of peroxides in the sample and q a constant.

In a series of measurements on PLPC-OOH the exponent (n) was fitted, by nonlinear regression, to 2.03 ± 0.27.

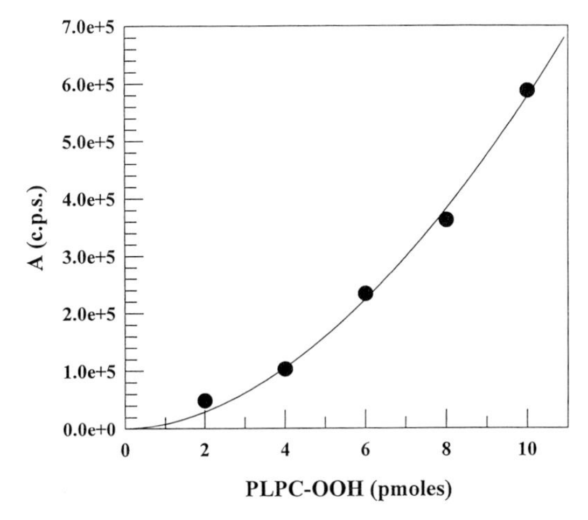

FIG. 2. Relationship between PLPC-OOH and A values [Eq. (5)], calculated from best fit of the kinetic equation [Eq. (4)].

When plasma or lipoprotein samples are analyzed, increasing amounts of PLPC-OOH are added to the sample and the values of A are calculated by best fitting of Eq. (4).

The relationship linking A to the total amount of peroxides in the sample is therefore

$$A = q(x + z)^n \quad (6)$$

where x is the amount of internal standard added and z is the endogenous plasma hydroperoxides. Both z and n can be estimated by fitting the model Eq. (6) to the experimental data. A typical analysis of a plasma sample is reported in Fig. 3.

The value of the exponent of Eq. (6) in different plasma samples ranges from 2 (as in the case of PLPC-OOH) to 1.2. This apparently indicates that not all hydroperoxides are available for oxidizing luminol, but some are lost in some side reactions. This is perfectly in agreement with the fact that the crude sample contains molecules able to compete with luminol for oxidizing species. Moreover, it supports the validity of the adopted internal calibration procedure.

This interpretation is validated by experiments where antioxidants decrease photon emission in the presence of PLPC-OOH, without significantly affecting the constants, but decreasing the value of the exponent of Eq. (5). For instance, in a typical experiment 10 pmol of ascorbate in the reaction mixture decreases by about 33% the A values of 2–10 pmol of PLPC-OOH and shifts the exponent of the calibration plot from 2.03 ± 0.26 to 1.57 ± 0.12.

Similar results have been observed with Trolox C.

Analytical Procedure

Materials

Palmitoyllinoleylphosphatidylcholine (PLPC); luminol, 3-[cyclohexylamino]-1-propanesulfonic acid (CAPS); peroxide-free Triton X-100; hemin; and glutathione are from Sigma (St. Louis, MO).

Luminol is purified as follows as previously described.[10]

The hydroperoxy derivative of PLPC (PLPC-OOH) is enzymatically prepared and quantified as previously described.[12]

Human Plasma and Lipoproteins

Human plasma is prepared in EDTA and immediately frozen at −70°C.

Low molecular weight antioxidants of plasma (mainly ascorbate and urate) are removed using a "desalting" cartridge Econo-Pac 10 DG from

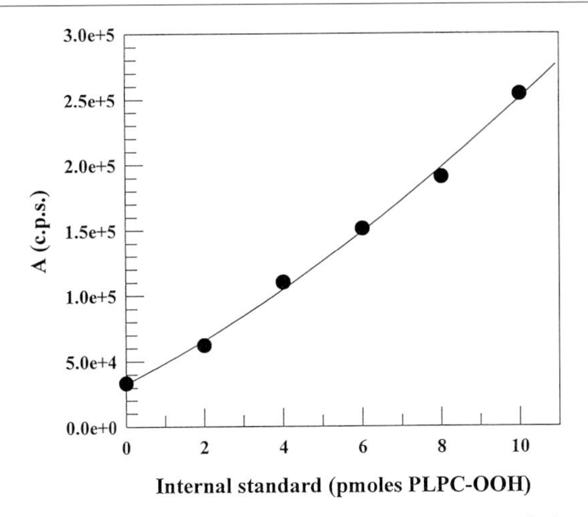

FIG. 3. Analysis of plasma hydroperoxide content. The photon emission of 20 μl plasma plus increasing amounts of internal standard hydroperoxide was analyzed to obtain A values from the best fitting of Eq. (4). The content of hydroperoxide in plasma is then calculated by fitting A values to Eq. (6).

Bio-Rad (Richmond, CA), as suggested by the manufacturer. This procedure leads to a 1:3 dilution of plasma, as tested by cholesterol content measurement.

Human plasma lipoproteins are prepared by routine centrifugation[27] on fresh plasma samples.

All buffers are Chelex-treated.

Chemiluminescence Measurements

The luminometer based on photon counting technology has been developed and patented (Italian patent MI 92A001815) by SEAS (Milano, Italy). The spontaneous photoemission of the detector is reduced by cooling at −20°C. Efficient stirring is obtained using a built-in minivortex. Injection of the sample, stirring, data acquisition, and processing are computer controlled.

The reaction mixture contains, in a final volume of 1.2 ml: 30 μM luminol, 0.3% (w/v) peroxide-free Triton X-100, 0.12 ml of methanol, and 4 μM hemin in 0.1 M CAPS buffer, pH 10.

After recording the dark current, the vial, containing the reaction mixture and hemin, is placed in a holder which rotates the vial in to front of the photomultiplier, and then the background photon emission is recorded.

[27] J. R. Havel, A. E. Howard, and J. H. Bragdon, *J. Clin. Invest.* **34,** 1345 (1955).

One milliliter of reaction mixture is aspirated from the vial and immediately reinjected, passing through the loop of a Rheodyne valve previously loaded with the hydroperoxide-containing sample (up to 50 μl).

The whole injection system is washed with the reaction mixture after each measurement.

For internal calibration, the sample, usually 20 μl of plasma after gel permeation, is injected from six to eight times alongside increasing amounts (2–10 pmol) of standard hydroperoxide (PLPC-OOH) in methanol.

Photon counting is integrated at 1-second intervals.

The row data produced are processed to fit the integrated equation [Eq. (4)] using the Levenberg–Marquardt algorithm.[28] This produces A values for plasma and each point of the internal calibration. The final calculation of lipid hydroperoxide content in the sample is done by best fitting Eq. (6). For both fittings, the commercially available software TableCurve 2D, Version 3.0 (SPPS Science ASC Gmbh, Erkrath, Germany, 1996) is used.

Conclusions

The measurement of plasma or lipoprotein lipid hydroperoxides requires a very sensitive, specific and accurate analytic procedure. The described procedure is definitely sensitive enough, detecting with sufficient precision 1 pmol of substrate. This means a threshold of 20–50 pmol/ml of plasma, where the lowest hydroperoxide concentration we found is 250 pmol/ml.

The specificity is ensured by the complete disappearance of the signal following enzymatic reduction of lipid hydroperoxides.

Finally, the accuracy is optimized by the use of the best fitting of the kinetic equation.

This kinetic approach ensures that no side reactions are taking place and reduces the influence of signal noise and background. It is worth noting, on the other hand, that kinetic analysis is the most appropriate way for measuring a product—the excited form of aminophthalate—which is transient and does not accumulate like usual reaction products.

So far this analytical approach has been utilized to show that human plasma contains lipid hydroperoxides in all lipoprotein classes,[10] but particularly the more electronegative LDL minus subfraction,[29] and that plasma lipid hydroperoxides increase in the postprandial state,[30] thus highlighting

[28] D. W. Marquart, *J. Soc. Ind. Appl. Math.* **11,** 431 (1963).

[29] A. Sevanian, H. Hodis, J. Hwang, G. Bittolo-Bon, G. Cazzolato, A. Zamburlini, M. Maiorino, and F. Ursini, *J. Lipid Res.* **38,** 419 (1997).

[30] F. Ursini, A. Zamburlini, G. Cazzolato, M. Maiorino, G. Bittolo-Bon, and A. Sevanian, *Free Rad. Biol. Med.,* in press (1988).

the role of alimentation in oxidative damage of lipoproteins, in turn related to atherosclerosis.

As a note of caution, this analysis clearly showed that a substantial amount of lipid hydroperoxides of lipoproteins are most likely produced during ultracentrifugation,[10] and it is well known how much this could affect lipoprotein peroxidability measurements.

[6] High-Performance Liquid Chromatography Analysis of Cholesterol Linoleate Hydroperoxide in Oxidized Low Density Lipoproteins: Calibration by Conjugated Diene Internal Standard

By GARRY J. HANDELMAN

Introduction

The susceptibility of low density lipoproteins (LDL) to oxidation *in vitro* has become a major technique in the study of interactions between LDL oxidation and atherogenesis. Substantial findings since 1985 indicate that dietary factors that increase the resistance of LDL to oxidation *in vitro* likewise decrease the occurrence of atherosclerotic heart disease. The original mechanism for the interaction between oxidized LDL and heart disease was proposed by Steinberg, Witzum, and Parthasarathy[1]; oxidized LDL formed in the vasculature was thought to accumulate in the cells of the arterial wall, setting up a cascade that led through foam cell formation and finally to accelerated plaque formation. The theory has undergone considerable refinement,[2] especially with the understanding of the role of adhesion factors that might cause accumulation of macrophages and neutrophils in the developing plaque. Pathways have been explored whereby oxidized LDL activates adhesion molecule expression in the vascular endothelium.

Based on the proposed role of oxidized LDL in atherosclerosis, the oxidation of LDL *in vitro* by reactive oxygen species has been extensively investigated.[3] In many individuals, the oxidative resistance of LDL increases

[1] D. Steinberg, S. Parthasarathy, T. E. Carew, J. C. Khoo, and J. L. Witztum, *New Engl. J. Med.* **320,** 915–924 (1989).
[2] M. N. Diaz, B. Frei, J. A. Vita, and J. F. J. Keaney, *New Engl. J. Med.* **327,** 408–416 (1997).
[3] H. Esterbauer, J. Gebicki, H. Puhl, and G. Jurgens, *Free Rad. Biol. Med.* **13,** 341–390 (1992).

0076-6879/99 $30.00

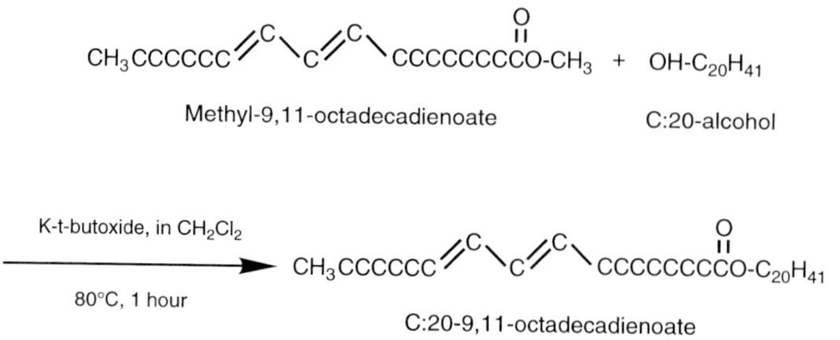

$CH_3CCCCCC{=}C{-}C{=}C{-}CCCCCCCC(=O){-}O{-}CH_3$ + $OH{-}C_{20}H_{41}$

Methyl-9,11-octadecadienoate C:20-alcohol

K-t-butoxide, in CH_2Cl_2, 80°C, 1 hour → $CH_3CCCCCC{=}C{-}C{=}C{-}CCCCCCCC(=O){-}O{-}C_{20}H_{41}$

C:20-9,11-octadecadienoate

FIG. 1. Synthesis of conjugated diene internal standard for HPLC analysis of cholesterol linoleate hydroperoxide in LDL.

following addition of vitamin E to the diet. The most reliable increases in LDL oxidative resistance occur following a 400–800 mg daily supplement.[4]

The measurement of damage to LDL *in vitro* has generally been limited to the late stage of oxidative damage, where considerably amounts of conjugated dienes, thiobarbituric-acid reactive substances (TBARS), fluorescent trienes, and volatile aldehydes can all be detected.[3] However, it is also appreciated that during the very early stages of LDL oxidation (corresponding to the formation of "minimally modified LDL"), there is the formation of hydroperoxides and the loss of antioxidants.[5,6] Much of the hydroperoxide formed is cholesteryl linoleate hydroperoxide (chol-18:2-OOH). It may be of considerable value to monitor the rate of hydroperoxide formation during this early phase of oxidative attack on LDL. The method described here[6] uses an isocratic HPLC system, with detection at 234 nm, to measure the formation of chol-18:2-OOH within a few minutes following oxidative attack on LDL.

Methods

Preparation of Conjugated Diene Internal Standard

The conjugated diene internal standard is prepared by transesterification of methyl-9,11-octadecadienoate (the methyl ester of conjugated linoleic acid) with *n*-arachidyl alcohol (C20:0). The reaction pathway is shown in Fig. 1. The synthetic precursors are obtained from Nu-Check Prep (Elysian,

[4] I. Jialal, C. J. Fuller, and B. A. Huet, *Arteriosclerosis, Thrombosis, and Vascular Biology* **15,** 190–198 (1995).

[5] B. Frei, R. Stocker, and B. N. Ames, *Proc. Natl. Acad. Sci. USA* **85,** 9748–9752 (1988).

[6] G. J. Handelman, E. N. Frankel, R. Fenz, and J. B. German, *Biochem. Mol. Biol. Internat.* **31,** 777–788 (1993).

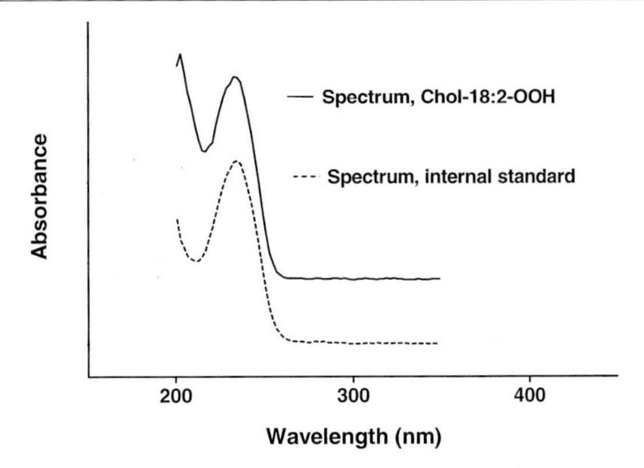

FIG. 2. The spectra of the conjugated diene internal standard (lower trace) and cholesterol linoleate hydroperoxide (upper trace). Both spectra were collected from the HPLC analyses of oxidized LDL shown in Fig. 3.

MN), or from Sigma (St. Louis, MO). The conjugated diene-methyl ester (5 mg) is mixed with the arachidyl alcohol (20 mg), in 1 ml of dichloromethane, and 100 mg of potassium *tert*-butoxide (Aldrich Milwaukee, WI) is added. After incubation for 60 min at 80°, about 10% of the methyl ester is converted to the arachidyl ester product.

The newly synthesized arachidyl ester is purified by reversed-phase HPLC on a C_{18} column, using a mobile phase prepared from 40% acetonitrile, 10% methanol, 50% 2-propanol, and 0.1% (w/v) ammonium acetate. The use of an HPLC diode-array detector (DAD) is very useful at the purification step, to aid in the reliable identification of the reaction product.

Because the arachidyl ester is far more hydrophobic than the methyl ester starting material, it should elute far later on the C_{18} column. Both the starting material and the synthesis product can be monitored by the presence of the distinct conjugated diene chromophore in the spectrum on the HPLC diode-array detector (Fig. 2, lower trace). The spectrum of chol-18:2-OOH (Fig. 2, upper trace) and the spectrum of the conjugated diene internal standard are very similar.

The concentration of the arachidyl ester product is determined using $\varepsilon_{234} = 25{,}000$.[7] The reaction product is stored in a tightly capped vial at −20° in HPLC mobile phase and is stable under these storage conditions.

Other long-chain alcohols (such as C18:0-OH or C22:0-OH) can be substituted during the transesterification reaction, to provide an HPLC retention time optimal for the specific chromatographic conditions employed.

[7] H. W.-S. Chan and G. Levett, *Lipids* **12,** 99–104 (1976).

Isolation of LDL

Human LDL is obtained by ultracentrifugation of EDTA–plasma at 4°.[8] If Cu^{2+} is used as oxidant, EDTA must be removed before oxidative stress. To eliminate EDTA, the LDL are dialyzed for 18 hours at 4° in 500 volumes of phosphate-buffered saline (PBS), which is changed once after 9 hours of dialysis. Protein in the dialyzed LDL is determined with a modification of the Lowry method,[9] or with the bicinchoninic acid reagent (BCA),[10] using bovine serum albumin (BSA) as standard. For both protein assays, LDL should be solubilized before measurement in 1% SDS.

LDL Oxidation in Vitro

The LDL are diluted to a final concentration of 100 μg protein/ml in 0.01 *M* phosphate buffer, pH 7.4, 0.1 *M* NaCl. $CuSO_4$ (5 μM) is added, and the mixture incubated under air at 37° in a shaking water bath. Experiments are typically conducted in a volume of 30 ml in a 100 ml Teflon-capped glass bottle. At timed intervals, 1 ml aliquots are collected into vials, diluted with 100 μl of 0.01 *M* EDTA (which prevents further Cu^{2+}-catalyzed oxidation) and stored at −80°.

Several other protocols have also been employed to initiate LDL oxidation, such as peroxyl radical, heme catalysts, and UV light. Chol-18:2-OOH is a major product during all of these LDL oxidation protocols.

Extraction of Chol-18:2-OOH

The LDL samples at −80° are thawed, and 250 μl aliquots dispensed into 4 ml screw-cap borosilicate glass vials. To the sample is added the diene internal standard (25 μl, containing typically 1.0 nmol of the standard), which is dispensed with a positive-displacement micropipette (available from several suppliers). The sample is extracted with a modified Folch extraction protocol.[11] To the sample is added 1 ml of methanol and 1 ml of dichloromethane; the vial is capped and vortexed for 30 sec. Additional dichloromethane (1 ml) and water (750 μl) are added, the sample is vortexed for 30 more seconds, and centrifuged at 2000*g* for 1 min at ambient temperature. The lower dichloromethane layer is collected and evaporated at 40°

[8] V. N. Schumaker and D. L. Puppione, Methods in Enzymology, Vol.128, pp. 155–169 (1986).

[9] M. A. Markwell, S. M. Haas, L. L. Bieber, and N. E. Tolbert, *Anal. Biochem.* **87,** 206–210 (1978).

[10] P. K. Smith, R. I. Krohn, G. T. Hermanson, A. K. Mallia, F. H. Gartner, M. D. Provenzano, E. K. Fujimoto, N. M. Goeke, B. J. Olson, and D. C. Klenk, *Anal. Biochem.* **150,** 76–85 (1985).

[11] F. J. G. M. van Kuijk, G. J. Handelman, and E. A. Dratz, *Free Rad. Biol. Med.* **1,** 421–427 (1985).

under nitrogen. The residue is dissolved in 40 μl of HPLC mobile phase for analysis.[6]

The organic solvent extract of LDL is also suitable for other analyses, including vitamin E[6] and carotenoids.[6]

HPLC Analysis of Chol-18:2-OOH Formed during Oxidative Attack on LDL

A sample of normal human LDL (1 mg protein/ml) was oxidized with 5 μM Cu^{2+} at 37°. Aliquots were taken at time points, mixed with EDTA (1 mM) to suppress further oxidation, and extracted for HPLC analysis. The HPLC results are shown in Fig. 3, for the 0, 20, and 180 min time points. At 20 min, a distinct new peak can be observed (retention time 4 min) which was identified as chol-18:2-OOH (see below). This peak becomes very large after 180 min of oxidation with Cu.$^{2+}$ Under optimal conditions, this newly formed chol-18:2-OOH can be measured within 5 min after addition of Cu^{2+} to the sample.

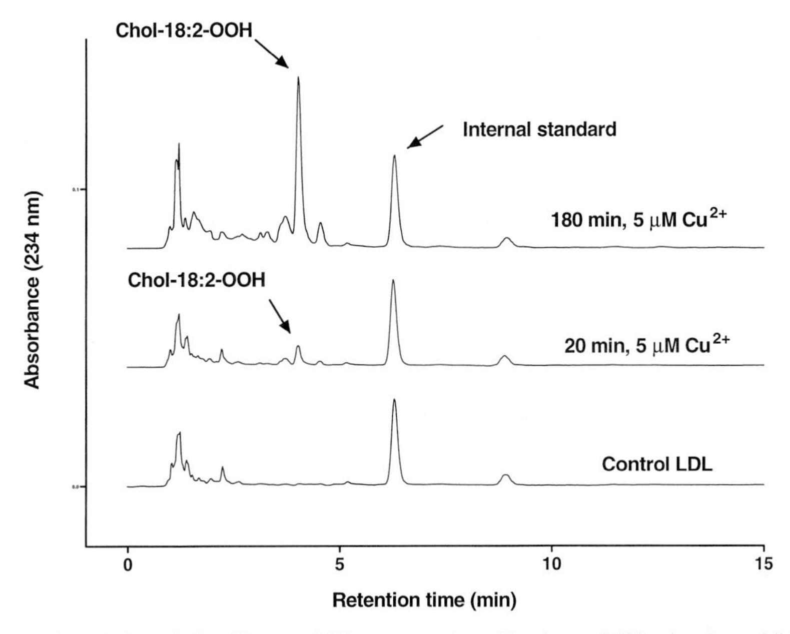

FIG. 3. HPLC analysis of human LDL, at zero time, 20 min, and 180 min after addition of 5 μM Cu^{2+}. The internal standard is the conjugated diene internal standard shown in Fig. 1. HPLC conditions: C_{18} HS Adsorbosphere (Alltech), 3 μm particle size, 15 cm × 0.46 cm. Mobile phase: 40% acetonitrile, 10% methanol, 50% 2-propanol, 0.01% ammonium acetate. Flow rate: 1 ml/min. Detection wavelength: 234 nm.

The spectrum of this peak, taken directly from the signal output of the diode-array detector, is shown in Fig. 2 (upper trace).

HPLC is done with an Alltech C_{18} Adsorbosphere-HS column, 3 μm particle size, 15 cm $\times$ 0.46 cm. Other C_{18} columns are satisfactory, but high chromatographic resolution is desirable for maximum sensitivity. The mobile phase is 50% 2-propanol, 40% acetonitrile, 10% methanol, and 0.1% (w/v) ammonium acetate, with 1.5 ml/min flow rate. Measurements are made at 234 nm.

The analyses described here were carried out with an HP Model 1100 diode-array detector and solvent delivery module (Hewlett-Packard, Avondale, PA).

Identification of Chol-18:2-OOH Peak

For identification of the new peak that forms during LDL oxidation (Fig. 3), chol-18:2-OOH can be prepared. Cholesteryl linoleate (Sigma), 1 mg/ml in hexane, is allowed to air-oxidize in a tightly sealed vial, at 37°. At suitable time points (6, 12, 24, and 48 hour intervals), 5 μl aliquots are injected directly on the HPLC. With HPLC detection at 234 nm, a new peak will gradually appear on the chromatogram; this peak has the same retention time and spectrum as the peak that accumulates during oxidative attack on LDL.

If the analysis can be monitored with a diode-array HPLC detector, the diene chromophore (Fig. 2, upper trace) is readily characterized.

Accelerated oxidation of cholesterol linoleate in hexane can be obtained with 1 m*M* 2,2′-azobis-(2-amidinopropane) dihydrochloride (AAPH) at 37°, where substantial amounts of this new peak can be seen within 30 min[12] after addition of peroxyl radical initiator. Standards of chol-18:2-OOH should be stored at −20°, with 1 m*M* BHT added to inhibit further radical-driven decomposition of the hydroperoxides to secondary products.

For definitive results, it is recommended that the product of cholesterol linoleate autoxidation be run on the HPLC in parallel with the extract of oxidized LDL. The identity of chol-18:2-OOH in the synthetic lipid, and in the LDL extract, will be readily apparent, since both peaks will have the same HPLC retention time and spectrum.

Calibration for Chol-18:2-OOH Content

The measurement of chol-18:2-OOH, containing a diene chromophore, is readily accomplished by direct comparison of the peak area with the

[12] I. Maitra, L. Marcocci, M. T. Droy-Lefaix, and L. Packer *Biochem. Pharmacol.* **49,** 1649–1655 (1995).

internal standard peak area, since both compounds contain a conjugated diene which absorbs at 234 nm. Both of these chromophores have a comparable molar extinction at 234 nm (25,000 ± 10%, for most conjugated dienes).[7]

For example, the integregated peak areas are determined for the HPLC analysis (Fig. 3, upper trace) of LDL after a 180-min exposure to 5 μM Cu^{2+}. The peak area of chol-18:2-OOH is twice the peak area of the internal standard, and we conclude that this sample contains twice as much chol-18:2-OOH as conjugated diene internal standard. Because 1 nmol of internal standard was added to this LDL sample prior to lipid extraction, the calculation yields 2 nmol of cholesterol ester hydroperoxide present in this aliquot of oxidized LDL.

Discussion

HPLC analysis of chol-18:2-OOH in LDL, with conjugated diene internal standard, provides a simple quantitative approach to *in vitro* LDL oxidation during the initial period of free-radical attack on the LDL (often referred to as the *lag phase*). The method has been employed[6] to charcterize changes to LDL *in vitro* during the early phases of oxidant stress with Cu^{2+}, and to show the protective effects of components in *Gingko biloba* extract against oxidative damage to LDL by water-soluble peroxyl radical initiator.[12]

The method described here has the virtue of simple calibration. The yield of chol-18:2-OOH is determined by the ratio of its peak area to the internal standard peak area.

By use of this HPLC technique, it was observed[6] that substantial chol-18:2-OOH forms in LDL while the sample still contains an abundance of α-tocopherol. This observation has been reported previously,[5] with analysis of chol-18:2-OOH carried out with the HPLC chemiluminescence detector.

Because many compounds can be detected during HPLC analysis at 234 nm, the use of the HPLC diode-array detector (DAD) is strongly recommended to aid in the identification of the conjugated diene internal standard, and the chol-18:2-OOH reaction oxidation product. The conjugated diene spectrum, with maximum absorption at 234 nm, is highly distinctive (Fig. 2).

Use of the DAD also allows concurrent measurements of carotenoids (at 450 nm) and cholesterol esters (at 206 nm), extending the range of information obtained during LDL oxidation *in vitro*.[6]

Previous investigators have also resolved these lipid hydroperoxides by

HPLC: Piotrowski *et al.*[13] utilized reversed-phase HPLC to isolate cholesterol linoleate hydroperoxides, which were detected at 234 nm.

Chol-18:2-OOH can also be measured by HPLC with the use of a chemiluminescence detector.[14] This method provides the sensitivity needed to detect chol-18:2-OOH in plasma samples directly after collection from human subjects,[15] without use of *in vitro* oxidative stress. The chemiluminescence detector approach is very sensitive, but requires the use of highly sophisticated HPLC equipment and calibration procedures.[5,14]

The limitations of the method must be addressed. The HPLC method does not resolve between chol-18:2-OOH and its reduction product, since these compounds have similar retention times in the HPLC system employed here. Some of the forms of chol-18:2-OOH formed by singlet oxygen attack on cholesteryl linoleate lack a conjugated diene chromophore[7] and will not be detected by analysis at 234 nm. By contrast, virtually all isomers of chol-18:2-OOH formed by free-radical attack on cholesteryl linoleate contain the diene chromophore.

The method described here requires a C_{18} column with good resolution, a single HPLC pump and HPLC detector, and the preparation of the conjugated diene internal standard from commercially available synthetic precursors. Therefore, this method should be readily adapted by the typical biochemistry laboratory.

Summary

The cholesterol linoleate hydroperoxides formed in LDL after oxidant stress are measured by HPLC, with UV detection at 234 nm. Calibration is performed with a conjugated diene internal standard. This internal standard is synthesized by the transesterification of the methyl ester of conjugated diene linoleic acid with a long-chain alochol, such as arachidyl alcohol (C_{20}). Different long-chain alcohols can be used during the transesterification, to achieve internal standards with variable HPLC retention times. The method allows measurement of cholesterol linoleate hydroperoxide in LDL very early during attack with Cu^{2+} or other initiator, so that the kinetics of antioxidant loss and hydroperoxide formation can be concurrently monitored.

[13] J. J. Piotrowski, GC, H., CD, E., MA, D., and VN., B. *Life Sciences 1990,* 715–721 (1990).
[14] Y. Yamamoto, M. H. Brodsky, J. C. Baker, and B. N. Ames, *Anal. Biochem.* **160,** 7–13 (1987).
[15] M. C. Polidori, B. Frei, G. Rordorf, C. S. Ogilvy, W. J. Koroshetz, and M. F. Beal, *Free Rad. Biol. Med.* **23,** 762–767 (1997).

[7] Determination of Phospholipid Oxidation in Cultured Cells

By GLENN T. SHWAERY, JASON M. SAMII, BALZ FREI, and JOHN F. KEANEY, JR.

Introduction

Lipid peroxidation and oxidative damage have been implicated in a number of disease processes such as atherosclerosis, neoplasia, and aging.[1] At the cellular level, lipid peroxidation is primarily manifested as oxidation of phospholipids. Phospholipid oxidation has been extensively studied in lipoproteins and model membranes, but the precise cellular events associated with intact cellular membrane phospholipid oxidation are less well understood.[2]

Peroxidation of polyunsaturated fatty acids in membrane phospholipids proceeds via a free radical chain reaction that is associated with formation of lipid hydroperoxides.[2] A number of methods have been used to detect phospholipid oxidation, such as oxygen consumption, diene conjugation,[3] or the formation of secondary reaction products such as malondialdehyde and 4-hydroxynonenal.[4] The measurement of these secondary reaction products as an index of lipid peroxidation works well in defined systems (e.g., liposomes, microsomes); however, their use in biologic systems is fraught with difficulty.[5] For example, the measurement of malondialdehyde in the thiobarbituric acid (TBA) assay is dependent on the sample content of lipid,[6] trace metals, and antioxidants.[5]

The purpose of this chapter is to illustrate the use of high-performance liquid chromatography (HPLC) with postcolumn chemiluminescence detection for the measurement of lipid peroxidation in cultured cells. This method also has several advantages over the more common TBA assay discussed above. Specifically, this assay measures the hydroperoxide groups directly, and it is less prone to interference such as is observed with the TBA

[1] B. Frei, *Am. J. Med.* **97**(supplement 3A), 5s (1994).
[2] E. H. Pacifici, L. L. McLeod, H. Peterson, and A. Sevanian, *Free Rad. Biol. Med.* **17,** 285 (1994).
[3] H. Esterbauer, H. Striegl, H. Puhl, and M. Rotheneder, *Free Rad. Res. Commun.* **6,** 67 (1989).
[4] H. Esterbauer, G. Jürgens, O. Quehenberger, and E. Koller, *J. Lipid Res.* **28,** 495 (1987).
[5] D. R. Janero, *Free Rad. Biol. Med.* **9,** 515 (1990).
[6] H. Esterbauer, G. Striegl, H. Puhl, S. Oberreither, M. Rotheneder, M. El-Saadani, and J. Jürgens, *Ann. NY Acad. Sci.* **570,** 254 (1989).

0076-6879/99 $30.00

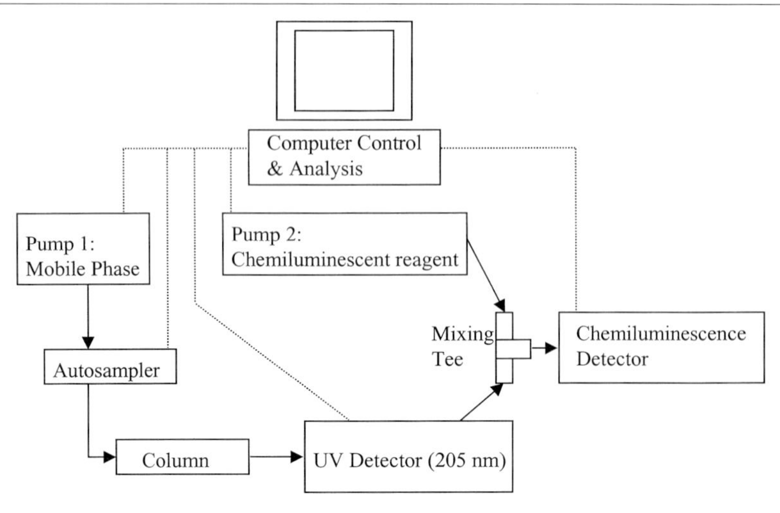

FIG. 1. Schematic diagram of HPLC system with postcolumn chemiluminescence. An isocratic pump (Hewlett Packard 1100 series) delivers mobile phase to the autosampler (Hewlett Packard) that injects samples across the reversed-phase LC-Si column (25 × 4.6 mm, 5 μm particle size, Supelco). Separated lipids are detected by UV absorbance at 205 nm (Hewlett-Packard series 1100) prior to entering a mixing tee (Supelco) where the sample is combined with the chemiluminescence reaction solution that is pumped at a flow rate of 1.5 ml/min. The mixture then passes through a chemiluminescence detector (Soma Chemi Lumi Detector S-3400, Soma Optics Ltd).

assay.[5] In addition, the assay can be adapted for individual identification of phospholipid hydroperoxides (PLOOHs) from the different phospholipid classes.[7]

HPLC Apparatus and Luminol Chemiluminescence

A schematic diagram of the HPLC system used for this assay is contained in Fig. 1. A standard Hewlett-Packard 1100 series ChemStation forms the basis of our HPLC system. For chemiluminescence detection we use a Soma Chemi Lumi Detector S-3400 (Soma Optics Ltd., Tokyo, Japan). This detector has a spiral flow cell which is readily removable to aid in cleaning should the chemiluminescence chamber become clogged. Good results have also been reported with a CLD-110 chemiluminescence detec-

[7] Y. Yamamoto, B. Frei, and B. N. Ames, Methods in Enzymology, Vol. 186, p. 371 (1990).

tor (Tohoku Electronic Ind. Co., Sendai, Japan)[8] or Shimadzu CLD-10A chemiluminescence detector (Shimadzu Corp., Tokyo, Japan).[9]

In this assay, extracted phospholipids are separated by HPLC over a silica column (Fig. 1) and individual lipid classes are detected by UV absorption. In the postcolumn chemiluminescence assay, microperoxidase catalyzes the decomposition of hydroperoxides to the corresponding alkoxyl radicals. This alkoxyl radical formation facilitates the oxidation and decomposition of isoluminol, yielding visible light that is maximal at a wavelength of 430 nm.[7]

Reagents and Materials

We take meticulous care to exclude traces of adventitious metal ions from all aqueous buffers. Accordingly, buffers are treated overnight with approximately 100 g/liter Chelex 100 resin (Bio-Rad, Richmond, CA) followed by filtration through an acid-washed Büchner funnel to remove the Chelex. All solvents are of HPLC grade, and care must be exercised in choosing vendors for HPLC-grade reagents, as we have found some solvents from some vendors to contain significant amounts of contaminating hydroperoxides. This problem can also be alleviated by treating organic solvents with HPLC-grade water (approximately 100 ml per liter of solvent). Alternatively, other investigators have reduced contaminating hydroperoxides by storing solvents at 4°C over a 4 Å molecular sieve (Aldrich, Milwaukee, WI).[8]

We also carefully treat all glassware to eliminate contaminating hydroperoxides. Glassware is sequentially sonicated for 15 minutes in each of the following solutions: hot water containing 2% Micro concentrated cleaning solution (International Products Corp., Burlington, NJ), double-distilled (DI) water, 1/1 (v/v) methanol/DI water, and HPLC-grade water. This final rinse is repeated twice followed by drying the glassware inverted at room temperature.

In order to prepare the postcolumn chemiluminescence reaction solution, we dissolve 76.3 g of sodium tetraborate decahydrate (reagent grade) in 2 liters of HPLC-grade water (100 m*M* final concentration) and adjust to pH 10.0 with the addition of 6.3 g solid sodium hydroxide. This aqueous borate buffer is filtered through a 0.45 μm Millipore filter (Millipore Corp., Bedford, MA) then combined with HPLC-grade methanol (Aldrich) 7/3 (v/v) followed by the addition of 5 mg per liter microperoxidase (MP-11,

[8] W. Sattler, D. Mohr, and R. Stocker, Methods in Enzymology, Vol. 233, p. 469 (1994).

[9] T. Miyazawa, K. Fujimoto, T. Suzuki, and K. Yasuda, Methods in Enzymology, Vol. 233, p. 324 (1994).

Sigma, St. Louis, MO) and 177.2 mg per liter (1 mM) isoluminol (6-amino-2,3-dihydro-1,4-phthalazinedione, Sigma). This solution is filtered and stored in an amber bottle where it is stable at room temperature for 2 to 3 weeks.

Procedure for Analysis of Phospholipid Hydroperoxides in Cultured Cells

Cell culture medium is removed and cells are washed twice with Hanks' balanced salt solution (HBSS) containing 50 μM diethylenetriaminepentaacetic acid (DTPA). Approximately 2.5×10^5 cells from a confluent 35 mm dish are removed with a cell scraper into a glass Kimax screw-top tube (Fisher Scientific, Pittsburgh, PA) with a PTFE-lined cap. The use of rubber or plastic-capped tubes should be avoided as their solvent sensitivity tends to produce high background levels of chemiluminescence. The culture dish is rinsed with HBSS containing DTPA, the rinses are pooled, and the sample is centrifuged at 250g for 10 minutes. The supernatant is discarded and the pellet is resuspended in 1 ml of balanced salt solution. The cells are then extracted with 4 volumes of 2/1 (v/v) chloroform/methanol. The aqueous and organic layers are separated by centrifugation at 4° for 10 minutes at 1500g. The lower chloroform layer is collected and evaporated to dryness under a stream of nitrogen at room temperature. The sample is resuspended in 200 μl of HPLC-grade methanol and placed in an amber injection vial that is purged with nitrogen and loaded into the autosampler. We then inject 50 to 100 μl of the sample onto an LC-Si column (5 μm, 4.6×250 mm, Supelco, Bellefonte, PA) with a mobile phase of methanol/*tert*-butanol/40 mM monobasic sodium phosphate (6/3/1 by volume) at a flow rate of 1.0 ml per minute. The eluent from the column passes through a UV detector with the wavelength set at 205 nm before passing into a mixing tube (low volume static mixer, Supelco) where it is combined with the chemiluminescence reaction solution at a flow rate of 1.5 ml per minute. Using this technique, we find that phosphatidylcholine (16:0, 18:2) hydroperoxide (PCOOH) elutes at approximately 12 minutes, as shown in Fig. 2 (*top*). Confirmation of the peak at 12 minutes as a hydroperoxide is readily performed by treating the sample with sodium borohydride (Sigma) and repeating the injection. The borohydride reagent reduces the lipid hydroperoxides to their corresponding alcohols, which are no longer detectable by chemiluminescence (Fig. 2, *bottom*).

We use chloroform/methanol for lipid extraction because it yields a quantitative recovery of 96 $\pm$ 2%. Extraction with methanol/hexane has

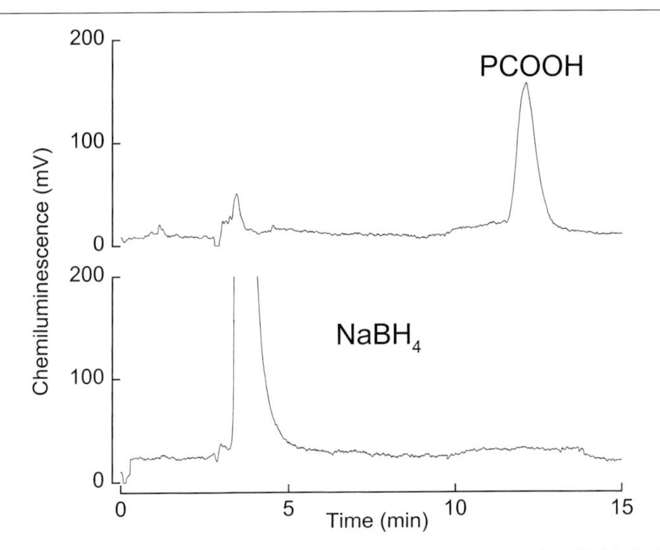

FIG. 2. *Top:* Chromatogram of a phosphatidylcholine hydroperoxide (PCOOH) injection showing a retention time of approximately 12.1 minutes. The measured chemiluminescent units are in millivolts (mV). *Bottom:* Chromatogram of a PCOOH injection after treatment with sodium borohydride. The hydroperoxides were reduced to their respective alcohols and no longer detectable by chemiluminescence.

also been described,[10] although the reported recovery rate is only 76%.[11] We specifically do not filter samples or dry them using a rotary evaporator, as we have found these procedures promote the formation of phospholipid hydroperoxides during the workup (unpublished observations, 1995).

We normalize the amount the phospholipid hydroperoxides formed based on the amount of total phospholipid in the sample as determined by the phosphate content. An aliquot of the chloroform extract is dried under a stream of nitrogen in a disposable 13 × 100 mm borosilicate glass tube (Fisher). We use new tubes exclusively for this assay in order to avoid artificial contamination by phosphate which is present in most detergents. The dried extract is then subjected to acid hydrolysis using 70% perchloric acid (Fisher) in a sand bath at 120°C for a minimum of 2 hours. Evaporation of perchloric acid is prevented during the acid hydrolysis by placing an acid-washed glass bead at the top of each tube. After hydrolysis, the tubes are allowed to cool for 20 minutes in the sand bath and then for 20 minutes at room temperature. To each tube, 4.6 ml of aqueous 0.22% ammonium

[10] Y. Yamamoto, M. H. Brodsky, J. C. Baker, and B. N. Ames, *Anal. Biochem.* **160,** 7 (1987).
[11] B. Frei, Y. Yamamoto, D. Niclas, and B. N. Ames, *Anal. Biochem.* **175**(1), 120 (1988).

molybdate tetrahydrate (Sigma) is added and samples are vortexed. The samples are reduced by the addition of 0.2 ml Fiske & Subbarow Reducer (1-amino-2-naphthol-4-sulfonic acid, Sigma) which is prepared fresh using 1 gram reagent per 6.3 ml HPLC-grade water. Samples are vortexed and incubated in a boiling water bath for 8 minutes covered by an acid-washed glass bead. Samples are allowed to cool to room temperature and the absorbance is read at 830 nm against a standard curve of potassium phosphate using a Cary 3E UV-visible spectrophotometer (Varian Analytical Instruments, Sugarland, TX).

Standardization of the Chemiluminescence Assay

To determine the retention time of phosphatidylcholine hydroperoxide, we use commercially available egg yolk phosphatidylcholine (Sigma) that contains palmitate in the *sn*-1 position and linoleate in the *sn*-2 position. If left exposed to light and air overnight, this standard contains enough autoxidized phospholipid for detection using the HPLC chemiluminescence assay. Alternatively, we have found that exposing the phosphatidylcholine to 2.5 m*M* AAPH results in a 23-fold increase in oxidized phospholipid over a 1 hour time period. Injection of approximately 1 nmol of phosphatidylcholine allows for detection in the UV chromatogram and the determina-

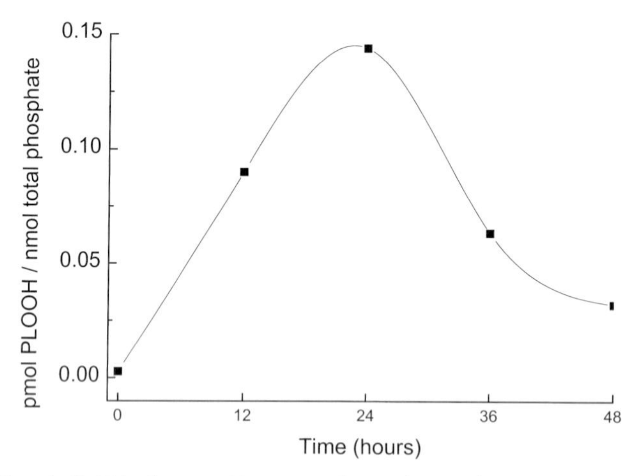

FIG. 3. Phospholipid hydroperoxide content in porcine aortic endothelial cells incubated with AAPH. Porcine aortic endothelial cells were incubated with 10 m*M* AAPH for 0–48 hours. Cells were washed, scraped into a Kimax tube, and extracted with chloroform/methanol as described in the text. Two-thirds of the chloroform phase is dried, reconstituted in methanol, and injected into the HPLC chemiluminescence system for the determination of PLOOH. The remaining one-third was used to quantify total phospholipid by phosphate determination.

tion of a retention time for PCOOH (typically 12–13 minutes as demonstrated in Fig. 2, *top*). To generate a standard curve relating hydroperoxide content to chemiluminescence we use authentic 15(*S*)-hydroperoxyeicosatetraenoic acid [15(*S*)-HpETE, Cayman, Ann Arbor, MI]. The concentration of 15(*S*)-HpETE can be determined spectrophotometrically using an extinction coefficient of 27,000 M^{-1} cm^{-1}.[11] The 15(*S*)-HpETE standard (0.1–0.4 pmol) is injected into the HPLC chemiluminescence system generating a peak with a retention time of approximately 2.8 minutes. A standard curve can then be constructed relating peak area to hydroperoxide content that is then used for the determination of PLOOH content in samples.

As an illustration of the application of this method to cultured cells, we exposed porcine aortic endothelial cells to a steady flux of aqueous peroxyl radicals generated by thermal decomposition of 10 m*M* 2,2′-azobis(2-amidinopropane) dihydrochloride (AAPH, Wako Chemicals, Richmond, VA) for 0–48 hours. The cells were washed, scaped into a Kimax tube, and extracted with chloroform/methanol as described above. Two-thirds of the chloroform phase is dried under nitrogen, reconstituted in methanol, and injected into the HPLC chemiluminescence system for the determination of PLOOH. The remaining one-third is used to quantify total phospholipid by phosphate determination. The relationship between time of incubation and PLOOH content is shown in Fig. 3.

Conclusions

We have demonstrated that by using HPLC with postcolumn chemiluminescence, we can directly measure minute levels of phospholipid hydroperoxides that can serve as a useful tool in measuring oxidation of cellular lipids *in vitro*. This method offers advantages over other assays as it uses a more direct measurement of lipid peroxidation, offers greater sensitivity than other assays, and is not subject to sample interference based upon sample lipid content, antioxidant content, or availability of trace metals.

Acknowledgments

This work was supported by the American Heart Association (96006120), the Council for Tobacco Research (#4073), and National Institutes of Health grants HL55854 and HL59346 (to J.F.K) and HL49954 (to B.F.). J. F. Keaney, Jr., is a recipient of a Clinical Investigator Development Award (HL03195) from the National Institutes of Health.

[8] Ferrous Ion Oxidation in Presence of Xylenol Orange for Detection of Lipid Hydroperoxides in Plasma

By JAFFAR NOUROOZ-ZADEH

Introduction

Hydroperoxides (ROOHs) are the initial stable products formed during peroxidation of unsaturated lipids such as fatty acids or cholesterol. A battery of assays are available for the measurement of ROOHs in liposomes, whole plasma, or plasma lipoprotein classes. These techniques are generally divided into two categories, including total and individual classes. The latter is measured by high-performance liquid chromatography linked with chemiluminescence detection (HPLC-CL) techniques. Total ROOHs have been estimated to be between 0.5 and 5.5 μM using colorimetric, chemiluminescence, or enzymatic techniques.[1–6] Using HPLC-CL assays, plasma levels of phospholipid hydroperoxides (PC-OOH) have been reported to range from 10 to 500 nM.[7,8] Plasma cholesteryl ester hydroperoxide levels are reported to be as low as 3 nM[9] or as high as 920 nM.[10]

In this laboratory, the ferrous oxidation in xylenol orange (FOX) version 2 assay was originally developed to measure ROOHs in liposome preparations or low density lipoprotein (LDL) suspensions.[11] The FOX assay is based on the oxidation of ferrous to ferric ions by ROOHs under acidic conditions [Eq. (1)]. The dye xylenol orange [*o*-cresolsulfonphthalein-3,3-bis(methyliminodiacetic acid sodium salt)] complexes with an equal molar concentration of ferric ion to produce a color (blue-purple) with an apparent extinction coefficient at 560 nm ($\varepsilon_{\lambda 560}$) of 1.5×10^4 M^{-1} cm^{-1}. In the presence

[1] P. J. Marshall, M. A. Warso, and W. E. M. Lands, *Anal. Biochem.* **145,** 192 (1985).

[2] G. L. Cramer, J. F. Miller, R. B. Pendleton, and W. E. M. Lands, *Anal. Biochem.* **193,** 204 (1991).

[3] A. Zamburlini, M. Maiorino, P. Barbera, A. Roveri, and F. Ursini, *Anal. Biochem.* **232,** 107 (1995).

[4] J. Nourooz-Zadeh, J. Tajaddini-Sarmadi, and S. P. Wolff, *Anal. Biochem.* **220,** 403 (1994).

[5] J. Nourooz-Zadeh, J. Tajaddini-Sarmadi, S. McCarthy, D. J. Betteridge, and S. P. Wolff, *Diabetes* **44,** 1054 (1995).

[6] J. Nourooz-Zadeh, A. Rahimi, J. Tajaddini-Sarmadi, H. J. Tritschler, P. Rosen, B. Halliwell, and D. J. Betteridge, *Diabetologia* **40,** 647 (1997).

[7] Y. Yamamoto and B. N. Ames, *Free Rad. Biol. Med.* **3,** 359 (1987).

[8] T. Miyazawa, *Free Rad. Biol. Med.* **7,** 209 (1989).

[9] Y. Yamamoto and E. Niki, *Biochem. Biophys. Res. Comm.* **165,** 988 (1989).

[10] A. E. Holly and T. F. Slater, *Free Rad. Res. Comm.* **15,** 51 (1991).

[11] Z. Y. Jaing, C. S. Woollard, and S. P. Wolff, *Lipids* **26,** 853 (1991).

0076-6879/99 $30.00

of ROOHs, the yield of ferric ion–xylenol orange complex, however, is higher than 1 : 1 because of limited chain oxidation of ferrous ion by ROOHs. Undesirable chain oxidation [Eqs. (2)–(4)] is prevented by inclusion of the lipid-soluble chain-breaking antioxidant butylated hydroxytoluene (BHT), which repairs the alkyl radicals produced by the reaction of alkoxyl radicals with unsaturated lipids [Eq. (5)]. The apparent $\varepsilon_{\lambda 560}$ for a number of ROOHs is $4.5 \times 10^4\ M^{-1}\ cm^{-1}$.

$$ROOH + Fe^{2+} \rightarrow Fe^{3+} + RO\cdot + OH^- \quad (1)$$
$$RO\cdot + RH \rightarrow ROH + R\cdot \quad (2)$$
$$R\cdot + O_2 \rightarrow ROO\cdot \quad (3)$$
$$ROO\cdot + RH \rightarrow ROOH + R\cdot \quad (4)$$
$$R\cdot + BHT \rightarrow RH + BH\cdot \quad (5)$$

The FOX2 assay in conjugation with triphenylphosphine (TPP) has been implemented for the measurement of plasma ROOHs.[4–6] TPP reduces ROOHs to their corresponding alcohols while itself being converted to triphenylphosphine oxide. This maneuver was also necessary to generate a proper control, since plasma contains interfering components, mainly ferric ions, that are detected by xylenol orange. There are other advantages of the FOX2 assay over existing techniques: (a) the kinetics of the reaction are independent of the chemical structure of ROOHs; (b) no extraction step is normally needed for analysis of liposomes and lipoprotein suspensions because of the use of 90% methanol/25 mM H_2SO_4, which denatures proteins sufficiently for access of the ferrous ions to available ROOHs.

Preparation of FOX2 Reagent

The FOX2 reagent comprises two stock solutions, A and B. Solution A is prepared by dissolving ammonium ferrous sulfate (98.03 mg) in 100 ml of 250 mM H_2SO_4. Subsequently, xylenol orange (76.06 mg) is added to the ammonium ferrous sulfate solution and the mixture is kept under stirring for about 10 min at room temperature. Solution B is prepared by dissolving 969.76 mg BHT in 900 ml methanol (HPLC-grade). A working FOX2 reagent is prepared by mixing one volume of solution A with nine volumes of solution B. The final FOX2 reagent is composed of xylenol orange (100 μM), BHT (4.4 mM), sulfuric acid (25 mM), and ammonium ferrous sulfate (250 μM). This solution is stable for 1 month at 4°C in the dark. The molar extinction coefficient for each batch of the FOX2 reagent is determined by calibration against varying concentrations of freshly prepared hydrogen peroxide (H_2O_2). Concentration of the stock H_2O_2 solution is determined

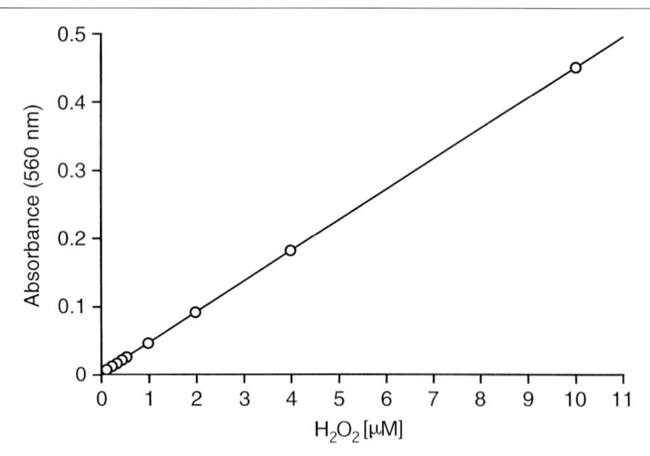

FIG 1. Calibration curve for H_2O_2 using the FOX2 assay.

using the molar $\varepsilon_{\lambda 240}$ 43.6 M^{-1} Cm^{-1}. Figure 1 shows the calibration curve for H_2O_2 in the concentration range 0–10 μM.

FOX2 Assay

Ninety microliter aliquots of plasma or lipoprotein suspensions (100 μg/ml protein) are transferred into 8 (1.5 ml) microcentrifuge reaction vials (L.I.P. Ltd, West Yorkshire, England). TPP (10 mM in methanol; 10 μl) is added to 4 vials to remove ROOHs. Methanol alone (10 μl) is added to the remaining 4 vials. The samples are vortexed and subsequently incubated at room temperature for 30 min. FOX2 reagent (900 μl) is added and the samples are vortexed. After incubation at room temperature for 30 min, the samples are centrifuged at 15,000g at 20°C for 10 min, the supernatant is carefully decanted into a cuvette, and absorbance is read at 560 nm.

Absorbance values in plasma, without TPP treatment, range between 0.050 and 0.070. The background is less if there is a very small degree of plasma hemolysis. If there is more hemolysis, the background is greater. The ROOH content in the plasma samples is determined as a function of the mean absorbance difference of samples with and without elimination of ROOHs by TPP. Absorbance differences between pre- and posttreatment with TPP range from 0.002 to 0.05. The standard deviation is taken as the larger of the standard deviations of the measurements obtained with or without TPP treatment. Figure 2 shows typical data for native plasma and plasma contaminated with a defined amount of synthetic ROOHs.

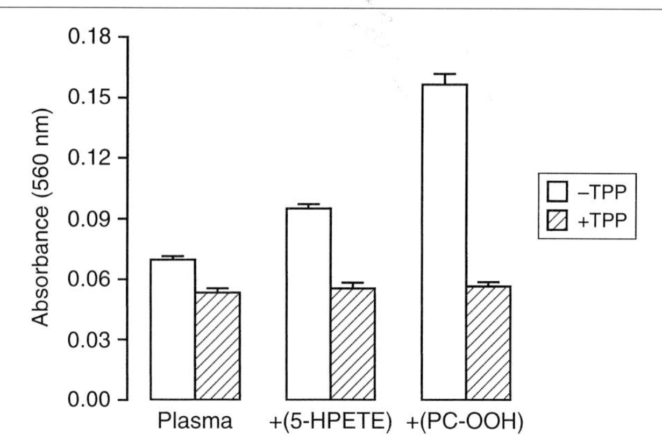

FIG 2. Detection of authentic ROOHs in plasma. PC-OOH: phospholipid hydroperoxides; 5-HPETE: 5-hydroperoxyeicosatetraenoic acid.

Application

The FOX2 assay has been successfully used to examine circulating ROOH levels in apparently healthy volunteers as well as in a variety of clinical conditions associated with oxidative stress. The mean ROOH content in fresh plasma from healthy volunteers from three separate studies was 3.02 ± 1.85 μM (n = 23), 3.76 ± 2.48 μM (n = 23), and 4.1 ± 2.2 μM (n = 41).[4–6] Values of 9.04 ± 4.3 μM and 9.4 ± 3.3 μM were found in fresh plasma from type II diabetic patients from two separate studies (n = 22 and 81, respectively).[5,6] For 67 patients with chronic renal failure, a mean level of 5.95 ± 3.04 μM of ROOHs was recorded. The mean level of plasma ROOHs in hypercholesterolemic patients (n = 52) was 4.4 ± 2.1 μM. The coefficient of variation for individual plasma using the FOX2 assay is <10%.[4–6] The assay is not influenced by diurnal, fasting, or nonfasting conditions.[5] It has also been shown that more than 65% of total ROOHs in native plasma is confined to the LDL fraction.[12]

Precautions

Precision of the FOX2 assay for the measurement of ROOHs in whole plasma or isolated lipoprotein fraction relies on accurate assessment of the amount of ferric ions formed during oxidation of ferrous ions by ROOHs. Thus, a prerequisite for the implementation of the assay is to eliminate the following confounding factors.

[12] J. Nourooz-Zadeh, J. Tajaddini-Sarmadi, K. L. E. Ling, and S. W. Wolff, *Biochem. J.* **313,** 781 (1996).

Spontaneous Oxidation. Ammonium ferrous sulfate should be dissolved in 250 m*M* H_2SO_4 immediately after weighing to prevent spontaneous oxidation of ferrous ions to ferric ions during preparation of the FOX2 reagent.

Reagent Decay with Time. A freshly prepared FOX2 reagent should be yellow with an absorbance at 560 nm < 0.005, against methanol as blank. Higher values suggest dirty glassware or the presence of substantial amounts of ferric ion in the reagent derived from oxidation ferrous ions to ferric ions. A high background absorbance (<0.05 at 560 nm) does not interfere with the assay when it is used for *in vitro* studies (lipoprotein suspensions or liposomes). However, FOX2-reagent with high background absorbance values is not suitable for the measurement of plasma ROOHs.

Metal-Chelating Agents. EDTA and DETAPAC are often used as anticoagulants during blood collection or as metal-chelating agents during dialysis of isolated lipoprotein fractions. These agents interfere with the accurate determination of ROOHs when present in the assay. Therefore, blood samples should be collected in heparinized tubes and chelating agents should be removed prior to the FOX assay.

Ferric Ions. Circulatory iron is generally bound to the protein transferrin. Iron may, however, be released in plasma as a result of hemolysis of red blood during sample collection. This would give rise to a high background absorbance as consequence of liberation of ferric ion during the assay. Again, plasma samples with absorbance values >0120, pretreatment with TPP, will not be suitable for the assay.

Antioxidants. Ascorbic acid (vitamin C) undergoes oxidation in the presence of transition metals (e.g., iron and copper ions) generating H_2O_2 and hydroxyl radicals as by-products. Vitamin C does not interfere with the FOX assay under physiological concentrations (12–100 μM). However, contamination of plasma with vitamin C at concentrations higher than 200 μM would give rise to a high background signal. Vitamin E or uric acid does not interfere with the assay.

Acknowledgment

We thank the British Heart Foundation for financial support.

[9] Purification and Characterization of Phospholipid for Use in Lipid Oxidation Studies

By TAMMY R. DUGAS and DANIEL F. CHURCH*

Introduction

Phospholipid vesicles are widely used as models for biological membranes in order to study both the physical characteristics of membranes and chemical processes such as lipid peroxidation.[1-3] Careful, thorough kinetic studies of lipid peroxidation and antioxidant activity can require large amounts of phospholipids that are devoid of peroxidic impurities. Because commercially available phospholipids can become expensive when large amounts are required, and, more importantly, since these products can vary widely with respect to the presence of peroxidic impurities, a reproducible, cost-effective method for the extraction and purification of these materials would be highly desirable.

We present here a rapid method for the preparation of large quantities of highly purified phospholipid (PL) from crude soy lecithin using flash column chromatography. The PL thus purified is especially suitable for oxidation experiments, as the oxidizability of vesicles formed from this purified PL is shown to be the same as that of vesicles formed from commercially available, 99% pure soy phosphatidylcholine (soy PC). The overall purity of the PL prepared using this method is characterized using reversed-phase HPLC, while the lipid content of the material is examined using fast-atom bombardment mass spectrometry (FAB-MS).

Experimental

Materials

L-α-Phosphatidylcholine (99%) from soybean, type III-S, and diethylenetriaminepentaacetic acid (DTPA) are obtained from Sigma (St. Louis,

* Address all correspondence to this author at: Radical Technologies, 12525 N. Oak Hills Pkwy, Baton Rouge, LA 70810.

[1] "Physical Methods on Biological Membranes and Their Model Systems" (F. Conti, W. E. Blumberg, J. D. Gier, and F. Pocchiari, Eds.), Vol. 71. Plenum, New York, 1985.

[2] R. R. C. New, "Liposomes: A Practical Approach." Oxford University Press, New York, 1990.

[3] J. P. Reeves, *in* "Biological Membranes" (R. M. Dowden, Ed.), pp. 223–254. Little and Brown, Boston, 1980.

0076-6879/99 $30.00

MO). Lecithin capsules, a source of crude L-α-phosphatidylcholine (soy PC), are purchased from a local pharmacy. Silica gel, grade 62, 60–200 mesh, is obtained from EM Science (Cincinnati, OH). Granular 2,2′-azobis-2-amidinopropane dihydrochloride (ABAP), type V-50, is obtained from Wako Pure Chemicals (Richmond, VA). All other incidental reagents used are of analytical grade or better.

Purification of Crude Soy Lecithin

Preparation of Column. The flash chromatography column apparatus can be obtained from Ace Glass, Inc. It consists of a column with a 10 mm I.D. and a 45.7 cm effective length, a 250 ml reservoir, and a flow control adapter.[4] A slurry of 40 g of silica gel in chloroform is poured into the column such that the resulting column bed is 35–37 cm in length.

Flash Chromatography. One lecithin capsule is dissolved in 10 ml chloroform and loaded onto the column. The column is then pressurized using nitrogen such that the flow rate is approximately 10 ml/min. The column is first eluted with 400 ml ethyl acetate/methanol (50 : 50, v/v). This fraction presumably contains less polar compounds such as α-tocopherol, β-carotene, and diglycerides. The PC is then eluted off the column using 200 ml methanol. The methanolic solution is evaporated to dryness under vacuum leaving the purified lipid as a thin film on the walls of the flask. The yield is approximately 200 $\pm$ 75 mg, depending on the source of lecithin. The purified material is dissolved in chloroform and can be stored under nitrogen at $-20°$ for several months with little degradation.

Analysis of Purified Phosphatidylcholine

HPLC. The purity of the purified phospholipid samples is examined using reversed-phase HPLC and is compared to commercially available 99% soy PC sample. The HPLC system used consists of a Perkin-Elmer (Norwalk, CT) Series 410 gradient pump with a Perkin-Elmer Series 235C diode array detector. The column used is a Spherisorb 25 cm $\times$ 4.6 mm I.D. C_8 reversed phase column with 5 μm particle size (Regis, Morton Grove, IL 60053). An isocratic solvent system consisting of 90% acetonitrile and 10% water (v/v) is formed using the HPLC pump at a 1.0 ml/min flow rate. The column temperature is 25°. Components are detected by UV absorption at 215 nm.

Mass Spectrometry. Fast-atom bombardment-mass spectra (FAB-MS) are acquired using a Finnigan (San Jose, CA) TSQ-70 triple quadrupole

[4] W. C. Still, M. Kahn, and A. Mitra, *J. Org. Chem.* **43,** 2923 (1978).

mass spectrometer with an Antek (Houston, TX) PS-4 cesium ion gun. Samples are applied to 3-nitrobenzyl alcohol by placing the matrix material on a stainless steel probe tip and applying the sample dissolved in chloroform (ca. 20–100 mg/ml) to the matrix. The heater current for the Cs pellet is run at 150 μA, the extractor at 2.5 kV, the focusing lens at 5.5 kV, and the acceleration voltage at 6.4 kV.

Measurement of Oxidizability of Phosphatidylcholine Liposomes. Phospholipid is purified from lecithin capsules using the flash column chromatography method described above. Stock solutions of the purified phospholipid or 99% pure commercial soy PC in chloroform are evaporated to a thin film by rotary evaporation. The evaporated PC is resuspended in 50 m*M* phosphate buffer, pH 7.4, containing 88.3 μM DTPA (diethylenetriaminepentaacetic acid), such that the final concentration of the phospholipid in buffer is 20 mg/ml. To produce multilamellar vesicles (MLVs), the phospholipid suspension is sonicated in a low energy bath-type sonifier for 8 minutes. Once formed, 3 ml of the MLV solution is placed in a Clark-type oxygen electrode cell thermostatted to 37°. Oxidation is initiated by addition of sufficient ABAP to give a final concentration of 20 m*M*. A Biological Oxygen Monitor, Model YSI 5300 (Yellow Springs Instrument Company, Yellow Springs, OH), interfaced to Labtech Aquire computer software via a 12 bit A/D converter is used to monitor and acquire oxygen concentration over time. Plots of decrease in oxygen concentration vs [ABAP] are constructed, and least squares treatment of the data gives the rate of oxidation in *M*/sec.

Results and Discussion

The HPLC chromatograms obtained for the phospholipid purified as described in the Experimental section above and for a commercial sample of soy PC (99% purity) are shown in Fig. 1. Both chromatograms show a similar group of peaks that elute from 14 to 23 min. These peaks reflect the mixture of phosphatidylcholines having various fatty acid composition (separating by fatty acid chain lengths and by varying degrees of unsaturation). A significant additional unidentified peak was observed at 26.5 min in the commercial soy PC sample that was not observed in the purified PL. In addition, several small peaks were observed between 5 and 15 min in the commercial sample that were not observed for the chromatographically purified PL. These peaks probably represent oxidation products in the commercial PC, including hydroperoxides. The peak eluting at 3 min in both samples was shown to be that of chloroform, the solvent (data not shown), probably along with any other small polar compounds present. The different sizes of these peaks is due to the different concentrations of

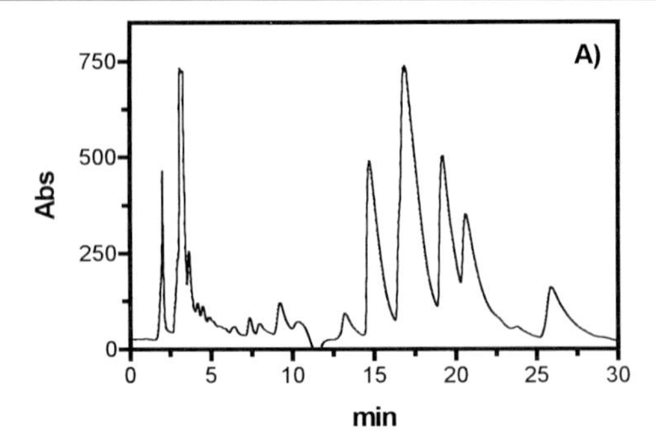

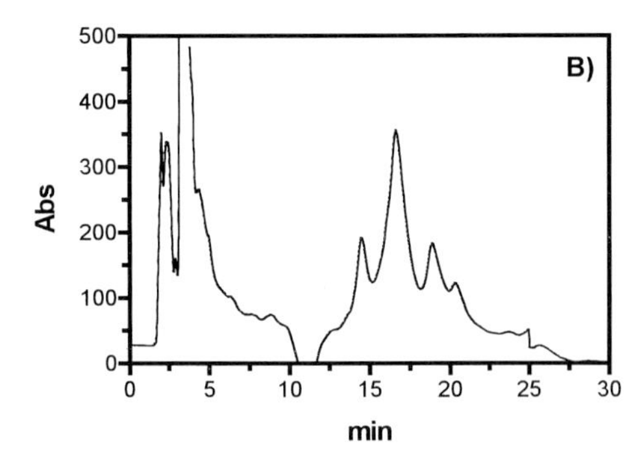

FIG. 1. HPLC chromatogram for L-α-phosphatidylcholine. Samples contained (A) 100 mg/ml chloroform solution of 99% L-α-phosphatidylcholine, and (B) 20 mg/ml chloroform solution of L-α-phosphatidylcholine purified from lecithin.

the phospholipid samples. It appears that on the whole the phospholipid purified by this method is of higher purity than PC obtained commercially, especially in terms of crucial impurities such as hydroperoxides.

The chromatographically purified PL gives a fast-atom bombardment-mass spectrum (FAB-MS) that shows mass clusters centered at 695.9, 717.9, 739.0, 759.1, 783.2, and 805.2 *m/e,* whereas the commercial soy PC sample shows mass clusters centered at 759.1 and 783.1 *m/e* (Fig. 2). Thus, it appears that although the purified PL is not contaminated with the oxidation product impurities that are present in commercial soy PC, the purified PL may represent a more complex lipid mixture. Table I shows the fatty acid composition of some commonly used phospholipids, including PC extracted from soybean. Possible combinations for fatty acyl side chains attached to a phosphatidylcholine headgroup were then determined for the three highest mass clusters observed in the FAB-MS of the purified PL (Table II). The

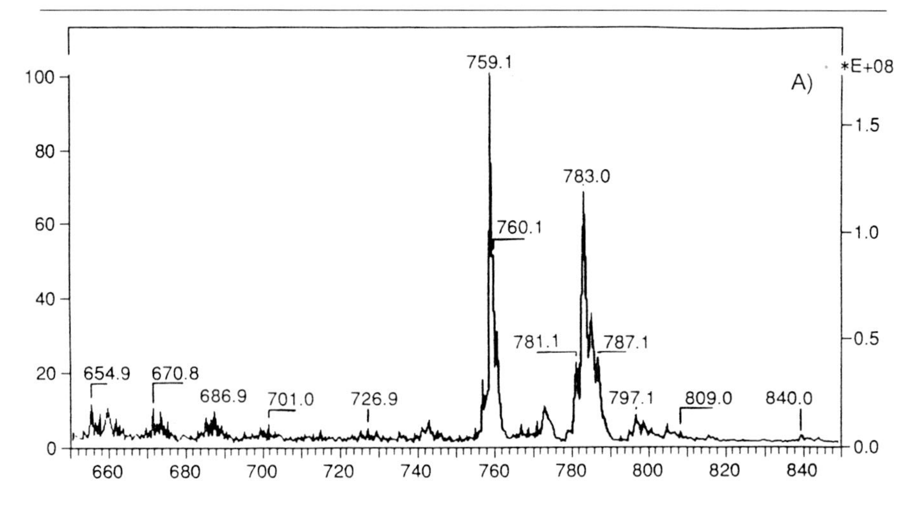

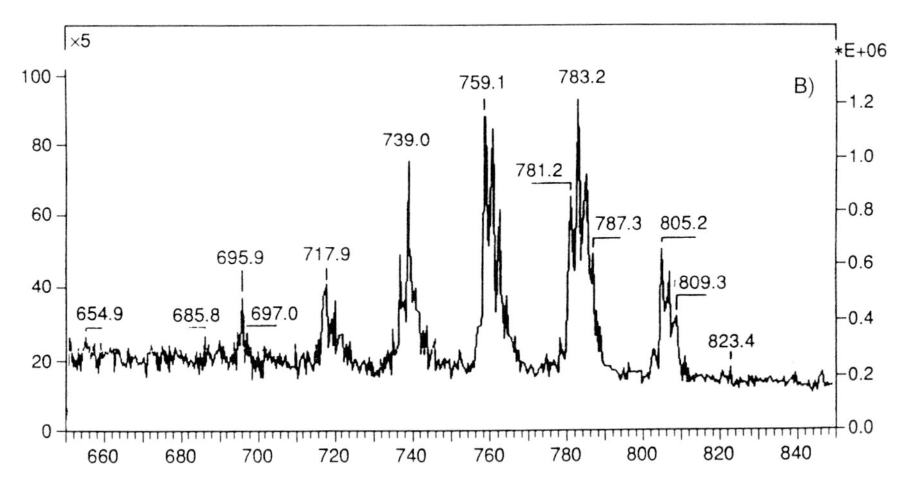

FIG. 2. FAB-MS obtained for (A) 99% L-α-phosphatidylcholine, and (B) phospholipid purified from lecithin capsule.

presence of the cluster at 805.2 *m/e* in the FAB-MS of the purified PL is particularly interesting. This mass range can only be accommodated by a phosphatidylcholine headgroup with one *20-carbon* fatty acid chain (e.g., arachidonic acid). Since C_{20} fatty acids are not naturally present in soy PC (Table I), the data suggest that the lecithin capsules may contain PC from some other source. The clusters at 739.0 and 717.9 *m/e* also cannot be accounted for by fatty acids that would normally be found in soy lipids. It is likely that rather than being due to exceptionally short-chained fatty

TABLE I

FATTY ACID COMPOSITION OF COMMONLY USED PHOSPHOLIPIDS[a]

Fatty acid abbreviation	Common name	Percent abundance (%)			
		Soy PC	Egg PC	Cod liver PC	Rat liver PC
14:0	Myristic			1.7	
16:0	Palmitic	17.2	35.3	23.0	28.8
16:1	Palmitoleic			2.7	
18:0	Stearic	3.8	13.5	3.4	18.0
18:1	Oleic	22.6	26.8	14.9	8.4
18:2	Linoleic	47.8	5.7	7.5	19.4
18:3	Linolenic	8.6	0.2	0.7	
18:4				0.9	
20:1	Gadoleic			3.1	
20:3					1.0
20:4	Arachidonic		1.0	1.7	17.0
20:5			3.6	14.5	2.4
22:1				0.3	
22:5			1.3	0.7	1.5
22:6			12.6	21.2	3.5

[a] See reference 1.

TABLE II

POSSIBLE COMBINATIONS OF R_1, R_2 FOR OBSERVED FAB-MS MASS CLUSTERS

Mass cluster	Possible R_1, R_2 combinations	Mass
759.1	16:0/18:0	761
	16:0/18:1	759
	16:0/18:2	757
	16:0/18:3	755
783.1	18:0/18:1	789
	18:0/18:2	787
	18:0/18:3	785
	18:1/18:2	783
	18:1/18:3	785
	18:2/18:3	783
	18:0/18:0	781
	18:1/18:1	781
	18:2/18:2	779
	18:3/18:3	777
805.2	20:4/18:0	809
	20:4/18:1	807
	20:4/18:2	805
	20:4/18:3	803

TABLE III
RATES OF OXYGEN CONSUMPTION FOR OXIDATION OF MULTILAMELLAR LIPOSOMES PREPARED FROM DIFFERENT PHOSPHATIDYLCHOLINE SOURCES

Lipid source	$-d[O_2]/dt \times 10^6$ (M/sec)
99% commercial soy PC[a]	0.0405 ± 0.0084
Purified phospholipid[b]	0.0412 ± 0.0153

[a] Oxidation was initiated by addition of 20 mM ABAP to 20 mg/ml soy PC MLV liposomes in 50 mM phosphate buffer (pH 7.4) containing 88.3 μM DTPA. Oxygen consumption was measured by oxygen electrode method.

[b] Same as *a*, except MLVs were prepared from soy PC purified using the column chromatography method presented here.

acids (C_{14}, for example), these clusters are due to a phospholipid having a lower mass headgroup, most likely phosphatidylethanolamines. The peak with mass 717.9 *m/e* from the purified lecithin sample may also be due to the presence of a small amount of phosphatidic acid that did not separate from the phosphatidylcholine. (Note that the masses reported by FAB-MS are one mass unit greater than the actual mass because the analyzed species are protonated. Also, as is true of UV detection,[5,6] size of the peak produced by FAB-MS is not necessarily proportional to the quantity present.[7])

Measurements of rates of oxygen consumption following ABAP-initiated oxidation of multilamellar vesicles (MLVs) prepared from either purified PL or from 99% pure soy PC showed that the oxidizabilities of the two were indistinguishable (Table III). Thus, the phospholipid purified using this method should be suitable for oxidation experiments.

We have presented a method for purifying phospholipid very inexpensively. Evidence from HPLC indicates that the PL purified by this method might actually be of higher purity than the 99% pure soy PC obtained commercially in terms of oxidation products such as hydroperoxides. From the MS data, however, the fatty acid and headgroup composition of the two samples may be somewhat different, although liposomes prepared from the two samples are shown here to have equivalent oxidizabilities.

[5] M. Pryszczewska and T. Lipiec, *Roczniki Chem.* **29,** 985 (1955).

[6] M. G. Tsyurupa and V. M. Peshkova, *Vestn. Mosk. Univ., Ser. II, Khim.* **19,** 60 (1964).

[7] S. Crosland, *Diss. Abstr. Int. B.* **49,** 5262 (1989).

Acknowledgments

We wish to acknowledge technical assistance in the chromatographic purification of phospholipids by Ms. Jherie Blazier and in the acquisition of the FAB-MS data by Mr. Rollie Singh.

[10] Assay of Malondialdehyde and Other Alkanals in Biological Fluids by Gas Chromatography-Mass Spectrometry

By HELEN C. YEO, JIANKANG LIU, HAROLD J. HELBOCK, and BRUCE N. AMES

Introduction

Oxidatively damaged lipids have been linked to cancer, coronary heart disease, and the degenerative diseases of aging.[1] The extent of lipid peroxidation has been quantified by measuring various analytes, including LOOH,[2] conjugated dienes,[3] expired hydrocarbons,[4] and carbonyls.[5]

Carbonyls, in addition to being a marker of lipid oxidation, have generated biochemical interest because of their intrinsic potential for toxicity.[6–10] Malondialdehyde (MDA) is a well-known carbonyl in rancid foods. In biological systems, it is a product of pathologic lipid oxidation and normal prostaglandin biosynthesis. MDA is commonly assayed by the thiobarbituric acid (TBA) assay as an indirect measure of lipid oxidation.[11,12] How-

[1] B. N. Ames, M. K. Shigenaga, and T. M. Hagen, *Proc. Natl. Acad. Sci. USA* **90,** 7915–7922 (1993).
[2] H. J. Helbock, P. A. Motchnik, and B. N. Ames, *Pediatrics* **91,** 83–87 (1993).
[3] A. E. Holley and T. F. Slater, *Free Rad. Res. Commun.* **15,** 51–63 (1991).
[4] A. Cailleux and P. Allain, *Free Rad. Res. Commun.* **18,** 323–327 (1993).
[5] H. C. Yeo, H. J. Helbock, D. W. Chyu, and B. N. Ames, *Anal. Biochem.* **220,** 391–396 (1994).
[6] L. J. Marnett, *IARC Sci. Publ.* **125,** 151–163 (1994).
[7] M. Nagao, Y. Fujita, T. Sugimura, and T. Kosuge, *IARC Sci. Publ.* **70,** 283–291 (1986).
[8] M. Nagao, Y. Fujita, K. Wakabayashi, H. Nukaya, T. Kosuge, and T. Sugimura, *EHP, Environ. Health Perspect.* **67,** 89–91 (1986).
[9] R. A. Woutersen, L. M. Appelman, A. Van Garderen-Hoetmer, and V. J. Feron, *Toxicology* **41,** 213–231 (1986).
[10] W. D. Kerns, K. L. Pavkov, D. J. Donofrio, E. J. Gralla, and J. A. Swenberg, *Cancer Res.* **43,** 4382–4392 (1983).
[11] D. R. Janero, *Free Rad. Biol. Med.* **9,** 515–540 (1990).
[12] H. H. Draper, E. J. Squires, H. Mahmood, J. Wu, S. Agarwal, and M. A. Hadley, *Free Rad. Biol. Med.* **15,** 353–363 (1993).

0076-6879/99 $30.00

ever, the TBA test lacks specificity, and the HPLC version, although addressing the specificity issue, nevertheless permits temperature-dependent oxidation artifacts that severely limit the validity of the method.[13]

In this chapter we describe a gas chromatography-mass spectrometry (GC/MS) assay for MDA capable of high throughput analysis.[5] Further, the method combines the specificity of mass spectral information with mild sample preparation, thereby avoiding the heat-generated artifacts that are an integral component of the TBA method.

Sample Preparation

Human Plasma. Preheparinized Vaccutainer blood collection tubes (Becton Dickinson, Rutherford, NJ) are used. Plasma is isolated by centrifuging the samples at 4°C for 10 min at 2000 rpm on an SS34 rotor and frozen at −20°C until analysis. Repeated freezing and thawing may cause an artificial increase in MDA levels and should be avoided.

Sperm Cells. Sperm samples provided by healthy individuals are centrifuged in an SS34 rotor at 3000 rpm and 4°C for 20 min. The supernatant is removed and the sperm pellet suspended in 10 ml of phosphate-buffered saline (PBS). The solution is then vortexed and centrifuged for another 20 min as above. After this washing step is repeated three times, the washed cells are resuspended in PBS equal to the original volume of sample and stored at −20°C until analysis.

Cell Culture. This method can be used to detect MDA in most types of cells. In our experiments, we measure MDA levels in HT-4 cells (a neuronal cell line) and also in primary rat cortical cells. HT-4 cells are derived from mouse neuronal tissue with hippocampal characteristics and are maintained in Dulbecco's modified Eagle's medium (DMEM) supplemented with 10% fetal calf serum at 33°C.[14] In all experiments, cells are grown for 3–5 days after passage, collected after trypsinization, centrifuged, washed, and sonicated with PBS buffer containing 1 m*M* deferoxamine mesylate (Sigma, St. Louis, MO) with a Sonifier Cell Disruptor (Heat Systems-Ultrasonics, Inc., NY) at 3 W for 15 s.[15]

Solid Organs. The brain, liver, kidney, and heart are obtained either from rats or mice. If the samples cannot be processed immediately, they are frozen in liquid nitrogen and stored at −80°C until analysis. The samples

[13] H. C. Yeo, H. J. Helbock, and B. N. Ames. "Toxic Carbonyl Compounds." In: Chromatogr. Science Series: Chromatographic Analysis of Toxicants in the Environment and Food" T. Shibamoto, ed. (Marcel Dekker, NY) **77,** 289–322 (1998).

[14] B. H. Morimoto and D. E. Koshland, Jr., *Proc. Natl. Acad. Sci. USA* **87,** 3518 (1990).

[15] J. Liu, H. C. Yeo, S. J. Doniger, and B. N. Ames, *Anal. Biochem.* **245,** 161 (1997).

are homogenized with PBS or any other buffer at physiological pH. We have found that adding 1 m*M* deferoxamine mesylate in homogenizing buffer is preferable to the use of EDTA and prevents lipid peroxidation during sample processing (data not shown).

Synthesis of Internal Standard

An internal standard is critical for the analysis of trace levels of biomolecules in methods which require several preparatory steps. A stable-isotope internal standard is used in the GC-MS technique described here to correct for derivatization and extraction efficiencies and permit the positive identification of the analyte of interest.

For this study, the stable isotope standard of MDA has been kindly provided by Prof. L. Marnett (Vanderbilt University, Tennessee). The standard may be obtained commercially through Cambridge Isotope Laboratories (Andover, MA) on custom order. Another source is Prof. J. R. Wagner [Address: Research Center in Geronto-geriatrics, Sherbrooke Geriatrics, University Institute (d'Youville), 1036 Belvedere Sud, Sherbrooke, Quebec, Canada J1H 4C4. Tel: 819-829-7131 ext. 2286/2291. Fax: 829-7141. Email: rwagner@courrier.usherb.ca].

Malonaldehyde Standard Preparation

Reference standards of MDA may be obtained by hydrolyzing 100 μmol of malonaldehyde bis(diethyl acetal) (Aldrich Chemicals Co., Milwaukee, MI) with 10 ml of 0.01 *M* HCl at room temperature for 6 hr. This will result in a 10 m*M* stock solution of MDA standard. Standards of $[^2H_2]$MDA are also prepared in this manner. The precise concentration of the stock solution may be verified by measuring its absorbance at 245 nm (see below). Note that the stock solution must be diluted appropriately for the sensitivity of the specific instrument used.

Derivatization

Preliminary Steps

1. Check the exact concentration of the MDA standard by UV:
   (a) Make a 10 μM solution from 10 m*M* stock using 0.01 *M* HCl.
   (b) Measure absorbance of the solution at 245 nm. Calculate the exact concentration of MDA using Beer's law where the coefficient constant is 13,700 cm^{-1} M^{-1}.
2. Thaw biological samples and store in ice.

Derivatizing Steps

1. Pipette 250 μl samples into 1.5 ml Eppendorf tubes.
2. Spike samples with 50 pmole [$^2H_2$]MDA as an internal standard.
3. Add 10 μl of BHT to prevent oxidation.
4. Add 10 μl 6.6 *N* H_2SO_4 to hydrolyze the aldehydes from the proteins. Vortex. Allow to incubate for 10 min at room temperature.
5. Add 75 μl 0.3 *M* Na_2WO_4 to precipitate the proteins. Vortex. Centrifuge for 5 min.
6. Transfer the supernatant into a clean Eppendorf tube.
7. Add 300 μl citrate–phosphate buffer, pH 5.5, followed by 50 μl PFPH reagent. Incubate the samples at room temperature for 30 min.
8. Add 10 μl 9 *N* H_2SO_4, followed by 125 μl isooctane, and vortex to extract the derivative.
9. Centrifuge for 5 min to separate the two phases.
10. Pipette approximately 60 μl of anhydrous isooctane (upper layer) into autosampler vials for GC-MS analysis.

List of Reagents

10 m*M* stock MDA standard
1 μ*M* stock [$^2H_2$]MDA internal standard
BHT (2 m*M* BHT solution containing 1 m*M* desferoxamine mesylate)
6.6 *N* H_2SO_4
9 *N* H_2SO_4
0.3 *M* Na_2WO_4
Citrate–phosphate buffer, pH 5.5 (0.58 *M* citric acid monohydrate/0.21 *M* Na_2HPO_4)
PFPH reagent (5 mg/ml aqueous solution)

Importance of pH. The formation of PFPH–aldehyde adduct occurs over the narrow range from pH 4 to pH 6. Therefore, care should be taken to ensure that the pH of the solution during derivatization is properly controlled.

Protein Assay. The levels of MDA are normalized to the protein content in the sample. Other options including lipid content may be used. Our studies measured protein by using bicinchoninic acid.[16]

[16] P. K. Smith, R. L. Krohn, G. T. Hermanson, A. K. Mallia, F. H. Gartner, M. D. Provenzano, K. K. Fujimoto, N. M. Goeke, B. J. Olson, and D. C. Klenk, *Anal. Biochem.* **150,** 76–85 (1985).

TABLE I
SET OF STANDARD SOLUTIONS

Final concentration (pmol/ml)	Volume of 1 μM MDA standard (μl)	Volume of PBS (μl)
0	0	250
10	2.5	247.5
20	5	245
40	10	240
80	20	230
100	25	225

Quantitation

Standard Curve

To generate a standard curve for this assay, varying amounts of stock MDA are added to phosphate-buffered saline in Chelex (PBS) solution to produce a final volume of 250 μl. The standard solutions are each spiked with 50 pmol of [$^2H_2$]MDA serving as an internal standard. The resulting samples are then derivatized as described above.

A typical set of standard solutions is described in Table I. For example, to prepare a standard of final concentration 100 pmol/ml, pipette 25 μl of a 1 μM MDA stock solution into 225 μl of PBS. The final concentration of the standard curve may be increased to 1 nmol/ml or higher if necessary, especially for tissue samples where the levels of MDA are much higher than in plasma.

Calculations

For each standard sample, two mass spectral traces are monitored in a single run: at $m/z = 234$ for MDA and $m/z = 236$ for the internal standard. A typical chromatogram is shown in Fig. 1. The peak areas for each trace are obtained and their ratios are calculated. Then a plot of $\text{MDA}_{(\text{Peak Area})}/[^2H_2]\text{MDA}_{(\text{Peak Area})}$ versus pmol/ml MDA is generated. A linear response curve should be achieved with the concentrations given above, and therefore a linear equation may be executed as shown in Fig. 2.

For each sample of interest, the above two mass spectral traces are also obtained where the peak area ratio is calculated (y value). Assuming that the linear equation is $y = mx + c$ (where y is the peak area ratio and x is pmol/ml MDA), the concentration of MDA in the sample, c, may be calculated accordingly.

In order for this calculation to be valid, care must be taken to make sure that the final volume of the standard and sample (in this case, 250 μl)

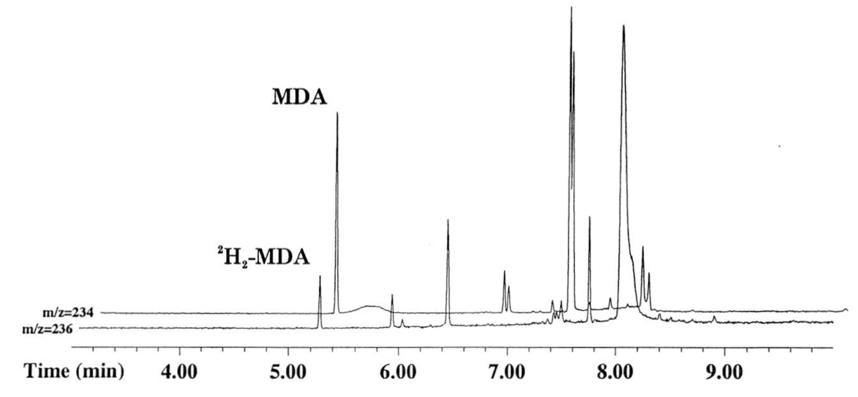

FIG. 1. Gas chromatographic profiles of a sperm sample monitoring *m/z* 234 and 236 for MDA and [$^2H_2$]MDA, respectively. From Yeo *et al., Anal. Biochem.* **220,** 391–396 (1994).

is constant and that the same amount of internal standard is added to both standard and sample (in this case, 50 pmol [$^2H_2$]MDA).

Instrumentation

A gas chromatograph (Hewlett-Packard 5890 Series II) interfaced to an HP 5989A mass spectrometer* was employed for the analysis. A DB-5

* Bench-top models of the GC-MS with NCI capabilities are now commercially available at considerably lower cost than the full-size models used here.

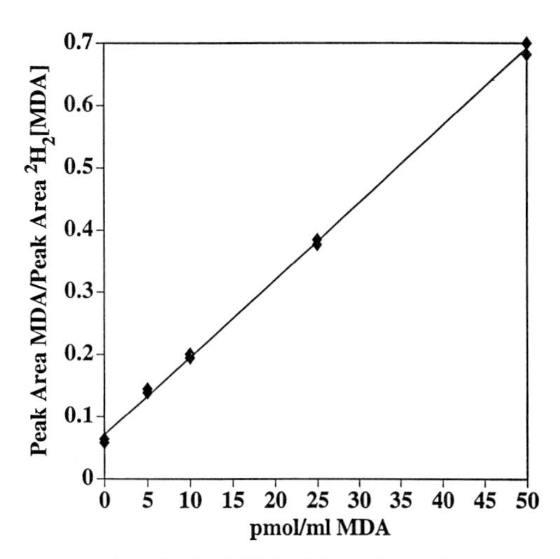

FIG. 2. Calibration curve for MDA–PFPH adduct. From Yeo *et al., Anal. Biochem.* **220,** 391–396 (1994).

(J&W Scientific, Folsom, CA) column or its equivalent of dimensions 15m × 0.25 mm I.D., 0.25 μm thickness, is used. Other less routine columns, such as DBWAX and DB17 or their equivalent, have been successfully applied with good resolution. The GC conditions include holding the initial temperature at 50°C for 1 min, ramping at 20°C/min to 280°C, and holding for 1 min at final temperature. Helium is used as a carrier gas with a flow of 58 cm/sec. A split mode injection with a split ratio of 50:1 is applied. Samples of 1–2 μl are injected. The mass spectrometer operates in the negative chemical ionization (NCI) mode using the following parameters: electron energy 230 eV, source pressure for methane 1.8 torr, source temperature 150°C. Selected ion monitoring of signals m/z 234 and 236 for derivatives of MDA and [$^2H_2$]MDA, respectively, are used for this analysis. Analyses are executed in the NCI mode to enhance sensitivity and specificity. Figure 3 shows the EI and NCI spectra of the PFPH adduct of MDA, demonstrating the differences in the degree of fragmentation.

Application of Method to Other Aldehydes

It has been shown that in addition to MDA, other aldehydes derived from lipid oxidation are also present in significant amounts.[17] Such aldehydes include the saturated alkanals. We have extended the MDA assay to encompass these aldehydes with little modification as described below.

Standard Preparation of Alkanals. As an internal standard for this analysis, acetaldehyde-d_4 (Cambridge Isotopes Laboratories, Inc.) is used. Typical concentration of the internal standard is 33 μM. Standard solutions of the alkanals are made, including acetaldehyde, propionaldehyde, butyraldehyde, valeraldehyde, hexanal, heptanal, octyl aldehyde, and nonyl aldehyde (Aldrich, Milwaukee, WI). The concentrations range from 10 to 500 μM, and higher if necessary, depending on the alkanal levels in the samples assayed. Because of the volatility of the standard solutions, the aldehyde vials must be chilled thoroughly before opening and they must be stored under nitrogen.

Alkanal Assay. The assay protocol is very similar to that of the MDA assay except for the following: (1) Spike 50 μl of 33 μM acetaldehyde-d_4 instead of [$^2H_2$]MDA; (2) after the 1 hr incubation with PFPH, add 150 μl of 200 mM $NaCNBH_3$ and incubate for another 1 hr to reduce the Schiff base. In developing the method, several reducing agents including sodium borohydride ($NaBH_4$), lithium aluminum hydride ($LiAlH_4$), and sodium cyanoborohydride ($NaCNBH_3$) were used. We also experimented with various conditions: incubation time, incubation temperature, whether the reduc-

[17] S. Toyokuni, X. P. Luo, T. Tanaka, K. Uchida, H. Hiai, and D. Lehotay, *Free Rad. Biol. Med.* **22,** 1019–1027 (1997).

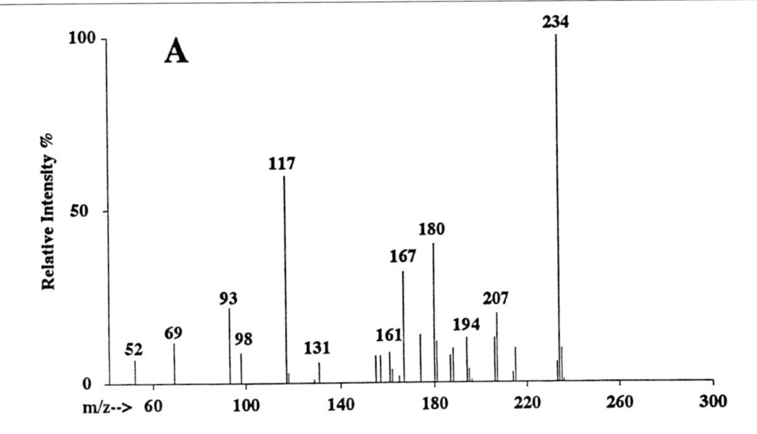

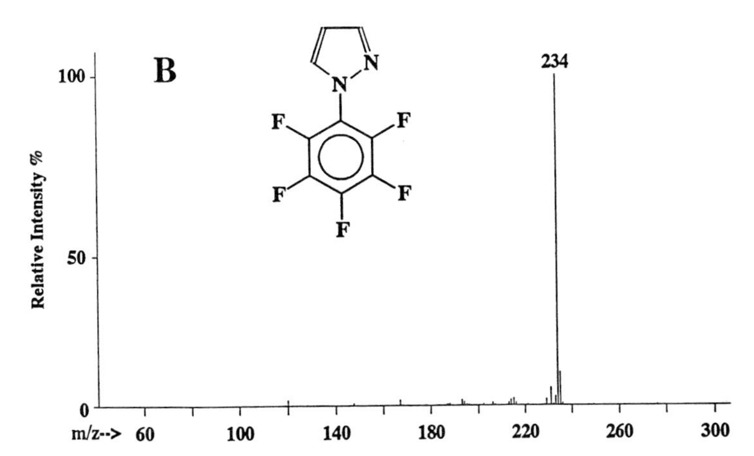

FIG. 3. (A) Electron impact mass spectrum of MDA-PFPH; (B) negative chemical ionization mass spectrum of MDA–PFPH. From Yeo *et al., Anal. Biochem.* **220,** 391–396 (1994).

ing agent is added as solid or liquid, and the reduction performed in aqueous or in organic phases. We concluded that $NaCNBH_3$ solution in aqueous phase for 1 hr is the preferred method because the reduction is efficient and the reducing agent is easy to handle.

It is possible to detect MDA and the other alkanals in the same sample. This may require more sample at the beginning; adding [$^2H_2$]MDA and acetaldehyde d_4 internal standards; splitting the PFPH–MDA and PFPH–

alkanal samples after the 1 hr incubation with PFPH; and adding $NaCNBH_3$ to the PFPH–alkanal portion for reduction.

Summary

The method described in this chapter allows the accurate measurement of MDA in diverse biological samples and can be extended to measurements of other alkanals. The use of GC/MS-NCI ensures specificity and sensitivity, and the ability to prepare samples without heating limits oxidation artifact. Although the widely used TBA assay for MDA does not require sophisticated equipment, its results may be of limited value as the assay is hindered by the possibility of cross reactivity and by heat-induced oxidation artifact. This GC-MS technique offers the additional advantages of efficient processing of large numbers of samples and the elimination of recovery errors by inclusion of an internal standard.

[11] Assessment of Physiological Interaction between Vitamin E and Vitamin C

By PETER A. GLASCOTT, JR. and JOHN L. FARBER

Introduction

Vitamin E (α-tocopherol) is the major, if not the only, antioxidant found in membranes, whereas vitamin C (ascorbate) is one of the major water-soluble antioxidants. α-Tocopherol (α-T) donates a hydrogen from the 6 position on its chromanol ring to free radicals,[1,2] thereby neutralizing the radical and forming the tocopheroxyl radical. Compared to the hydroxyl radical, which has a half-life of approximately 1 nanosecond,[3] the tocopheroxyl radical is relatively persistent with a half-life of milliseconds.[4] Ascorbate provides its reducing equivalents by donating a hydrogen from the 3 position to produce the ascorbate anion radical before oxidizing to form dehydroascorbate.[4,5] Dehydroascorbate has a half-life of approximately 6 min at physiological pH[5] and can be reduced back to ascorbate by an

[1] R. S. Parker, *Adv. Food Nutr. Res.* **33,** 157 (1989).
[2] D. C. Liebler, *Crit. Rev. Toxicol.* **23,** 147 (1993).
[3] G. R. Buettner and B. A. Jurkiewicz, *Free Rad. Biol. Med.* **14,** 49 (1993).
[4] P. B. McKay, *Ann. Rev. Nutr.* **5,** 323 (1985).
[5] R. C. Rose and A. M. Bode, *FASEB J.* **7,** 1135 (1993).

0076-6879/99 $30.00

excess of reducing equivalents, such as those provided by homocysteine or dithiothreitol.

It has been speculated that a mechanism exists to regenerate α-T. Cellular α-T is present in pmol/mg protein concentrations and is obtained solely through the diet. Tocopherol has a persistent radical upon donation of a hydrogen atom. Regeneration of α-T is thought to involve donation of reducing equivalents from the millimolar concentrations of vitamin C.

Numerous *in vitro* experiments utilizing homogeneous solutions, liposomes, plasma, and cellular organelles suggest that vitamin C reduces the tocopheroxyl radical to α-T (discussed in[1,2,4–6]). In brief, a lag occurs between the decline of α-T and the initiation of lipid peroxidation. Addition of ascorbate prolongs this lag, suggesting that there is an interaction between vitamin E and vitamin C. The tocopheroxyl radical appears upon the depletion of ascorbate concentrations or the decay of the ascorbate anion radical. Other studies demonstrate the disappearance of the tocopheroxyl radical after the addition of ascorbate.

Living, intact systems, such as cell preparations and whole animal, yield conflicting results with regard to the ability of vitamin C to regenerate vitamin E (discussed in[7–9]). For example, ODS rats are mutant Wistar rats which lack the enzyme L-gulono-γ-lactone oxidase and thus cannot synthesize ascorbate. When fed diets enriched in ascorbate, ODS rats had greater tissue concentrations of α-T.[10] By contrast, the turnover of deuterated α-T in guinea pig tissues is not different among animals on either a high or low tocopherol diet supplemented with high, normal, and low ascorbate. It is concluded that ascorbate had little effect on the sparing of α-T.[11]

The use of primary cultures of rat hepatocytes provides a useful tool in detailing the interaction between vitamins E and C.[7–9,12] The characteristics of cultured hepatocytes provide a model that is greatly reduced in vitamin E after only 16–18 hr (Table I). In addition, cultures from mutant rats have added to the usefulness of this model system. As previously mentioned, ODS rats are deficient in gulonolactone oxidase, the last enzyme in the biosynthesis of ascorbate. Thus, with cultures from ODS rats, it is

[6] R. E. Beyer, *J. Bioenerg. Biomembr.* **26,** 349 (1994).

[7] P. A. Glascott, Jr., E. Gilfor, and J. L. Farber, *Mol. Pharmacol.* **48,** 80 (1995).

[8] P. A. Glascott, Jr., M. Tsyganskaya, E. Gilfor, M. A. Zern, and J. L. Farber, *Mol. Pharmacol.* **50,** 994 (1996).

[9] P. A. Glascott, Jr., E. Gilfor, A. Serroni, and J. L. Farber, *Biochem. Pharmacol.* **52,** 1245 (1996).

[10] O. Igarashi, Y. Yonekawa, and Y. Fujiyama-Fujihara, *J. Nutr. Sci. Vitaminol.* **37,** 359 (1991).

[11] G. W. Burton, U. Wronska, L. Stone, D. O. Foster, and K. U. Ingold, *Lipids* **25,** 199 (1990).

[12] P. A. Glascott, Jr., E. Gilfor, and J. L. Farber, *Mol. Pharmacol.* **41,** 1155 (1992).

TABLE I
CELLULAR VITAMIN E AND VITAMIN C CONTENT IN HEPATOCYTES FROM NORMAL AND ODS RATS[a]

Hepatocytes	Supplemental α-TP	Supplemental ascorbate	Cellular α-tocopherol (pmol/mg protein)	Cellular ascorbate (nmol/mg protein)
Freshly isolated	—	—	170 ± 49 (24)	4.62 ± 0.92 (13)
Cultured				
Normal rat	−	−	27 ± 16 (18)	4.87 ± 1.42 (13)
	+	−	193 ± 43 (21)	4.94 ± 1.67 (11)
ODS rat	−	−	16 ± 2 (3)	0.8 ± 0.7 (4)
	+	−	154 ± 35 (6)	0.5 ± 0.4 (4)
	−	+	22 ± 2 (3)	13.4 ± 0.7 (7)
	+	+	166 ± 32 (6)	13.0 ± 1.4 (7)

[a] Hepatocytes from Sprague-Dawley or ODS rats were isolated and cultured for 16–18 hr in 5 ml of complete Williams' E medium in 5% CO_2/95% air at 37°C. Cells were cultured in the presence or absence of 1–1.2 μM α-TP and/or 100 μM sodium ascorbate. The next day, the cells were washed and assayed for vitamin E and vitamin C by HPLC. The vitamin content of isolated hepatocytes was determined on the day of isolation. Values are the mean ± standard deviation. The number of determinations is indicated in parentheses. (Reprinted with permission of Am. Soc. Pharmacol. Exp. Ther., *Mol. Pharmacol.* **41,** 1155, 1992 and *Mol. Pharmacol.* **50,** 994, 1996.)

possible to have a living system that is not only deficient in vitamin E content, but also limited in its ability to synthesize vitamin C (Table I). Cultured cells also provide the flexibility of supplementing the cells with a wide concentration of extracellular vitamin E and/or C that results in a predictable cellular concentration (Fig. 1) and protect the cell against oxidative damage such as that produced by *tert*-butyl hydroperoxide (TBHP)[7,8,12] or allyl alcohol.[9] Finally, the measurements of cellular levels of vitamins E and C from experiments are only of the endogenous contents of vitamins, because any extracellular supplementation is eliminated from the culture medium prior to the initiation of the experiment. This elimination of extracellular supplementation negates any confounding effects produced in the presence of exogenous vitamins. Thus, it is possible to obtain a profile of the utilization of vitamins E and C during oxidative injury and to explore their interaction.

Methods

Animals

Male Sprague-Dawley rats (150–200 g; Charles River, Wilmington, MA) or the ODS rats (4-week-old; Clea, Tokyo, Japan) are housed for at least

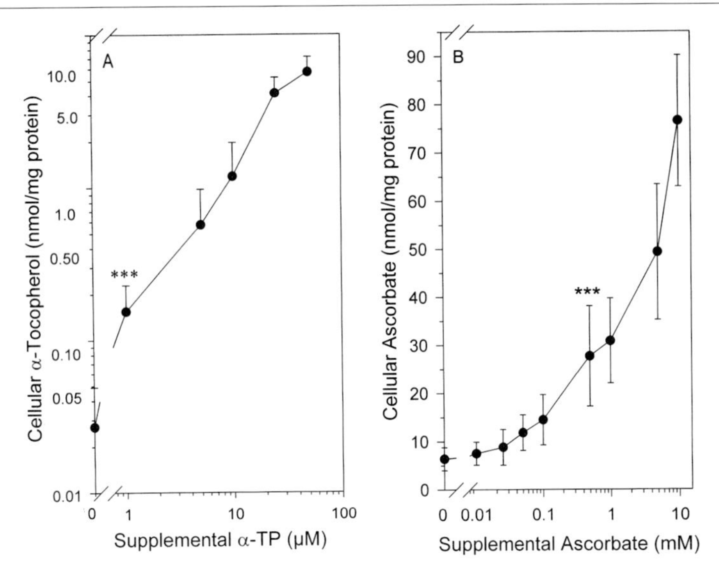

FIG. 1. Effect of overnight supplementation with (A) α-TP or (B) ascorbate on the preexperiment content of vitamins in hepatocytes. Hepatocytes were prepared and incubated overnight in the presence or absence of 1–50 μM α-TP or 0.01–10 mM ascorbate. After 16–18 hr, cells were washed and assayed for α-T or ascorbate. Values are the mean ± standard deviation for the results of 4–6 experiments. ***, At least $p < 0.005$ compared to respective control group. Reprinted with permission of Elsevier Science, Inc. *Biochem. Pharmacol.* **52,** 1245, 1996, and Am. Soc. Pharmacol. Exp. Ther., *Mol. Pharmacol.* **48,** 80, 1995.

1 week in quarters approved by the American Association of Laboratory Animals, allowed food (Purina Rodent Laboratory Chow 5001) and water *ad libitum*, and fasted overnight prior to use. This brand of rat chow contains 40 IU/kg vitamin E and no ascorbate. Clinical symptoms of scurvy[13] in ODS rats are largely prevented by supplementing the drinking water with 200 mg/liter Ester C (72% calcium ascorbate, 9% dehydroascorbate, and 1% calcium threonate; Inter-Cal Corp., Prescott, AZ). Occasionally, when ODS rats became scorbutic, additional supplementation with sodium ascorbate (100 mg/liter in the water and/or by dusting the food pellets) alleviated the symptoms. ODS rats on this regimen for 14 weeks gained weight at a rate comparable to wild-type rats. Ester C is not removed from the water prior to preparation of hepatocytes.

Isolation and Culturing of Rat Hepatocytes

Isolated hepatocytes are prepared by collagenase (Boehringer Mannheim Biochemicals, Indianapolis, IN) perfusion as described by Seglen.[14]

[13] A. J. Verlangieri, M. J. Fay, and A. E. Bannon, *Life Sci.* **48,** 2275 (1991).

[14] P. O. Seglen, *Methods Cell Biol.* **13,** 29 (1976).

Briefly, rats are anesthetized with sodium pentobarbital. After ligating the inferior vena cava, the liver is perfused through the portal vein at 30 ml/min for 5–6 min with an EGTA buffer (137 m*M* NaCl, 4.7 m*M* KCl, 1.2 m*M* KH_2PO_4, 0.65 m*M* $MgSO_4$, and 10 m*M* HEPES, 1 m*M* EGTA, pH 7.4). The liver is then perfused at the same rate for 3–4 min with 130 units/ml collagenase B (Boehringer Mannheim Biochemicals, Indianapolis, IN) in a buffer composed of 67 m*M* NaCl, 6.7 m*M* KCl, 5.1 m*M* $CaCl_2$, and 100 m*M* HEPES, pH 7.6. The liver is removed from the rat and then half of the hepatic capsule is removed from the tissue. Hepatocytes are suspended in ice-cold HEPES buffer (142 m*M* NaCl, 6.7 m*M* KCl, 1.2 m*M* $CaCl_2$, and 10.1 m*M* HEPES, pH 7.4) with gentle shaking over a plastic comb. The isolated hepatocytes are transferred to a 50 ml conical tube and allowed to settle for 15 min after which they are passed through gauze. The cells are then counted. Yields of 2–4 $\times$ 10^8 cells/liver with 90–95% viability (as determined by trypan blue exclusion) are routinely obtained.

After counting, the hepatocytes are plated on 25 cm^2 plastic flasks (Corning Glass Works, Corning, NY) at a density of 1.33 $\times$ 10^6 cells per flask in 3 ml of Williams' E medium (GIBCO Laboratories, Grand Island, NY) containing 9.1 IU/ml penicillin, 9.1 μg/ml streptomycin, 47 μg/ml gentamicin sulfate, 0.018 U/ml insulin, and 9% heat-inactivated (55°C for 15 min) fetal calf serum (JRH Biosciences, Kansas City, MO) (complete Williams' E medium). After a 2–3 hr incubation at 37°C in an atmosphere of 5% CO_2–95% (v/v) room air, the cultures are rinsed twice with 3 ml warmed HEPES buffer to remove any unattached or dead cells. Fresh complete Williams' E medium (5 ml) is added, and the hepatocytes incubated overnight.

Supplementation of Hepatocytes with Vitamin E and Vitamin C

Depending upon the experiment, cells are cultured for 16–18 hr in the presence or absence of vitamin E and/or vitamin C in a final concentration of 1–50 μM *dl*-α-tocopherol phosphate, disodium (α-TP), 1 μM *d*-α-tocopherol succinate (α-TS), *d*-α-tocopherol acetate (α-TA), or 1 μM non-esterified *dl*-α-tocopherol (Sigma, St. Louis, MO). The compounds are dissolved and diluted in dimethyl sulfoxide (DMSO) with the exception of α-TP, which is dissolved in water. The tocopherols are added to flasks in a volume of 0.5%. This concentration of dimethyl sulfoxide has no effect on the hepatocytes. In experiments involving vitamin C, 0.01–10 m*M* sodium ascorbate (Sigma Chemical Co., St. Louis, MO) dissolved in water, is added to overnight flasks at 1% volume.

After 16–18 hr, cultures are washed twice with warmed HEPES buffer, and 5 ml of serum-free Williams' E medium (complete Williams' E medium without fetal calf serum) without additional tocopherol is added. Cells are then treated with the compound of interest and incubated under 5% CO_2–95% air at 37°C for the desired time.

Cellular Tocopherol Determinations

α-, γ-, and δ-Tocopherol (Sigma) are determined by HPLC by the method previously described.[12] Briefly, at the designated times, the flasks are washed twice with 3 ml warmed HEPES buffer. Two ml of 50% ethanol containing 0.1 mg/ml butylated hydroxytoluene are added to the flasks, and the cells scraped from the plates. The cells are sonicated on ice for 15–20 seconds and 250 pmol of the internal standard, δ-tocopherol (90%), added. The sonicated cells are extracted with 2 ml of hexane, vortexed for 1 minute, and centrifuged for 2 minutes × 1640*g* at room temperature. The top layer of hexane is removed, and the extraction procedure repeated a second time. The hexane is evaporated to dryness under nitrogen, and the residue dissolved in 500 μl of methanol. Hexane and methanol are of the highest grade (Optima brand from Fisher Scientific Co., Pittsburgh, PA).

The tocopherols are determined by injection of 100 μl of the methanol resuspension onto an HPLC (Perkin-Elmer LC 410 with an ISS 100 autoinjector) with a fluorometric detector (Perkin-Elmer LC 240: excitation 205 nm, emission 325 nm; fixed factor 10; response 4). The conditions for analysis consist of a mobile phase of 96% methanol: 4% water (v/v), a flow rate of 2 ml/min, and a Supelcosil LC-18 column (150 mm × 4.6 mm; 3 μm packing) with a C_{18} precolumn (Supelco Inc., Bellefonte, PA). Quantitation of α and γ tocopherol is performed with a Nelson Series 900 Interface with Model 2100 PC Integrator (ver 5.1) by using the area ratio of the unknown tocopherol to δ-tocopherol. The standard curve is based on the area ratios of known concentrations of tocopherols to δ-tocopherol. The standard curve for γ-tocopherol is corrected for any γ-tocopherol contributed by the addition of internal standard; correction for α-T is not necessary. δ-, γ-, and α-Tocopherols have retention times of 5.2, 6.2, and 8.1 minutes, respectively.

Cellular Ascorbate Determinations

Ascorbic acid (reduced form) is determined by HPLC by the method previously described.[7] Briefly, the flasks are washed twice with 3 ml of warm HEPES buffer. Two ml of 0.85% *m*-phosphoric acid containing 25 μg/ml L-tyrosine (internal standard) are added to the flasks. The cells are scraped from the plates and kept on ice until being sonicated, also on ice, for 10–15 seconds. An aliquot is removed for protein determination and the remaining sample placed in a microfuge tube. The microfuge tube is spun at 12,000g for 6 minutes at 4° to pellet the protein. Approximately 600 μl of the acid supernatant is removed with a 1 ml syringe and filtered through a 0.5 μm filter (Millipore Millex-LCR_4, 4 mm) into an autosampler vial. Ten μl is injected onto a high-performance liquid chromatography system (Perkin-Elmer, Norwalk, CT, LC 410 solvent delivery system with an ISS 100 autoinjector) with an ultraviolet/visible spectrophotometer (Perkin-Elmer LC-95; 264 nm).

Either a 4.6 × 150 mm Supelcosil LC-ABZ C_{18} column (5 μm packing) with an LC-ABZ C_{18} precolumn or a 4.6 × 150 mm Supelcosil LC-18 C_{18} column (3 μm packing) with an LC-18 C_{18} precolumn (Supelco Inc., Bellefonte, PA) is used. The conditions for analysis employed a mobile phase of 0.34% glacial acetic acid (60 m*M*), 0.0825% octylamine (5 m*M*), and 10% methanol (v/v). The mobile phase is adjusted to a final pH of 4.8 with 10 *N* sodium hydroxide, filtered (0.45 μm), and purged with helium prior to use. The flow rate is 1 ml/min for sample determination (approximately 6 min), after which the flow rate is increased to 1.6 ml/min for 7 min to elute acid-extractable peaks that appear after tyrosine and ascorbate. The retention times for tyrosine and ascorbate with the LC-ABZ column are approximately 2.7 and 3.3 min, respectively. The retention times for tyrosine and ascorbate with the LC-18 column are approximately 3.1 and 4.7 min, respectively. Ascorbate is quantified with a Nelson series 900 Interface and 2100 PC Integrator (version 5.1) software using the ratio of the peak height of unknown ascorbate to the peak height of tyrosine. The standard curve is based on the ratio of the peak height of known ascorbate to the peak height of tyrosine. Endogenous tyrosine from cultured hepatocytes is not detectable under experimental conditions.

Other Assays

Cell killing is determined by release of lactate dehydrogenase (LDH) into the medium as described previously.[12] Lipid peroxidation, as determined by the accumulation of thiobarbituric acid (TBA)-reactive products released into the medium, is measured as described previously.[12] Protein is determined by the bicinchoninic acid method using bovine serum albumin as the standard.[15]

Discussion

Cultured hepatocytes provide a simple living model to examine the interaction between vitamins E and C.[7–9,12] Hepatocytes cultured for 16–

[15] P. K. Smith, R. I. Krohn, G. T. Hermanson, A. K. Mallia, F. H. Gartner, M. D. Provenzano, E. K. Fujimoto, N. M. Goeke, B. J. Olson, and D. C. Klenk, *Anal. Biochem.* **150,** 76 (1985).

18 hr overnight lose 85% of their endogenous α-T due to normal hepatic secretion into the culture medium.[12] This time is relatively short compared to the months required to achieve vitamin E deficiency in whole animals on special diets. The complete Williams' E medium in which the hepatocytes are incubated does not contain sufficient vitamin E to maintain physiological concentrations of α-T.[12] Figure 2 illustrates that culturing cells with at least 1 μM tocopherol or tocopherol ester maintains physiological concentrations of vitamin E and protects cells against TBHP.[12]

Experiments with cultured hepatocytes from normal and ODS rats allow the examination of the relationship between vitamins E and C.[7–9] Figure 3 shows the cellular vitamin E and vitamin C content, the cell killing and accumulation of TBA-reactive products during oxidative stress induced by TBHP in normal hepatocytes. These data demonstrate that physiological

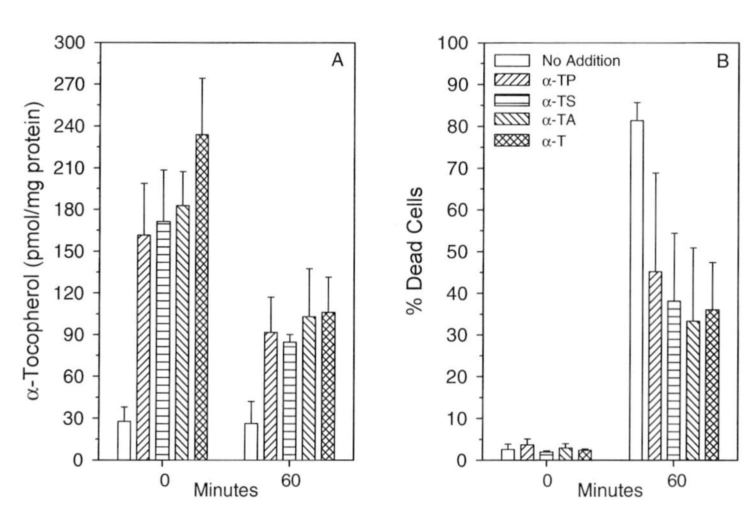

FIG. 2. Effect of 500 μM TBHP on the (A) α-T content and (B) cell killing of hepatocytes cultured overnight with 1 μM α-T or one of its esters. Hepatocytes were prepared and incubated overnight without supplementation (open), or with 1 μM α-TP (diagonal rising right), 1 μM α-TS (horizontal), 1 μM α-TA (diagonal rising left) or 1 μM nonesterified α-T (diagonal cross-hatched). After 16–18 hr, cells were washed, placed in Williams' E medium without serum and without tocopherol equivalents, and then incubated with 500 μM TBHP for 60 min. Values are the mean ± standard deviation for the results of 4 experiments. (B) While exposure to TBHP caused significant cell killing in all groups ($p < 0.03$), cells given tocopherol or one of its esters showed lower killing compared to unsupplemented cells ($p < 0.0001$). (A) Cells supplemented with nonesterified α-T had higher concentrations of α-T than cells supplemented with tocopherol esters ($p < 0.0001$). While exposure to TBHP reduced α-T in all supplemented groups ($p < 0.003$), these groups were still greater than the unsupplemented group ($p < 0.005$). Reprinted with permission of Am. Soc. Pharmacol. Exp. Ther., *Mol. Pharmacol.* **41,** 1155, 1992.

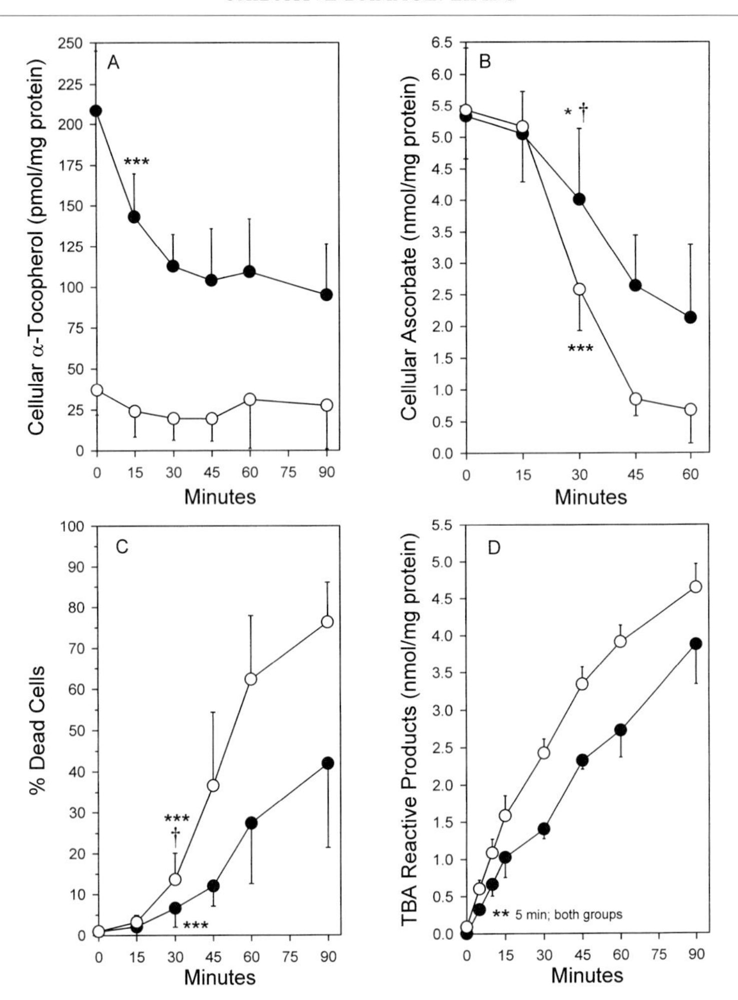

FIG. 3. Effect of 750 μM TBHP on the (A) cellular α-T content, (B) cellular ascorbate content, (C) cell killing, and (D) accumulation of TBA-reactive products in hepatocytes cultured overnight with (●) and without (○) 1 μM α-TP. Hepatocytes were prepared and incubated overnight with or without 1 μM α-TP. After 16–18 hr, cells were washed, placed in Williams' E medium without serum or additional α-TP, and then incubated with 750 μM TBHP. Values are the mean ± standard deviation for the results of 3–4 experiments. Control cells not treated with TBHP showed no significant change in the α-T or ascorbate concentration over the course of 90 min (data not shown). *, $p < 0.05$ from respective zero time; **, $p < 0.005$ from respective zero time; ***, $p < 0.001$ from respective zero time; †, $p < 0.05$ from same time in unsupplemented cells. Reprinted with permission of Am. Soc. Pharmacol. Exp. Ther., *Mol. Pharmacol.* **41,** 1155, 1992, and *Mol. Pharmacol.* **48,** 80, 1995.

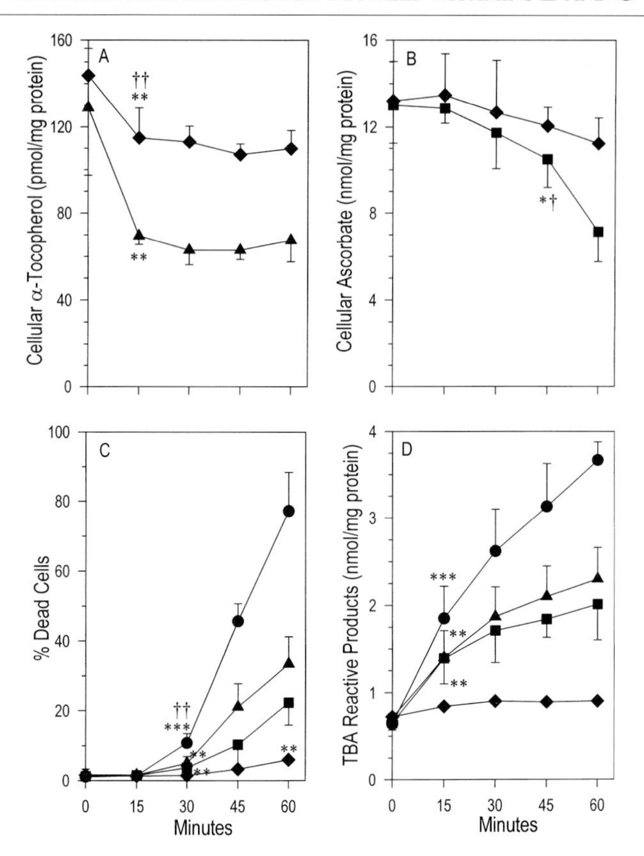

FIG. 4. Effect of 500 μM TBHP on the (A) cellular vitamin E content, (B) cellular vitamin C content, (C) cell killing, and (D) accumulation of TBA-reactive products of hepatocytes that were cultured overnight without α-TP and ascorbate (●), with either α-TP (▲) or ascorbate (■) alone, or with both vitamins (◆). Hepatocytes were prepared and incubated overnight with or without 1.2 μM α-TP and/or 100 μM ascorbate. After 16–18 hr, cells were washed and placed in Williams' E medium without serum and additional α-TP and/or ascorbate. The cells were then incubated with 500 μM TBHP for 60 min. Ascorbate values have been corrected for nonspecific release of ascorbate seen with supplementation in ODS cells.[8] (D) Hepatocytes supplemented with either vitamin alone were different from cells receiving both or no vitamins, but were not different from each other (p at least <0.01). Asterisks and daggers indicate the first point at which statistical differences are present. Values are the mean ± standard deviation for the results of 3 experiments. *, p at least <0.05 compared with respective time zero; **, p at least <0.01 compared with respective time zero; ***, p at least <0.001 compared with respective time zero; †, p at least <0.05 from other supplementation groups at the same time; ††, p at least <0.005 from other supplementation groups at the same time. Reprinted with permission of Am. Soc. Pharmacol. Exp. Ther., *Mol. Pharmacol.* **50,** 994, 1996.

amounts of vitamin E significantly protect cells against an oxidative stress induced by TBHP and cause a decrease in the formation of TBA-reactive products.[12] During oxidative stress, vitamin E declines within the first 15 minutes and prior to cell death.[7,12] The decline in vitamin E involves a cellular source of ferric iron.[12] The loss in vitamin E precedes the decline in vitamin C which occurs at 30 min.[7] The decrease in vitamin C parallels the increase in cell killing.[7] That is to say, the loss of ascorbate was more closely associated with the disruption of the plasma membrane than the regeneration of vitamin E.

Similarly, Fig. 4 shows the cellular vitamin E and vitamin C content, the cell killing and TBA-reactive products during oxidative stress induced by TBHP in ODS hepatocytes that were incubated in the presence or absence of vitamins E and/or C. The ability of vitamin C to protect against cell killing and lipid peroxidation did not depend on the content of vitamin E.[7,8] Vitamin E and vitamin C act as antioxidants independent of each other and protect cells when compared to cells lacking both vitamins E and C. As is seen in normal hepatocytes, the decline in vitamin C occurs after the initiation of cell death. Taken together, these and other supplementary data[7–12] argue against the regeneration of the tocopheroxyl radical by ascorbate. Rather, these data better support the position that vitamin C acts as an antioxidant independent of vitamin E and reacts with radicals prior to their reaction with vitamin E. Thus, vitamin C prevents the loss of vitamin E in an indirect manner in an intact living model.

[12] Modification of Proteins and Lipids by Myeloperoxidase

By STANLEY L. HAZEN, FONG F. HSU, JOSEPH P. GAUT, JAN R. CROWLEY, and JAY W. HEINECKE

Oxidative modification of lipids, nucleic acids, and proteins is thought to be of central importance in the pathogenesis of disease. In atherosclerosis, the leading cause of death in industrialized nations, one important oxidative target may be low density lipoprotein (LDL), the major carrier of blood cholesterol.[1–5] LDL is a complex of lipid and protein, predominantly apoli-

[1] M. S. Brown and J. L. Goldstein, *Science* **232,** 34 (1986).
[2] U. P. Steinbrecher, F. F. Zhang, and M. Lougheed, *Free Radical Biol. Med.* **9,** 155 (1990).
[3] J. L. Witztum and D. Steinberg, *J. Clin. Invest.* **88,** 1785 (1991).

0076-6879/99 $30.00

poprotein B100.[1] Although many lines of evidence suggest that oxidation converts LDL into an atherogenic form,[2–5] it has been difficult to identify the relevant reaction pathways *in vivo* because reactive intermediates are short-lived and difficult to detect directly.

One pathway for LDL oxidation may involve myeloperoxidase,[6] a heme protein secreted by activated phagocytes.[7,8] The enzyme generates oxidants that play key roles in defenses against invading pathogens, but it may also damage host tissues.[7–13] It is present and active in human atherosclerotic lesions and, in intermediate lesions, colocalizes with macrophages.[6,11,13]

Myeloperoxidase uses hydrogen peroxide (H_2O_2), also produced by activated phagocytes, to generate potent oxidants.[8–10,14] We have investigated several reaction pathways to identify unique and stable products of this system.[11,15–17] When these markers are detectable at sites of inflammation and vascular disease, they implicate myeloperoxidase as the agent of damage.

At physiological concentrations of chloride ion, myeloperoxidase converts H_2O_2 to hypochlorous acid (HOCl; refs. 8, 10):

$$Cl^- + H_2O_2 + H^+ \rightarrow HOCl + H_2O \quad (1)$$

Moreover, it is the only human enzyme known to convert chloride to HOCl at plasma concentrations of halides.[8,18] Therefore, detection of chlorinated or chloride-dependent oxidation products can serve as a powerful tool for identifying sites where myeloperoxidase has inflicted oxidative damage *in vivo*.

---

[4] J. A. Berliner and J. W. Heinecke, *Free Rad. Biol. Med.* **20,** 707 (1996).
[5] G. Jurgens, H. F. Hoff, G. M. Chisolm, and H. Esterbauer, *Chem. Phys. Lett.* **45,** 315 (1987).
[6] A. Daugherty, J. L. Dunn, D. L. Rateri, and J. W. Heinecke, *J. Clin. Invest.* **94,** 437 (1994).
[7] S. J. Klebanoff, *Ann. Intern. Med.* **93,** 480 (1980).
[8] J. E. Harrison and J. Schultz, *J. Biol. Chem.* **251,** 1371 (1976).
[9] J. M. Albrich, C. A. McCarthy, and J. K. Hurst, *Proc. Natl. Acad. Sci. USA* **78,** 210 (1981).
[10] C. S. Foote, T. E. Goyne, and R. I. Lehrer, *Nature* **301,** 715 (1981).
[11] S. L. Hazen and J. W. Heinecke, *J. Clin. Invest.* **99,** 2075 (1997).
[12] C. Leeuwenburgh, M. M. Hardy, S. L. Hazen, P. Wagner, S. Oh-ishi, U. P. Steinbrecher, and J. W. Heinecke, *J. Biol. Chem.* **272,** 1433 (1997).
[13] L. J. Hazell, L. Arnold, D. Flowers, G. Waeg, E. Malle, and R. Stocker, *J. Clin. Invest.* **97,** 1535 (1996).
[14] S. L. Hazen, F. F. Hsu, D. M. Mueller, J. R. Crowley, and J. W. Heinecke, *J. Clin. Invest.* **98,** 1283 (1996).
[15] J. W. Heinecke, W. Li, D. M. Mueller, A. Bohrer, and J. Turk, *Biochem.* **33,** 10127 (1994).
[16] S. L. Hazen, J. P. Gaut, F. F. Hsu, J. R. Crowley, A. d'Avignon, and J. W. Heinecke, *J. Biol. Chem.* **272,** 16990 (1997).
[17] S. L. Hazen, F. F. Hsu, K. Duffin, and J. W. Heinecke, *J. Biol. Chem.* **271,** 23080 (1996).
[18] S. J. Weiss, S. T. Test, C. M. Eckmann, D. Ross, and S. Regiani, *Science* **234,** 200 (1986).

One target for chlorination is tyrosine, a phenolic amino acid.[11,14,19–21] HOCl or HOCl-derived oxidants halogenate protein-bound tyrosyl residues through an electrophilic reaction, generating 3-chlorotyrosine, a stable end product. Our *in vitro* studies have shown that 3-chlorotyrosine is a very specific marker of protein oxidation by myeloperoxidase.[11,14] Using stable isotope dilution gas chromatography-mass spectrometry (GC/MS), we also have demonstrated that 3-chlorotyrosine levels are much higher in LDL isolated from human atherosclerotic lesions than in circulating LDL, implicating myeloperoxidase as one pathway for LDL oxidation *in vivo*.[11] Our method for detecting 3-chlorotyrosine in proteins and tissues is described in an accompanying chapter.[22]

Whereas the myeloperoxidase–H_2O_2–Cl^- system converts protein-bound tyrosine residues to 3-chlorotyrosine, it generates the amphipathic aldehyde *p*-hydroxyphenylacetaldehyde (*p*HA) from free tyrosine.[23] The reaction pathway involves initial chlorination of the α-amino group of tyrosine; the resulting unstable monochloramine then decomposes into *p*HA,[24] which is the major product of optimally stimulated phagocytes at plasma concentrations of L-tyrosine and chloride. In turn, *p*HA can covalently modify proteins because it reacts with the ε-amino moiety of lysine residues to form a Schiff base. This base can be detected readily by isotope dilution GC/MS after it is reduced to *p*HA–lysine, which is acid-stable.[16]

This chapter describes our methods for identifying and quantifying oxidative damage to proteins, lipids, and LDL that has resulted from myeloperoxidase activity. We first discuss general procedures for working with HOCl and other reactive halogenating agents. Then we describe methods for isolating LDL from plasma and LDL-like particles from human atherosclerotic tissue. We also describe methods for synthesizing, detecting, and quantifying *p*HA–lysine[16] and chlorinated sterols,[15,17] lipid oxidation products generated by myeloperoxidase. In an accompanying chapter in this volume, we outline our mass spectrometric procedures for quantifying the oxidation products 3-chlorotyrosine, 3-nitrotyrosine, *o*-tyrosine, *m*-tyrosine, and *o,o'*-dityrosine in proteins, LDL, lesion LDL, and tissue.[22]

[19] N. M. Domigan, T. S. Charlton, M. W. Duncan, C. C. Winterbourn, and A. J. Kettle, *J. Biol. Chem.* **270,** 16542 (1995).

[20] A. J. Kettle, *FEBS Lett.* **379,** 103 (1995).

[21] S. L. Hazen, J. R. Crowley, D. M. Mueller, F. F. Hsu, and J. W. Heinecke, *Free Rad. Biol. Med.* **23,** 909 (1997).

[22] J. W. Heinecke, F. F. Hsu, J. R. Crowley, S. L. Hazen, C. L. Leeuwenburgh, J. E. Rasmussen, and J. Turk, *Methods in Enzymology* **300** [16] (1998) (this volume).

[23] S. L. Hazen, F. F. Hsu, and J. W. Heinecke, *J. Biol. Chem.* **271,** 1861 (1996).

[24] S. L. Hazen, A. d'Avignon, F. F. Hsu, and J. W. Heinecke, *J. Biol. Chem.* **37,** 6864 (1998).

Obtaining Chlorine-Demand Free Conditions

When assaying chlorinated compounds, it is important to ensure that all glassware is rendered chlorine-demand free and that reaction buffers consume an insignificant fraction of reagent HOCl. Otherwise, HOCl will react with the glassware and buffers instead of the target molecules, artifactually lowering the yield of the desired reaction products.

We operationally define chlorine-demand free conditions as no detectable (<3%) consumption of a 1 m*M* HOCl solution during a 10 min incubation at 37°. Glassware is rendered chlorine-demand free by thorough rinsing with sodium hypochlorite (NaOCl) and H_2O; the water for this and the other procedures should be deionized and glass-distilled. The glassware then is pyrolyzed at 500° overnight. Quartz cuvettes used for HOCl or H_2O_2 determinations are washed with a concentrated stock (>10 m*M*) of the oxidant, exhaustively rinsed with H_2O, and dried prior to use.

Reaction buffers should be demonstrated to be chlorine-demand free prior to use by incubating them with HOCl in a gas-tight vial and monitoring HOCl concentration. HOCl concentration can be determined either by the taurine chloramine assay[25] or by the oxidation of iodide to triiodide.[26] Of the buffers we have tested, only phosphate has consistently proved to be chlorine-demand free over a wide range of pH values. A physiological buffer that is generally chlorine-demand free and can be used for cells is phosphate-buffered saline (PBS) containing 10 m*M* phosphate. When chloride-free conditions are required, reactions can be performed in sodium phosphate. In experiments where trace metal ions need to be avoided, we supplement phosphate buffers with 100 μM diethylenetriaminepentaacetic acid (DTPA). Under these conditions, buffers are first treated with Chelex resin (Bio-Rad, Richmond, CA) to remove metal ions; DTPA is then added to the reaction mixture.

We determine concentrations of H_2O_2 and hypochlorite (ClO^-, the conjugate base of HOCl) spectrophotometrically ($\varepsilon_{240} = 39.4\ M^{-1}\ cm^{-1}$ and $\varepsilon_{292} = 350\ M^{-1}\ cm^{-1}$, respectively; refs. 27, 28). Note that ClO^- is in equilibrium with HOCl ($pK_a \sim 7.5$ at 30°; refs. 28, 29). The absorption spectra of HOCl and ClO^- differ.[28] It is therefore important for accurate

[25] A. J. Kettle and C. C. Winterbourn, *Methods in Enzymology* **233,** 502 (1994).

[26] M. El-Saadani, H. Esterbauer, M. El-Sayad, M. Goher, A. Y. Nassar, and G. Jurgens, *J. Lipid Res.* **30,** 627 (1989).

[27] D. P. Nelson and L. A. Kiesow, *Anal. Biochem.* **49,** 474 (1972).

[28] J. C. Morris, *J. Phys. Chem.* **70,** 3798 (1966).

[29] G. C. White, *in* "Handbook of Chlorination," p. 182. Van Nostrand Reinhold, New York, 1972.

determination of ClO^- concentration that the stock solution be alkaline (>pH 9.5), such as in NaOCl aliquoted into distilled H_2O.

Preparation of Chloride-Free Sodium Hypochlorite

To study mechanisms of chlorination by myeloperoxidase, ClO^- is often used in place of the enzymatic system. Reagent NaOCl is a strongly basic solution containing high concentrations of contaminating chloride. Because certain chlorination reactions mediated by $HOCl/ClO^-$ are chloride-dependent,[9,14,17] it is critical to prepare chloride-free reagent ClO^- to investigate reaction mechanisms.

We use the following procedure to prepare chloride-free NaOCl, working in a well-ventilated hood with frequent venting of the Cl_2 that is released during acid addition. In a separatory funnel, we mix reagent NaOCl with ethyl acetate (1 : 1, v/v) and then convert it to HOCl by dropwise addition of concentrated phosphoric acid. The final pH of the aqueous solution should be ≤5. Because HOCl is uncharged, it partitions into the organic solvent. This yellow organic phase is washed several times with H_2O to remove residual chloride. The HOCl then is converted to ClO^- by the dropwise addition of concentrated NaOH to the solution. Because ClO^- is charged, it partitions into the aqueous phase, reforming NaOCl. The final pH of the solution should be ≥10. Bubbling with N_2 removes the residual ethyl acetate in the yellow aqueous solution of chloride-free NaOCl. In all experiments using chloride-free conditions, it is important to avoid the use of chloride salts of buffers and reagents. Solutions containing HCl are another common source of contaminating chloride. Proteins, lipids, LDL, and myeloperoxidase also need to be free of chloride.

Myeloperoxidase Purification

A number of methods are useful for isolating myeloperoxidase. Typically, the enzyme is first extracted with cetyltrimethylammonium bromide from human leukocytes. We use the procedure of Rakita and Rosen[30] to obtain a reasonable yield from the detergent extract. In this method, solubilized myeloperoxidase is purified by lectin-affinity chromatography and size-exclusion chromatography (Superose 6 column; Pharmacia, Piscataway, NJ) as described.[30] The enzyme is then dialyzed against phosphate buffer, followed by H_2O, to remove chloride and other halides, and stored in 50% (v/v) glycerol at −20°. Myeloperoxidase prepared by this method is stable for long periods of time and has an $A_{430\,nm}/A_{280\,nm}$ ratio of ≥0.6.

[30] R. M. Rakita, B. R. Michel, and H. Rosen, *Biochemistry* **29,** 1075 (1990).

Enzyme concentration is determined spectrophotometrically (ε_{430} = 170 mM^{-1} cm^{-1}; ref. 31).

Isolation of LDL from Human Atherosclerotic Intima

We have isolated LDL from human arterial tissue to evaluate the roles of different pathways in the oxidative modification of LDL. In our hands, the following method yields greater amounts of LDL-like particles (d = 1.019–1.070 g/ml) from atherosclerotic tissue than previously described procedures.[32–34]

We perform all steps at 4° unless otherwise indicated. Atherosclerotic intima from fatty streaks and intermediate lesions is rapidly isolated by blunt dissection in thoracic aorta from a single subject. Grossly necrotic and calcified lesions (i.e., advanced lesions) are discarded. Intima is gently minced into small pieces with a pair of scissors and then pulverized into a fine powder in a stainless steel mortar and pestle at the temperature of liquid nitrogen. Tissue powder (up to 9 g) is placed into a 50 ml conical centrifuge tube, which then is filled completely with extraction buffer [0.15 *M* NaCl, 10 m*M* sodium phosphate, 300 μM EDTA, 100 μM DTPA, 50 μg/ml soybean trypsin inhibitor, 40 μg butylated hydroxytoluene (BHT), 10 m*M* 3-aminotriazole, pH 7.4] to avoid the formation of air bubbles. The tube is gently rocked end-over-end overnight. The next morning, the mixture is centrifuged at 5000*g* for 15 min to remove gross particulate material from the crude intimal extract and the pellet is discarded. If the amount of tissue is limited, the pellet may be washed with an equal volume of extraction buffer to recover additional LDL.

Lesion LDL is isolated from the intimal extract by a modification of the sequential flotation ultracentrifugation method described for plasma LDL by Schumaker and Puppione in an earlier volume of this series.[35] The extract containing lesion LDL is centrifuged at 100,000*g* for 30 min. The lipemic layer on the surface is carefully aspirated and discarded, the supernatant saved, and the pellet discarded. The density of the extraction buffer should be 1.006 g/ml as measured with a volumetric pipette and weighing. Otherwise, it is necessary to adjust the amount of KBr added in the next step accordingly. The density of the supernatant is adjusted to 1.019 g/ml

[31] Y. Morita, H. Iwamoto, S. Aibara, T. Kobayashi, and E. Hasegawa, *J. Biochem.* **99,** 761 (1986).

[32] A. Daugherty, B. S. Zweifel, B. E. Sobel, and G. Schonfeld, *Arteriosclerosis* **8,** 768 (1988).

[33] S. Yla-Hertualla, W. Palinski, M. E. Rosenfeld, S. Parthasarathy, T. E. Carew, S. Butler, J. L. Witztum, and D. Steinberg, *J. Clin. Invest.* **84,** 1086 (1989).

[34] U. P. Steinbrecher and M. Lougheed, *Arterioscler. Thromb.* **12,** 608 (1992).

[35] V. N. Schumaker and D. L. Puppione, *Methods in Enzymology* **128,** 155 (1986).

with anhydrous KBr. The amount of KBr added to each sample is calculated using the following formula: volume of solution (ml) × 17.14 (mg/ml) KBr = KBr (mg). The KBr is completely dissolved by rocking the sample gently end-over-end. The mixture then is carefully transferred into sealable ultracentrifuge tubes while avoiding bubble formation. A convenient method of transfer is to cut the needle off of a winged infusion needle attached to a syringe, thread the tubing into the centrifuge tube, pour the sample into the syringe, and slowly push the sample into the tube. Tubes are filled to the top with EDTA–saline buffer (1 m*M* EDTA, 0.15 *M* NaCl, pH 7.4) whose density has been adjusted to 1.019 g/ml with KBr (add 4.56 g of anhydrous KBr to 250 ml of EDTA–saline buffer). The sealed tubes are centrifuged at 45,000 rpm in a 50.2 Ti rotor (245,000*g*) for 18 hr at 15°, and the rotor is allowed to stop without braking. The tubes are cut with a Beckman Tube Slicer, and the pellet is harvested and transferred to a 50 ml conical tube using a small volume of EDTA–saline buffer adjusted to density 1.019 g/ml.

The LDL-containing pellet is gently resuspended in EDTA–saline buffer, and the solution is then adjusted to 1.070 g/ml by adding anhydrous KBr. The amount of KBr is calculated as follows: volume of pellet plus buffer (ml) × 68.09 (mg/ml) = KBr (mg). The KBr is dissolved carefully to avoid bubble formation. Solubilizing the pellet is facilitated by addition of the KBr and mashing the pellet with the wooden end of a cotton-tipped applicator. The dissolved pellet is then transferred into a sealable ultracentrifuge tube, again avoiding air bubbles. Residual air in the tubes is displaced with EDTA–saline buffer that previously has been adjusted to a density of 1.070 g/ml (add 22.67 g of anhydrous KBr to 250 ml of EDTA–saline buffer). The samples are again centrifuged at 45,000 rpm in a 50.2 Ti rotor (245,000*g*) for 18 hr at 15° and allowed to stop without braking. LDL is recovered from the lipemic layer floating in the top of the tube. This "lesion LDL" is dialyzed for 24 hr at 4° under nitrogen against three changes of 50 m*M* sodium phosphate buffer (pH 7.4) containing 100 μM DTPA. BHT (100 μM final) is added again at this step to avoid *ex vivo* oxidation of the lipoprotein. To avoid loss of sample, final LDL concentration may be estimated spectrophotometrically ($A_{280} - A_{315}$) employing a difference extinction coefficient ($\Delta\varepsilon_{280-315}$) of 1.05 $(\text{mg/ml})^{-1}$ cm^{-1}; the absorbance of the solution at 315 nm is subtracted from that at 280 nm (to correct for light scattering) and then divided by 1.05 $(\text{mg/ml})^{-1}$ cm^{-1}. A more accurate protein determination can be performed by the Markwell-modified Lowry protein assay[36] with bovine serum albumin as the standard.

[36] M. A. Markwell, S. M. Haas, L. L. Bieber, and N. E. Tolbert, *Anal. Biochem.* **87,** 206 (1978).

This method yields from 0.1 to 2 mg of LDL per 9 g of aortic intima. When we analyze preparations of LDL-like particles isolated by this procedure, we found that the vast majority of its cholesterol and apolipoprotein B100 migrate on FPLC (fast protein liquid chromatography) size-exclusion chromatography (tandem Superose 6 and 12 columns; Pharmacia LKB) with an apparent M_r indistinguishable from that of native LDL.[11] SDS–PAGE and Western blot analysis with a rabbit antihuman apolipoprotein B100 polyclonal antibody revealed aggregated, intact and fragmented forms of immunoreactive protein in lesion LDL preparations,[11] as previously reported by others.[13,32–34]

Isolation of Plasma Low Density Lipoprotein

We isolate human LDL from healthy male and female subjects by sequential density ultracentrifugation of plasma. The blood is collected into an EDTA-containing syringe [1 ml of 10% (w/w) EDTA (potassium salt; pH 7.4) per 50 ml of blood] and centrifuged at 500*g* at 4° for 10 min. LDL (d = 1.019–1.070 g/ml) is then isolated from the plasma supernatant by two ultracentrifugation steps as outlined above for lesion LDL preparations.

LDL preparations are rendered chloride-free by dialysis at 4° under nitrogen against 65 m*M* sodium phosphate buffer (pH 7.4) containing 100 μM DTPA. Protein concentration is determined using the Markwell-modified Lowry protein assay[36] with bovine serum albumin as the standard.

Modification of Proteins in Vitro and in Vivo with pHA

We have shown that activated phagocytes use the myeloperoxidase–H_2O_2–Cl^- system to convert L-tyrosine into the amphipathic aldehyde *p*HA.[23] This tyrosine-derived aldehyde can then form a Schiff base with the N^ε-amino moiety of protein lysine residues.[16] The reduced Schiff base, N^ε-[2-(*p*-hydroxyphenyl)ethyl]lysine (*p*HA–lysine), is acid-stable and serves as a specific marker for the modification of protein by myeloperoxidase.[16] We determine the formation of protein-bound Schiff base by reduction, followed by acid hydrolysis to free the amino acids, and then quantifying *p*HA–lysine by isotope dilution negative-ion electron capture GC/MS.

Synthesis of N^α-Acetyl[$^{13}C_6$]pHA–Lysine. [$^{13}C_6$]*p*HA is prepared from reagent NaOCl and [$^{13}C_6$]tyrosine. NaOCl (1 : 1, mol : mol, NaOCl : tyrosine) is added dropwise with constant mixing to an ice-cold solution of 2 m*M* L-tyrosine in 20 m*M* sodium phosphate (pH 7). The solution is then warmed to 37° for 60 min and analyzed by reverse-phase high-performance liquid chromatography (HPLC) with monitoring of absorbance at 276 nm.[16] Products are applied to a Beckman C_{18} column (Ultrasphere, 5 μm particle size, 4.6 × 250 mm) equilibrated with solvent A (0.1% trifluoroacetic acid in H_2O) at a flow rate of 1 ml/min. Products are analyzed with a gradient generated with solvent B (0.1% trifluoroacetic acid in methanol) as follows: 0–5 min, 0–35% solvent B; 5–25 min, isocratic elution with 35% solvent B; 25–35 min, 35–100% solvent B. *p*HA is baseline-resolved from other products as a single major peak with a longer retention time than L-tyrosine. Yield of the aldehyde is typically ≥95%. *p*HA can react with the primary amine on residual L-tyrosine, but using *p*HA isolated by HPLC avoids potential interference from these products with the quantification of N^{α}-acetyl-[$^{13}C_6$]*p*HA–lysine. Ensuring that all glassware and buffers are chlorine-demand free as described above prevents the *p*HA yield from being decreased by the reaction of HOCl with trace quantities of organic material on glassware or in buffer.

To prepare the Schiff base adduct of *p*HA and the ε-amino group of lysine, we add N^{α}-acetyllysine (4–10 m*M* final concentration) to a solution of *p*HA (2 m*M*). The mixture is incubated at 37° for 4 hr, and the Schiff base is reduced to N^{α}-acetyl-*p*HA-lysine (the acid-stable adduct) by overnight incubation with 10 m*M* $NaCNBH_3$. The overall synthesis is shown in Scheme I.

We quantify N^{α}-acetyl[$^{13}C_6$]*p*HA-lysine by reversed-phase HPLC using an external calibration curve constructed with L-tyrosine.[16] The conditions for HPLC analysis are similar to those outlined above for *p*HA analysis. *p*HA-lysine is baseline-resolved from L-tyrosine and *p*HA by use of a C_{18} Porasil column and the following gradient: 0–5 min, 0–10% solvent B; 5–25 min, isocratic elution with 10% solvent B; 25–35 min, 10–100% solvent B. N^{α}-Acetyl[$^{13}C_6$]*p*HA-lysine is stored in aliquots at −20° under argon. A typical HPLC chromatogram from a synthesis of *p*HA-lysine, along with

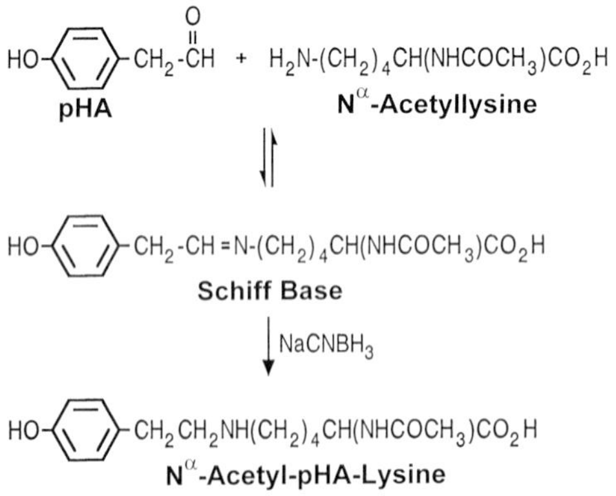

HO–C_6H_4–CH_2–CH(=O) + H_2N-$(CH_2)_4$CH($NHCOCH_3$)CO_2H

pHA **N$^{\alpha}$-Acetyllysine**

⇅

HO–C_6H_4–CH_2-CH=N-$(CH_2)_4$CH($NHCOCH_3$)CO_2H

Schiff Base

↓ $NaCNBH_3$

HO–C_6H_4–$CH_2CH_2NH(CH_2)_4$CH($NHCOCH_3$)CO_2H

N$^{\alpha}$-Acetyl-pHA-Lysine

SCHEME I

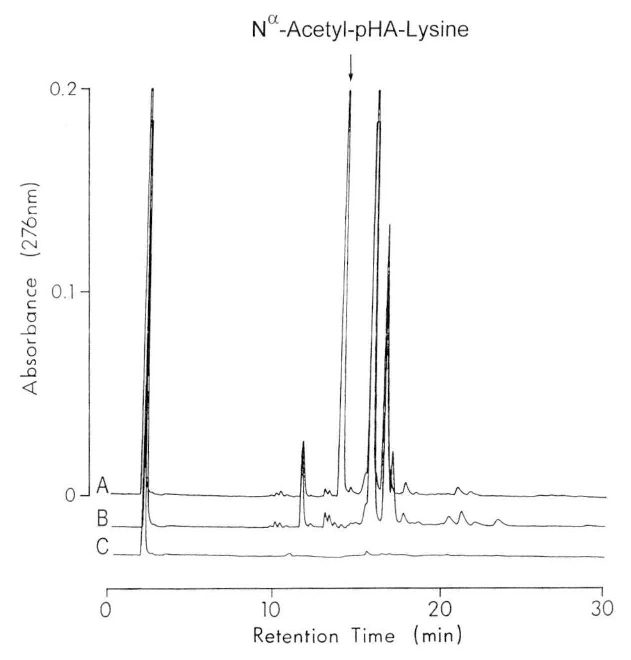

FIG. 1. Reversed-phase HPLC detection of the reduced Schiff base adduct formed in the reaction between *p*HA and N^{α}-acetyllysine. Reaction mixtures containing both 2 m*M* *p*HA and 4 m*M* N^{α}-acetyllysine (A), *p*HA alone (B), or N^{α}-acetyllysine alone (C) were incubated for 4h at 37°C in 20 m*M* sodium phosphate, 100 μM DTPA, pH 7.4. $NaCNBH_3$ (10 m*M*) was then added, and the reaction mixtures were incubated at 37°C overnight. Products were subjected to HPLC analysis on a C_{18} column as described in the text. The single major acid-stable compound formed in the presence of both *p*HA and N^{α}-acetyllysine (retention time 14.0 min) is N^{α}-acetyl-*p*HA-lysine (reprinted with permission from ref. 16).

control mixtures to aid in peak identification, are shown in Fig. 1. The identity of the peak representing N^{α}-acetyl[$^{13}C_6$]*p*HA–lysine is confirmed by mass spectrometry as outlined below.

Tissue Collection and Storage. For practical purposes, sample processing and analyses are best performed in batches. Freshly harvested tissue is rinsed in buffer supplemented with antioxidants and inhibitors of myeloperoxidase (catalase, a scavenger of peroxide; azide, a heme poison). We typically use 65 m*M* sodium phosphate (pH 7.4), 100 μM DTPA (an inhibitor of metal-catalyzed reactions; added from a 100 m*M* stock solution, pH 7.0), 100 μM butylated hydroxytoluene (an inhibitor of lipid peroxidation; added from a 10 m*M* stock solution prepared in 95% ethanol), 10 μg/ml catalase, and 10 m*M* aminotriazole. Tissue is stored frozen under nitrogen at −80° until use. For analysis of *p*HA–lysine, the reducing agent

$NaCNBH_4$ (20 mM) is included in all buffers used to store and prepare samples.

Tissue Processing. Tissue is thawed in sealed plastic bags under cool running water. All subsequent tissue processing steps are performed at 4° unless otherwise indicated. Specimens of interest (e.g., aortic intima) are rapidly isolated from the tissue, frozen in liquid nitrogen, and then pulverized in a stainless steel mortar and pestle at the temperature of liquid nitrogen. Tissue powder (~10–20 mg wet weight) is transferred to a 13 × 100 mm heavy-walled threaded glass hydrolysis tube and suspended in solution A (100 μM DTPA and 10 mM 3-aminotriazole, pH 7.0). Samples are delipidated twice using a single-phase extraction mixture comprised of H_2O : methanol : H_2O-washed diethyl ether (1 : 3 : 7; v/v/v). For each extraction, mixtures are vigorously vortexed, incubated for 30 min, and the protein precipitate is isolated by centrifugation at 5,000g for 5 min. Following the addition of $^{13}C_6$-labeled internal standards (typically 50 pmol N^{α}-acetyl[$^{13}C_6$]pHA-lysine and 100 nmol L-[$^{13}C_6$]lysine), the samples are dried in a rotary vacuum device and then hydrolyzed with acid.

Protein Hydrolysis. Proteins (0.1–2 mg) are hydrolyzed at 120° for 24 hr in glass hydrolysis tubes containing 0.5 ml of 6 N HBr supplemented with 1% phenol. Prior to heating, the tubes are capped with Mininert gas-tight valves, and the samples are alternately degassed and purged five times with argon. HBr is added to the reduced reaction mixture or tissue protein in a well-ventilated hood because cyanide is released when $NaCNBH_4$ is exposed to acid.

C_{18} Solid-Phase Extraction of Protein Hydrolyzates. Protein hydrolyzates are adjusted to a final volume of 2 ml with 0.1% trifluoroacetic acid and applied to a C_{18} minicolumn (Supelclean, 3 ml; Supelco Co.) equilibrated with 0.1% trifluoroacetic acid. Following a wash with 2 ml of 0.1% trifluoroacetic acid, products are recovered with 2 ml of 20% methanol in 0.1% trifluoroacetic acid.

Derivatization of Amino Acids. Prior to derivatization, samples are evaporated to dryness either under vacuum or under a stream of anhydrous nitrogen. n-Propyl esters are prepared by adding 200 μl of 3.5 M HBr in n-propanol followed by heating at 65° for 1 hr. The products are dried under nitrogen, and pentafluoropropionyl (PFP) derivatives are generated by adding 50 μl pentafluoropropionic acid anhydride (Pierce Chemical Co., Rockford, IL) in ethyl acetate (1 : 3; v/v) and heating the sealed vial for 30 min at 65°. Alternatively, heptafluorobutyryl (HFB) derivatives of the n-propylated amino acids may be used; these are generated by adding 50 μl of 4 : 1 (v/v) ethyl acetate/heptafluorobutyric acid anhydride (Pierce Chemical Co.) and incubating the sealed vials at 65° for 30 min. Comparable results are obtained with quantification of amino acids by GC/MS with selected ion monitoring of *n*-propyl, per PFP derivatives and *n*-propyl, per HFB derivatives; methods for both are described since it may be helpful to prepare the alternative derivative if difficulties arise from interfering ions.

Mass Spectrometric Analysis. Prior to mass spectrometric analysis, we separate derivatized amino acids on a 30 m DB-17 capillary column (P. J. Cobert, St. Louis, MO; 0.25 mm i.d., 0.25 μm film thickness) with helium as the carrier gas. For *p*HA–lysine and lysine determinations, the column is typically run with the following temperature gradient: 175° for 0–3 min, then 175° to 270° at 40°/min. The injector, transfer line and source temperatures are set at 250°, 250°, and 130°, respectively.

We analyze the derivatives with a Hewlett-Packard 5890 Gas Chromatograph interfaced with a Hewlett-Packard 5988A mass spectrometer equipped with extended mass range. The amino acids are quantified as either their *n*-propyl, per-PFP derivatives or their *n*-propyl, per-HFB derivatives using selected ion monitoring in the negative-ion electron capture mode.[16,22] The *n*-propyl, PFP derivative of *p*HA–lysine is monitored at mass-to-charge ratio (m/z) 726 $(M - HF)^-$ and m/z 598 $(M - CF_3\text{–}CF_2\text{–}CHO)^-$, and their corresponding isotopically labeled internal standard ions at m/z 732 and m/z 604. The *n*-propyl, PFP derivative of lysine is monitored using the base ion at m/z 460 $(M - HF)^-$, another major fragment ion at m/z 440 $(M - 2\ HF)^-$, and their corresponding isotopically labeled internal standard ions at m/z 466 and m/z 446.

To quantify the derivatives, we use an external calibration curve constructed with each authentic compound and its isotopically labeled internal standard.[22] The ratio of ion currents of two characteristic ions of each compound and its internal standard are monitored to ensure that no interfering ions coeluted with the analyte. Chromatography conditions are adjusted to ensure that all amino acids are baseline separated, that peaks contain at least 8 data points for accurate integration, and that target analytes coelute with their ^{13}C-labeled internal standards. The limit of detection (signal/noise $>$ 10) is $<$1 pmol for all of the compounds.

Detection of pHA–Lysine in Vivo by Gas Chromatography-Mass Spectrometry. Figure 2 illustrates the detection of *p*HA–lysine in human inflammatory fluid by GC/MS with selected ion monitoring.[16] Immediately following tissue collection, samples were mixed 1:1 (v/v) with reduction buffer (20 m*M* $NaCNBH_3$, 100 m*M* ammonium acetate, 1 m*M* NaN_3, 10 μg/ml catalase, 50 m*M* sodium phosphate, pH 7.4) and incubated at 37° for 1 hr. Ammonium acetate was included in the buffer to scavenge any free *p*HA and thereby prevent artifactual generation of *p*HA–lysine during reduction of the Schiff base. Catalase and NaN_3 were included to inhibit myeloperoxidase-dependent reactions. Preliminary experiments demonstrated that *p*HA–lysine content did not increase when tissue samples

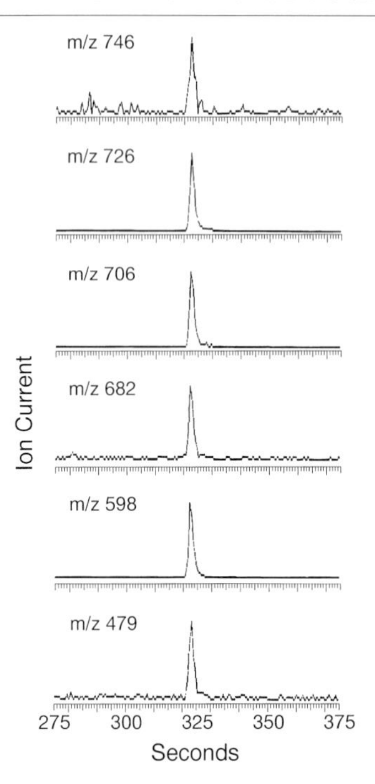

FIG. 2. Detection of *p*HA–lysine in human inflammatory tissue by GC/MS with selected ion monitoring. Residual fluid from an intraabdominal abscess was reduced in the presence of ammonium acetate and inhibitors of the myeloperoxidase–H_2O_2 system and then prepared for GC/MS analysis as described in the text. Ions from the *n*-propyl-per PFP derivative of *p*HA–lysine were monitored at *m/z* 746 ($M^{\cdot-}$), *m/z* 726 $(M - HF)^-$, *m/z* 706 $(M - 2HF)^-$, *m/z* 682 $(M - HF - CO_2)^-$, *m/z* 598 $(M - CF_3CF_2CHO)^-$ and *m/z* 479 $(M - CF_3CF_2COO\text{–}C_6H_4\text{–}CH_2CH_2)^-$ (reprinted with permission from ref. 16).

were supplemented with either 20 n*M* myeloperoxidase or 100 μM H_2O_2,[16] strongly suggesting that *p*HA–lysine was not being generated *ex vivo*. Reduced tissue was then delipidated, the pellet washed with 10% trichloroacetic acid, and the protein residue hydrolyzed with acid as described above. Amino acids in the acid hydrolyzate were isolated with a C_{18} column, eluted, derivatized, and analyzed by selected ion monitoring electron capture chemical ionization GC/MS. Ions from the *n*-propyl per-PFP derivative of *p*HA–lysine were monitored at *m/z* 746 ($M^{\cdot-}$), *m/z* 726 $(M - HF)^-$, *m/z* 706 $(M - 2\ HF)^-$, *m/z* 682 $(M - HF - CO_2)^-$, *m/z* 598 $(M - CF_3CF_2CHO)^-$, and *m/z* 479 $(M - CF_3CF_2COO\text{–}C_6H_4\text{–}CH_2CH_2)^-$. All of the ions coeluted (Fig. 2) and had the same relative abundance as authentic, isotope-labeled *p*HA–lysine, confirming *p*HA–lysine is present in acid hydrolysates of reduced human inflammatory fluid.[16]

Specificity and Sensitivity of pHA–Lysine as a Marker of Myeloperoxidase-Mediated Protein Modification

Several features, including specificity, stability, and ease of detection, make *p*HA–lysine a valuable indicator of myeloperoxidase-mediated protein modification.

Specificity. To establish the specificity of *p*HA–lysine as a marker for protein modification by myeloperoxidase, we performed *in vitro* experiments to evaluate the ability of a variety of oxidation systems to generate *p*HA adducts on BSA.[16] All oxidation systems were supplemented with plasma concentrations of L-tyrosine.

Following reduction of the Schiff base adduct, acid hydrolysis and derivatization, *p*HA–lysine was readily detected by isotope dilution GC/MS in reaction mixtures of BSA exposed either to the complete myeloperoxidase–H_2O_2–Cl^- system or reagent HOCl.[16] Interestingly, HOBr could not replace HOCl. *p*HA–lysine also was undetectable in BSA exposed to myeloperoxidase and H_2O_2 supplemented with other halides (Br^-, I^-, F^-) or with the pseudohalide SCN^-. Furthermore, no detectable *p*HA–lysine formed in BSA incubated with L-tyrosine and the following *in vitro* oxidation systems: a hydroxyl radical-generating system (copper plus H_2O_2), lactoperoxidase plus H_2O_2, horseradish peroxidase plus H_2O_2, or peroxynitrite.[16] Collectively, these results demonstrate that *p*HA–lysine is a highly specific marker for protein modification by the myeloperoxidase–H_2O_2–Cl^- system in the presence of free tyrosine.

Stability and Artifactual Generation. Markers of protein oxidation *in vivo* also need to be stable during analysis. Therefore, the acid-labile Schiff base that forms when *p*HA reacts with the ε amino group of lysine is not suitable for this purpose. However, the reduced product, *p*HA–lysine, is quantitatively recovered following acid hydrolysis.[16] Importantly, control experiments failed to detect ^{13}C-labeled *p*HA–lysine in inflamed human tissue samples supplemented with [$^{13}C_6$]tyrosine under the conditions described above for sample storage and processing. Thus, the adduct does not appear to form *ex vivo* under our experimental conditions.

Sensitivity. Selected-ion monitoring negative-ion chemical ionization GC/MS is an extremely sensitive and versatile method for detecting and quantifying derivatized amino acids.[21,22] Femtomole quantities of *p*HA–lysine are readily detected in a single analysis.[16]

Myeloperoxidase Generation of Molecular Chlorine

The products we have discussed so far result from the chlorinating action of HOCl (or its conjugate base, ClO^-). However, HOCl is in equilibrium with Cl_2 via a reaction that requires H^+ and chloride (ref. 29):

$$HOCl + Cl^- + H^+ \rightleftharpoons Cl_2 + H_2O \tag{2}$$

This suggested that Cl_2 might be a chlorinating intermediate in reactions catalyzed by myeloperoxidase. Because molecular chlorine, the solvated form of Cl_2, is in equilibrium with Cl_2 gas, we analyzed the headspace gas above the myeloperoxidase–H_2O_2–Cl^- reaction system.[14] Electron ionization mass spectrometric analysis revealed a gas with the expected mass-to-charge ratio of Cl_2. Both the retention time and isotopic distribution of the compound were identical to those of authentic Cl_2. These observations, together with previous studies of the formation of HOCl,[8] provide evidence that myeloperoxidase generates Cl_2 via a reaction pathway that involves HOCl as an intermediate.

Molecular Chlorine Oxidation of LDL Cholesterol

Activated phagocytes lyse phospholipid liposomes in a reaction that also requires halide and H_2O_2, implicating HOCl in the pathway.[37,38] Reagent HOCl reacts with fatty acid acyl groups to form chlorohydrins,[39] suggesting that polar chlorohydrins disrupt membrane structure. These electrophilic addition compounds appear stable and may therefore represent specific markers for myeloperoxidase-mediated damage.

Oxidation of Cholesterol in Phospholipid Liposomes. Because chlorohydrins are oxygenated as well as chlorinated, and because oxygenated sterols have been isolated from human vascular lesions,[40,41] we were interested in the idea that HOCl also might react with cholesterol, a major component of plasma membranes and circulating LDL. Therefore, we incorporated cholesterol into phospholipid liposomes and exposed the liposomes to the myeloperoxidase–H_2O_2–Cl^- system. Analysis of the reaction mixture by normal phase chromatography revealed three major products. We identified these by GC/MS as cholesterol α- and β-chlorohydrins, cholesterol α- and β-epoxides, and a novel cholesterol chlorohydrin.[15]

The myeloperoxidase-initiated chlorination of cholesterol was optimal under acid conditions.[15] This pH dependence was not due to enzymatic activity because it was also seen when reagent HOCl was used. HOCl itself

[37] S. M. Sepe and R. A. Clark, *J. Immunol.* **134,** 1888 (1985).

[38] S. M. Sepe and R. A. Clark, *J. Immunol.* **134,** 1896 (1985).

[39] C. C. Winterbourn, J. J. M. VandenBerg, E. Roitman, and F. A. Kuypers, *Arch. Biochem. Biophys.* **296,** 547 (1992).

[40] H. N. Hodis, A. Chauhan, S. Hashimoto, D. W. Crawford, and A. Sevanian, *Atherosclerosis* **96,** 125 (1992).

[41] G. M. Chisolm, G. Ma, K. C. Irwin, L. L. Martin, K. G. Gunderson, L. F. Linberg, D. W. Morel, and P. E. DiCorleto, *Proc. Natl. Acad. Sci. USA* **91,** 11452 (1994).

did not appear to be the chlorinating intermediate, however, since the pK_a for $HOCl/ClO^-$ is ~7.5.[28,29]

Because HOCl is in equilibrium with Cl_2 via a reaction that requires H^+ and chloride [Eq. (2)], we suspected that Cl_2 might be the chlorinating intermediate. To test this hypothesis we examined the reaction requirements for the oxidation of LDL cholesterol by HOCl.[17] HOCl failed to chlorinate cholesterol in the absence of chloride ion, consistent with the requirement for the halide in the generation of Cl_2 as an intermediate (Fig. 3). The reaction also was optimal under acid conditions, consistent with a requirement for H^+ in Cl_2 formation. Finally, at neutral pH and in the absence of chloride, molecular chlorine readily generated cholesterol chlorohydrins in LDL. These results strongly suggest that Cl_2—not HOCl—is the chlorinating intermediate when myeloperoxidase oxidizes cholesterol.[17] They also illustrate the importance of using chloride-free conditions when evaluating the mechanism of sterol oxidation by HOCl.

Oxidation of Cholesterol in LDL. To study oxidation of LDL cholesterol by reagent HOCl and by the myeloperoxidase–H_2O_2–Cl^- system, we incubate reagents in sealed, chlorine-demand-free and chloride-free reaction vials.[17] We prepare [^{14}C]cholesterol-labeled LDL by adding tracer quanti-

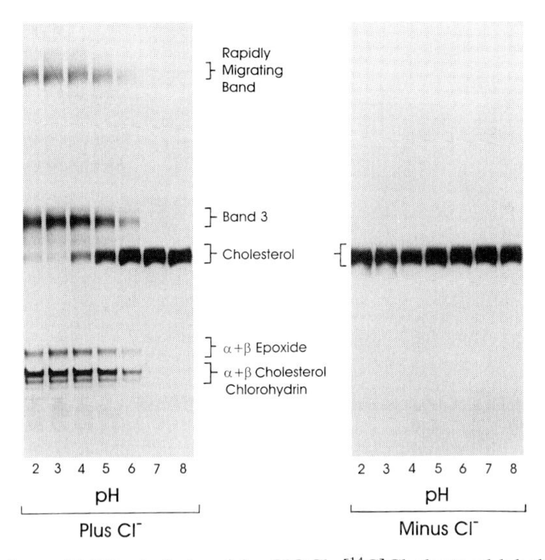

FIG. 3. Oxidation of LDL cholesterol by HOCl. [^{14}C]Cholesterol-labeled LDL (510 μg protein/ml) and chloride-free HOCl (500 μM) were incubated in chlorine-demand free sodium phosphate buffer (50 mM) at the indicated pH in the presence (*left*) or absence (*right*) of NaCl (100 mM) within sealed reaction vials for 30 min at 37°C. Reaction products were subsequently extracted into ethyl acetate, resolved by normal phase TLC, and visualized by autoradiography (reprinted with permission from ref. 17).

ties of [^{14}C]cholesterol (DuPont NEN Research Products) in ethanol (<0.2% final concentration, v/v) to chloride-free LDL at 37°. Following incubation for 5 min, we subject the LDL to size exclusion chromatography on an Econopac 10-DG column (Bio-Rad) equilibrated with H_2O. Under these conditions, >99% of [^{14}C]cholesterol is incorporated into the LDL particle as determined by scintillation spectrometry and protein analysis of fractions prepared from the column.

We oxidize LDL in 65 m*M* sodium phosphate buffer using chloride-free HOCl at the indicated pH in the presence and absence of supplemental chloride. Following a 30 min incubation at 37°, 0.5 ml of a NaCl-saturated aqueous solution is added to 0.5 ml of reaction mixture, and lipid soluble products are extracted by three sequential additions of 2 ml of ethyl acetate.[17] The combined organic extracts are brought to near dryness under

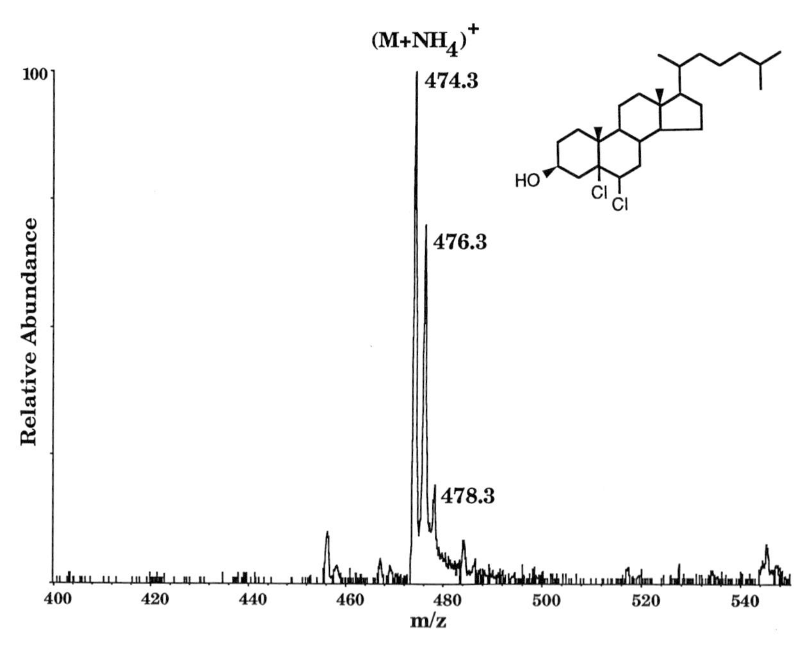

FIG. 4. Positive-ion electrospray ionization mass spectrum of the dichlorinated sterol in the rapidly migrating band. Cholesterol oxidation products generated by the myeloperoxidase–H_2O_2–Cl^- system were separated by normal phase TLC as described in the text. The rapidly migrating band (Fig. 3) was extracted into ethyl acetate, cationized with ammonium acetate, and analyzed by electrospray mass spectrometry as described in the text. The predicted *m/z* of an ammoniated, dichlorinated ion derived from cholesterol (*inset*) is 474 $(M + NH_4)^+$. The mass spectrum demonstrates the isotopic cluster expected for a dichlorinated compound, with ions of relative intensity of 10:6:1 at M^+, $M^+ + 2$, and $M^+ + 4$, respectively (reprinted with permission from ref. 17).

anhydrous nitrogen and immediately analyzed by normal-phase thin-layer chromatography (TLC) on silicic acid (Silica Gel 60A Linear-K with preabsorbent strips; 250 μm; Whatman, Clifton, NJ) with hexane : diethyl ether (40 : 60; v : v) as the mobile phase. TLC plates are developed twice, and the reaction products are subsequently quantified by phosphor imaging (Molecular Dynamics, Sunnyvale, CA). Under these conditions, radioactivity is a linear function of [^{14}C]cholesterol, and recovery of chlorinated sterols is $\geq$95%.

Characterization of Cholesterol Oxidation Products by Mass Spectrometry. We confirm the identity of chlorinated and oxygenated sterols by electrospray mass spectrometry, as shown in Fig. 3. A complete description of electrospray mass spectrometric analysis is presented in an earlier volume of this series.[42] The utility and power of the method is illustrated in Fig. 4. The positive-ion electrospray mass spectrum of the rapidly migrating band (Fig. 3) is shown. The spectrum demonstrates the isotopic cluster expected for a dichlorinated sterol such as that formed by direct addition of Cl_2 to the C5–C6 double bond of LDL cholesterol. The electrospray mass spectrum was acquired on a Sciex API III+ triple quadrupole mass spectrometer (Sciex Inc., Thornhill, Ontario) in positive-ion mode. Thin-layer chromatography (TLC) extracts of the band were diluted in chloroform : methanol (1 : 1; v : v) containing 10 m*M* ammonium acetate and infused into the mass spectrometer at a rate of 4 μl/min. The electrospray interface was maintained at 5.0 kV with respect to ion entrance of the mass spectrometer. Air was the nebulizing gas.

Conclusions

The identification of protein and lipid oxidation products that are specifically generated by the myeloperoxidase system, together with the development of methods to detect low levels of these markers in tissues, enables one to probe the role of activated phagocytes in the many diseases associated with oxidative damage. The specificity, sensitivity, and reproducibility of mass spectrometry offers a remarkably powerful technique for this purpose. Identifying additional oxidation products, and developing further mass spectrometric methods for detecting them *in vivo,* should elicit important insights about the involvement of phagocytes in a variety of human diseases.

[42] J. A. McClosky, *Methods Enzymol.* **193,** 825 (1990).

[13] Protein Carbonyl Measurement by Enzyme-Linked Immunosorbent Assay

By CHRISTINE C. WINTERBOURN and I. HENDRIKJE BUSS

Introduction

Protein carbonyls are formed by a variety of oxidative mechanisms and are sensitive indices of oxidative injury.[1] The conventional assay for protein carbonyls is a colorimetric procedure that measures binding of dinitrophenylhydrazine (DNP).[2–4] Protein-bound DNP can also be measured, with increased sensitivity, either by HPLC,[3] or using an anti-DNP antibody with Western blots[5] or tissue sections.[6] We have adapted the immunological approach to an ELISA method that enables carbonyls to be measured quantitatively.[7] The advantages of this method are that it requires only microgram quantities of protein and it avoids the high and sometimes variable blanks due to unbound DNP that are limitations for the colorimetric method.[4,8] We have found it to be highly sensitive in analyzing plasma and lung aspirates from both critically ill adult patients and premature infants. Results correlate well with the colorimetric assay but are more discriminatory. Absolute values in both assays are subject to some uncertainty, and the method is best for comparing samples analyzed using a standard system.

Principle

Samples containing protein are reacted with DNP, then the protein is nonspecifically adsorbed to an ELISA plate. Unconjugated DNP and nonprotein constituents are easily washed away and give minimal interfer-

[1] E. R. Stadtman, *Free Rad. Biol. Med.* **9,** 315 (1990).
[2] R. L. Levine, D. Garland, C. N. Oliver, A. Amici, I. Climent, A. Lenz, B. Ahn, S. Shalteil, and E. R. Stadtman, *Methods Enzymol.* **186,** 464 (1990).
[3] R. L. Levine, J. A. Williams, E. R. Stadtman, and E. Shacter, *Methods in Enzymol.* **233,** 346 (1994).
[4] A. Z. Reznick and L. Packer, *Methods Enzymol.* **233,** 357 (1994).
[5] E. Shacter, J. A. Williams, M. Lim, and R. L. Levine, *Free Rad. Biol. Med.* **17,** 429 (1994).
[6] M. A. Smith, G. Perry, L. M. Sayre, V. E. Anderson, M. F. Beal, and N. Kowall, *Nature* **382,** 120 (1996).
[7] H. Buss, T. P. Chan, K. B. Sluis, N. M. Domigan, and C. C. Winterbourn, *Free Rad. Biol. Med.* **23,** 361 (1997).
[8] L. Lyras, P. J. Evans, P. J. Shaw, P. G. Ince, and B. Halliwell, *Free Rad. Res.* **24,** 397 (1996).

0076-6879/99 $30.00

ence. The adsorbed protein is probed with a commercial biotinylated anti-DNP antibody followed by streptavidin-linked horseradish peroxidase. Absorbances are related to a standard curve prepared for bovine serum albumin (BSA) containing increasing proportions of HOCl-oxidized protein that is calibrated colorimetrically.

Procedure

Reagents

- 2,4-Dinitrophenylhydrazine (DNP; Riedel-de-Haen, Seelze-Hannover, Germany). Solution: 10 m*M* DNP in 6 *M* guanidine hydrochloride, 0.5 *M* potassium phosphate buffer, pH 2.5
- Biotin-conjugated rabbit immunoglobulin G (IgG) polyclonal antibody raised against a DNP conjugate of keyhole limpet hemocyanin (anti-DNP) (Molecular Probes, Eugene, OR); 1:1000 dilution in PBS, 0.1% (w/v) Tween 20 solution
- Streptavidin-biotinylated horseradish peroxidase (Amersham International, Buckinghamshire, UK); 1:3000 in PBS, 0.1% Tween 20 solution
- Phosphate-buffered saline (PBS): 10 m*M* sodium phosphate buffer, pH 7.4, in 0.14 *M* sodium chloride
- Blocking solution: PBS containing 0.1% Tween 20
- *o*-Phenylenediamine (0.6 mg/ml) and hydrogen peroxide (30% stock diluted 1:2500) in 50 m*M* Na_2HPO_4 plus 24 m*M* citric acid, mixed directly before use

A microplate reader and washer are required. We use a Dynatech reader and AM60 washer. Nunc Immuno Plate Maxisorp ELISA plates.

Unless noted otherwise, biochemicals are obtainable from Sigma Chemical Co. (St. Louis, MO).

Albumin Standards

Bovine serum albumin (BSA) as purchased already contains carbonyls. Fully reduced BSA is prepared by reacting a 0.5 g/100 ml solution in PBS with 0.1 g sodium borohydride for 30 min, followed by neutralizing with 2 *M* HCl, added slowly. Since this reaction produces hydrogen, it should be carried out in a fume hood. After overnight dialysis against PBS, the protein concentration is checked by measuring A_{280} and adjusted to 4 g/liter. Oxidized BSA containing additional carbonyls is prepared for use as a reference by reacting (at 50 mg/ml in PBS) with hypochlorous acid (final concentration 5 m*M*). Albumin solutions do not change in carbonyl content when stored at −80°.

Standard Curves and Calibration of Assay

Standard curves are constructed by mixing varying proportions of HOCl-oxidized BSA with fully reduced BSA at a constant total protein concentration. A range of 0–2 nmol carbonyl/mg protein is appropriate for most purposes. The carbonyl content of the oxidized BSA is determined from A_{375} in the colorimetric assay performed as described below. In our hands the fully reduced BSA consistently gives an A_{375} of about 0.13 per 10 mg (equivalent to 0.6 nmol carbonyls/mg), which was unaffected by further treatment with sodium borohydride. It is assumed to be nonspecific and not due to carbonyls.

Analysis of Plasma and Lung Aspirates

Samples can be used fresh or stored at −80°. A total protein assay is performed using the Bio-Rad assay (Bio-Rad Laboratories, Richmond, CA) on all standards, plasmas, and aspirates, which are then diluted in PBS to a protein concentration of 4 mg/ml. Because aspirates generally contain lower protein concentrations than this, they need to be concentrated by mixing a volume containing 60 μg protein with 0.8 volume of 28% (w/v) trichloroacetic acid (TCA), centrifuging at 10,000*g* for 2 min and discarding the supernatant. PBS (15 μl) is added to the protein precipitate, which dissolves completely after addition of the DNP reagent (see below).

ELISA Assay

Protein derivatization is carried out with 45 μl of DNP solution added to 15 μl of sample (4 mg/ml) to give a final protein concentration of 1 mg/ml. The blank consists of DNP reagent added to PBS without protein. After vortex mixing and incubating at room temperature for 45 min, 5 μl of each solution is added to 1 ml PBS. Triplicate 200 μl aliquots (containing 1 μg protein) are added to wells of the ELISA plate which is incubated overnight at 4°, then washed 5 times with PBS before each of the following incubations:

Blocking solution (250 μl/well) for 1.5 h at room temperature
Biotinylated anti-DNP antibody (200 μl/well) for 1 h at 37°
Streptavidin-biotinylated horseradish peroxidase (200 μl) for 1 h at room temperature

o-Phenylenediamine/peroxide solution (200 μl) is then added and the color allowed to develop for 25 min before stopping with 100 μl of 2.5 *M* sulfuric acid and reading absorbances with a 490 nm filter. A six point standard curve of reduced and oxidized BSA is included with each plate. The background absorbance for DNP reagent (typically about 0.08) is subtracted from all other absorbances.

Colorimetric Assay for Calibration of Standard

The carbonyl content of the stock oxidized BSA is calibrated using a modification of the standard colorimetric method[2] in which 10 mg protein in 250 μl PBS is reacted with 1 ml 10 m*M* DNP in 2 *M* HCl for 45 min with occasional mixing, precipitated with 1 ml 28% TCA and washed three times with 2.5 ml ethanol/ethyl acetate (1 : 1, v/v). Pellets should be broken up mechanically and by sonicating during the washing steps, and dissolved at 37° in 1 ml 6 *M* guanidine hydrochloride, 0.5 *M* potassium phosphate, pH 2.5, to measure A_{375}. The protein concentration of the final extract is determined by diluting 1 : 5 and measuring A_{280} (ε_{280} 5.31 for a 1 g/100 ml solution) and the absorbance adjusted for the protein loss (about 10%) that occurs with this method.[4] A blank with the protein reacted with 2 *M* HCl containing no DNP is carried through the procedure and its absorbance is subtracted. Carbonyl content is determined as nmol/mg protein using ε_{375} 22,000 M^{-1} cm^{-1} after subtracting the value for reduced albumin.

Comments

Sensitivity and Reproducibility

Standard curves for the ELISA are linear up to at least 10 nmol carbonyl/mg protein. With the development time described, 1.0 nmol/mg gives a net absorbance of about 0.8 above the blank. Repeat analyses of 10 different plasma samples, carried out on three separate occasions, gave variations of 1.7–21.4% with a mean of ±8.8%.[7] Most of this variation arose from analyses at the lower end of the standard curve (mean 17% for values <0.06 nmol/mg compared with a mean of 4.5% for values over 0.8 nmol/mg). Intraassay variation for 9 samples analyzed in triplicate was 0.27–2.53% with a mean of 1.4%.

Blanks

A key requirement for sensitivity is a low reagent blank. This is typically about 0.04 absorbance units above PBS alone. We originally used reduced BSA in the blocking solution to achieve this, but have subsequently found that PBS containing 0.1% Tween 20 is also satisfactory. There are minor differences in sensitivity and absolute values for plasma samples, but results with both solutions correlate closely. Reduced BSA typically gives an absorbance about 0.02 units above the reagent blank.

Validation

As described elsewhere,[7] the conditions described for the assay have been chosen after testing different ELISA plates, protein concentrations, and antibodies. Reactivity can be outcompeted by free DNP, it requires protein, DNP, and the primary antibody, and it decreases to baseline if the albumin or plasma is pretreated with sodium borohydride.

Clinical Samples

Plasmas from normal adults give a mean protein carbonyl of about 0.1 nmol/mg protein by the ELISA method. Higher levels (up to 4 nmol/mg) have been measured in plasma from critically ill patients and lung aspirates from adults and premature infants. The TCA-precipitation step before DNP derivatization that is generally necessary to concentrate the protein from aspirates causes a small (20%) increase in response.

Correlation with Colorimetric Assay

A good correlation has been found between the ELISA and colorimetric methods for plasma samples over a range of 0.01–2.2 nmol/mg ($r^2 = 0.70$, $n = 26$).

Absolute V Relative Values

We use BSA treated with hypochlorous acid to generate stable DNP-reactive carbonyls to calibrate the ELISA. Carbonyls can also be generated with iron and ascorbate. For an equivalent number of carbonyls measured colorimetrically, the response in the ELISA to iron/ascorbate-oxidized BSA was 65% of that for HOCl-treated protein.[7] Thus, carbonyl values are not absolute and depend to some extent on the oxidized protein standard used. Absolute values are also affected by whether the blank is taken as the reagents without protein, or fully reduced BSA. The absorbances of these solutions are normally within 0.01 absorbance units, but this can make a difference for samples with low carbonyl content.

Although results correlate well with the colorimetric assay, absolute values for plasma samples in the two assays differ.[7] This is partly due to the standards and blanks described above. However, absolute values from the colorimetric assay are also subject to uncertainties, with different versions giving different results.[8] The assay is highly dependent on blank values and protein loss, which vary with different washing procedures.[9] With a mixture such as plasma, it is possible that some proteins adhere better to the ELISA plate than others and therefore are selected for in the assay. The main strength of the assay, therefore, is for comparing samples analyzed using a standard system.

[9] G. Cao and R. G. Cutler, *Arch. Biochem. Biophys.* **320,** 106 (1995).

Effects of DNA and RNA

Although we have not used the ELISA for tissue extracts, it should be suitable for this purpose. However, it has been reported[9] that nucleic acids in tissue extracts can contribute to carbonyl measurements so that the assay is not specific for protein carbonyls. We found that adding 3–30 μg RNA to 30 μg reduced BSA progressively increased carbonyls detected by ELISA (from zero to 0.6 nmol/mg protein), whereas DNA at the same concentration had no effect.[7] Removal of RNA before analysis of tissue extracts may therefore be necessary.

Acknowledgments

This work was supported by the Health Research Council of New Zealand.

[14] Detection of 3-Chlorotyrosine in Proteins Exposed to Neutrophil Oxidants

By ANTHONY J. KETTLE

Introduction

Neutrophils have the greatest potential of any cell type to inflict oxidant-dependent tissue injury. These phagocytic white blood cells accumulate in substantial numbers at sites of inflammation where they generate an array of destructive oxidants.[1] When stimulated, they assemble an oxidase in their cell membrane which produces superoxide by using NADPH to reduce molecular oxygen.[2] The superoxide can give rise to a number of reactive oxidants, including hydrogen peroxide, hypochlorous acid, peroxynitrite, hydroxyl radical, and singlet oxygen.[3] Neutrophil oxidants are strong contenders for contributing to the pathophysiology of inflammatory diseases. However, their involvement in inflammation remains equivocal. This is primarily due to the difficulty of detecting short-lived oxidants, and the lack of availability of unique markers for these reactive intermediates.

[1] S. J. Weiss, *N. Engl. J. Med.* **320,** 365 (1989).
[2] A. W. Segal and A. Abo, *TIBS* **18,** 43 (1993).
[3] A. J. Kettle and C. C. Winterbourn, *Redox Report* **3,** 3 (1997).

0076-6879/99 $30.00

Hypochlorous acid is the strongest oxidant produced by neutrophils in appreciable amounts.[4–6] Myeloperoxidase, the most abundant protein in neutrophils, catalyzes its production from hydrogen peroxide and chloride.[3] This heme enzyme is also present in monocytes. Chlorination reactions of hypochlorous acid have the potential to unmask the role neutrophil oxidants play in inflammation because chlorine is incorporated into products.[7] Chlorinated biomolecules provide definitive footprints that can then be used to implicate hypochlorous acid in the etiology of disease processes.

Hypochlorous acid reacts with peptides and proteins to convert tyrosyl residues to 3-chlorotyrosine [reaction (1)].[8–10]

$$\text{Tyrosine (R–C}_6\text{H}_4\text{–OH)} + \text{HOCl} \longrightarrow \text{3-Chlorotyrosine (R–C}_6\text{H}_3\text{Cl–OH)} + H_2O \quad (1)$$

At physiological pH and at low ratios of hypochlorous acid to peptide or protein, chlorination occurs by the initial formation of chloramines which undergo intramolecular reactions with tyrosyl residues to form 3-chlorotyrosine.[8] Below pH 4, where hypochlorous acid exists predominantly as molecular chlorine, the aromatic ring of tyrosine is chlorinated directly.[10] 3-Chlorotyrosine is a minor product when hypochlorous acid reacts with protein.[9] However, it is the only modification of proteins identified to date that is unique to hypochlorous acid. It is also a stable reaction product that is not destroyed by acid hydrolysis of proteins.[11] Halogenation of aromatic rings is an enzymatic activity that is specific to peroxidases, since myeloperoxidase is the only human enzyme capable of producing hypochlorous acid.

[4] C. S. Foote, T. E. Goyne, and R. I. Lehler, *Nature* **301,** 715 (1983).

[5] S. J. Weiss, R. Klein, A. Slivka, and M. Wei, *J. Clin. Invest.* **70,** 598 (1982).

[6] A. J. Kettle, C. A. Gedye, and C. C. Winterbourn, *Biochem. Pharmacol.* **45,** 2003 (1993).

[7] C. C. Winterbourn, J. J. M. van den Berg, E. Roitman, and F. A. Kuypers, *Arch. Biochem. Biophys.* **296,** 547 (1992).

[8] N. M. Domigan, T. S. Charlton, M. W. Duncan, C. C. Winterbourn, and A. J. Kettle, *J. Biol. Chem.* **270,** 16542 (1995).

[9] A. J. Kettle, *FEBS Lett.* **379,** 103 (1996).

[10] S. L. Hazen, F. F. Hsu, D. M. Mueller, J. R. Crowley, and J. W. Heinecke, *J. Clin. Invest.* **98,** 1283 (1996).

[11] S. K. Chowdhury, J. Eshraghi, H. Wolfe, D. Forde, A. G. Hlavac, and D. Johnston, *Anal. Chem.* **67,** 390 (1995).

Detection of 3-chlorotyrosine has the potential to identify unequivocally the contribution myeloperoxidase-containing cells make to inflammation.

In this chapter I have described two HPLC assays for detecting 3-chlorotyrosine in proteins. They are useful for determining if tyrosyl residues have been chlorinated in proteins that have been treated with reagent hypochlorous acid, myeloperoxidase, hydrogen peroxide, and chloride, or stimulated neutrophils. Both assays detect at least one molecule of 3-chlorotyrosine per 1000 tyrosines and can be used to assay as little as 0.5 pmol of 3-chlorotyrosine. They are superior to conventional amino acid analysis because they are specific for aromatic amino acids and consequently better at identifying 3-chlorotyrosine when it is present at low levels compared to other amino acids. The first assay described involves derivatization of hydrolyzed amino acids with 1-nitroso-2-naphthol to give fluorescent products.[9] Its major advantage is that it measures only tyrosine, 3-chlorotyrosine, and 3-bromotyrosine. Brominated tyrosines may be formed from the reaction of proteins with hypobromous acid, which is produced by eosinophil peroxidase.[12] Other potential products of tyrosyl residues in proteins, including 3-iodotyrosine, *o*-tyrosine, *m*-tyrosine, 3-nitrotyrosine, and DOPA, do not give compounds with appreciable fluorescence.

A second assay has also been developed, using electrochemical detection to measure 3-chlorotyrosine. Given that it is selective for aromatic amino acids and also detects other substituted tyrosines, including 3-bromotyrosine, 3,5-dichlorotyrosine, and 3,5-dibromotyrosine, without the need for derivatization, it is preferred to the fluorescence method. It is prudent to use both assays to check results for consistency. The major limitation of these assays is that at best only one chlorotyrosine per few thousand tyrosines can be detected. When greater sensitivity is required, gas chromatography with mass spectrometry is the preferred option.[13]

Chemicals and Biochemicals

3-Chlorotyrosine, tyrosine, phorbol myristate acetate, pronase E, 1-nitroso-2-naphthol, phenol, xanthine oxidase, and glucose oxidase can be purchased from the Sigma Chemical Co. (St. Louis, MO). Concanavalin A-Sepharose, CM-Sepharose, and phenyl-Sepharose are available from Pharmacia (Piscataway, NJ). Constant boiling HCl is obtained from Pierce (Rockford, IL). We use commercially available chlorine bleach for our hypochlorous acid reagent. At pH 10, its concentration is determined by

[12] S. J. Weiss, S. T. Test, C. M. Eckmann, D. Roos, and S. Regiani, *Science* **234,** 200 (1986).

[13] S. L. Hazen, J. R. Crowley, D. M. Mueller, and J. W. Heinecke, *Free Rad. Biol. Med.* **23,** 909 (1997).

measuring its absorbance at 292 nm (ε_{292} 350 M^{-1} cm^{-1}). Alternatively, the concentration of hypochlorous acid can be determined by reacting it with 5-thio-2-nitrobenzoic acid and measuring the loss in absorbance at 412 nm (ε_{412} 14,100 M^{-1} cm^{-1}).[14] Hydrogen peroxide should be prepared fresh each day from a 30% stock and its concentration determined using ε_{240} 43.6 M^{-1} cm^{-1}.[15]

Purification of Myeloperoxidase

Granules are isolated from neutrophils as described by Mathewson *et al.*[16] Myeloperoxidase is purified from them as outlined previously[16–18] with the following modifications. The enzyme is extracted from neutrophil granules by sonicating them in 50 ml of 100 m*M* phosphate buffer, pH 7.0, containing 1% cetyltrimethylammonium bromide (CTAB). The sonicated material is then centrifuged at 20,000*g* for 30 minutes and 1 m*M* each of magnesium chloride, manganese chloride, and calcium chloride are added to the supernatant. If the solution becomes turbid it should be filtered through Whatman (Clifton, NJ) 1 filter paper. Concanavalin A bound to Sepharose gel (5 ml) is added and the mixture is rotated overnight at 4°C to allow for binding of myeloperoxidase. The green gel is allowed to settle, the supernatant is discarded, and the gel is then resuspended in 0.1 *M* sodium acetate, pH 6.0, containing 0.5 *M* sodium chloride and 0.05% CTAB. The gel is then poured into a column (1 cm diameter × 6 cm height) and washed with the same buffer at 10 ml/min until the A_{280} of the eluant is zero. The myeloperoxidase is then eluted with 0.5 *M* α-*D*-mannoside in wash buffer at a flow rate of 1 ml/min. Fractions with RZ (A_{430}/A_{280}) values greater than 0.4 are pooled and dialyzed against 50 m*M* phosphate buffer pH 6.0 containing 0.2 *M* sodium chloride and loaded onto a column (2 cm diameter × 20 cm height) of CM-Sepharose equilibrated in the same buffer at a flow rate of 25 ml/hr. The column is washed with binding buffer until A_{280} reaches zero and then the myeloperoxidase is eluted with a linear gradient from 0.2 to 1 *M* sodium chloride. Fractions with RZ values greater than 0.5 are pooled, dialyzed against 1.7 *M* ammonium sulfate in 100 m*M* phosphate buffer, pH 7.0, and loaded onto a phenyl-Sepharose column (2 cm diameter × 20 cm height) equilibrated in the same buffer. After washing

[14] A. J. Kettle and C. C. Winterbourn, *Methods Enzymol.* **233,** 502 (1994).
[15] R. J. Beers and I. W. Sizer, *J. Biol. Chem.* **195,** 133 (1952).
[16] N. R. Matheson, P. S. Wong, and J. Travis, *Biochemistry* **20,** 325 (1981).
[17] A. J. Kettle and C. C. Winterbourn, *Biochem. J.* **252,** 529 (1988).
[18] R. M. Rakita, B. R. Michel, and H. Rosen, *Biochemistry* **29,** 1075 (1990).

the column with loading buffer until A_{280} drops to zero, myeloperoxidase is eluted with a linear gradient from 1.7 M down to 0.1 M ammonium sulphate. Fractions with RZ values greater than 0.8 are pooled, dialyzed against 5 mM phosphate buffer, pH 7.0, and stored at −80°C. Pure myeloperoxidase has an RZ value of 0.85.[19] The concentration of myeloperoxidase per heme group is determined by using ε_{430} 89,000 M^{-1} cm^{-1}.[20]

Preparation of Neutrophils

Neutrophils are isolated from the blood of healthy donors by Ficoll–Hypaque centrifugation, dextran sedimentation, and hypotonic lysis of red cells.[21] After isolation, neutrophils are suspended in phosphate-buffered saline (PBS) containing 1 mM calcium chloride, 0.5 mM magnesium chloride, and 1 mg/ml of glucose.

Chlorination of Proteins

Chlorination with Reagent Hypochlorous Acid

It is essential to add hypochlorous acid while vortexing the protein solution. Commercial hypochlorous acid has a pH > 10 and should be neutralized before it is added to reaction systems. It is best to chlorinate proteins in phosphate buffers. Tris, Good, and HEPES buffers, as well as cell media, should not be used because the amino groups scavenge hypochlorous acid. The amount of hypochlorous acid required to convert tyrosyl residues in a protein will depend on the particular protein under investigation. For albumin, 5 mol of hypochlorous acid per mol of protein promotes the chlorination of about 1 in 200 tyrosyl residues (Fig. 1). Above 20 mol of hypochlorous acid per mol of albumin, the level of 3-chlorotyrosine declines due to the formation of secondary chlorination products.

The treated protein should be allowed to react for about an hour at 37°C to obtain maximum formation of 3-chlorotyrosine. This is because chlorination occurs via initial formation of chloramines which react slowly with tyrosyl residues.[8] For time course studies, reactions can be stopped by adding a slight molar excess of methionine over hypochlorous acid to quench reactive chloramines. It is prudent to spike samples to establish

[19] A. R. J. Bakkenist, R. Wever, T. Vulsma, H. Plat, and B. F. van Gelder, *Biochim. Biophys. Acta* **524,** 45 (1978).

[20] K. Agner, *Acta Physiol. Scand.* **2,** 1 (1941).

[21] A. Boyum, *Scand. J. Clin. Lab. Invest.* **21** (suppl 97), 77 (1968).

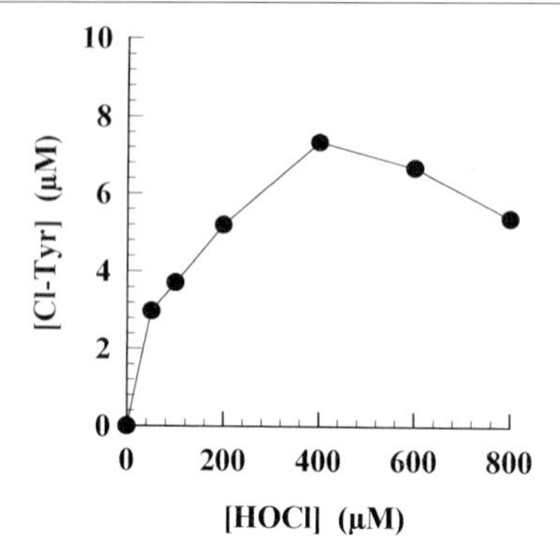

FIG. 1. Chlorination of albumin by hypochlorous acid. Increasing concentrations of HOCl were added to 500 μg/ml of bovine serum albumin in 10 m*M* phosphate buffer, pH 7.4, containing 140 m*M* NaCl. After an hour at 37°C, reactions were stopped by adding 1 mM methionine and then pronase was added to hydrolyze albumin. 3-Chlorotyrosine was detected using HPLC with electrochemical detection.

that the HPLC peak assigned to 3-chlorotyrosine coelutes precisely with the authentic molecule.

Chlorination with Myeloperoxidase, Hydrogen Peroxide, and Chloride

When chlorinating proteins with myeloperoxidase, the enzyme should be present at about 10–50 n*M*. Myeloperoxidase is active over a pH range from pH 5 to pH 8, with the pH optimum decreasing as the ratio of hydrogen peroxide concentration to chloride concentration increases.[22] Hydrogen peroxide should be added in aliquots no higher than 50 μM and a chloride concentration of 100 m*M* is desirable. At higher concentrations of hydrogen peroxide the enzyme is rapidly inactivated. Alternatively, glucose oxidase and glucose (1 mg/ml) or xanthine oxidase and acetaldehyde (10 m*M*) can be used to generate a flux of hydrogen peroxide. Hydrogen peroxide generation can be measured using the FOX assay or a hydrogen peroxide electrode.[14,23] A hydrogen peroxide flux of about 5 μM/min is optimal so as to maximize chlorination and minimize inactivation of myeloperoxidase. Acetaldehyde is the preferred substrate of xanthine oxidase because hypoxanthine or xanthine are oxidized to urate, which scavenges hypochlorous acid and interferes with myeloperoxidase activity.[17,24] It is preferable to

[22] J. M. Zgliczynski, R. J. Selvaraj, B. B. Paul, T. Stelmaszynska, P. K. F. Poskitt, and A. J. Sbarra, *Proc. Soc. Exp. Biol. Med.* **154,** 418 (1977).

[23] S. P. Wolff, *Methods Enzymol.* **233,** 182 (1994).

[24] C. C. Winterbourn, *Biochim. Biophys. Acta* **840,** 204 (1985).

redistill acetaldehyde before use. Reactions can be stopped by adding a 10-fold excess of catalase over myeloperoxidase and 1 m*M* methionine to scavenge hypochlorous acid and chloramines. When using xanthine oxidase, allopuriniol (100 μ*M*) should also be added to stop chlorination. To confirm that the HPLC peak is correctly assigned to 3-chlorotyrosine, it is necessary to show that it is diminished by inhibiting chlorination with azide (100 μ*M*), catalase (10 μg/ml), or methionine (1 m*M*).

Chlorination with Stimulated Neutrophils

Neutrophils (about 2×10^6/ml) are suspended in PBS containing 1.0 mM calcium chloride, 0.5 mM magnesium chloride and 1 mg/ml of glucose, plus the protein or proteins of interest. After five minutes at 37°C, the cells should then be stimulated with 100 ng/ml of phorbol myristate acetate. An optimal time for chlorination is about 1 hour. To stop chlorination, add 10 μg/ml of catalase, place cells on melting ice, and then pellet them at 1000 *g* for 5 minutes.

Hydrolysis of Proteins

Proteins are dried down in 6 × 50 mm Kimble tubes and placed in a hydrolysis container with 200 μl of constant-boiling HCl containing 1% phenol. The container is then alternately evacuated and flushed with nitrogen and finally sealed under vacuum and placed in an oven at 110°C overnight. It is imperative that the HCl/phenol solution be fresh and that a good vacuum be obtained to avoid artifactual chlorination of tyrosine.[11] Acid hydrolysis is the preferred method of liberating 3-chlorotyrosine from proteins because complete hydrolysis is achieved. Untreated proteins should always be included as controls to ensure that any artifactual chlorination is negligible. Alternatively, artifactual chlorination of tyrosine can be avoided by hydrolyzing with 50 μg/ml of pronase at pH 7.4 and 37°C overnight. The disadvantage of pronase, however, is that incomplete hydrolysis can lead to noise in the HPLC chromatograms when the signal from 3-chlorotyrosine is small.

Acid-hydrolyzed proteins should be dissolved in 100 μl of 0.3 *M* trichloroacetic acid if they are to be derivatized and assayed by fluorescence detection. For electrochemical detection, hydrolyzates are dissolved in the HPLC eluant. When proteins have been hydrolyzed with pronase, they are diluted 1:1 with 0.6 *M* trichloroacetic acid and after 10 minutes centrifuged at 10,000 rpm to pellet any undigested protein.

HPLC Analysis

Fluorescence Detection

Derivatization with 1-Nitroso-2-naphthol. Tyrosine and chlorotyrosine are derivatized with 1-nitroso-2-naphthol essentially as described by Waalkes and Udenfriend [reaction (2)].[25] An ortho carbon on phenols has to be unsubstituted for this reaction to occur. Hence, when both ortho positions are halogenated, derivatization is blocked. To 100 μl of amino acids, 167 μl each of 1.5 *M* nitric acid, 0.1 *M* sodium nitrate, and 7.5 m*M* 1-nitroso-2-naphthol in 95% ethanol are added. Solutions are then mixed and heated at 60°C for 20 minutes. After allowing the solutions to cool, the derivatized tyrosines are extracted by adding 0.4 ml of deionized water and 4 ml of dichloromethane. The mixtures are shaken vigorously and centrifuged for 5 minutes at 1000*g*. The upper aqueous phase is removed for analysis.

3-Chlorotyrosine + **1-Nitroso-2-naphthol** → **Fluorescent Product**

(2)

HPLC Separation. Derivatized tyrosine and 3-chlorotyrosine are separated by reversed-phase high performance chromatography (HPLC) using a 250 × 4.6 mm Nucleosil 5 μm C_{18} AB column (Macherey-Nagel, Düren, Germany) with fluorescence detection (λ_{ex} 375 nm and λ_{em} 530 nm). Typically, 50 to 100 μl of derivatized tyrosine is injected onto the column and is eluted with water/acetonitrile/acetic acid (77:23:0.1) at a flow rate of 0.8 ml/min. After 15 min of elution the column is washed for 5 min with 100% acetonitrile, then reequilibrated with the eluant for 10 min. Fluorescence of the 3-chlorotyrosine and 3-bromotyrosine products is about 25% and 10%, respectively, of that of the tyrosine derivative.[9] Because of tailing of the tyrosine peak, a maximum of 1 3-chlorotyrosine per 1000 tyrosines can be

[25] T. P. Waalkes and S. Udenfriend, *J. Lab. Clin. Med.* **50,** 733 (1957).

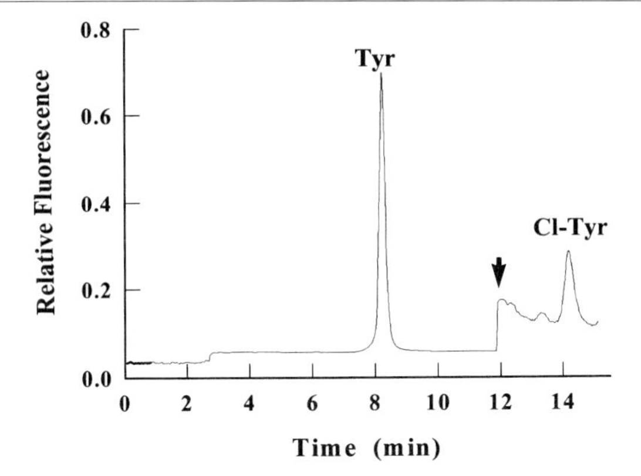

FIG. 2. Measurement of 3-chlorotyrosine using HPLC with fluorescence detection. Human serum albumin (500 μg/ml) was incubated with stimulated neutrophils as described in text. Albumin was hydrolyzed with pronase and 3-chlorotyrosine was detected by HPLC with fluorescence detection. The arrow indicates an increase in sensitivity of 128-fold.

detected in this assay (see Fig. 2). Bromotyrosine elutes immediately after chlorotyrosine.

Electrochemical Detection

Samples (20 μl) are separated by HPLC using a 12.5 cm $\times$ 4.6 mm Hypersil H% ODS column (Hichrom Ltd., UK) (Fig. 3). They are eluted from the column with 100 mM phosphate buffer, pH 3.0, with 10% methanol

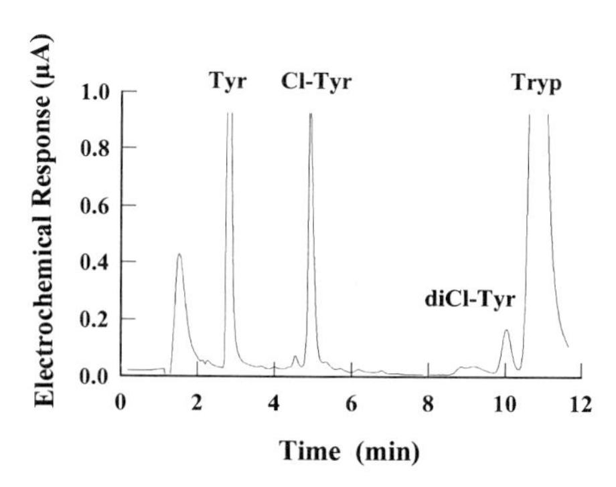

FIG. 3. Measurement of 3-chlorotyrosine by HPLC with electrochemical detection. Bovine serum albumin (500 μg/ml) was treated with 100 μM HOCl and chlorinated tyrosines were detected using HPLC with electrochemical detection. At 4 min the sensitivity was increased 10-fold. Other conditions were as described in Fig. 1.

at a flow rate of 1 ml/min. We use a Coulochem II electrochemical detector (ESA Inc., Chelmsford, MA) to detect eluted tyrosines. The guard cell potential is set at 700 mV while the E_1 and E_2 potentials for the 5011 analytical cell are set at 400 and 650 mV, respectively.

Acknowledgment

This work was supported by the Health Research Council of New Zealand. I would like to thank Christine Winterbourn for help in the preparation of this manuscript.

[15] Determination of 2-Oxohistidine by Amino Acid Analysis

By SANDRA A. LEWISCH and RODNEY L. LEVINE

Introduction

Oxidative modification of proteins gives rise to a large number of covalent modifications, many of which might serve as qualitative or quantitative markers of oxidative damage.[1] Histidine frequently functions as a ligand for divalent cations at metal-binding sites in proteins, precisely where metal-catalyzed oxidative modifications will occur. Multiple products are likely to be produced by such caged reactions,[2] and 2-oxohistidine is known to be one of them (Fig. 1). Thus, 2-oxohistidine may serve as a marker of oxidative modification, especially that occurring at metal binding sites within the protein.[3,4]

Uchida and Kawakishi developed a reversed-phase high-pressure liquid chromatography (HPLC) method employing electrochemical detection to identify 2-oxohistidine at the active site of oxidatively modified Cu,Zn-superoxide dismutase.[3] They also demonstrated the site-specific character of the oxidation of histidine to 2-oxohistidine in angiotensin.[5] Using the same method, Frei and colleagues demonstrated 2-oxohistidine in apolipoprotein B-100 of human low-density lipoprotein exposed to copper and ascorbate (B. Frei and K. Chen, personal communication, 1998). Apffel and colleagues identified 2-oxohistidine in modified glutamine synthetase

[1] E. R. Stadtman, *Free Rad. Biol. Med.* **9,** 315 (1990).
[2] M. Tomita, M. Irie, and T. Ukita, *Biochemistry* **8,** 5149 (1969).
[3] K. Uchida and S. Kawakishi, *FEBS Lett.* **332,** 208 (1993).
[4] S. A. Lewisch and R. L. Levine, *Anal. Biochem.* **231,** 440 (1995).
[5] K. Uchida and S. Kawakishi, *Arch. Biochem. Biophys.* **283,** 20 (1990).

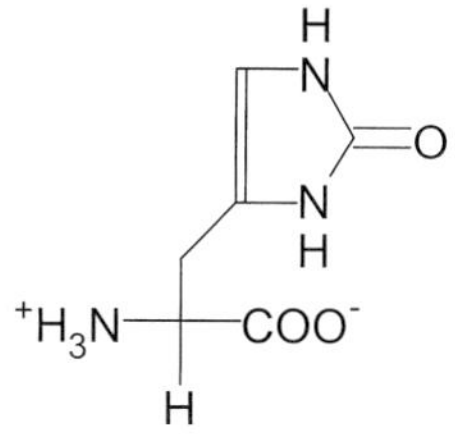

FIG. 1. Structure of 2-oxohistidine.

(glutamate–ammonia ligase) using automated Edman protein sequencing and mass spectroscopy.[6] 2-Oxohistidine was among the products of metal-catalyzed oxidation of growth hormone.[7,8] Schöneich and colleagues also established that the oxygen derives from molecular oxygen, not water.[8]

It would be convenient to assay 2-oxohistidine with standard methods for amino acid analysis, following hydrolysis of the protein. We found that 2-oxohistidine is unstable during the usual hydrolysis with 6 *M* hydrochloric acid, breaking down to products which include aspartate and ammonia.[4] This finding explains why 2-oxohistidine was previously thought to coelute with ammonia during classical amino acid analysis.[5] Through empirical trials, we found that inclusion of very high concentrations of dithiothreitol stabilized 2-oxohistidine during acid hydrolysis. The hydrolyzate can then be analyzed by the typical methods of amino acid analysis, including classical ion exchange with postcolumn derivatization and precolumn derivatization with either *o*-phthaldialdehyde (OPA) or 6-aminoquinoyl-*N*-hydroxysuccinimidyl carbamate (AQC).[4]

Preparation of 2-Oxohistidine

At present, 2-oxohistidine is not commercially available. However, it is a simple matter to oxidatively modify superoxide dismutase, glutamine synthetase, or other proteins according to published procedures,[9,10] hydrolyze the protein, and utilize the hydrolyzate as a 2-oxohistidine-containing reference.

[6] A. Apffel, J. Sahakian, and R. L. Levine, *Protein Sci.* **3,** 99 (1994).
[7] S. H. Chang, G. M. Teshima, T. Milby, B. Gillece-Castro, and E. Canova-Davis, *Anal. Biochem.* **244,** 221 (1997).
[8] F. Zhao, E. Ghezzo-Schoneich, G. I. Aced, J. Hong, T. Milby, and C. Schoneich, *J. Biol. Chem.* **272,** 9019 (1997).
[9] K. Uchida and S. Kawakishi, *J. Biol. Chem.* **269,** 2405 (1994).
[10] R. L. Levine, Methods in Enzymology, Vol. 107, p. 370 (1984).

If a purified standard is desired, it must be synthesized. The procedure given here is that of Uchida and Kawakishi[9] with minor modifications.

Materials

N-Benzoyl-L-histidine (Sigma, St. Louis, MO)
Sodium phosphate buffer, 50 m*M*, pH 7.2
Sodium phosphate buffer, 100 m*M*, pH 2.0
Phosphoric acid
Sodium hydroxide, 1 *M*
Ascorbic acid, 250–1000 m*M* in phosphate buffer, using NaOH to adjust the pH to 7.2; prepare this solution just before use.
Copper sulfate
Dowex AG 50W-X8 resin, 60-140 mesh, H^+ form (Bio-Rad Laboratories, Hercules, CA)
Chelex 100 resin, if desired (Bio-Rad Laboratories, Hercules, CA)

Procedure

Place a beaker on a magnetic stirrer at room temperature and add the desired volume (at least 50 ml) of 50 m*M* *N*-benzoyl-L-histidine. With vigorous mixing, bring the solution to 50 m*M* ascorbate and then 0.5 m*M* copper sulfate. Keep the solution well stirred to ensure adequate oxygenation. Every 30 min for 6 hr, use 1 *M* NaOH to readjust to pH 7.2. After 6 hr, the pH may be adjusted every 4–8 hr. Allow to react for 18 to 48 hr. After purification, overall yield will only be a few percent of the starting material, although the yield can be increased slightly by complexing the copper with Chelex 100 resin (Bio-Rad Laboratories, Hercules, CA) before adding it to the reaction beaker.

To begin the purification, bring the solution to pH 2 with 100 m*M* phosphoric acid and pass it over the Dowex 50 resin, previously equilibrated with 100 m*M* sodium phosphate, pH 2.0. *N*-Benzoyl-2-oxohistidine elutes in the void volume while the starting material will bind to the column. At this stage, *N*-benzoyl-2-oxohistidine is the major component, but it is not pure. If additional purification is desired, concentrate the solution to about one-third of the original volume in a vacuum centrifuge (Savant Speed Vac, Farmingdale, NY). Load on a C_{18} reversed-phase column (Vydac 218TP54, 0.46 mm × 25 cm, Separations Group, Hesperia, CA) equilibrated in 0.05% trifluoroacetic acid (Pierce, Rockford, IL). Develop a gradient from 0 to 10% methanol over 20 min. *N*-Benzoyl-2-oxohistidine will elute near the end of the gradient and can be monitored at 210 and 270 nm. Dry the product in the vacuum centrifuge and store at −20°C.

Hydrolysis of *N*-Benzoyl-2-oxohistidine or Proteins

The free amino acid 2-oxohistidine is prepared from the *N*-benzoyl-2-oxohistidine by acid hydrolysis. However, 2-oxohistidine is not very stable to storage, so small amounts should be prepared as needed. In contrast, 2-oxohistidine is quite stable in protein hydrolyzates stored at −20°C.

Materials

Hydrochloric acid, constant boiling (Pierce, Rockford, IL)
Dithiothreitol, 1 *M*
Borosilicate 4 ml vials (Wheaton 224882, Millville, NJ)
PTFE-lined vial caps (Kimble 738013425)
Argon or other inert gas

Procedure

Hydrolysis is conveniently carried out with a rapid 155°C benchtop method.[11] Place the sample into a Wheaton vial and add 200 μl 6 *M* HCl plus 8 μl 1 *M* dithiothreitol (40 m*M* final). Addition of this unusually high concentration of dithiothreitol substantially increases the recovery of 2-oxohistidine from both purified *N*-benzoyl-2-oxohistidine and proteins. Flush the vial with argon for 1 min before capping. The cap should be snug, but avoid overtightening which may crack the neck of the vial. Then place in a benchtop heater in the hood at 155°C. Hydrolyze the *N*-benzoyl-2-oxohistidine for 20 min and proteins for 45 min. After cooling, dry promptly in the vacuum centrifuge.

Amino Acid Analysis

We have quantitated 2-oxohistidine by three different methods of amino acid analysis: (1) classical Moore–Stein ion-exchange chromatography with postcolumn derivatization; (2) precolumn derivatization with OPA followed by reverse phase chromatography; and (3) precolumn derivatization with AQC.[4] Presumably, other methods would also be successful. It is important to note that when precolumn OPA is used, one must decrease the pH of the derivatizing solution from the usual ~10 to 8.0, again to stabilize the 2-oxohistidine.[4]

2-Oxohistidine elutes between serine and glutamate in a Moore–Stein analysis, between threonine and arginine with precolumn OPA derivatiza-

[11] R. L. Levine, D. Garland, C. N. Oliver, A. Amici, I. Climent, A. G. Lenz, B. W. Ahn, S. Shaltiel, and E. R. Stadtman, Methods in Enzymology, Vol. 186, p. 464 (1990).

tion, and between histidine and ammonia with precolumn AQC derivatization. Optimal separation may require adjustment of the standard gradient program, with the specific changes dependent on the particular HPLC being used. See ref. 4 for guidelines.

[16] Detecting Oxidative Modification of Biomolecules with Isotope Dilution Mass Spectrometry: Sensitive and Quantitative Assays for Oxidized Amino Acids in Proteins and Tissues

By JAY W. HEINECKE, FONG FU HSU, JAN R. CROWLEY, STANLEY L. HAZEN, CHRISTIAAN LEEUWENBURGH, DIANNE M. MUELLER, JANE E. RASMUSSEN, and JOHN TURK

Oxidative modification of biomolecules has been implicated in the pathogenesis of diseases including atherosclerosis, ischemia–reperfusion injury, and cancer.[1,2] Many lines of evidence suggest that oxidative stress also plays a causal role in aging.[3] Protein oxidation may be a key event in these processes because proteins play fundamental roles as biological catalysts, gene regulators, and structural components of cells. One widely studied model of protein oxidation involves the generation of hydroxyl radical-like species by metal-catalyzed oxidation reactions.[3] Other mechanisms of protein oxidation involve reactive nitrogen species, including peroxynitrite and nitrogen dioxide radical derived from nitric oxide.[4–6]

Because white blood cells use oxidative pathways to kill invading microorganisms, they also generate reactive species that can damage host proteins, including lipoproteins.[7–9] Our studies have focused on the potential

[1] B. N. Ames, M. K. Shigenaga, and T. M. Hagen, *Proc. Natl. Acad. Sci.* **90,** 7915 (1993).
[2] J. A. Berliner and J. W. Heinecke, *Free Rad. Biol. Med.* **20,** 707 (1996).
[3] E. R. Stadtman, *Science* **257,** 1220 (1992).
[4] W. H. Koppenol, J. J. Moreno, W. A. Pryor, H. Ischiropoulos, and J. S. Beckman, *Chem. Res. Toxicol.* **5,** 834 (1992).
[5] J. P. Eiserich, C. E. Cross, A. D. Jones, B. Halliwell, and A. van der Vleit, *J. Biol. Chem.* **271,** 19199 (1996).
[6] S. Moncada and A. Higgs, *N. Engl. J. Med.* **329,** 2002 (1993).
[7] S. J. Klebanoff, *Ann. Intern. Med.* **93,** 480 (1980).
[8] J. K. Hurst and W. C. Barrette, *CRC Crit. Rev. Biochem. Mol. Biol.* **24,** 271 (1989).
[9] S. J. Weiss, *N. Engl. J. Med.* **320,** 365 (1989).

0076-6879/99 $30.00

threat posed by myeloperoxidase,[10] a heme enzyme secreted by activated phagocytes. Using hydrogen peroxide, which phagocytes also secrete, myeloperoxidase generates several potent oxidants.[7,9] These include hypochlorous acid,[11] tyrosyl radical,[12] and nitrating species.[5]

The oxidative pathways that damage biomolecules *in vivo* are difficult to identify because the toxic intermediates are short-lived and difficult to measure directly.[1–3] One approach to studying the effects of oxidative pathways that operate *in vivo* is to analyze normal and pathological tissues for stable end products of oxidative reactions that have been identified through *in vitro* studies (Fig. 1). For example, *o*-tyrosine and *m*-tyrosine, the unnatural isomers of *p*-tyrosine, represent stable, posttranslation modifications of protein-bound phenylalanine residues that can be generated by hydroxyl radical.[13–15] Similarly, *o,o'*-dityrosine is formed when hydroxyl radical cross-links tyrosine residues.[13,15,16] A metal-catalyzed oxidation system might therefore produce all three of these compounds *in vivo*. The compound *o,o'*-dityrosine is also formed by reaction of free or protein-bound tyrosine with tyrosyl radical,[17] which is produced from tyrosine and H_2O_2 by myeloperoxidase.[12] In contrast to *o,o'*-dityrosine formation caused by hydroxyl radical,[13,15] that caused by tyrosyl radical is not accompanied by formation of *o*-tyrosine or *m*-tyrosine.[15,18,19] 3-Chlorotyrosine is formed by reaction of tyrosine with HOCl or with species such as molecular chlorine derived from HOCl.[20,21] At concentrations of chloride which occur in plasma, HOCl is a specific product of the myeloperoxidase–H_2O_2–Cl^- system.[11,22,23] 3-Nitrotyrosine is formed by reactive nitrogen intermediates that

[10] A. Daugherty, J. L. Dunn, D. L. Rateri, and J. W. Heinecke, *J. Clin. Invest.* **94,** 437 (1994).

[11] J. E. Harrison and J. Schultz, *J. Biol. Chem.* **251,** 1371 (1976).

[12] J. W. Heinecke, W. Li, H. L. Daehnke, and J. A. Goldstein, *J. Biol. Chem.* **268,** 4069 (1993).

[13] T. G. Huggins, M. C. Wells-Knecht, N. A. Detorie, J. W. Baynes, and S. R. Thorpe, *J. Biol. Chem.* **268,** 12341 (1993).

[14] H. Kaur and B. Halliwell, *Methods Enzymol.* **233,** 67 (1994).

[15] C. Leeuwenburgh, J. E. Rasmussen, F. F. Hsu, D. M. Mueller, S. Pennathur, and J. W. Heinecke, *J. Biol. Chem.* **272,** 3520 (1997).

[16] K. J. Davies, *J. Biol. Chem.* **262,** 9895 (1987).

[17] R. C. Prince, *Trends Biochem. Sci.* **13,** 286 (1988).

[18] J. S. Jacob, D. P. Cistola, F. F. Hsu, S. Muzaffar, D. M. Mueller, S. L. Hazen, and J. W. Heinecke, *J. Biol. Chem.* **271,** 19950 (1996).

[19] J. W. Heinecke, W. Li, G. A. Francis, and J. A. Goldstein, *J. Clin. Invest.* **91,** 2866 (1993).

[20] E. L. Thomas and M. B. Grisham, *Methods Enzymol.* **132,** 569 (1986).

[21] S. L. Hazen, F. F. Hsu, D. M. Mueller, J. R. Crowley, and J. W. Heinecke, *J. Clin. Invest.* **98,** 1283 (1996).

[22] S. L. Hazen and J. W. Heinecke, *J. Clin. Invest.* **99,** 2075 (1997).

[23] S. J. Weiss, S. T. Test, C. M. Eckmann, D. Ross, and S. Regiani, *Science* **234,** 200 (1986).

(a) Metal ions/hydroxyl radical

OH
+
OH
R R R
Phenylalanine o-Tyrosine m-Tyrosine

(b) Tyrosyl radical

OH OH OH OH
R R R R
Tyrosine Tyrosyl radical o,o′-Dityrosine

(c) Myeloperoxidase/hypochlorous acid

OH OH
Cl
R R
Tyrosine 3-Chlorotyrosine

(d) Reactive nitrogen species

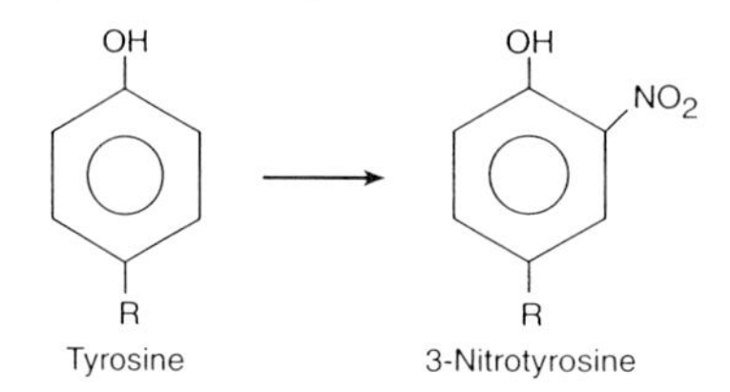

FIG. 1. Stable products of protein oxidation *in vitro*.

oxidize tyrosine.[4,5,24,25] All of these amino acid products are stable to acid hydrolysis, making them potentially useful markers for protein oxidation *in vivo*.[13,15,22,25]

In this chapter, we describe sensitive and quantitative assays for measuring *o*-tyrosine, *m*-tyrosine, *o,o′*-dityrosine, 3-nitrotyrosine, and 3-chloro-

[24] J. S. Beckman, T. W. Beckman, J. Chen, P. A. Marshall, and B. A. Freeman, *Proc. Natl. Acad. Sci. USA* **87,** 1620 (1990).

[25] C. Leeuwenburgh, M. M. Hardy, S. L. Hazen, P. Wagner, S. Oh-ishi, U. P. Steinbrecher, and J. W. Heinecke, *J. Biol. Chem.* **272,** 1433 (1997).

tyrosine in proteins and tissues. The methods combine gas chromatography (GC) with isotope dilution negative-ion electron capture mass spectrometry (MS). GC/MS has enabled us to determine the relative amounts of oxidized amino acids in proteins and lipoproteins oxidized *in vitro*.[15,19,21,22,25,26] We also have used this approach to quantify levels of modified amino acids in atherosclerotic lesions,[15,22,25] inflammatory tissues,[26] and tissue proteins of aging animals.[27,28] Our observations indicate that the quantitative evaluation of protein oxidation products by GC/MS is a powerful tool for identifying pathways that promote oxidative protein modification *in vivo*.

Advantages of Isotope Dilution GC/MS Analysis

Although immunohistochemical methods are widely used to study oxidative stress *in vivo*, they are only semiquantitative.[29,30] Moreover, cross-reacting but structurally distinct molecules may confound immunodetection methods. Existing high-performance liquid chromatographic (HPLC) approaches to quantifying oxidation products[14,31] can be more sensitive and more specific than immunochemistry, but they provide no structural information about the analyte and can be confounded by coeluting but structurally distinct molecules. In contrast, GC/MS provides specific structural information,[32,33] thereby reducing the potential for confusion with extraneous compounds that coelute with the target analyte during chromatography. In addition, GC/MS analyses permit the use of a stable, isotopically labeled internal standard which, apart from its heavy isotope, is structurally identical to the target analyte and therefore behaves identically during extraction, processing, and chromatographic analyses.[32] Including such a standard corrects for analyte loss during processing and increases the precision of quantitative measurements. Although internal standards can be incorporated into the HPLC-based methods, they must be chromatographically separable and therefore structurally distinct from the target analyte. This raises the

[26] S. L. Hazen, J. P. Gaut, F. F. Hsu, J. R. Crowley, A. d'Avignon, and J. W. Heinecke, *J. Biol. Chem.* **272,** 16990 (1997).

[27] C. Leeuwenburgh, P. Wagner, J. O. Holloszy, R. S. Sohal, and J. W. Heinecke, *Arch. Biochem. Biophys.* **346,** 74 (1997).

[28] C. Leeuwenburgh, P. Hansen, A. Shaish, J. O. Holloszy, and J. W. Heinecke, *Am. J. Physiol.* **274,** R453 (1998).

[29] M. E. Rosenfeld, W. Palinski, S. Yla-Herttuala, S. Butler, and J. L. Witztum, *Arterioscler.* **10,** 336 (1990).

[30] J. L. Witztum and D. Steinberg, *J. Clin. Invest.* **88,** 1785 (1991).

[31] M. K. Shigenaga, H. H. Lee, B. C. Blout, S. Christen, E. T. Shigeno, H. Yip, and B. A. Ames, *Proc. Natl. Acad. Sci.* **94,** 3211 (1997).

[32] J. R. Chapman, "Practical Organic Mass Spectrometry." Wiley, New York, 1985.

[33] J. T. Watson, *Methods Enzymol.* **193,** 86 (1990).

possibility of differential extraction or interference from extraneous substances that coelute with either the internal standard or the target analyte.

Synthesis of Internal Standards

Using isotope dilution GC/MS, it is possible to quantify trace amounts of analyte in a complex biological mixture.[32,33] It is necessary to add a known amount of internal standard to the protein or tissue sample prior to analysis. A variety of isotopically labeled compounds can be used to prepare the internal standard, including ^{2}H-, ^{13}C-, ^{15}N-, and ^{18}O-enriched compounds. For tyrosine- and phenylalanine-derived compounds, we typically use $^{13}C_6$-ring-labeled amino acids which exhibit several advantageous properties. First, ^{13}C-labeled compounds behave essentially like nonisotopically labeled compounds during sample processing and chromatography. In contrast, chromatographic separation of ^{2}H-labeled internal standards and their ^{1}H natural counterparts sometimes occurs. Second, the increase in six mass units minimizes the contribution of nonisotopically labeled internal standard to the ion current of the target analyte. Finally, ^{13}C-labeled compounds are intrinsically stable to a wide range of chemical conditions and do not readily undergo isotopic exchange. In contrast, ^{2}H- or ^{18}O-labeled compounds can undergo isotope exchange under some conditions.[34] When this occurs, it may be impossible to accurately quantify the amount of material present in the sample. ^{13}C-labeled amino acids are, however, generally more expensive than their ^{2}H-labeled counterparts.

Sigma Chemical Company (St. Louis, MO) provides standard *o*-tyrosine, *m*-tyrosine, 3-nitrotyrosine, and 3-chlorotyrosine. Cambridge Isotope Laboratories supplied the ^{13}C-labeled amino acids used in our studies. We synthesize isotopically labeled *o*-tyrosine and *m*-tyrosine using [$^{13}C_6$]phenylalanine, copper, and H_2O_2.[13,15] The isomers are then isolated by reversed-phase HPLC.[35] Standard *o,o'*-[$^{13}C_{12}$]dityrosine and *o,o'*-dityrosine are prepared from [$^{13}C_6$]tyrosine and tyrosine, respectively, using horseradish peroxidase and H_2O_2.[36] We use 10 m*M* borate (pH 9) as the reaction buffer to facilitate the isolation of *o,o'*-dityrosine by cellulose phosphate column chromatography[36,37] or by anion-exchange FPLC (fast protein liquid chromatography).[12] Standard 3-nitro[$^{13}C_6$]tyro-

[34] J. A. McCloskey, *Methods Enzymol.* **193,** 329 (1990).

[35] S. L. Hazen, F. F. Hsu, and J. W. Heinecke, *J. Biol. Chem.* **271,** 1861 (1996).

[36] R. Amado, R. Aeschbach, and H. Neukom, *Methods Enzymol.* **107,** 377 (1984).

[37] G. A. Francis, A. J. Mendez, E. L. Bierman, and J. W. Heinecke, *Proc. Natl. Acad. Sci. USA* **90,** 6631 (1993).

sine is prepared by treating tyrosine with tetranitromethane.[38,39] The product is then isolated by reversed-phase HPLC.[35] Standard 3-chloro[$^{13}C_6$]tyrosine is prepared from [$^{13}C_6$]tyrosine with reagent hypochlorous acid (HOCl)[21,22,40] and then isolated by reversed-phase HPLC.[35] The concentrations of these synthetic amino acids are determined spectrophotometrically by comparison with unlabeled standards using HPLC and monitoring UV absorbance at 276 nm. Alternatively, their concentrations can be determined by liquid scintillation spectrometry when a starting material is used that contains tracer amounts of a radioactive isotope at a known specific activity.

Tissue Collection and Preparation

Tissue resected at surgery or autopsy is immediately placed in ice-cold antioxidant buffer [100 μM diethylenetriaminepentaacetic acid (DTPA), 1 mM butylated hydroxytoluene, 1% (v/v) ethanol, 140 mM NaCl, 10 mM sodium phosphate, pH 7.4], and then frozen at $-80°$C until analysis. DTPA (prepared as a 10 mM stock solution, pH 7.4) and butylated hydroxytoluene [prepared as a 100 mM stock solution in 95% (v/v) ethanol] are included to inhibit metal-catalyzed oxidation reactions and lipid peroxidation, respectively.[41] Depending on the oxidation products of interest, the antioxidant buffer is supplemented with inhibitors of other pathways (e.g., 10 mM 3-aminotriazole to inhibit myeloperoxidase[12,19]). In animal studies, it is desirable to perfuse tissues with antioxidant buffer to remove red blood cells that might contribute oxidation products to the samples either directly or via hemoglobin-catalyzed reactions.

Two procedures are used to prepare tissue for hydrolysis. Most tissues (50–100 mg wet weight) are suspended (1 : 10, w/v) in ice-cold homogenization buffer (50 mM sodium phosphate (pH 7.4) supplemented with 100 μM DTPA, 1 mM butylated hydroxytoluene, 1% (v/v) ethanol (from the stock solution of butylated hydroxytoluene in 95% ethanol), 10 mM 3-aminotriazole) and then homogenized using a motor-driven Teflon pestle in a glass receptacle. Aortic tissue, which is rich in collagen and elastin, is difficult to homogenize by this procedure. Instead, it is frozen in liquid nitrogen and then pulverized with a stainless steel mortar and pestle that had been previously immersed in liquid nitrogen. All subsequent procedures are performed at 4°. Tissue powder (10–20 mg wet weight) is suspended in

[38] G. E. Means and R. E. Feeney, in "Chemical Modification of Proteins," p. 183. Holden-Day, Oakland, CA, 1971.

[39] M. Solkolovsky, J. F. Riordan, and B. L. Vallee, *Biochemistry* **5,** 3582 (1966).

[40] S. L. Hazen, F. F. Hsu, D. M. Mueller, J. R. Crowley, and J. W. Heinecke, *Free Rad. Biol. Med.* **23,** 909 (1997).

[41] J. W. Heinecke, L. Baker, H. Rosen, and A. Chait, *J. Clin. Invest.* **77,** 757 (1986).

1 ml of buffer B (100 μM DTPA, pH 7) and then dialyzed for 24 hr against that buffer to remove free amino acids, nitrite, and other low molecular weight contaminants. The protein solution is dried under vacuum, resuspended in 0.5 ml of H_2O, and delipidated by incubation for 10 min with 6.5 ml of H_2O/methanol/water-washed diethyl ether (1:3:7, v/v/v). For the analysis of 3-chlorotyrosine, an additional wash with 1 ml of H_2O and 13 ml of methanol/water-washed diethyl ether (3:10, v/v) is performed to remove residual chloride that might contribute to chlorination of tyrosine during acid hydrolysis. To recover tissue protein, the sample is centrifuged at 5000*g* for 10 min, the supernatant removed, and the fluffy protein powder again delipidated with 10 ml of water-washed diethyl ether.

Protein and Tissue Hydrolysis

Proteins and tissues must be enzymatically or chemically hydrolyzed before their content of oxidized amino acid residues is measured by GC/MS. Although enzymatic hydrolysis is gentler than chemical hydrolysis and therefore less likely to result in artifactual generation of oxidized amino acids, it is difficult to ensure that enzymatic digestion of proteins is quantitative, especially with biological samples. Moreover, autodigestion of protease enzymes can add free amino acids to those derived from tissue, lowering the apparent relative concentration of oxidation products (J. R. C. and J. W. H., unpublished observation). We therefore routinely use acid to hydrolyze tissues because it quantitatively releases amino acids from proteins. This approach requires that the oxidation product of interest be stable under conditions of acid hydrolysis (generally a 24 hr exposure of the protein to 6 *N* HCl at 110°). Acid hydrolysis also can contribute to protein oxidation *ex vivo*. Therefore, we add phenol (0.1–1%) and benzoic acid (0.1–1%) to the reaction mixture to suppress such reactions.

Exposure of proteins to HCl generates low levels of 3-chlorotyrosine, which can contribute significantly to the apparent level of 3-chlorotyrosine in biological material.[22,40,42] As noted below, we avoid this artifact by using HBr for acid hydrolysis. It also is possible to generate 3-nitrotyrosine from nitrate, nitrite, and other nitrogen species under acidic conditions.[4,31] Many commercial sources of HCl promote the nitration of tyrosine under the conditions used for acid hydrolysis (J. R. C. and J. W. H., unpublished observation), and nitrate and nitrite also can be present in biological material.[4,6] It therefore is important to use reagent-quality HCl for hydrolysis and to perform control experiments to exclude artifactual amino acid oxidation. Such experiments include subjecting model proteins to the hydrolysis

[42] F. Sanger and E. O. P. Thompson, *Biochim. Biophys. Acta* **71,** 471 (1963).

procedure and including ^{13}C-labeled amino acids in the reaction mixture so that any generation of ^{13}C-labeled oxidation products can be identified.[15,22] It also is useful to dialyze tissue proteins to remove low molecular weight contaminants such as nitrate from biological material prior to hydrolysis. We also extract tissue samples and lipoproteins with methanol and water-washed diethyl ether to remove lipids that might undergo peroxidation during acid hydrolysis.

o,o'-Dityrosine, o-Tyrosine, m-Tyrosine, and 3-Nitrotyrosine. Following the addition of ^{13}C-labeled internal standards (~50 pmol oxidized amino acid and 50 nmol precursor amino acid), the protein or tissue residue (~0.5 mg protein) is concentrated to dryness under nitrogen and then hydrolyzed at 110° for 24 hr in 0.5 ml of 6 *N* HCl (Sequenal Grade, Pierce Chemical, Rockford, IL) supplemented with 1% benzoic acid and 1% phenol. Prior to hydrolysis, tubes are capped with gas-tight valves and alternately degassed and purged with argon five times.

3-Chlorotyrosine. Trace amounts of 3-chlorotyrosine are formed in proteins hydrolyzed with HCl.[22,40,42] We avoid this artifact by substituting HBr for HCl and including 1% phenol in the acid.[22,40] Internal standards (~10 pmol 3-chlorotyrosine and 100 nmol tyrosine) are added to the protein or tissue residue (~0.5 mg protein). The sample is then concentrated to dryness under nitrogen and hydrolyzed at 120°C for 24 h in 0.5 ml of 6 *N* HBr supplemented with 1% phenol. Prior to hydrolysis, tubes are capped with gas-tight valves and alternately degassed and purged with argon as described above. Stoichiometric protein hydrolysis under these conditions was demonstrated by quantitative recovery of a variety of amino acids (L-tyrosine, L-lysine, and L-phenylalanine) by isotope dilution GC/MS. The compound 3-chlorotyrosine is stable under these hydrolysis conditions and is recovered in greater than 98% yield, as determined by reverse-phase HPLC analysis.[40]

Oxidation of Amino Acids during Acid Hydrolysis. It is important to determine whether acid hydrolysis itself oxidizes amino acid residues because this would inflate estimates of their content in proteins. We therefore perform preliminary experiments to quantify the amount of oxidized amino acid in an authentic precursor of the amino acid (e.g., phenylalanine for *o*-tyrosine) and in the precursor amino acid subjected to the analytical workup.[15,22,25] For example, we have demonstrated that the amount of *o*-tyrosine in standard phenylalanine subjected to the acid-hydrolysis procedure increases by less than 10 μmol oxidation product/mol precursor amino acid.[15] This represents less than 5% of the amount of *o*-tyrosine observed in tissue. Under these conditions, there is no detectable conversion of tyrosine to *o,o'*-dityrosine or 3-nitrotyrosine. Levels of oxidation products in model proteins (bovine serum albumin, ribonuclease, immunoglobulin

G) subjected to acid hydrolysis, derivatization, and GC/MS analysis vary with the identity of the analyte.[15] Relative concentrations range from 100 to 1100 μmol for *o*-tyrosine per mol tyrosine. Less than 5 μmol of *o,o'*-dityrosine per mol of tyrosine is observed. The relative concentrations of oxidized amino acids observed on acid hydrolysis of different model proteins remain consistent during repeated analyses, suggesting that they may reflect endogenous content. These observations indicate that the acid hydrolysis procedure itself is not a major contributor to the amounts of oxidized amino acids that are observed in biological materials.

Solid-Phase Extraction of Oxidized Amino Acids

We isolate amino acids from protein hydrolyzates by solid-phase extraction with octadecylsilicic (ODS) columns. This procedure removes contaminants that interfere with derivatization of amino acid analytes and with their analysis by negative-ion electron capture mass spectrometry. ODS extraction columns (Supelclean LC-18 SPE tube, 3 ml; Supelco Inc., Bellefonte, PA) are activated prior to use by washing with 3 ml of methanol.

o,o'-Dityrosine, o-Tyrosine, m-Tyrosine, and 3-Nitrotyrosine. After activation with methanol, ODS extraction columns are conditioned prior to use by washing with 10 ml of 50 m*M* sodium phosphate (pH 7.4) supplemented with 100 μM DTPA, followed by 12 ml of 0.1% trifluoroacetic acid. Washing the ODS column with DTPA inhibits oxidation of phenylalanine and tyrosine during isolation of the amino acids. The volume of the amino acid hydrolyzate (~0.5 ml) is adjusted to 2 ml with 0.1% trifluoroacetic acid and passed over the ODS solid-phase extraction column. Columns are washed with 2–4 ml of 0.1% trifluoroacetic acid, and amino acids are eluted with 2 ml of H_2O : methanol (4 : 1, v/v). The amino acid solution is concentrated to dryness under vacuum and immediately derivatized (see below). Studies with standard compounds indicate that >80% of tyrosine oxidation products are recovered from the ODS column using this procedure.[15,25]

3-Chlorotyrosine. After activation with methanol, ODS extraction columns are conditioned prior to use with 12 ml of 0.1% trifluoroacetic acid. The volume of protein hydrolyzates (~0.5 ml) is adjusted to 2 ml with 0.1% trifluoroacetic acid and the amino acid solution is passed over the ODS column. Columns are washed with 2 ml of 0.1% trifluoroacetic acid, and amino acids are eluted with 2 ml of H_2O : methanol (1 : 1, v/v) containing 0.1% trifluoroacetic acid. The amino acid solution is concentrated to dryness under vacuum and immediately derivatized. Under these conditions, 3-chlorotyrosine liberated by HBr hydrolysis and isolated by passage over a solid-phase ODS minicolumn is recovered quantitatively.[22,40]

Derivatization of Amino Acids

Negative-ion electron capture mass spectrometry is more sensitive than electron impact mass spectrometry for quantifying trace levels of organic compounds, but can be used only when the target analyte contains an electron-capturing moiety.[32,33,43] Conversion of amino acids to their heptafluorobutyryl derivatives with heptafluorobutyric acid anhydride (HFBA) generates compounds whose high volatility and electron-capturing properties make them suitable for negative-ion electron capture GC/MS analysis.[43]

HFBA reacts with several functional groups, including aliphatic or aromatic hydroxyl moieties and primary or secondary amines.[43] For example, it converts *n*-propyl-*o*-tyrosine to a bis-HFB derivative bearing a fluoroacyl moiety on both the amino group and phenol hydroxyl.[15] This derivative yields detectable negative-ion electron capture GC/MS signal at subpicomole levels. We therefore routinely prepare HFBA derivatives of oxidized amino acids for negative-ion electron capture GC/MS analysis.[12,15,19,22,25]

Amino acids in protein hydrolyzates that have been concentrated to dryness (~0.5 mg amino acids) are first converted to their *n*-propyl carboxylic acid esters by addition of 200 μl of 3.5 *M* HCl (or 3.5 *M* HBr for 3-chlorotyrosine determination) in *n*-propanol, followed by heating at 65° for 1 hr. Reaction products are then concentrated to dryness under nitrogen, and their heptafluorobutyryl (HFB) derivatives are generated by addition of 50 μl of 3:1 (v/v) ethyl acetate/heptafluorobutyric acid anhydride and incubation at 65° for 5 min.

Gas Chromatography/Mass Spectrometry

We have used commercially available quadrupole mass spectrometers with chemical ionization sources and negative ion capabilities for GC/MS analysis of amino acids, including Hewlett-Packard 5988A and Finnigan SSQ-7000 systems. We typically employ 10–30 m, 0.2–0.3 mm i.d. fused-silica capillary columns with helium as the carrier gas for GC analysis. Both J & W Scientific Inc. (Folsom, CA) DB-1 columns (15 m, 0.33 μm methyl silicone film thickness, 0.32 mm i.d.) and DB-17 columns (30 m, 0.25 μm phenyl methyl silicone film thickness, 0.25 mm i.d.) have proven useful.[12,15,19,22,25] These columns provide excellent separation of HFBA-derivatized amino acids. Precursor amino acids (*p*-tyrosine, phenylalanine), which are present at high concentrations in the amino acid hydrolyzate, are injected into the GC with a 50–100:1 split to avoid column overload and saturation of the detector of the mass spectrometer. Oxidized amino

[43] D. R. Knapp, "Handbook of Analytical Derivatization Reactions." Wiley, New York, 1979.

acids, which are present at low concentrations in biological material, may be injected in the splitless mode to increase the sensitivity of the analysis. The injector temperature and transfer lines of the GC are maintained at 200–250°, and the mass spectrometer ion source temperature is 100–150°. The mass spectrometer is operated in negative-ion electron capture mode with methane as moderating gas at a source pressure of 0.6–1.0 torr. Emission current is set at 300 μA, and the electron energy is 240 ev.

Selected Ion Monitoring for Quantification of Trace Quantities of Oxidized Amino Acids

In the selected ion monitoring mode, a limited number of ions of defined mass-to-charge (m/z) values are monitored by the mass spectrometer as a function of GC retention time.[33] This greatly increases the sensitivity of the analysis compared to full-scan mode because a greater proportion of the duty cycle of the instrument is used to detect the ions of interest.

To perform GC/MS analysis with selected ion monitoring, it is first necessary to determine the GC retention time and the mass spectrum of the oxidized amino acid derivatives.[12,19,21,25] To determine these parameters, the *n*-propyl ester, HFBA derivative of authentic material is prepared and analyzed by GC/MS in full-scan mode. When a relatively pure amino acid is used, there typically is a single major peak in the total ion current chromatogram. Alternatively, selected ion monitoring can be used to detect ions that are likely to be present in the mass spectrum of the derivatized target analyte in order to determine its GC retention time so that a mass spectrum of the appropriate peak can be determined. The molecular ion (M^-) of HFBA-derivatized amino acids is often of low abundance, but intense ions are generally observed at M − 20 (M^- − HF) and M − 198 (M^- − $CF_3(CF_2)_2CHO$). After tentatively identifying the retention time of the derivatized amino acid, a full-scan mass spectrum of the peak is obtained to confirm that the material exhibits a mass spectrum consistent with the proposed derivative.

An example of this approach is illustrated in Fig. 2. We used horseradish peroxidase and H_2O_2 to convert tyrosine to *o,o'*-dityrosine[36] and isolated that product from the reaction mixture by anion exchange chromatography.[12] The isolated product then was converted to its *n*-propylcarboxylic acid ester derivative. Treatment of that derivative with HFBA yielded a tetra-HFB derivative in which the fluoracyl moiety modified both α-amino and both phenolic hydroxyl groups. The structure and negative-ion chemical ionization mass spectrum of this derivative are illustrated in Fig. 2. The molecular anion (m/z 1228) is of low abundance, but ions reflecting neutral

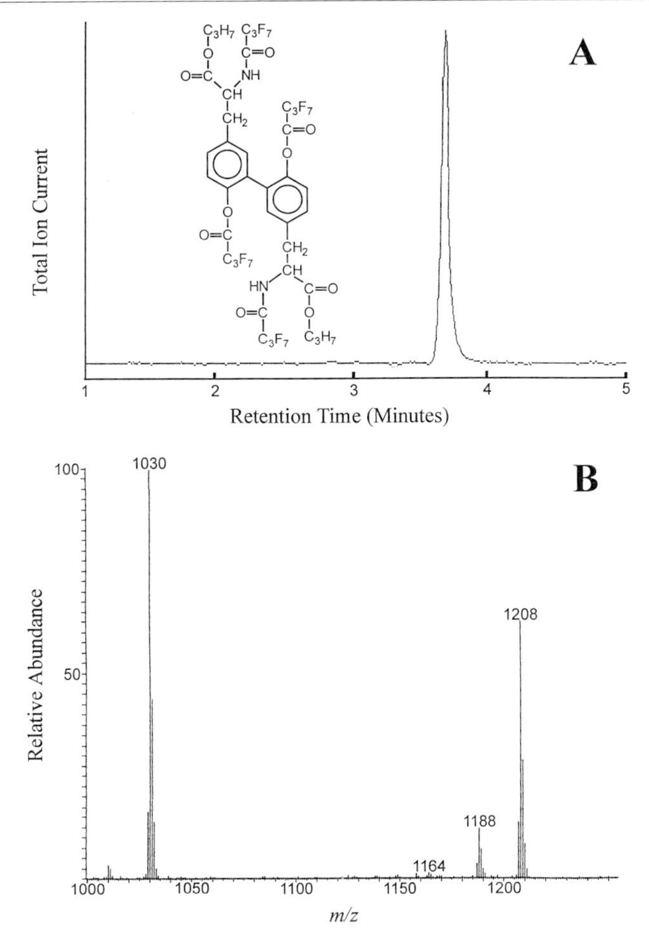

FIG. 2. Total ion chromatogram (A) and negative-ion electron capture mass spectrum of the *n*-propyl-HFB derivative of *o,o'*-dityrosine (B). *o,o'*-Dityrosine was prepared from tyrosine by oxidation with H_2O_2 and horseradish peroxidase and then isolated by anion-exchange chromatography. The isolated amino acid was converted to its *n*-propyl-HFB derivative and then analyzed by negative-ion electron capture GC/MS as described in the text. The molecular weight of the derivative is 1228 (*Inset*). The loss of HF and $CF_3(CF_2)_2CHO$ from a molecular ion of *m/z* 1228 yields ions with *m/z* values of 1208 and 1030, respectively.

losses of HF (*m/z* 1208) and heptafluorobutyric acid (*m/z* 1030) are abundant.

Once the GC retention time and mass spectrum of the analyte are established, we determine whether the presence of the analyte can be demonstrated qualitatively in biological materials. After hydrolysis to release the analyte from proteins, it is derivatized and analyzed by negative-ion electron capture GC/MS.[32,33] Structurally informative ions of high abundance which occur in the negative-ion electron capture mass spectrum of the derivatized standard compound are detected by selected ion monitoring. Coelution of multiple ions with the expected relative abundance at the appropriate GC retention time indicates that the compound of interest is present in the material. The identities of compounds in biological extracts can be confirmed by comparing the GC/MS behavior of other derivatives (e.g., pentafluoropropionyl derivatives[15,25]) of the compound with that of standards.

For quantitative analyses, a standard curve is constructed so that the concentration of the analyte can be established by isotope dilution.[32] Determination of the standard curve involves addition of a constant amount of heavy isotope-labeled internal standard and varied but known amounts of standard target analyte to a series of reaction vessels. The materials are then derivatized and analyzed by negative-ion electron capture GC/MS with selected monitoring of an ion arising from the target analyte and the analogous ion arising from the internal standard. The ratio of the integrated ion current peaks for these ions is then plotted as a function of the amount of target analyte initially added to the reaction vessel. For the analytes studied here, the resultant regression line is typically linear over at least two orders of magnitude.

The mass of the internal standard is shifted to a higher mass-to-charge ratio owing to the heavy isotope content of the compound. Because the chemical and physical properties of natural and ^{13}C-labeled compounds are otherwise essentially identical, the ions of the isotopically labeled standard coelute with the corresponding ions of the target analyte. In contrast, deuterated internal standards are slightly less polar than hydrogenated compounds and typically elute slightly earlier than the authentic material on GC.

For measurements of the analyte from biological materials, internal standards are added in sufficiently small amounts that the blank value of the internal standard, which reflects its intrinsic content of unlabeled analyte, represents less than 5% of the total current of ions that are monitored to quantitate the target analyte. To ensure that interfering ions from extraneous substances are not coeluting with the analyte, we monitor the ratio of ion currents of the two most abundant structurally informative ions of each compound and its internal standard. Quantitation of *o,o'*-dityrosine using this approach is shown in Fig. 3.

Phenylalanine and p-Tyrosine. An aliquot of derivatized amino acids is diluted 1:100 (v/v) with ethyl acetate. A 1 μl sample is injected (1:100 split) into the gas chromatograph and analyzed on a 12 m DB-1 capillary

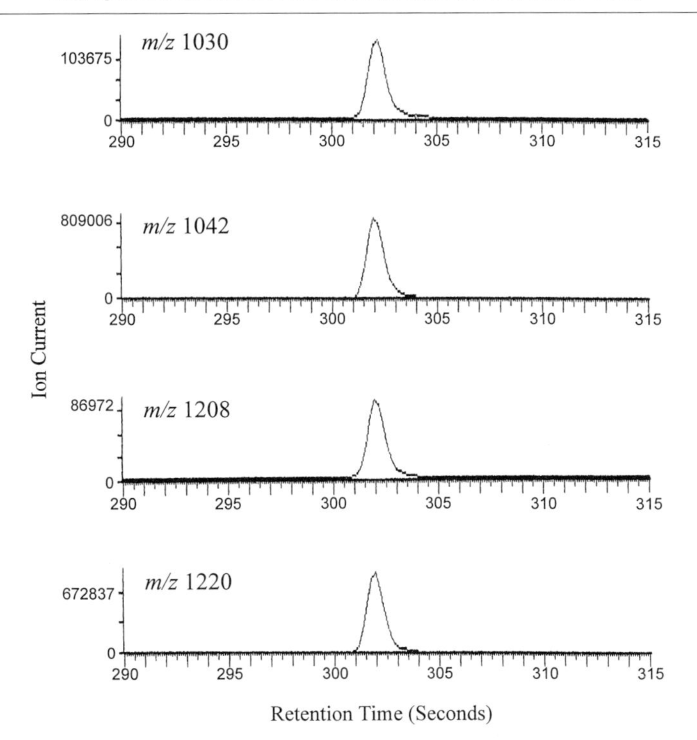

FIG. 3. Detection of the *n*-propyl-HFB derivative of *o,o'*-dityrosine in mitochondrial proteins by negative-ion electron capture GC/MS with selected ion monitoring. Rat heart mitochondrial proteins were homogenized, dialyzed, and delipidated. Following the addition of *o,o'*-[$^{13}C_{12}$]dityrosine as internal standard, the proteins were subjected to acid hydrolysis, and liberated amino acids were extracted with an ODS solid-phase column. The isolated amino acids were derivatized, and the *n*-propyl-HFB derivatives of the amino acids were analyzed by negative-ion electron capture GC/MS with selected ion monitoring. The major ions derived from *o,o'*-dityrosine (*m/z* 1208 and 1030; Fig. 2B) coelute with analogous ions from *o,o'*-[$^{13}C_{12}$]dityrosine (*m/z* 1220 and 1042).

column. The initial GC oven temperature of 120° is maintained for 1 min and then increased at a rate of 10°/min to a final temperature of 220°. The mass spectrum of the *n*-propyl-HFB derivative of phenylalanine includes a low abundance molecular ion (M^-) at *m/z* 403 and a prominent ion at *m/z* 383 ($M^- - HF$). Phenylalanine is quantified by monitoring the intensity of the ion at *m/z* 383 ion relative to that at *m/z* 389 for the derivatized [$^{13}C_6$]phenylalanine internal standard. The mass spectrum of the *n*-propyl-HFB derivative of *p*-tyrosine reveals prominent ions at *m/z* 595 ($M^- - HF$) and 417 ($M^- - CF_3(CF_2)_2CHO$). The ion at *m/z* 417 is used to quantify tyrosine relative to the ion at *m/z* 423 for the derivatized [$^{13}C_6$]tyrosine internal standard.

o,o′-Dityrosine, o-Tyrosine, and m-Tyrosine. The *n*-propyl-HFB derivatives of the amino acids are prepared, concentrated to dryness under nitrogen, and redissolved in 50 μl of ethyl acetate. Aliquots (1–2 μl) are then analyzed on a 12 m DB-1 capillary GC column after injection in the splitless mode. The initial GC oven temperature of 70° is increased at a rate of 40°/min to a final temperature of 300°. The mass spectrum of the *n*-propyl-HFB derivative of *o,o′*-dityrosine includes a low intensity molecular ion at *m/z* 1228 (M^-) and prominent ions at *m/z* 1208 (M^- − HF) and 1030 (M^- − $CF_3(CF_2)_2CHO$). The ion at *m/z* 1208 is used to quantify *o,o′*-dityrosine relative to the ion at *m/z* 1220 for the derivatized *o,o′*-[$^{13}C_{12}$]dityrosine internal standard. The mass spectrum of the *n*-propyl-HFB derivative of *o*-tyrosine exhibits prominent ions at *m/z* 595 (M^- − HF) and 417 (M^- − $CF_3(CF_2)_2CHO$). The ion at *m/z* 595 is used to quantitate this analyte relative to the ion at *m/z* 601 for the derivatized *o*-[$^{13}C_6$]tyrosine internal standard. The mass spectrum of the *n*-propyl-HFB derivative of *m*-tyrosine exhibits prominent ions at *m/z* 595 (M^- − HF) and 417 (M^- − $CF_3(CF_2)_2CHO$). The ion at *m/z* 417 is used to quantitate this analyte relative to the ion at *m/z* 423 for the derivatized *m*-[$^{13}C_6$]tyrosine internal standard.

3-Nitrotyrosine. The *n*-propyl-HFB derivatives of the amino acids are prepared, concentrated to dryness under nitrogen, and redissolved in 50 μl of ethyl acetate. Aliquots (1–2 μl) are then analyzed on a 12 m DB-1 capillary GC column. The mass spectrum of the *n*-propyl-HFB derivative of 3-nitrotyrosine includes prominent ions at *m/z* 464 (M^-) and *m/z* 444 (M^- − HF). The base ion (*m/z* 464) and the corresponding ion (*m/z* 470) from the derivatized 3-nitro[$^{13}C_6$]tyrosine internal standard are used for quantification. A splitless injection is typically used for analysis of 3-nitrotyrosine, though 1:10 split injections are sometimes possible and prolong column life. The initial column temperature of 70° is increased to 180° at 60°/min and then raised to 205° at 4°/min.

As with many compounds containing a nitro group or other polar moiety, the *n*-propyl-HFB derivative of 3-nitrotyrosine tends to yield tailing peaks on GC.[34,43] This may reflect the fact that the phenolic hydroxyl moiety is not fluoracylated in the derivative owing to stereoelctronic effects of the nitro group on the aromatic ring. We have found that the assay is improved by first reducing the nitro group to an amine (J. R. C. and J. W. H., unpublished observation). This permits facile fluoroacylation of the phenolic hydroxyl, the aromatic amino group, and the α-amino group of the compound. The resulting derivative is substantially more volatile than the mono-fluoroacylated derivative of 3-nitrotyrosine, and it exhibits excellent chromatographic properties on a GC column. Moreover, it yields a greater signal on negative-ion electron capture mass spectrometric analysis, reflecting the enhanced electron capturing properties attributable to its threefold higher fluoroacyl content. Using this approach, we have successfully measured the 3-nitrotyrosine content of acid hydrolyzates of a standard nitrated protein and native tissue proteins.

3-Chlorotyrosine. 3-Chlorotyrosine is analyzed as its *n*-propyl-per-HFB derivative. For quantitation, the intensities of analyte ions at *m/z* 451 ($M^- - CF_3(CF_2)_2CHO$) and *m/z* 629 ($M^- - HF$) are compared to those of the analogous ions (*m/z* 457 and 635) from the derivatized 3-[$^{13}C_6$]chlorotyrosine internal standard. GC analyses are performed on a 30 m DB-17 capillary column. Samples are diluted 50-fold with ethyl acetate and injected using a 1:50 split. The initial GC oven temperature of 150° is increased at a rate of 20°/min to a final temperature of 250°.

Specificity of Products as Markers of Oxidative Damage by Different Reaction Pathways

The goal of our research is to identify the molecular mechanisms of oxidative injury that trigger human disease. We have a particular interest in atherosclerosis because many lines of evidence indicate that oxidation renders low density lipoprotein (LDL) atherogenic.[2,30,44] *In vitro*, myeloperoxidase, metal ions, hydroxyl radical and reactive nitrogen species all promote LDL oxidation,[2] but the physiologically relevant pathways are not established.[44] We selected *o,o'*-dityrosine, *o*-tyrosine, *m*-tyrosine, 3-nitrotyrosine, and 3-chlorotyrosine as markers for oxidative injury because they might represent unnatural amino acids generated by each of these oxidation systems.[15,22,25] To evaluate the potential utility of these compounds as markers of oxidative reactions that occur *in vivo*, we have performed *in vitro* studies to establish the product yields of each of the amino acids in LDL and model proteins exposed to different oxidation systems.

Myeloperoxidase. Activated phagocytes use an NADPH oxidase to produce superoxide, which dismutates to H_2O_2.[7,8] The H_2O_2 serves as an oxidizing substrate for the heme protein myeloperoxidase, which the phagocytes also secrete. Myeloperoxidase greatly amplifies the toxic potential of H_2O_2, generating potent microbicidal and cytotoxic oxidants.[7,8,11] The NADPH oxidase plays a critical role in host defenses against invading pathogens. The ability of phagocytes to generate oxidants has a deleterious aspect, however, because reactive species can damage tissue at sites of inflammation. Indeed, many lines of evidence implicate this process in the pathogenesis of diseases including atherosclerosis, ischemia–reperfusion injury, and cancer.[1,7,9,44]

[44] J. W. Heinecke, *Curr. Opin. Lipid.* **8,** 268 (1997).

Myeloperoxidase cannot directly damage large macromolecules because its active site is buried in a hydrophobic cleft.[8,19] Instead, myeloperoxidase relies on low molecular weight intermediates to convey oxidizing equivalents from its heme group to the target for damage. One such intermediate is the long-lived tyrosyl radical,[17] which the enzyme generates from the phenolic amino acid tyrosine.[12,19] The productive interaction of two tyrosyl radicals yields *o,o'*-dityrosine, an intensely fluorescent compound.[45]

Tyrosyl radical generated by myeloperoxidase promotes peroxidation of LDL lipids[46]; it also cross-links protein tyrosine residues into *o,o'*-dityrosine.[19,37] When we exposed BSA or LDL to tyrosyl radical generated by the myeloperoxidase–H_2O_2–tyrosine system, there was a dramatic, selective increase in the *o,o'*-dityrosine content of the proteins.[15,19] In contrast, there was no increase in the levels of *o*-tyrosine or *m*-tyrosine.[15]

Myeloperoxidase is the only human enzyme known to generate the potent oxidation HOCl at concentrations of chloride which occur in plasma.[11,23] The detection of chlorinated molecules *in vivo* would therefore constitute presumptive evidence that myeloperoxidase represents one pathway for oxidative damage. Most oxidation products generated by HOCl are either nonspecific or decompose to uninformative compounds.[20] However, studies have demonstrated that myeloperoxidase converts tyrosine into 3-chlorotyrosine,[21,47] a stable product that may serve as a molecular fingerprint of the action of the enzyme.

We detected significant amounts of 3-chlorotyrosine in LDL that had been exposed to the complete myeloperoxidase–H_2O_2–Cl^- system or to reagent HOCl.[22] In contrast, there was little increase in 3-chlorotyrosine of LDL oxidized with a variety of other oxidation systems. The systems we examined included copper or iron with molecular oxygen, which oxidize LDL lipids by poorly understood mechanisms[2]; copper plus H_2O_2, which generates hydroxyl radical or a species of similar reactivity[13]; H_2O_2 and chloride ion plus lactoperoxidase or horseradish peroxidase, which will directly oxidize proteins and lipids (these peroxidases, however, do not generate HOCl); glucose, which autoxidizes with the generation of dicarbonyls and promotes glycoxidation[48]; and soybean lipoxygenase plus phospholipase A_2 and molecular oxygen, which generates high levels of peroxidized polyunsaturated fatty acids.[49] These results demonstrate that 3-chlorotyrosine is a highly specific marker of LDL oxidation by the myeloperoxidase–

[45] S. O. Anderson, *Acta. Physiol. Scand.* **66,** 1 (1966).

[46] M. I. Savenkova, D. M. Mueller, and J. W. Heinecke, *J. Biol. Chem.* **269,** 20394 (1994).

[47] N. M. Domigan, T. S. Charlton, M. W. Duncan, C. C. Winterbourn, and A. J. Kettle, *J. Biol. Chem.* **270,** 16542 (1995).

[48] J. W. Baynes, *Diabetes* **40,** 405 (1991).

[49] C. P. Sparrow, S. Parthasarathy, and D. Steinberg, *J. Lipid Res.* **29,** 745 (1988).

H_2O_2–Cl^- system and is not produced by a variety of other oxidation systems.[22]

Metal Ions. One important pathway for LDL oxidation *in vitro* requires metal ions.[2] The oxidizing intermediates involved in this reaction have not been conclusively identified. LDL oxidation by cultured arterial wall smooth muscle cells requires micromolar concentrations of iron or copper,[50] and metal chelators block LDL oxidation by most other types of cells.[51,52] Moreover, high concentrations of free metal ions oxidize LDL even in the absence of cells.[50,52]

LDL oxidized by copper in the presence of molecular oxygen exhibited large increases in contents of *o*-tyrosine and *m*-tyrosine compared to native LDL that had not been exposed to this oxidizing system.[15] In contrast, there was little change in the level of *o,o'*-dityrosine in LDL under these conditions. These results indicate that LDL oxidized by copper *in vitro* exhibits a selective increase in its content of *o*-tyrosine and *m*-tyrosine. In contrast, there was little change in the *o*-tyrosine and *m*-tyrosine content of BSA exposed to copper alone, strongly suggesting that LDL lipids are an important substrate for oxidation by copper.

Hydroxyl Radical. Incubation of copper with high concentrations of H_2O_2 generates hydroxyl radical,[13] an extremely potent oxidant that reacts at a diffusion-controlled rate with a wide variety of biological molecules.[1,3] Systems which generate hydroxyl radical are widely studied, although it is uncertain whether the free metal ions required to generate this oxidant exist *in vivo*. When LDL was exposed to the hydroxyl radical-generating copper–H_2O_2 system, there were large increases in the content of *o*-tyrosine, *m*-tyrosine, and *o,o'*-dityrosine.[15] Levels of *o,o'*-dityrosine in different model proteins exposed to this system varied, but there was a striking increase in *o*-tyrosine in all cases examined.[15,27,28] Huggins *et al.* reported similar results when they oxidized ribonuclease and lysozyme with hydroxyl radical-generating systems involving metal ions or ionizing radiation.[13] Thus, LDL oxidized by two different types of metal-catalyzed reactions exhibits clear increases in *o*-tyrosine and *m*-tyrosine but a variable increase in *o,o'*-dityrosine. Collectively, these results indicate that LDL and model proteins damaged *in vitro* by three distinct oxidizing systems (free copper, copper and H_2O_2, and the myeloperoxidase–H_2O_2–tyrosine system) yield distinguishable patterns of formation of phenylalanine and tyrosine oxidation products.

[50] J. W. Heinecke, H. Rosen, and A. Chait, *J. Clin. Invest.* **74,** 1890 (1984).

[51] D. W. Morjel, P. E. DiCorletto, and G. M. Chisolm, *Arterio.* **4**, 357 (1984).

[52] U. P. Steinbrecher, S. Parthasarathy, D. S. Leade, J. L. Witztum, and D. Steinberg, *Proc. Natl. Acad. Sci. USA* **81,** 3883 (1984).

3-Nitrotyrosine. Nitric oxide (NO) is a relatively stable free radical produced by nitric oxide synthase expressed in endothelial and other cells.[4,6] NO plays a central role in the regulation of vasomotor tone. NO generated by the inducible isoform of nitric oxide synthase contributes to inflammation. *In vitro*, NO reacts with superoxide ($O_2^{\cdot-}$) to form the reactive nitrogen species peroxynitrite ($ONOO^-$)[24]:

$$NO + O_2^{\cdot-} \rightarrow ONOO^-$$

In vitro studies demonstrate that $ONOO^-$ (or species derived from $ONOO^-$ such as ONOOH) promotes oxidation of proteins, nucleic acids, and lipids and is a potent cytotoxin.[4] Murine phagocytes generate high levels of NO, which reacts to form intermediates that appear to be important in certain types of microbicidal activity.[53] Among the protein oxidation products formed by $ONOO^-$ is 3-nitrotyrosine.[4,24]

To evaluate the specificity of 3-nitrotyrosine as a marker for protein modification by reactive nitrogen species, we examined a variety of *in vitro* oxidation systems for their ability to generate 3-nitrotyrosine in LDL.[25] Significant levels of 3-nitrotyrosine were formed when we exposed LDL to reagent $ONOO^-$. In contrast, there was little change in 3-nitrotyrosine content of LDL oxidized by free copper ions, by free iron ions, by a hydroxyl radical-generating system (H_2O_2 plus copper), by lactoperoxidase or horseradish peroxidase and H_2O_2, by glucose, or by lipoxygenase together with phospholipase A_2. Collectively, these results indicate that 3-nitrotyrosine is a highly specific marker for LDL oxidation by reactive nitrogen species derived from $ONOO^-$.[25]

Application of Isotope Dilution GC/MS to Study of Oxidative Pathways in Human Atherosclerosis

LDL oxidation may play a critical role in atherogenesis, but the mechanisms by which LDL are oxidized *in vivo* have not yet been identified. We have used the isotope dilution GC/MS methods outlined above to investigate the mechanisms that promote LDL oxidation in the human artery wall.[15,22,25]

Myeloperoxidase. Catalytically active myeloperoxidase is a component of human atherosclerotic tissue, where it colocalizes with foamy macrophages in the cell-rich regions of lesions.[10] The enzyme is a potent catalyst for LDL oxidation *in vitro*, and oxidation products of the enzyme have

[53] D. L. Granger, J. B. Hibbs, J. R. Perfect, and D. T. Durack, *J. Clin. Invest.* **81,** 1129 (1988).

been detected by immunohistochemistry in atherosclerotic vascular lesions, suggesting that myeloperoxidase promotes LDL oxidation *in vivo*.[54]

To explore this hypothesis, we used isotope dilution GC/MS to quantify 3-chlorotyrosine levels in atherosclerotic lesions.[22] In vascular tissue freshly harvested at surgery, the 3-chlorotyrosine level was six times higher in atherosclerotic lesions than in normal aortic tissue. Moreover, the content of 3-chlorotyrosine (per mol of its precursor amino acid tyrosine) was 30 times higher in LDL isolated from atherosclerotic lesions than in circulating LDL. These results suggest that chlorination by myeloperoxidase constitutes one mechanism for protein oxidation *in vivo*. They also suggest that myeloperoxidase may play a role in converting LDL into an atherogenic form.

Tyrosyl radical is another reactive intermediate generated by myeloperoxidase. To explore the role of tyrosyl radical in promoting LDL oxidation *in vivo*, we used isotope dilution GC/MS to quantify the level of *o,o'*-dityrosine in lesion LDL and human atherosclerotic tissue.[15] We detected a remarkable 100-fold increase in *o,o'*-dityrosine levels in LDL isolated from atherosclerotic lesions compared with those in circulating LDL. Analysis of fatty streaks, the earliest lesion of atherosclerosis, revealed a similar pattern of oxidation products. Compared with normal tissue, there was a 10-fold increase in *o,o'*-dityrosine content of such lesions with no difference in either *o*-tyrosine or *m*-tyrosine. These results suggest that tyrosyl radical, which can be generated by myeloperoxidase, is one agent for LDL oxidation in the human artery wall.[15]

Metal Ions and Hydroxyl Radical. To explore the possible role of free metal ions in oxidizing LDL *in vivo*, we applied our GC/MS methods to detect products in human vascular tissue that can be formed by metal ion catalyzed oxidation.[15] Such products include *m*-tyrosine and *o*-tyrosine. We were unable to demonstrate elevated levels of either compound in tissue proteins contained in fatty streak or intermediate atherosclerotic lesions. Moreover, LDL isolated from human atherosclerotic lesions contained no more *o*-tyrosine or *m*-tyrosine than did circulating LDL. These observations suggest that metal ions are unlikely to serve as catalysts of LDL oxidation early in atherogenesis.

In contrast to fatty streaks and lesion LDL, we did observe a trend toward higher levels of *o*-tyrosine and *m*-tyrosine in advanced atherosclerotic lesions.[15] Perhaps cellular dissolution, a prominent feature of the complex atherosclerotic lesion, promotes the release or activity of free metal ions at this late stage of the disease process.

[54] L. J. Hazell, L. Arnold, D. Flowers, G. Waeg, E. Malle, and R. Stocker, *J. Clin. Invest.* **97,** 1535 (1996).

Reactive Nitrogen Species. Immunohistochemical studies have detected 3-nitrotyrosine in human atherosclerotic lesions,[55] suggesting that reactive nitrogen species also may promote LDL oxidation. To explore oxidation products derived from reactive nitrogen species *in vivo*, we quantified the level of 3-nitrotyrosine by isotope dilution GC/MS in LDL isolated from human atherosclerotic lesions.[25] There was a striking 90-fold increase in the 3-nitrotyrosine content of lesion LDL per mol of tyrosine when compared with circulating LDL. These observations raise the possibility that nitric oxide may promote atherogenesis under certain conditions by virtue of its ability to form reactive nitrogen species.

It is noteworthy that myeloperoxidase generates nitrating species in the presence of nitrite and generates chlorinating species in the presence of chloride.[5,56] The fact that elevated levels of 3-nitrotyrosine and 3-chlorotyrosine are observed in LDL isolated from atherosclerotic tissue raises the possibility that myeloperoxidase contributes both to nitration and chlorination reactions in the inflammatory milieu of the evolving atherosclerotic lesion.[22,25]

Conclusions

The ability to quantify low levels of distinct amino acid oxidation products in proteins and tissues enables one to explore the roles of various pathways in the diseases which may be initiated or propagated by oxidative injury. The specificity, sensitivity, and precision of isotope dilution GC/MS to identify products of specific oxidation pathways makes this approach a remarkably powerful technique for this purpose. Development of further mass spectrometric methods for detecting specific molecules *in vivo* should provide important insights into the involvement of particular oxidative pathways in the onset and progression of a variety of human diseases.

[55] S. J. Beckman, Y. Z. Ye, P. G. Anderson, J. Chen, M.-A. Accaviti, M. M. Tarpey, and R. White, *Biol. Chem. Hoppe-Seyler* **375,** 81 (1994).

[56] A. van der Vliet, J. P. Eiserich, B. Halliwell, and C. E. Cross, *J. Biol. Chem.* **272,** 7617 (1997).

[17] Measurement of Protein Carbonyls in Human Brain Tissue

By PATRICIA EVANS, LEONIDAS LYRAS, and BARRY HALLIWELL

Introduction

Oxidative modification of proteins *in vivo* may affect a variety of cellular functions involving proteins: receptors, signal transduction mechanisms, transport systems, and enzymes. It could also contribute to secondary damage to other biomolecules, for instance, inactivation of DNA repair enzymes, loss of fidelity of DNA polymerases in replicating DNA,[1] and the development of new antigens provoking autoimmune responses. The chemical reactions resulting from attack of reactive oxygen and nitrogen species (ROS/RNS) on proteins are complex[2] and lead to a variety of products, many as yet uncharacterized. With 20 amino acids present in proteins and a number of possible oxidizing species, the range of products is likely to be enormous. The carbonyl assay has been developed as a general assay of oxidative protein damage to assess steady-state protein damage in animal tissues and body fluids.[3,4] The assay is based on the fact that several ROS can attack amino acid residues in proteins (particularly histidine, arginine, lysine, and proline) to produce carbonyl functions[5] that can react with 2,4-dinitrophenylhydrazine (DNPH) to generate chromophoric dinitrophenylhydrazones. This reaction can therefore be used to estimate the carbonyl content of proteins in human tissues and body fluids. Western blotting assays based on the use of anti-dinitrophenol antibodies have also been developed to identify oxidatively damaged proteins in tissues and body fluids[6] after their extraction and derivatization with DNPH.

The carbonyl assay has become widely used and many laboratories have developed individual protocols for it.[3,4,7] Sometimes the assay procedures used are not precisely specified, and when they are, they may differ from

[1] H. Wiseman and B. Halliwell, *Biochem. J.* **313,** 17 (1996).

[2] R. T. Dean, S. Fu, R. Stocker, and M. J. Davies, *Biochem. J.* **324,** 1 (1997).

[3] R. L. Levine, D. Garland, C. N. Oliver, A. Amici, I. Climent, A. G. Lenz, B. W. Ahn, S. Shaltiel, and E. R. Stadtman, Methods in Enzymology, Vol. 186, p. 464 (1990).

[4] R. L. Levine, J. A. Williams, E. R. Stadtman, and E. Schacter, Methods in Enzymology, Vol. 233, p. 346 (1994).

[5] A. Amici, R. L. Levine, L. Tsai, and E. R. Stadtman. *J. Biol. Chem.* **264,** 3341 (1989).

[6] J. Keller, N. C. Halmes, J. A. Hinson, and N. R. Pumford, *Chem. Res. Toxicol.* **6,** 430 (1993).

[7] A. Z. Reznick and L. Packer, Methods in Enzymology, Vol. 233, p. 357 (1994).

0076-6879/99 $30.00

those originally proposed by the group of Stadtman *et al.* (e.g., refs. 8–10). This point is important, because there is considerable variation in the baseline levels of protein carbonyls in certain tissues, depending on how the assay is performed.[11,12] For example, levels reported for human brain cortex range from 1.5 to 6.4 nmol/mg protein (Table I).[13–15] By contrast, broadly comparable values for protein carbonyls (<1 nmol/mg protein) have been reported by several laboratories assaying human plasma (reviewed in ref. 16).

To investigate the reasons for the wide variation in reported values in human brain cortex, we analyzed postmortem cortex samples from patients suffering from amyotrophic lateral sclerosis (ALS) or progressive muscular atrophy (PMA, both diseases commonly known as motor neuron disease, MND), together with 10 normal controls (NC) and 11 disease controls (DC) from individuals suffering from other neurodegenerative disorders not affecting the cortex. This disease was chosen for study since there is evidence for the involvement of a mutant superoxide dismutase (SOD) in one subtype of the familial disease[17,18] which could lead to oxidative stress in the cortex and consequent protein damage. There have also been two reports of elevated protein carbonyls in brain or spinal cord from MND patients with normal SOD activity.[8,10]

In this study, we investigated the level of carbonyl groups in proteins extracted from these tissues by two different methods described below and use these methods to illustrate technical difficulties in the assays which can

[8] P. J. Shaw, P. G. Ince, G. Falkous, and D. Mantle, *Ann. Neurol.* **38,** 691 (1995).

[9] C. N. Oliver, B. A. Ahn, E. J. Moermann, S. Goldstein, and E. R. Stadtman, *J. Biol. Chem.* **262,** 5488 (1987).

[10] A. C. Bowling, J. B. Schultz, R. H. Brown, Jr., and M. F. Beal, *J. Neurochem.* **61,** 2322 (1993).

[11] G. Cao and R. G. Cutler, *Arch. Biochem. Biophys.* **320,** 106 (1995).

[12] L. Lyras, P. J. Evans, P. J. Shaw, and B. Halliwell, *Free Rad. Res.* **24,** 397 (1996).

[13] C. D. Smith, J. M. Carney, P. E. Stark-Reed, C. N. Oliver, E. R. Stadtman, R. A. Floyd, and W. A. Markesbury, *Proc. Natl. Acad. Sci. USA* **88,** 10540 (1991).

[14] R. E. Stafford, I. T. Mak, J. H. Kramer, and W. B. Weglicki, *Biochem. Biophys. Res. Commun.* **196,** 596 (1993).

[15] C. N. Oliver, P. E. Stark-Reed, E. R. Stadtman, G. J. Liu, J. M. Carney, and R. A. Floyd, *Proc. Natl. Acad. Sci. USA* **87,** 5144 (1990).

[16] B. Halliwell, *Free Rad. Res.* **25,** 57 (1996).

[17] D. R. Rosen, T. Siddique, D. Paterson, D. A. Figlewicz, P. Sapp, A. Hentati, D. Donaldson, J. Goto, J. P. O'Regan, H.-Z. Deng, Z. Rahmani, A. Krizus, D. McKenna-Yasek, A. Cayabyab, S. M. Gaston, R. Berger, R. E. Tanzi, J. J. Halperin, B. Herzfeldt, R. Van den Burgh, W.-Y. Hung, T. Bird, G. Deng, D. W. Mulder, C. Smyth, N. G. Laing, E. Soriano, M. A. Pericak-Vance, J. Haines, G. A. Rouleau, J. S. Gusella, H. R. Horvitz, and R. H. Brown, *Nature* **363,** 59 (1991).

[18] W. Robberecht, P. Sapp, M. K. Viaene, D. Rosen, D. McKenna-Yasek, J. Haines, R. Horvitz, P. Theys, and R. Brown Jnr., *J. Neurochem.* **62,** 383 (1994).

TABLE I
SELECTED LITERATURE VALUES FOR BRAIN PROTEIN CARBONYLS

Samples studied	Carbonyl value (nmol/mg)	Comments	Ref.
Human brain, frontal and occipital cortex	Alzheimer 4.51–7.14 Age-matched controls 4.00–6.43	Not clear whether DNPH reagent was added to protein in solution or after acid precipitation. Pierce bicinchoninic acid (BCA) method used to measure protein, not A_{280}	*a*
Human brain, frontal cortex	1.51 ± 0.80 controls 2.79 ± 1.13 MND	Method of Levine *et al.* (3) used with unstated "minor modifications"	*b*
Rat brain	1.6 ± 0.6 control 3.2 ± 0.8 Mg deficient	Method of Oliver *et al.* (9) used. DNPH reagent added to precipitated protein. Pierce BCA assay used to measure protein instead of A_{280}	*c*
Gerbil brain	~6.8 (control) 8.8–11.8 after 10 min ischemia–reperfusion	Method of Oliver *et al.* (9) used	*d*
Human lumbar spinal cord	0.73 ± 0.63 controls 1.60 ± 1.23 MND 0.85 ± 0.62 other neurological disease controls	Modified version of assay used in which protein is measured by Lowry method in initial supernatant	*e*

[a] C. D. Smith, J. M. Carney, P. E. Stark-Reed, C. N. Oliver, E. R. Stadtman, R. A. Floyd, and W. A. Markesbury, *Proc. Natl. Acad. Sci. USA* **88,** 10540 (1991).
[b] A. C. Bowling, J. B. Schultz, R. H. Brown Jnr., and M. F. Beal, *J. Neurochem.* **61,** 2322 (1993).
[c] R. E. Stafford, I. T. Mak, J. H. Kramer, and W. B. Weglicki, *Biochem. Biophys. Res. Commun.* **196,** 596 (1993).
[d] C. N. Oliver, P. E. Stark-Reed, E. R. Stadtman, G. J. Liu, J. M. Carney, and R. A. Floyd, *Proc. Natl. Acad. Sci. USA* **87,** 5144 (1990).
[e] P. J. Shaw, P. G. Ince, G. Falkous, and D. Mantle, *Ann. Neurol.* **38,** 691 (1995).

lead to variations in baseline levels of carbonyl groups. Both methods have been used in literature reports describing protein carbonyl levels in normal and diseased brain cortex and spinal cord.[8,10]

Materials and Methods

Reagents

Coomassie blue protein reagent is from Pierce (Chester, UK). All other reagents are of the highest quality available from Sigma Chemical Company, Poole, Dorset, UK.

Brain Tissue Sampling

Brain cortex samples were excised postmortem from 22 patients with MND (15 with ALS and 7 with PMA), 10 NC, and 11 DC. At autopsy, the brain is dissected as soon as possible to obtain tissues for rapid freezing. The brain stem is removed by a horizontal incision through the upper pons. The diencephalon is bisected and the cerebral hemispheres separated by incision in the sagittal plane. The left cerebral hemispheres are then further subdissected by slicing 1 cm coronal sections, after removing the mid-brain. The motor cortex is identified in the appropriate coronal slices and then the slices are sealed in polyethylene and snap-frozen by immersion in melting Arcton surrounded by a liquid nitrogen bath. They are stored at −80°C until required.

All samples were generously provided by the Brain Bank, Royal Victoria Infirmary, Newcastle, UK.

Patient Details

The patient groups comprise cases of motor neuron disease (MND, $n = 22$), normal controls ($n = 10$), and disease controls ($n = 11$). The MND group consists of 11 males and 11 females whose mean age was 61.5 ± 13.2 years (mean ± SD, range 40–89 years). Mean delay from death to freezing of tissue was 17.5 ± 8.7 hours (mean ± SD, range 9–48 hours). The normal control group comprises 8 males and 2 females whose mean age was 69.1 ± 9.3 years (mean ± SD, range 54–82 years). The mean delay from death to freezing of tissue for this group was 18.4 ± 6.7 hours (mean ± SD, range 10–31 hours). The causes of death in the normal control group were as follows: ischemic heart disease (5), respiratory failure (2), and carcinoma of the lung (1), rectum (1), or uterus (1).

The disease control group consists of patients with neurodegenerative disorders who had a similar mode of death and agonal status compared to the MND group. Further details are given in Lyras *et al.*[12] This group contains 5 males and 6 females, with a mean age of 65.6 ± 16.5 years (mean ± SD, range 35–85 years). The mean delay from death to freezing of tissue was 17.7 ± 7.5 hours (mean ± SD, range 9–34 hours).

Assay of Protein Carbonyls in Human Brain Extracts

Two different methods are carried out on the brain extracts. One set of measurements is made by the method of Reznick and Packer,[7] in which protein is reacted with DNPH before precipitation and carbonyl determination (method A). The samples are assayed in parallel as described by Levine *et al.*[3] (method B), in which protein is precipitated from the tissue homogenate and reacted with DNPH before being redissolved for carbonyl determination. We also measure the initial protein content of the homogenates using the Lowry assay, as this has also been used to measure protein concentration in protein carbonyl determinations.[8]

Method A. Brain tissue (200 mg) is homogenized in 2 ml of homogenizing buffer (100 m*M* KH_2PO_4–K_2HPO_4, pH 7.4, plus 0.1% (w/v) digitonin) in a 2 ml glass homogenizer. The homogenized tissue is transferred to a plastic tube, left for 15 min at room temperature, and then streptomycin sulfate solution (10% w/v) is added to a final concentration of 1% to precipitate any extracted DNA which could react with DNPH and contribute to the carbonyl level. The solution is mixed and left to stand a further 15 min at room temperature, after which it is centrifuged at 2800*g* for 10 min at room temperature. The supernatant is removed and 0.8 ml is divided equally between two 12 ml plastic centrifuge tubes with the remaining supernatant being reserved for assay by method B (stage 1). DNPH (1.6 ml, 10 m*M* in 2 *M* HCl) is added to one tube and 1.6 ml of 2 *M* HCl to the other tube (ratio of supernatant to DNPH solution should be 1 : 4, v/v). The tubes are then incubated for 1 h on a rotator at room temperature and then the protein is precipitated by adding an equal volume of 20% (w/v) trichloroacetic acid (TCA) to the tubes and leaving them for 15 min. The protein is spun down at 3400*g* (10 min, room temperature), the supernatant is discarded, and the pellet is washed with 1.5 ml of an ethyl acetate : ethanol mixture (1 : 1, v/v) to remove excess DNPH. This procedure is repeated three times. The final protein pellet is dissolved in 1.25 ml of 6 *M* guanidine hydrochloride and the absorbance of both solutions (DNPH and HCl) is measured at 370 nm from which the carbonyl content can be evaluated (carbonyl concentration in nmol/ml: $\Delta A_{370} \times 45.45$, where ΔA_{370} equals A_{370} of DNPH solution $- A_{370}$ of HCl solution). The protein concentration is calculated from the A_{280} of the HCl samples ($A_{280} \times 1.8$ gives protein concentration in mg/ml). Carbonyl values can then be calculated as in ref. 7:

Carbonyl concentration (nmol/mg of protein) = nmol/ml of carbonyl groups/protein concentration in mg/ml

The protein concentration of some of the final protein solutions is also determined by Coomassie Plus Protein Assay reagent[19] as described later.

Method B. From the remaining supernatant (stage 1, method A), 0.8 ml is divided equally into two 1.5 ml Eppendorf tubes and an equal volume of 20% TCA is added to both tubes to precipitate the protein. After 15 min standing at room temperature the protein is spun down at 2800*g* (10 min, room temperature) and the supernatant carefully removed. DNPH (0.5 ml, 10 m*M* in 2 *M* HCl) is added to the precipitate in one tube and 0.5 ml of 2 *M* HCl to the other tube. After mixing, the tubes are left

[19] M. Marcart and L. Gerbaut, *Clin. Chim. Acta* **122,** 93 (1982).

standing for 1 hr at room temperature and then a further 0.5 ml of 20% TCA is added to each tube. The protein is spun down at 3400*g* (10 min, room temperature), the supernatant is discarded, and the pellet is washed three times with 1.5 ml of ethyl acetate : ethanol (1 : 1 v/v) to remove excess DNPH. The remainder of the assay is performed as described in method A.

Measurement of Protein Concentration

Apart from measurement of protein concentration by A_{280} used at the end of carbonyl determination, protein concentration is also determined by two other methods, since this is obviously a critical measurement in estimating protein carbonyls.

In the Lowry assay,[20] brain homogenates are diluted in water to contain 25–500 mg/ml of protein (usually a 1 : 50 dilution) and assayed according to the standard procedure using bovine serum albumin as the standard protein.

Lowry Protein Assay. The following solutions are prepared:

Solution A: 2% (w/v) Na_2CO_3 in 0.1 *M* NaOH
Solution B: 0.5% (w/v) $CuSO_4 \cdot 5H_2O$ in 1% (w/v) trisodium citrate
Solution C: 50 ml of solution A is mixed with 1 ml of solution B
Solution D: Folin and Ciocalteau's reagent diluted with water to make it 1 *M* in acid (a 2-fold dilution is usually required)

One ml of solution C is added to 0.2 ml of sample and then the solutions are mixed and left at room temperature for 10 min. Solution D (0.1 ml) is then added, the solution immediately mixed and then left for 30 min after which the absorbance is read at 750 nm. An accurately prepared bovine serum albumin solution is used to construct a standard curve.

Coomassie Plus Protein Assay Reagent. Other protein assays are performed using Coomassie Plus Protein Reagent, an assay[16] which is based on the absorbance shift from 465 to 595 nm that occurs when Coomassie Blue G-250 binds to proteins in acidic solution. The standard assay procedure is given below.

A range of known protein concentrations is prepared by diluting a stock bovine serum albumin (BSA) standard solution in the same diluent as the protein sample whose concentration is to be determined (in this case 6 *M* guanidine hydrochloride). The protein standard series should cover the range of concentration between 75 and 1500 μg/ml. Diluted standard or unknown protein sample (0.1 ml) is pipetted into a test tube, and 3.0 ml of Protein Assay Reagent is added and the solution mixed well. Absorbance is then read against the blank (sample diluent) at 595 nm.

[20] O. H. Lowry, N. J. Rosebrough, A. L. Farr, and R. J. Randall, *J. Biol. Chem.* **193,** 265 (1951).

As everything is done in duplicate, a standard curve is prepared by plotting the average net absorbance at 595 nm for each diluted protein standard. Using the standard curve, the protein concentration for each unknown protein sample is determined.

Results

Protein Carbonyls in Human Brain Cortex

Figure 1 summarizes the results obtained by performing two different types of carbonyl assay (as detailed above, methods A and B) on 22 patients with MND, 10 NC, and 11 DC. It can be seen that the levels of protein carbonyls in cortex vary over a wide range depending on which assay procedure is used. Table II summarizes the mean and SD values. Method B* in Table II indicates the results obtained when protein carbonyl values are calculated using protein concentrations obtained using the Lowry protein assay on the original homogenate.

Timing of Protein Concentration Measurement

One approach is to use the A_{280} of the final protein solution in guanidine hydrochloride to calculate protein concentrations, but sometimes protein carbonyl values have been based on the protein content of the initial brain homogenate, e.g., as determined by the Lowry method[20] (see also Table II, B*). We found that it was also possible to determine protein content on the final solution using the Coomassie blue method. The results obtained correlated well with those obtained by A_{280} (Fig. 2), which suggests that the A_{280} measurement is a reliable way of determining final protein content. Interestingly, however, protein levels as determined by Coomassie blue (using a calibration curve of BSA in guanidine hydrochloride) were reproducibly less (50%) than those calculated from the A_{280} values. However, when the results of the final A_{280} protein or Coomassie blue determinations were compared with those of the Lowry measurements on the initial homogenate, the correlation was poor (Fig. 3). Guanidine hydrochloride interferes with the Lowry assay and so cannot be used to assess protein concentration at the final stage. We conclude, in agreement with Reznick and Packer,[7] that losses of protein are variable during the extraction and washing procedures, i.e., it is questionable to calculate protein carbonyl values on the basis of protein determinations in the initial tissue homogenate.

Discussion

The carbonyl assay is a widely used assay to measure oxidative protein damage, but some authors have criticized it (e.g., ref. 11). However, the

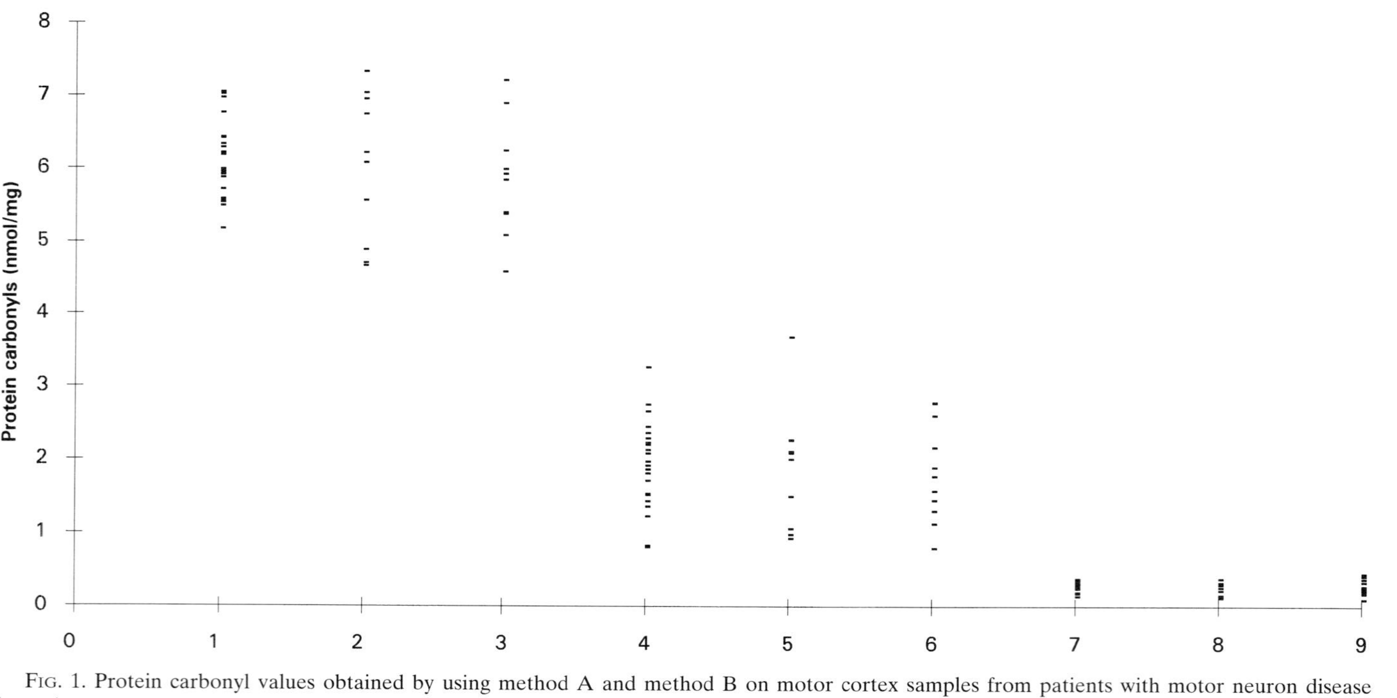

FIG. 1. Protein carbonyl values obtained by using method A and method B on motor cortex samples from patients with motor neuron disease (MND), normal controls (NC), and patients with other neurodegenerative diseases (DC). Key: method A, 1 MND, 2 NC, 3 DC; method B, 4 MND, 5 NC, 6 DC; method B, but carbonyl calculation based upon protein concentration of initial brain homogenate as determined by the Lowry method, 7 MND, 8 NC, 9 DC.

TABLE II
HUMAN BRAIN PROTEIN CARBONYLS IN MOTOR NEURON DISEASE (MND), NORMAL CONTROL (NC), AND DISEASE CONTROL (DC) SUBJECTS MEASURED BY DIFFERENT PROCEDURES

Method used	Carbonyls (nmol/mg protein) mean ± SD		
	MND	NC	DC
A	6.07 ± 0.56	6.02 ± 1.02	5.81 ± 0.77
B	1.93 ± 0.61	1.89 ± 0.83	1.84 ± 0.67
B*[a]	0.28 ± 0.06	0.25 ± 0.09	0.28 ± 0.11

[a] (*) based on protein measurement using Lowry assay in initial brain homogenate.

data in Table I suggests that more comparison and standardization of different methods used by various laboratories are required.

Protein Determination in the Carbonyl Assay

Carbonyl contents of proteins are expressed as nmol/mg protein. Therefore, it is necessary to determine the protein concentration of the samples. This can be done either on the initial supernatant by using a protein assay (e.g., Pierce BCA or Lowry [Table I]) or on the final guanidine hydrochlo-

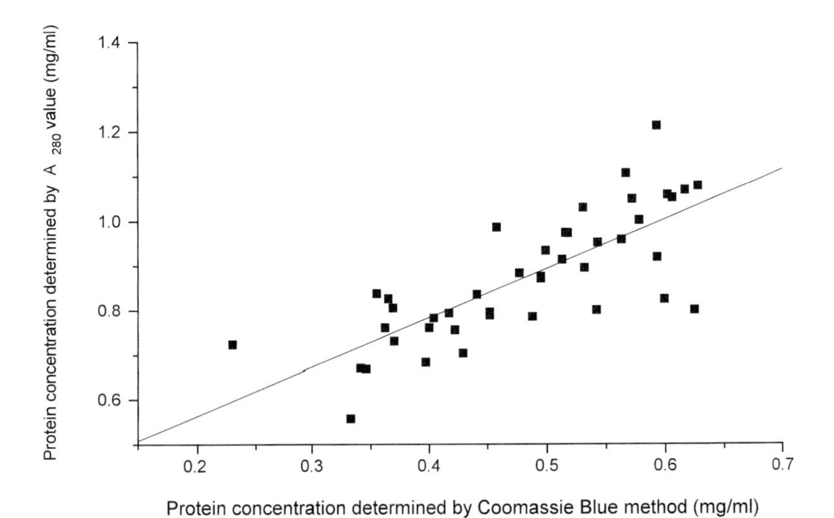

FIG. 2. Correlation graph of MND, NC, and DC brain cortex protein concentrations determined (on final solutions) by A_{280} and Coomassie blue assay on protein samples obtained from method A ($r = 0.77$, $n = 43$).

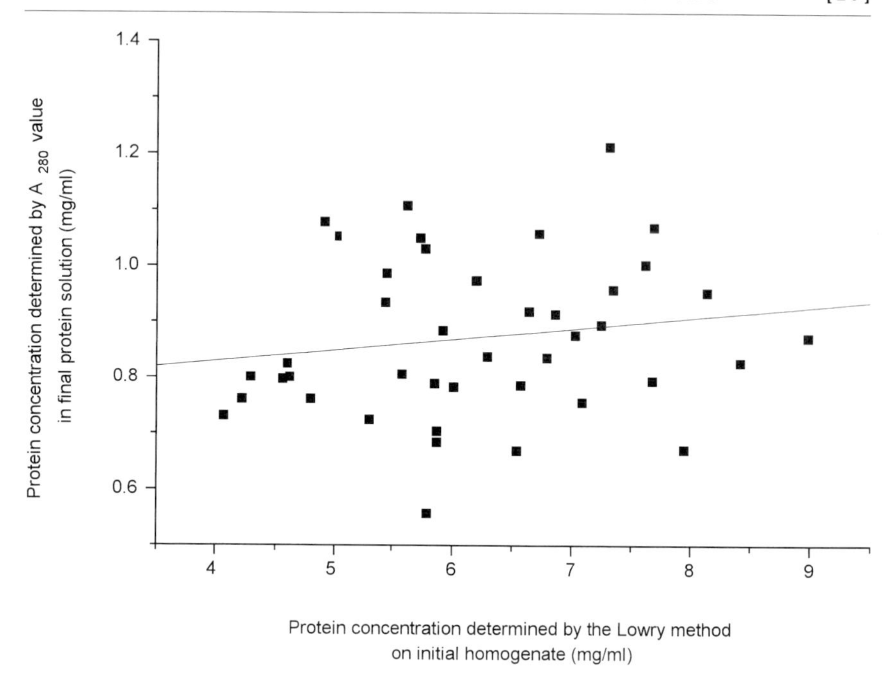

FIG. 3. Correlation graph between MND, NC, and DC brain cortex protein concentrations determined by Lowry assay on initial tissue homogenate and A_{280} values of the final protein solution ($r = 0.165$, $n = 43$).

ride solution by measuring A_{280} or by the Coomassie blue method used here. A_{280} measurements might be expected to be prone to interference by the many biologically important substances other than proteins which absorb at this wavelength. However, we would expect such substances to be removed by the multiple washing procedures. The Coomassie blue method applied to the final protein solutions gave results that correlated well with A_{280}, although they were about 50% lower. We do not know the reason for this, but the various protein assays measure different parameters of protein structure.

Whereas Levine *et al.*[3] state that protein recovery is generally excellent throughout the carbonyl assay procedure, our comparison of protein determinations in the initial supernatant with A_{280} values suggests that considerable and variable protein loss occurs (at least with brain samples); the several washing stages with ethyl acetate/ethanol may be particularly prone to variable losses of protein. Reznick and Packer[7] state that about 10–15% of protein is lost in the various washing steps and they recommend that protein levels be determined in the final protein solutions (by the A_{280}

value), after all washings are finished. Our data support this view and suggest that use of A_{280} value to determine protein concentration of the final protein solution is preferable to calculating carbonyl values based upon measuring protein concentration in the initial supernatant.

Precipitation of Protein before Addition of DNPH Reagent

In the method described in ref. 3 and our method B, protein is dried or precipitated from solution and DNPH reagent is added to a protein pellet. This is done when dealing with low amounts of protein in order to concentrate the proteins. By contrast, Reznick and Packer[7] use larger amounts of proteins and DNPH is reacted with soluble proteins—our method A. Our data (Fig. 1) show that precipitating the protein before addition of DNPH (method B), alters the results significantly compared to omitting the precipitation step. However, we found that the protein concentrations estimated by A_{280} were very similar and highly correlated for both methods (Fig. 4). This suggests that the difference in carbonyl values may relate to incomplete reaction with DNPH when the protein is in the solid form. We recommend reaction of the protein with DNPH in solution whenever possible.

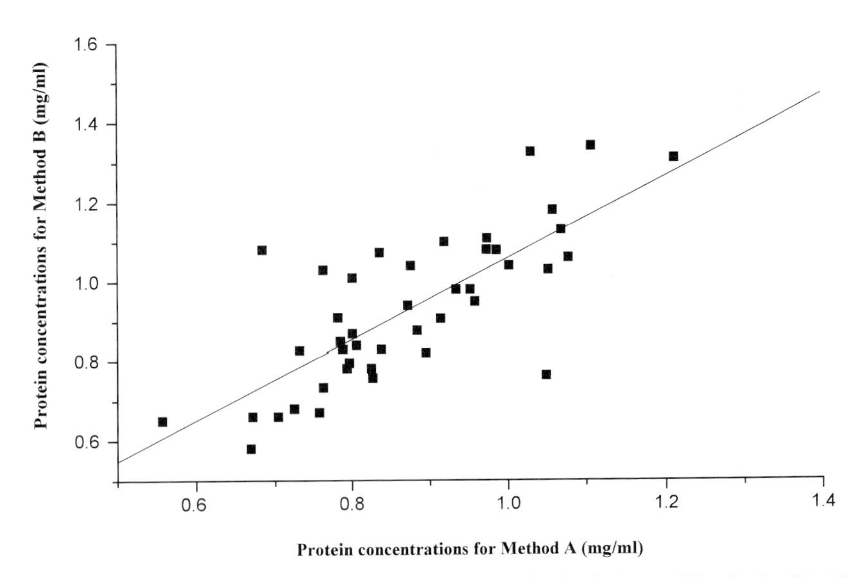

FIG. 4. Protein concentrations as determined by A_{280} in the final guanidine hydrochloride solution of brain samples from MND, NC, and DC patients. A comparison of method A and method B ($r = 0.76$, $n = 43$).

The protein carbonyl assay detects oxidation of several amino acid side chains to carbonyls, by a variety of reactive oxygen species, however generated,[2] and has been widely used to study oxidative protein damage in brain. The procedure we use to measure carbonyls removes lipids and nucleic acids that can interfere with the assay.

Our investigation into methods used and cited in the literature reveals the use of a variety of assays, sometimes incompletely described (Table I). These lead not only to differences in baseline levels of protein carbonyls observed in brain tissue, but also to differences between normal and diseased tissue, which seem dependent on the method of assay used (compare refs. 8 and 12). As with other assays of oxidative damage, there is a need to standardize procedures so that values observed in tissues are directly comparable.

[18] 8-Hydroxydeoxyguanosine and 8-Hydroxyguanine as Biomarkers of Oxidative DNA Damage

By HAROLD J. HELBOCK, KENNETH B. BECKMAN, and BRUCE N. AMES

Introduction

Oxidative damage to DNA is postulated to be an important contributor to aging and the degenerative diseases of aging, including cancer.[1,2] 8-Oxo-2′-deoxyguanosine (oxo^8dG) and its corresponding base (oxo^8Gua) are particularly useful markers of oxidative DNA damage because they represent approximately 5%[3] of the total oxidized bases that are known to occur in DNA, and because they can be measured with excellent sensitivity using electrochemical (EC) detection.[4] Initially, an high-performance liquid chromatography (HPLC)–EC assay was used to detect oxo^8dG in DNA hydrolyzates as a measure of steady-state DNA oxidation *in situ*. Later, the development of a monoclonal antibody, specific for oxo^8dG and oxo^8Gua, allowed purification of these adducts from urine and tissue culture medium and

[1] B. N. Ames, M. K. Shigenaga, and T. M. Hagen, *Proc. Natl. Acad. Sci. USA* **90,** 7915–7922 (1993).

[2] K. B. Beckman and B. N. Ames. *Physiol Rev.* **78,** No. 2, 1998.

[3] M. Dizdaroglu *Mutat. Res.* **275,** 331–342 (1992).

[4] R. A. Floyd, M. S. West, K. L. Eneff, J. E. Schneider, P. K. Wong, D. T. Tingey, and W. E. Hogsett *Anal. Biochem.* **188,** 155–158 (1990).

0076-6879/99 $30.00

provided what was hoped would be an integrated, noninvasive measure of the rate of endogenous DNA damage.[5]

Modifications and alternatives to the original HPLC-EC methods have produced a burgeoning literature on oxidative DNA damage. At the same time, they have resulted in discordant results and disagreements about the most appropriate techniques.[6–18] Because of the apparent importance of DNA damage, and the interest in oxo^{8}dG and oxo^{8}Gua as biomarkers for this damage, it is important to identify the best available assay methods. We previously evaluated various published methods and investigated sources of artifacts.[19] Here we review the HPLC–EC methods that we find to be the most reliable, indicate the common pitfalls associated with these techniques, and briefly review uncertainties of interpretation.

DNA Hydrolyzates

Tissue Homogenization

Two to 5 grams of tissue are diced with a pair of scissors and homogenized in 10 ml of homogenization buffer with 5–6 strokes of a Teflon–glass homogenizer at 200 rpm. The homogenization buffer is made up of 0.32 m*M* sucrose, 5 m*M* $MgCl_2$, 10 m*M* Tris-HCl, pH 7.5, and 0.1 m*M* Desferal (deferoxamine mesylate, Sigma Chemicals, St. Louis, MO). Homogenates are

[5] E. M. Park, M. K. Shigenaga, P. Degan, T. S. Korn, J. W. Kitzler, C. M. Wehr, P. Kolachana, and B. N. Ames, *Proc. Natl. Acad. Sci. USA* **89,** 3375–3379 (1992).

[6] H. G. Claycamp, *Carcinogenesis* **13,** 1289–1292 (1992).

[7] K. Frenkel, Z. J. Zhong, H. C. Wei, J. Karkoszka, U. Patel, K. Rashid, M. Georgescu, and J. J. Solomon, *Anal. Biochem.* **196,** 126–136 (1991).

[8] G. Harris, S. Bashir, and P. G. Winyard, *Carcinogenesis* **15,** 411–413 (1994).

[9] L. Wang, K. Hirayasu, M. Ishizawa, and Y. Kobayashi, *Nucleic Acids Res.* **22,** 1774–1775 (1994).

[10] H. Wiseman, H. Kaur, and B. Halliwell, *Cancer Lett.* **93,** 113–120 (1995).

[11] S. Adachi, M. Zeisig, and L. Moller, *Carcinogenesis* **16,** 253–258 (1995).

[12] D. Nakae, Y. Mizumoto, E. Kobayashi, O. Noguchi, and Y. Konishi, *Cancer Lett.* **97,** 233–239 (1995).

[13] M. Nakajima, T. Takeuchi, and K. Morimoto, *Carcinogenesis* **17,** 787–791 (1996).

[14] K. E. Herbert, M. D. Evans, M. T. Finnegan, S. Farooq, N. Mistry, I. D. Podmore, P. Farmer, and J. Lunec, *Free Radic. Biol. Med.* **20,** 467–472 (1996).

[15] M. T. Finnegan, K. E. Herbert, M. D. Evans, and J. Lunec, *Biochem. Soc. Trans.* **23,** 430S (1995).

[16] A. Collins, J. Cadet, B. Epe, and C. Gedik, *Carcinogenesis* **18,** 1833–1836 (1997).

[17] H. Kasai, *Mutation Res.* **387,** 147–163 (1997).

[18] M. Pflaum, O. Will, and B. Epe, *Carcinogenesis* **18,** 2225–2231 (1997).

[19] H. Helbock, K. Beckman, M. Shigenaga, P. Walter, A. Woodall, H. Yeo, and B. N. Ames, *Proc. Natl. Acad. Sci. USA* **95,** 288–293 (1998).

then centrifuged for 10 minutes at 4°C and 900*g*. The supernatant is aspirated and the nuclei from 1 g of tissue are resuspended in homogenization buffer to a final volume of 1 ml. The sample can then be frozen with liquid nitrogen and stored at −80°C. Some workers prefer to perform a second homogenization of the resuspended nuclear pellet prior to the final centrifugation, resuspension, and further processing.

DNA Extraction Methods

Two primary methods of DNA extraction from tissues will be reviewed. Other methods have been described, but have been found to offer several disadvantages when reviewed in our laboratory.[19] The most recent technique is the application, to the oxo^{8}dG assay by Nakae *et al.*, of the chaotropic NaI DNA isolation method of Wang *et al.*[9,12] This method, described below, is now marketed as the DNA Extractor WB kit (Wako BioProducts, Richmond, VA). The phenol extraction method of Gupta[20] as modified by Shigenaga[21] will also be reviewed, with appropriate references to earlier editions of this series. Thereafter, we will review the methods for nucleobase extraction from urine, with particular reference to modifications introduced since Volume 234 of this series.

Chaotropic Sodium Iodide Method

This description is for liver homogenates but may easily be adapted to other tissues and cultured cells (Wako BioProducts, technical bulletins). Frozen samples are thawed on ice, mixing frequently by gentle inversion. All reagents are maintained on ice with the exception of the enzyme solution and the NaI solution, both of which must be just warm enough during use, about 20°C, to produce a clear solution. It is well to have the water bath preheated, as the initial lysis steps should be carried out without delay in order to minimize the effects of endogenous DNase.

Preparation of RNase. RNase is not provided with the kit and is prepared as follows. 300 μg of RNase A (Boehringer Mannheim, Germany) is dissolved in 1 ml of water. The solution is then boiled in a water bath for 20 minutes to destroy contaminating DNase and thereafter cooled to room temperature, aliquoted into 0.5 ml microcentrifuge tubes, and stored at −20°C. Fifteen μl of the RNase solution, when added as described below, will give 20–21 μg/ml final concentration.

[20] R. C. Gupta, *Proc. Natl. Acad. Sci. USA* **81,** 6943–6947 (1984).

[21] M. K. Shigenaga, E. N. Aboujaoude, Q. Chen, and B. N. Ames, Methods in Enzymology, Vol. 234, pp. 16–33 (1994).

DNA Extraction. Resuspended nuclear pellet, 0.5 to 1.0 ml, is pipetted into a 2 ml microcentrifuge tube to which is added 1 ml of "lysis buffer" [1% (w/v) Triton X-100, 0.32 m*M* saccharose, 5 m*M* $MgCl_2$, 10 m*M* Tris-HCl, pH 7.5].* This is vortexed at moderate speeds for 30 sec and then centrifuged for 20 sec at 10,000*g* and 4°. The supernatant is discarded and replaced with 1 ml of fresh lysis buffer. The sample is again mixed and then transferred to a 1.5 ml Eppendorf tube and centrifuged for 20 sec at 10,000*g* and 4°C. The supernatant is discarded and 200 μl of enzyme solution [SDS 1% (w/v), 5 m*M* EDTA, 10 m*M* Tris-HCl, pH 8.0] is added, followed by 15 μl RNase. The sample is then incubated at 50°C for 10 min. This first incubation is followed by the addition of 10 μl of proteinase solution (17 mg/ml proteinase K) and incubation at 50°C for 50 min. During the second incubation the samples should be mixed by inversion every 10–15 minutes. The supernatant is now transferred to a fresh Eppendorf tube, 0.3 ml of NaI solution added (7.6 *M* NaI, 20 m*M* EDTA, 40 m*M* Tris-HCl, pH 8.0), and the sample mixed by repeated inversion. 2-Propanol (0.5 ml) is added and the sample mixed until 2–3 minutes after the first white DNA precipitate appears. The tube is then spun at 10,000*g* for 5 min at ambient temperature, following which the supernatant is removed and the tube drained by inversion on absorbant paper, taking care not to dislodge the pellet. The pellet is washed with 1 ml of "washing solution A" [40% (w/v) 2-propanol], including adequate mixing to dislodge the pellet from the wall of the tube. Following another centrifugation at 10,000*g* for 5 min at room temperature, the supernatant is aspirated and 1 ml of "washing solution B" [70% (w/v) ethanol] added. Again the sample is mixed and the pellet induced to float free in the wash solution. The sample may now be stored at −80°C in wash solution B, or may be centrifuged and the pellet immediately hydrolyzed as described below.

Alternative Method: Phenol Extraction

Preparation of RNase A and T Stock Solution. Add 20 mg of RNase A (Boehringer Mannheim) to 20 ml of 1 m*M* EDTA/10 m*M* Tris-HCl buffer, pH 7.4, and boil the resulting solution in a water bath for 20 min. To the cooled solution add 2000 units of RNase T (Boehringer Mannheim) and then freeze at −20°C in 0.5–1.0 ml aliquots.

DNA Extraction. Using 0.5–1 ml of the nuclear pellet suspension, add 20 μl of RNase stock solution and incubate for 1 hour at 50°C. Add 40 μl of 10% sarkosyl and 40 μl of a 5 mg/ml solution of pronase (Boehringer Mannheim) prepared in 10 m*M* Tris/1 m*M* EDTA, pH 7.4, and incubate for 1 hour at 50°C.

* Composition of the kit solutions are as described in the original publication of Wang *et al.*

Add 400 μl of high-purity phenol (Clontech, Palo Alto, CA), which has been saturated with buffer (0.1 *M* Tris, pH 8.0, 0.1 *M* NaCl, 20 m*M* EDTA), to the resulting sample and vortex vigorously for 1 min. Centrifuge the sample at 10,000*g* and room temperature for 3 min. Transfer the aqueous phase to a new Eppendorf tube, being careful not to carry over the interface macromolecules. Add 400 μl of the saturated phenol solution and 400 μl of Sevag (chloroform : isoamyl alcohol, 24 : 1, v/v). Vortex vigorously and then centrifuge at 10,000*g* for 10 min at room temperature. Transfer the upper aqueous phase to a new 1.5 ml Eppendorf tube and add 400 μl of Sevag. Centrifuge as above, transfer the aqueous phase to a new Eppendorf tube, and precipitate the DNA by adding 40 μl of 3 *M* sodium acetate, pH 5.1, and 800 μl of ice-cold ethanol. Precipitate the sample for at least 1 hour or overnight at −20°C. Centrifuge the sample at 10,000*g* for 15 min at 4°C, remove the supernatant, and wash the DNA pellet twice with 70% (v/v) ethanol, centrifuging as above. Then proceed to the hydrolysis step.

DNA Hydrolysis

If previously frozen, the sample is thawed on a bed of ice. It is then centrifuged for 5 min at 10,000*g* and 4°C, and the supernatant aspirated. The original method calls for partial drying under a stream of nitrogen or argon for 3 min. Other workers have used vacuum drying. Whatever method is used, it is important not to expose the sample to a stream of air or other oxygen-containing gases, as this may produce oxidation artifact. Also, the sample should not be taken to complete dryness as this will complicate dissolving the DNA. We have substituted a brief buffer wash for drying. Following centrifugation, Wash Solution B is aspirated and the tube drained by inversion on absorbant paper, taking care not to dislodge the pellet from the bottom of the tube. Next, 200 μl of sodium acetate buffer (see below) is added and immediately removed by aspiration and tube inversion. Fresh buffer is then added and the hydrolysis continued.

Dissolve the DNA sample in 200 μl of 0.1 m*M* Desferal/20 m*M* sodium acetate, pH 4.8. This is accomplished by using a micropipette to dislodge the pellet and break it up. Repeated aspiration with a wide-orifice 200 μl Pipetman tip then further disperses the DNA. Previously, we used Desferal at 1.0 m*M*, but we have reduced the concentration to 0.1 m*M* because the lower concentration is adequate to chelate the free iron in the sample[19] and, at the higher concentration, we noted interference with the oxo^{8}dG peak and baseline distortion that persisted despite extensive wash cycles (Fig. 1).

Continue the hydrolysis by adding 4 μl of nuclease P1 (Boehringer Mannheim), prepared by resuspending the contents of one vial of nuclease

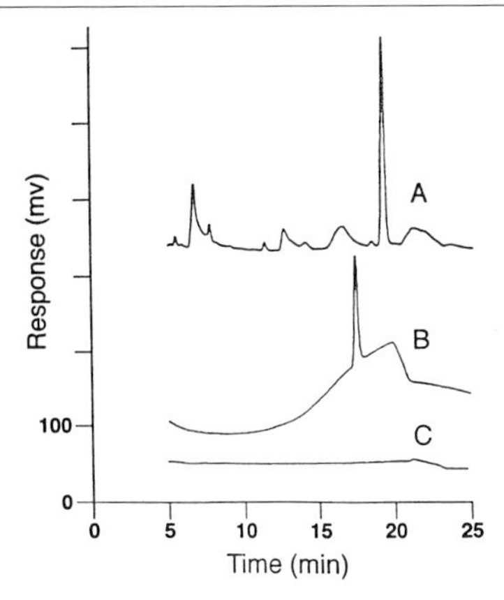

FIG. 1. Effect of Desferal. (A) Injection of buffer containing 1.0 m*M* Desferal, showing the interfering peaks. (B) Injection of oxo^{8}dG standard showing the baseline distortion that occurred after multiple injections of buffer containing 1.0 m*M* Desferal. These changes were seen despite wash cycles prior to this injection. (C) Baseline following injection of buffer not containing Desferal.

P1 in 300 μl of sodium acetate buffer. The final concentration of nuclease P1 in the sample will be 0.1 mg/ml. Continue to mix the suspension by inversion or hand agitation (do not vortex) for 5–10 minutes. Then incubate the sample for 12–15 minutes at 65–70°C. This step will digest the DNA to the corresponding nucleotide 5′-monophosphates. After the incubation is complete add 20 μl of 1 M Tris-HCl buffer, pH 8.0, to the sample mixture and mix by inversion. This step optimizes the pH for the action of alkaline phosphatase. Now add 4 μl of 1 U/μl of calf intestine alkaline phosphatase (Boehringer Mannheim: 10 μl diluted to 250 μl in water gives the desired 1 U/μl solution). Mix well by slow inversion and then incubate for 1 hour at 37°C. Following incubation, adjust the pH of the hydrolyzed DNA by adding 20 μl of 3 *M* sodium acetate buffer, pH 5.1. Transfer the sample to a 30,000-Da microfiltration tube (UltraFree, Millipore, Bedford, MA) and centrifuge for 5 min at 10,000*g* and 4°C. If more than one-third of the sample remains unfiltered, transfer the remaining unfiltered fraction to a new filter and centrifuge as above, combining the filtrates from both filters. The sample is then placed in autoinjector vials with low-volume inserts for HPLC–EC analysis.

HPLC-EC Analysis of Hydrolyzed DNA

A high quality solvent delivery system, coupled to flowthrough electrochemical and UV detection systems, is required. Autoinjectors should be refrigerated to maintain samples at 4°C. We use a Waters Associates (Milford, MA) Model 625 solvent delivery system and a refrigerated WISP Model 712 autoinjector. Chromatography is performed with a 15 cm × 4.6 mm, 3 μm LC-18-DB column (Supelco, Bellefonte, PA) using gradient elution and a flow rate of 1 ml/min. Elution and column washing require three solvents; solvent A is 2.5% methanol in 50 m*M* KH_2PO_4, pH 5.5, solvent B is 10% methanol in the same buffer, and solvent C is 50/50 (v/v) methanol/water. The column should be equipped with an LC-18-DB precolumn cartridge and the back pressure should not exceed 2900 psi, with a usual range of 2400 to 2700 psi. For these studies we place the electrochemical detector in line before the UV detector to improve peak shape and increase sensitivity. The electrochemical detector is an ESA (Bedford, MA) Model 5100 Coulochem equipped with a 5011 analytical cell. The oxidation potentials of electrodes 1 and 2 are set at 0.1 and 0.4 V, respectively, and gain at 500. These settings produce reliable oxidation of oxo^8dG and related species but do not interfere with the post EC, ultraviolet analysis of unmodified deoxy nucleosides. A Kratos (Westwood, NJ) model 773 UV detector is used at 260 or 290 nm, the setting depending on the amount of DNA injected. At 290 nm the absorbance is substantially diminished but remains a linear function of the nucleoside concentration. This permits the injection of larger (5×) amounts of DNA, thereby increasing the sensitivity of the method. Elution is accomplished with a linear gradient beginning with solvent A and progressing over 21 min to a 50/50 mixture of solvent A and solvent B. This mixture is maintained for 4 min and the system then changed to solvent C at a flow rate of 0.5 ml/min for 6 min. Following this washing cycle, solvent A is reintroduced at 0.5 ml/min and this lower flow continued for 8 min to prevent sharp rises in back pressure. The rate is then changed to 1 ml/min for a further 5 min, at which time the next sample is injected. Easily quantified oxo^8dG peaks can be obtained using this method, even at ratios of 0.04 $oxo^8dG/10^5dG$, by using a UV setting of 290 nm and injecting 50–100 μg of DNA hydrolyzate. Typical chromatograms have been published in Volume 234 of this series.[21]

Figure 2 emphasizes the danger of attempting to analyze <20 μg of DNA as starting material. Studies in our laboratory suggest that this problem is likely to be the result of carryover of sample from one injection to the next, so that at low oxo^8dG concentrations a small, constant carryover contributes increasingly significant area to the measured response. We have been able to overcome this limitation by complete disassembly and cleaning of the autoinjector prior to use, combined with meticulous attention to total system hygiene. Nevertheless, where possible, we recommend that DNA samples be 100 μg or more. Even with larger samples *it is imperative* that system hygiene be maintained and that buffer blanks and sample blanks

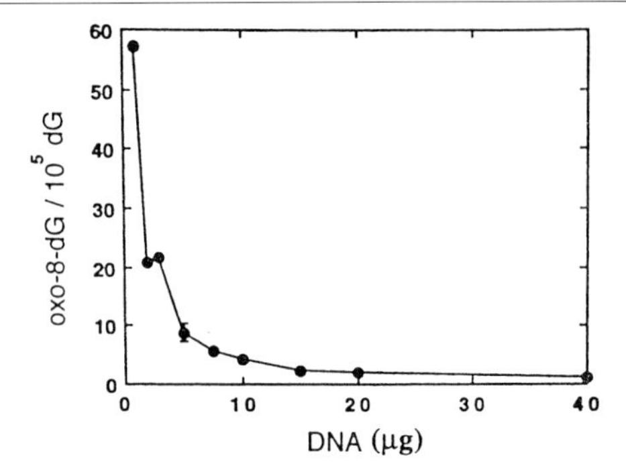

FIG. 2. Analyzing small quantities of DNA can result in aberrant results. See text for discussion.

(i.e., water blanks that have been carried through the entire extraction–hydrolysis process) be included in the analysis.

The results are expressed as the ratio of oxidized species (e.g., oxo^{8}dG) to dG as measured by UV detection. This technique uses dG as an internal standard and greatly improves the validity of the measurements. Typical results are in the range of 0.04 to 0.6 oxo^{8}dG/10^{5}dG. The values can be converted to total DNA analyzed by using the conversion factor 0.648 nmol dG per μg DNA, for DNA which is composed of 0.21 parts dG as is true for nuclear DNA of eukaryotes. Thus, the results can be expressed as a molar ratio of the nucleobases or as femtograms of nucleobase per microgram DNA.

Isolation of oxo^{8}dG, oxo^{8}G, oxo^{8}Gua, and Related Oxidation Products from Urine by Immunoaffinity Column Separation

Solid-Phase Extraction

Urine samples are stored at $-20°$ or $-80°$ prior to analysis. Before using the C_{18}OH solid-phase extraction (SPE) columns (Perkin Elmer/Analytichem, Harbor City, CA) they must be preconditioned by washing sequentially with 3 ml of methanol, water, and finally, 50 m*M* KH_2PO_4 buffer, pH 7.5. Gravity elution is preferred in all SPE steps. For the measurement of oxo^{8}dG or oxo^{8}G, 1 ml samples are diluted with an equal volume of 1 *M* NaCl, spiked with 3000–10,000 cpm of [^{3}H]oxo^{8}dG or [^{14}C]oxo^{8}G, and applied to the SPE column (see ref. 21 for preparation/storage of immunoaffinity columns and preparation of radiolabeled standards). The column is washed with 5 ml of 50 m*M* KH_2PO_4 buffer, pH 7.5, and then eluted with 3 ml of 15% methanol in the same buffer.

Urinary oxo^{8}Gua must be analyzed separately because special preparation of the urine samples is required to optimize the recovery of this species. Dissolving the precipitate, formed when urine is stored in the cold, improves and stabilizes the recovery of oxo^{8}Gua. Therefore, after the sample is thawed, adjust the pH to pH 6.9–7.2 using 1 *M* NaOH, following which the sample is warmed in a 37°C water bath for 10 min. A 1 ml aliquot is then *immediately* diluted with an equal volume of 1 *M* NaCl, spiked with 3000–10,000 cpm of [^{14}C]oxo^{8}Gua, and applied to SPE columns preconditioned as described above. Under these conditions oxo^{8}Gua is retained by the column but will not tolerate extensive washing. Therefore, the wash volume is limited to 1 ml, a volume that will not adequately remove interfering species for the analysis of oxo^{8}dG but does permit recovery of 65–70% of the oxo^{8}Gua without producing interference at the oxo^{8}Gua retention time. The SPE column is eluted as described above.

Immunoaffinity Separation

The SPE eluate is applied to the immunoaffinity column at 4°C and the column washed sequentially with 8 ml of water, 1 *M* NaCl, and finally with water again. Some workers then wash with 4 ml of acetonitrile to reduce the water content of the eluate and decrease drying time. The column is eluted with 5–7 ml of methanol and the eluate collected in 15-ml polypropylene culture tubes. The column is regenerated by the addition of a further 10 ml of methanol followed by two 10 ml aliquots of water. During the water wash, it is imperative that the bed be resuspended by stirring, with a Pasteur pipette, to remove air bubbles.

HPLC-EC Analysis of Immunoaffinity Column Prepared Samples

The immunoaffinity column eluate is taken to dryness under a stream of nitrogen or argon. The residue is resuspended in 200 μl of high-purity water, an aliquot removed for isotope analysis, and the balance placed in low volume insert autoinjector vials and analyzed. The equipment and chromatography are as described above for DNA hydrolyzates with the following exceptions. The UV component of the system is not used, the injection volume is 20 to 100 μl, depending on the concentration of analyte in the sample, and the eluting solvents are changed to avoid the interference of an artifact peak that intermittently occurs with the use of immunoaffinity columns.[19] Elution and column washing are accomplished using three solvents; solvent A is 0.75% acetonitrile and 0.75% methanol in 50 m*M* KH_2PO_4 buffer, pH 5.05, solvent B is 3% acetonitrile and 3% methanol in the same buffer, and solvent C is 50/50 methanol/water (v/v). Elution of oxo^{8}dG is accomplished with a linear gradient beginning with solvent A and progressing over 20 min to 35% solvent A and 65% solvent B. The flow rate for this first phase is 1 ml/min. The flow is now decreased to 0.6 ml/min and changed to solvent C for 9 min. Solvent A is then pumped for 5 min at 0.6 ml/min and for another 10 min at 1 ml/min, followed by the next sample injection. A typical chromatogram including the structures of the major analytes has been previously published.[21]

Interpretation

Oxo^8dG and oxo^8Gua have been used to estimate two different quantities: the number of oxidative adducts in DNA, which is sometimes referred to as the "steady-state" value, and the rate of DNA oxidation *in vivo*. The number of oxidative adducts is estimated by measuring the amount of oxo^8dG in extracted DNA, and is accurate to the extent that the estimates correctly reflect the state of DNA in the intact cell. The HPLC-EC estimates currently range from 0.04 to 3.0 $oxo^8dG/10^5$ dG for young rat liver, although many values lie in the range of 0.2 to 0.6 $oxo^8dG/10^5$ dG. The chaotropic NaI method gives some of the lowest values and, in our hands, low variability. Nevertheless, the phenol method has produced valuable information and gives the same ratio as the NaI method for the age-associated increase in oxo^8dG for rat liver nuclear DNA.[19] Several authors[6,17] have argued for the avoidance of light, air, and organic solvents, particularly phenol, during the isolation process. We have not found all of these restrictions to be necessary, although we do not use air, which can lead to oxidation, to dry DNA. Further, we have found that many of the alternative techniques give $oxo^8dG/10^5dG$ values that are higher than those obtained with the phenol method.[19] It is possible that the phenol procedure gives higher values of $oxo^8dG/10^5dG$ than the NaI method by including a wider range of DNA fractions. Nevertheless, at the time of writing the chaotropic NaI method appears to be the procedure of choice, with the caveat that further work may reveal a problem of underestimation due to lost DNA fractions or other anomalies.

In contrast to steady-state levels of adducts, estimates of the rate of oxidation of nuclear DNA are dependent on measurements of excreted adducts. The interpretation of adduct excretion in urine is complicated by the existence of other sources of, and pathways of degradation for, oxo^8dG and oxo^8Gua: (i) mitochondrial DNA is a target of oxidative damage and may contribute to the excretion rate through mitochondrial turnover and repair[22]; (ii) the apoptotic death of cells, characteristically associated with

[22] M. K. Shigenaga, T. M. Hagen, and B. N. Ames, *Proc. Natl. Acad. Sci. USA* **91,** 10771–10778 (1994).

the degradation of DNA, may also contribute adducts; (iii) the oxidation of cytosolic and circulating nucleobase pools and DNA fragments are potential sources of excreted adducts; (iv) large amounts of oxo^{8}Gua are of dietary origin; (v) oxo^{8}dG and oxo^{8}Gua are sensitive to oxidation, and may be destroyed *in vivo* before excretion, or by enzymatically catabolized via salvage pathways which operate on intact nucleotides, resulting in loss of these analytes from the urinary pathway under study. All of these confounders would change the estimate of excreted adducts, and therefore lead to an inaccurate estimate of the number of damage sites ("hits") repaired each day.

In conclusion, methods for DNA isolation continue to improve, with minor modification of the chaotropic NaI technique producing the lowest oxo^{8}dG values reported to date using HPLC-EC. Although the use of urinary oxo^{8}Gua adducts to monitor DNA oxidation *in vivo* remains attractive, the results can be difficult to interpret. Comparison of carefully controlled groups can minimize some of the confounders (such as dietary contributions), but the effect of experimental treatments on, for example, the rate of cell turnover, will remain unknown and uncontrolled. Thus, given our ignorance about the dynamics of nucleotides and their oxidative adducts in whole-body metabolism, studies of urinary excretion must be interpreted with considerable caution.

Acknowledgment

This work was supported by National Cancer Institute Outstanding Investigator Grant CA39910 and National Institute of Environmental Health Sciences Center Grant ES01896 to B.N.A.

[19] Markers of Oxidative Damage to DNA: Antioxidants and Molecular Damage

By STEFFEN LOFT and HENRIK ENGHUSEN POULSEN

Cells are constantly exposed to oxidants from both physiological processes, such as mitochondrial respiration, and pathophysiological conditions, such as inflammation, ischemia–reperfusion, foreign compound metabolism, and radiation.[1] Failure of the system of enzymatic, endogenous, and nutritional antioxidants may lead to carcinogenesis and degenerative

[1] B. N. Ames, L. S. Gold, and W. C. Willett, *Proc. Natl. Acad. Sci. USA* **92,** 5258 (1995).

0076-6879/99 $30.00

diseases of aging. Thus, the concept of supplementing nutritionally based antioxidants in order to prevent cancer and postpone aging appears rational.[1,2] Descriptive epidemiological studies support the notion by a highly reproducible close relationship between a high dietary intake of vegetables and fruits rich in antioxidants and a low risk of epithelial cancers, particularly in the airways and upper gastrointestinal tract.[3,4] So far, however, the large-scale intervention studies of the effects of single and combinations of antioxidants, vitamin E and β-carotene have been negative and even indicated a cancer-promoting effect of the latter in smokers.[5-7] The enormous costs and the very limited number of compounds or combinations and doses that can be tested in only a few high risk population groups in such studies may warrant an alternative approach. Small-scale human intervention studies with mechanistically based biomarkers as intermediate end points and supported by animal and *in vitro* experiments may target the optimum intervention strategy for the large-scale intervention.[8] In addition, the use of such biomarkers may provide further proof of a causal relationship between oxidative damage to DNA and cancer and aging. Accordingly, oxidative DNA adducts may serve as valuable biomarkers in this context.

More than 100 different oxidative modifications have been observed in DNA.[9,10] However, so far only a few of the base modifications have been used as biomarkers, and of these, the oxidative C-8 adduct of guanine is by far the most studied as either the nucleoside or base. In principle, the level in DNA from target or surrogate tissues or cells or the excretion of repair products into the urine can be measured. Under the usual steady-state conditions, the latter will reflect the rate of damage, whereas the former will reflect the balance between damage and repair. The present

[2] B. N. Ames, M. K. Shigenaga, and T. M. Hagen, *Proc. Natl. Acad. Sci. USA* **90,** 7915 (1993).

[3] G. Block, B. Patterson, and A. Subar, *Nutr. Cancer.* **18,** 1 (1992).

[4] G. Block, *Nutr. Rev.* **50,** 207 (1992).

[5] The Alpha-Tocopherol and Beta Carotene Cancer Prevention Group, *N. Engl. J. Med.* **330,** 1029 (1994).

[6] G. S. Omenn, G. E. Goodman, M. D. Thornquist, J. Balmes, M. R. Cullen, A. Glass, J. P. Keogh, F. L. Meyskens, B. Valanis, J. H. Williams, S. Barnhart, and S. Hammar, *N. Engl. J. Med.* **334,** 1150 (1996).

[7] C. H. Hennekens, J. E. Buring, J. E. Manson, M. Stampfer, B. Rosner, N. R. Cook, C. Belanger, F. LaMotte, M. Gaziano, P. M. Ridker, W. Willett, and R. Peto, *N. Engl. J. Med.* **334,** 1145 (1996).

[8] "Molecular Epidemiology. Principles and Practices" (P. A. Schulte and F. P. Perera, Eds.). Academic Press, London, 1993.

[9] M. Dizdaroglu, *Methods Enzymol.* **234,** 3 (1994).

[10] J. Cadet, J. L. Ravanat, G. W. Buchko, H. C. Yeo, and B. N. Ames, *Methods Enzymol.* **234,** 79 (1994).

review will focus on the involved assays, the interpretation of the methods, and some results from antioxidant intervention studies, particularly in humans.

Urinary Excretion of DNA Repair Products

The repair products from oxidative DNA damage, i.e., oxidized bases and nucleosides, are poor substrates for the enzymes involved in nucleotide synthesis; they are fairly water soluble and generally are excreted into the urine without further metabolism.[11,12] Among the possible repair products from oxidative DNA modifications, 8-oxo-7,8-dihydro-2′-deoxyguanosine (8-oxodG), 8-oxoguanine (8-oxoGua), thymine glycol (Tg), thymidine glycol (dTg), and 5-hydroxymethyluracil (5-OHmU) have so far been identified in urine.[11,13–18] Of these, 8-oxodG and the thymine derivatives are the most intensively studied. The levels of concentration and excretion of the oxidized bases and nucleosides obtained in different laboratories are in the same range (Table I).

The assays for the urinary DNA repair products include high-performance liquid chromatography (HPLC) with electrochemical detection (EC) for 8-oxodG and 8-oxoGua and with UV absorbance detection for dTg and Tg, whereas all the repair products can potentially be measured by gas chromatography-mass spectrometry (GC/MS).[9,14,16,17] The major problem with all these assays involves separation of the very small amounts of analyte from urine, which is a very complicated matrix. Thus, although several of the products are electrochemically active and high sensitivity is achievable, the HPLC methods require extensive cleanup procedures such as multiple solid-phase extractions, HPLC column switching techniques, or

[11] M. K. Shigenaga, C. J. Gimeno, and B. N. Ames, *Proc. Natl. Acad. Sci. USA* **86,** 9697 (1989).

[12] S. Loft, P. N. Larsen, A. Rasmussen, A. Fischer-Nielsen, S. Bondesen, P. Kirkegaard, L. S. Rasmussen, E. Ejlersen, K. Tornøe, R. Bergholdt, and H. E. Poulsen, *Transplantation* **59,** 16 (1995).

[13] R. Cathcart, E. Schwiers, R. L. Saul, and B. N. Ames, *Proc. Natl. Acad. Sci. USA* **81,** 5633 (1984).

[14] H. Faure, M. F. Incardona, C. Boujet, J. Cadet, V. Ducros, and A. Favier, *J. Chromatogr.* **616,** 1 (1993).

[15] J. Suzuki, Y. Inoue, and S. Suzuki, *Free Rad. Biol. Med.* **18,** 431 (1995).

[16] M. G. Simic and D. S. Bergtold, *Mutation Res.* **250,** 17 (1991).

[17] J. Teixeira, M. R. Ferreira, W. J. van Dijk, G. van de Werken, and A. P. de Jong, *Anal. Biochem.* **226,** 307 (1995).

[18] S. Loft and H. E. Poulsen, *J. Mol. Med.* **74,** 297 (1996).

TABLE I
RANGE OF PUBLISHED VALUES REGARDING URINARY BIOMARKERS OF OXIDATIVE DNA DAMAGE IN HUMANS[a]

Lesion	Assay	Range of averages	Total number of subjects	Total number of publications
dTg	HPLC	390–435 pmol/kg 24 h	19	2
Tg	HPLC	100–174 pmol/kg 24 h	19	2
dTg	GC/MS	110–250 pmol/kg 24 h	3	2
5-OHmU	GC/MS	74 ± 9 nmol/24 h	14	1[b]
8-oxodG	HPLC	170–600 pmol/kg 24 h	360	6
		1.0–3.0 nmol/mmol creatinine	206	8
8-oxodG	GC/MS	110–345 pmol/kg 24 h	26	3
8-oxodG	ELISA	1600–4800 pmol/kg 24 h	4	1
		7.7 ± 3.4 nmol/mmol creatinine	52	1[c]
8-oxoGua	HPLC	1.5–5.0 nmol/mmol creatinine	13	1

[a] Data from Loft and Poulsen, *J. Mol. Med.* **74,** 297 (1996).
[b] From Faure *et al.*, *Free Rad. Biol. Med.* **20,** 979 (1996).
[c] From Erhola *et al.*, *FEBS Lett.* **409,** 287 (1997).

immunoaffinity columns.[11,13,19–25] The complicated extraction procedures cause recovery problems in both HPLC and GC/MS-SIM methods and may require labeled internal standards. Moreover, the complicated procedures limit the analytical capacity. An enzyme-linked immunosorbent assay (ELISA) based on monoclonal antibodies has been developed for estimation of 8-oxodG in urine samples.[26] However, the values obtained in rat and human urine samples[26,27] were 3–5 times higher than other published values.[18] Similarly, in 4 smokers studied before and after smoking cessation,

[19] S. Loft, K. Vistisen, M. Ewertz, A. Tjønneland, K. Overvad, and H. E. Poulsen, *Carcinogenesis* **13,** 2241 (1992).
[20] C. Tagesson, M. Källberg, and P. Leanderson, *Toxicol. Meth.* **1,** 242 (1992).
[21] E.-M. Park, M. K. Shigenaga, P. Degan, T. S. Korn, J. W. Kitzler, C. M. Wehr, P. Kolachana, and B. N. Ames, *Proc. Natl. Acad. Sci. USA* **89,** 3375 (1992).
[22] M. K. Shigenaga, E. N. Aboujaoude, Q. Chen, and B. N. Ames, *Methods Enzymol.* **234,** 16 (1994).
[23] R. K. Brown, A. McBurney, J. Lunec, and F. J. Kelly, *Free Rad. Biol. Med.* **18,** 801 (1995).
[24] C. Tagesson, M. Kallberg, C. Klintenberg, and H. Starkhammar, *Eur. J. Cancer.* **31A,** 934 (1995).
[25] D. Germadnik, A. Pilger, and H. W. Rudiger, *J. Chromatogr.* **689,** 399 (1997).
[26] T. Osawa, A. Yoshida, S. Kawakishi, K. Yamashita, and H. Ochi, *in* "Oxidative Stress and Aging" (R. G. Cutler, L. Packer, J. Bertram, and A. Mori, Eds.), pp. 367–378. Birkhauser Verlag, Basel, 1995.
[27] M. Erhola, S. Toyokuni, K. Okada, T. Tanaka, H. Hiai, H. Ochi, K. Uchida, T. Osawa, M. M. Nieminen, H. Alho, and P. Kellokumpu-Lehtinen, *FEBS Lett.* **409,** 287 (1997).

the urinary 8-oxodG excretion values estimated by the ELISA method were 8 times higher than and showed only a weak correlation ($r = 0.42$) with the values obtained by HPLC.[28]

HPLC–EC Assay of 8-OxodG in Urine

Most data regarding urinary excretion of oxidative DNA repair products have been obtained on 8-oxodG using HPLC–EC assays with column switching techniques for both extraction and separation. In our laboratory an automated three-dimensional HPLC method with isocratic separation and electrochemical detection has been used for 6 years with minimum modifications for the analysis of 8-oxodG in several thousand urine samples.[19]

Urine can be stored frozen. For analysis, thawed or fresh urine samples e.g., 2 ml, are acidified with 40 μl 2 *M* HCl and frozen at −20° overnight for precipitation of uric acid and other solutes. After thawing and centrifugation for 5 min at 3000*g*, 34 μl NaOH 2 *M* is added to 1.7 ml of the supernatant. At this stage the samples can again be stored frozen for prolonged periods. To 95 μl aliquots of the treated urine, 5 μl of a solution containing 8-oxodG (Sigma, St. Louis, MO) 0, 800, or 8000 n*M* and 100 μl 1 *M* Tris buffer, pH 7.9, are added and 25 μl of the mixture is injected on the HPLC apparatus. This consists of the following Merck-Hitachi (San Jose, CA) components: L-6000 and L-6200 pumps, 655A-40 autosampler, column oven set at 40°, and D-6000 data handling and integration software, as well as an LP21 (Science Systems Inc., Mikrolab, Aarhus, DK) pulse dampener in the flow path, a Waters 440 UV absorbance detector at 254 nm, and an ESA Coulochem II electrochemical detector (Bedford, MA) equipped with a 5011 cell. The flow path diagram is shown in Fig. 1. In the first column (Spherisorb ODS2 15 cm 5 μm; Waters, Denmark), 8-oxodG is extracted from the urine sample with an alkaline mobile phase [2.5% acetonitrile and 1.5% methanol (v/v/v) in 10 m*M* borate buffer, pH 7.9]. The retention time of 8-oxodG in the extraction column is determined by UV absorbance after injection of a 4000 n*M* solution in the Tris buffer. Via a 6-port automatic Valco (Switzerland) valve, a 1 ml fraction of the effluent from urine samples containing the 8-oxodG is brought onto a 2 cm ion-exchange column PRP-X100 (Hamilton, Reno, NE). With a switch of the valve, this column is then flushed with an acidic eluent (100 m*M* phosphate buffer, pH 2.1) with 2.5–4% acetonitrile, bringing the contents onto a 25 cm Nucleosil C_{18} 3 μm column (Knaur, Germany). The effluent

[28] H. Priemé, S. Loft, R. G. Cutler, and H. E. Poulsen, *in* "Natural Antioxidants and Food Quality in Atherosclerosis and Cancer Prevention" (J. T. Kumpulainen and J. T. Salonen, Eds.), pp. 78–82. The Royal Society of Chemistry, London, 1996.

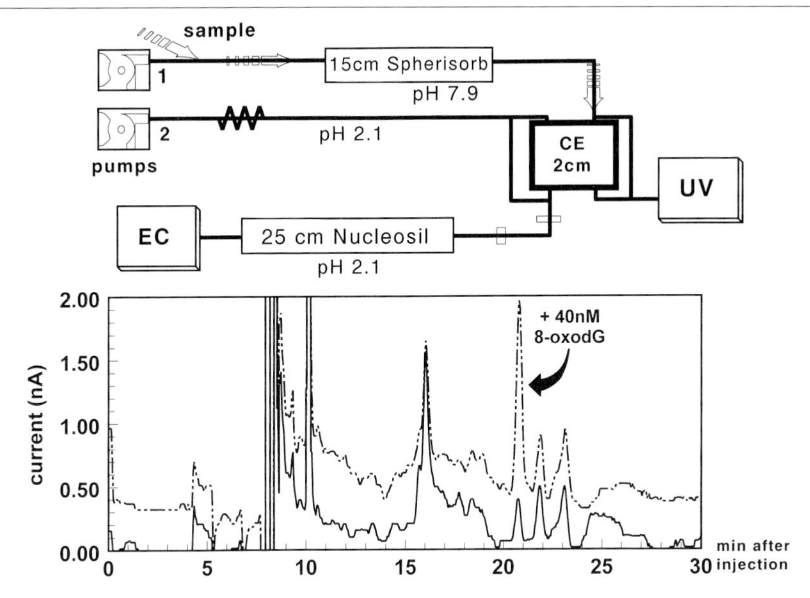

FIG. 1. Flow path (top) and chromatogram (bottom) of three-dimensional HPLC analysis of urinary 8-oxodG. The urine sample is introduced in the flow path of pump 1 and a fraction containing 8-oxodG is extracted by a Spherisorb column and brought on a cation exchange column (CE). The retained 8-oxodG is brought into the flow path of pump 2 by a six-port valve, separated on a Nucleosil column, and monitored by an EC-detector. In the chromatogram the broken and solid lines represent the tracing of a sample containing 11 n*M* 8-oxodG with and without addition of 40 n*M*.

of this column is monitored by the electrochemical detector in the oxidation mode set at 100–120 mV at electrode 1 and 200–300 mV at electrode 2 according to the voltammogram, which is the S-shaped relationship between the applied voltage and the response of the EC detector. The voltammogram is subject to change, particularly with wear of the cell, and should be checked frequently for optimum adjustment of the voltage to a response at the upper part of the S-shape (Fig. 2). The peak identity of 8-oxodG in urine samples has been ascertained by identical retention time and voltammogram as compared with the genuine compound.[19] Moreover, after acid hydrolysis in 0.5 *M* HCl at 140° for 1 hour, the 8-oxodG peak disappeared from the chromatogram of urine samples with and without the addition of 40 n*M* genuine compound (unpublished observations).

For quantification each urine sample is run with and without addition of at least two concentrations of genuine 8-oxodG and the concentrations calculated from the individual calibration curve based on the peak heights. The performance of the method includes linear calibration curves, an interassay coefficient of variation of 9–13%, and a limit of detection of 0.2

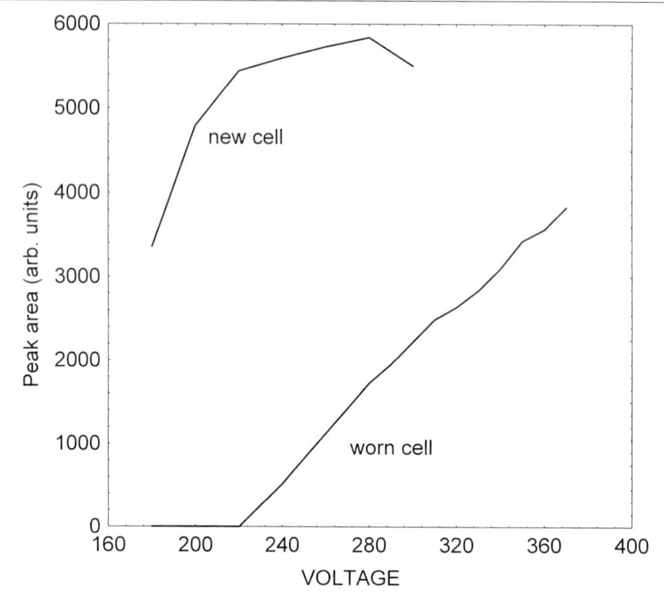

FIG. 2. Examples of voltammograms of 8-oxodG after HPLC analysis with an ESA Coulochem II detector equipped with a new or a worn 5011 cell.

n*M* as compared to concentrations ranging from 1 to 100 n*M* in human urine. We have reanalyzed 75 samples after 6 years of storage at −20° (Fig. 3). The two sets of values were highly correlated ($r^2 = 0.75$) and the residual standard deviation was 22% of the grand mean. The linear regression slope of 0.87 may be due to change of the source of calibration standard from solutions of 8-oxodG received as gifts from other researchers to the now commercially available crystallized compound; with the old calibration standards the slope would have been 1.09. The significant positive intercept may be related to a change in the mode of integration of the chromatographic peaks. Considering the long interval, the method appears reasonably robust and 8-oxodG is stable in storage for at least 6 years. Moreover, 20- to 25-year-old urine samples show concentrations in the range of fresh ones (unpublished observations). Although the method is relatively simple and automated, problems with separations frequently limit the number of samples that can be handled. In our hands, separation is primarily optimized by adjusting the acetonitrile concentration in the mobile phase of the Nucleosil column, and secondly by adjusting the corresponding pH. This column usually has a long lifetime, whereas the extraction columns wear out after only 4–8 weeks of use. In general, separation is more difficult with rat samples than with human samples.

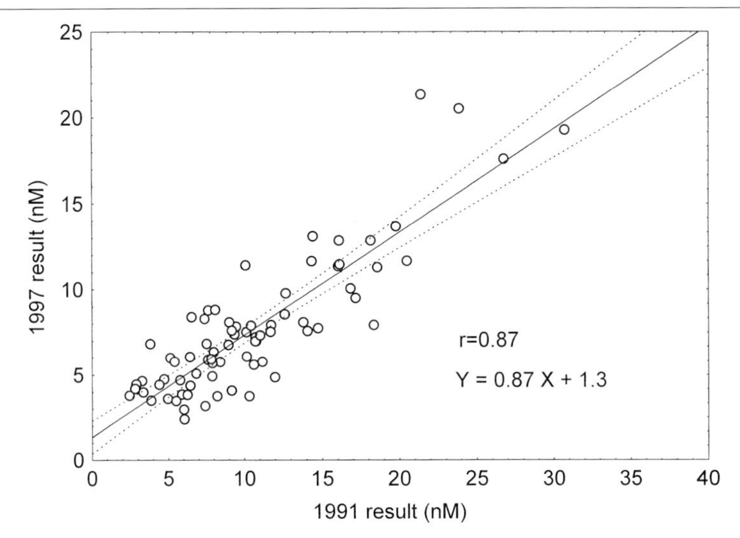

FIG. 3. Results of a three-dimensional HPLC analysis of 75 urine samples for 8-oxodG in 1991 and in 1997. Between the two assays the source of calibration standard and the method of peak integration were changed. The linear regression line and the equation are included.

A Swedish group has developed a similar column switching assay for 8-oxodG in urine.[24] However, in that assay the sample is concentrated on a Bondelut CH minicolumn before injection on a polymer reversed-phase column eluted with an acidic mobile phase (pH 4.0) with heptanesulfonic acid. A timed fraction of the effluent is brought on a C_{18} column eluted at pH 7.0 and monitored with a Waters 460 electrochemical detector. That assay seems to perform satisfactorily and has been used in some large-scale epidemiological studies.[29–31] The use of monoclonal antibody-based immunoaffinity columns for extraction of 8-oxodG, 8-oxoGua, and 8-oxoguanosine from biological fluids, including urine, has been extensively described in this series previously.[22]

For convenience, the use of spot urine samples corrected for creatinine would be simpler than 24 h collection of urine. However, creatinine production is variable, e.g., increased with exercise and decreased with age. Indeed, in 74 healthy subjects we collected urine for 24 h and a spot urine sample from the subsequent voiding. The correlation between the 8-oxodG to

[29] S. Lagorio, C. Tagesson, F. Forastiere, I. Iavarone, O. Axelson, and A. Carere, *Occup. Environ. Med.* **51,** 739 (1994).

[30] C. Tagesson, M. Kallberg, and G. Wingren, *Int. Arch. Occup. Environ. Health* **69,** 5 (1996).

[31] R. I. Nilsson, R. G. Nordlinder, C. Tagesson, S. Walles, and B. G. Jarvholm, *Am. J. Ind. Med.* **30,** 317 (1996).

creatinine ratio in the spot samples and the 24 h excretion of 8-oxodG was rather poor ($r = 0.50$; unpublished).

For measurement of 8-oxoGua in human urine only one HPLC–EC method appears to have been used for published data.[15] It involves complicated strong cation exchange and cellulose partition extraction procedures before separation on a C_{18} column. No further data involving the method than those included in the original paper appear to have been published.

Interpretation of Urinary Oxidized Nucleobases and Nucleosides

For the use of oxidized bases and deoxynucleosides as urinary biomarkers, the repair pathways of the modifications, in particular 8-oxoGua, in DNA and other sources may be debated.[21] Thus, two different DNA repair enzymes, one with glycosylase activity and one excising single 8-oxodG as a nucleotide, have been isolated from nuclear extracts of a human cell line.[32] The human 8-oxoGua glycosylase was cloned by several groups,[33,34] whereas nucleotide excision repair was shown to contribute to the repair of 8-oxodG in DNA.[35] Moreover, the 8-oxodGTP phosphatase and 8-oxodGMP nucleotidase will selectively and rapidly convert the liberated oxidized nucleotide to a nucleoside ready for excretion.[36,37] These enzymes will also sanitize oxidized dGTP from the cellular pool and allow its excretion as 8-oxodG. In addition, digestion of damaged DNA from cell renewal and mitochondrial turnover will liberate 8-oxodG. Unpublished data from rat studies indicate that the induction of 8-oxodG in nuclear DNA in target organs corresponds to the increase in urinary excretion after administration of the carcinogen 2-nitropropane, supporting the idea that 8-oxodG is the primary repair product *in vivo*. Animal experiments have shown that injected 8-oxodG is readily excreted unchanged into the urine, whereas 8-oxodG in the diet or oxidation of dG during excretion does not contribute.[11,12,21] In rats, at least, dietary purines are an important determinant of the excretion of 8-oxoGua, which is far larger than the excretion of 8-oxodG.[21] However, in humans the excretion of 8-oxoGua and 8-oxodG are in the same range and both are increased by smoking.[15,19] Accordingly,

[32] T. Bessho, K. Tano, H. Kasai, E. Ohtsuka, and S. Nishimura, *J. Biol. Chem.* **268,** 19416 (1993).

[33] J. P. Radicella, C. Dherin, C. Desmaze, M. S. Fox, and S. Boiteux, *Proc. Natl. Acad. Sci. USA* **94,** 8010 (1997).

[34] T. Roldan-Arjona, Y. F. Wei, K. C. Carter, A. Klungland, C. Anselmino, R. P. Wang, M. Augustus, and T. Lindahl, *Proc. Natl. Acad. Sci. USA* **94,** 8016 (1997).

[35] J. T. Reardon, T. Bessho, H. C. Kung, P. H. Bolton, and A. Sancar, *Proc. Natl. Acad. Sci. USA* **94,** 9463 (1997).

[36] J. Y. Mo, H. Maki, and M. Sekiguchi, *Proc. Natl. Acad. Sci. USA* **89,** 11021 (1992).

[37] H. Hayakawa, A. Taketomi, K. Sakumi, M. Kuwano, and M. Sekiguchi, *Biochemistry* **34,** 89 (1995).

although the exact relative importance of the repair pathways remain to be determined, the urinary excretion of 8-oxodG reflects the general average risk of a promutagenic oxidative adduct in DNA of all tissues and organs. Possibly, 8-oxoGua in the urine will be a valuable addition, allowing a complete account of the repair of 8-oxodG in DNA. The RNA oxidation product 8-oxoguanosine (8-oxoG) has also been used as a urinary biomarker.[38] The excretion is 3–4 times higher than of 8-oxodG.

For the use of the urinary excretion of repair products as biomarkers of oxidative DNA, extensive repair is assumed. Thus, after ionizing radiation the increase in urinary excretion of thymine glycol and 8-oxodG occurred within 24 hours in humans, whereas excess 8-oxodG was removed from mouse liver DNA after approximately 90 min.[39–41] In a study of 8-oxodG in human brain, the accumulation rate in the nuclear DNA corresponded to two lesions per cell per day.[42] In humans the reported values of the urinary excretion of the repair products, 8-oxodG, are in the range 15–50 nmol per 24 h[18] and the alternative repair product 8-oxoGua appears to be excreted in similar amounts.[15] The sum of these products thus corresponds to an average of 300–1000 lesions per day for each of the assumed 5×10^{13} cells in the body per day.[19,43] Accordingly, the calculated repair efficiency under these assumptions ranges from 99.4% to 99.8%.

Because of the extensive and rapid repair, the urinary excretion of the repair products will reflect the average rate of oxidative DNA damage in all the cells in the body. In contrast, the levels of oxidized bases in DNA lymphocytes or other accessible cells will reflect the steady-state levels, i.e., the balance between damage and repair, albeit only in a surrogate for target tissues. Accordingly, the two groups of biomarkers are supplementary.

Tissue/Cell Levels of Oxidized Bases in DNA

In tissue or cell samples, the level of oxidatively modified nucleobases can be measured by various techniques, including HPLC–EC (or MS or UV), GC/MS-SIM, TLC with ^{32}P-postlabeling, and various immunoassays

[38] H. Witt, A. Z. Reznick, C. A. Viguie, P. Starke-Reed, and L. Packer, *J. Nutr.* **122,** 766 (1992).

[39] D. S. Bergtold, C. D. Berg, and M. G. Simic, *Adv. Exp. Med. Biol.* **264,** 311 (1990).

[40] S. Blount, H. R. Griffiths, and J. Lunec, *Molec. Aspects Med.* **12,** 93 (1991).

[41] H. Kasai, P. F. Crain, Y. Kuchino, S. Nishimura, A. Ootsuyama, and H. Tanooka, *Carcinogenesis* **7,** 1849 (1986).

[42] P. Mecocci, U. MacGarvey, A. E. Kaufman, D. Koontz, J. M. Shoffner, D. C. Wallace, and M. F. Beal, *Ann. Neurol.* **34,** 609 (1993).

[43] S. Loft, E. J. M. V. Velthuis-te Wierik, H. van den Berg, and H. E. Poulsen, *Cancer Epidemiol. Biomarkers Prev.* **4,** 515 (1995).

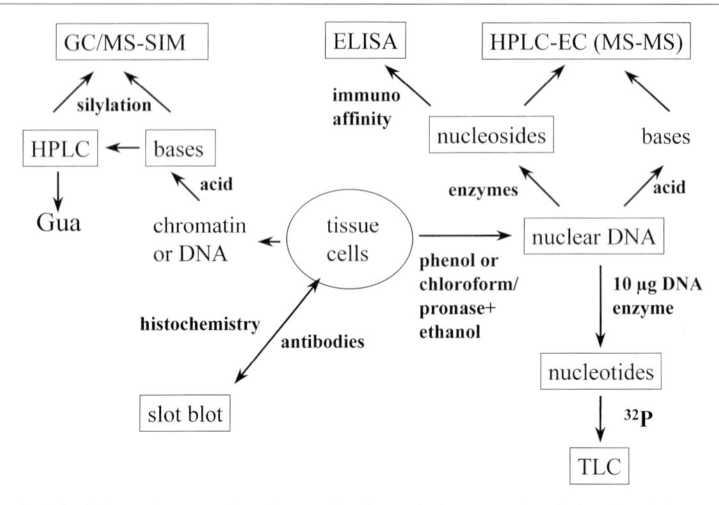

FIG. 4. Principles of assays for determination of tissue and cell levels of 8-oxodG and other modified nucleosides in DNA.

(Fig. 4). Except in the slot-blot technique and immunohistochemistry,[44,45] DNA or chromatin is isolated and hydrolyzed by enzymes or acid at high temperature. In all the assays the abundant unmodified nucleobases may be oxidized and thus cause artificially high values. Particularly, the derivatization with silyl groups for the GC/MS is prone to give rise to oxidation and should be carried out after removal of the unmodified base[46,47] or under controlled temperature and other conditions.[48] Similarly, the γ-radiation from the [^{32}P]phosphate used for postlabeling could oxidize guanine and thus explain the rather high values measured in human lymphocytes and rat organs by that method.[49,50] Even with the HPLC–EC method the reported values for 8-oxodG in leukocyte DNA varies from 0.3 to 13 per 10^5 dG.[18] There is no doubt that oxidation may occur during DNA extraction, particularly with the use of impure phenol and during drying the DNA after

[44] J. Musarrat and A. A. Wani, *Carcinogenesis* **15,** 2037 (1994).
[45] A. Yarborough, Y. J. Zhang, T. M. Hsu, and R. M. Santella, *Cancer Res.* **56,** 683 (1996).
[46] J.-L. Ravanat, R. J. Turesky, E. Gremaud, L. J. Trudel, and R. H. Stadler, *Chem. Res. Toxicol.* **8,** 1039 (1995).
[47] T. Douki, T. Delatour, F. Bianchini, and J. Cadet, *Carcinogenesis* **17,** 347 (1996).
[48] M. Hamberg and L. Y. Zhang, *Anal. Biochem.* **229,** 336 (1995).
[49] V. L. Wilson, B. G. Taffe, P. G. Shields, A. C. Povey, and C. C. Harris, *Environ. Health Perspec.* **99,** 261 (1993).
[50] U. Devanaboyina and R. C. Gupta, *Carcinogenesis* **17,** 917 (1996).

ethanol precipitation.[51–53] The lowest values have been obtained with anaerobic DNA extraction.[54,55] Some of the immunoassays are calibrated by HPLC-EC values,[44,56] whereas others appear to yield much higher 8-oxodG values than the HPLC-EC assays,[57] but whether that is due to insufficient specificity of the antibodies is unknown.

HPLC–EC Assays of 8-Oxo(d)G and Other Products in DNA from Tissues and Cells

Chemicals. 8-OxodG, 8-oxoGua, 8-aminoguanosine, 2′-deoxyguanosine (dG), guanine, pronase E, and nuclease P1 are from Sigma (St. Louis, MO), and alkaline phosphatase is from Boehringer Mannheim (Germany).

DNA Extraction and Hydrolysis

Tissues. Samples of tissues, e.g., liver, kidney, spleen, brain, or heart (200–600 mg) are homogenized in 10–30 ml HEPES 5 m*M* buffer [*N*-(hydroxyethyl)piperazine-*N*′-(2-ethanesulfonic acid)], sucrose 70 m*M*, manitol 250 m*M*, pH 7.4, on ice. After centrifugation at 1000*g*, the pellet containing the nuclei is resuspended in 0.15 *M* NaCl, 1 μl per mg tissue, on ice. Two hundred μl of suspension is transferred to 1.8 ml ice-cold TE buffer (150 m*M* NaCl, 10 m*M* Tris, 10 m*M* Na_2EDTA, pH 8.0), and 200 μl 10% sodium dodecyl sulfate (SDS) is added. After vortexing for 30 s, followed by rotation in an extraction bench for 15 min and incubation in a water bath at 37°C for 10 min, 200 μl 3 *M* sodium acetate (pH 5.2), 550 μl 5 *M* sodium perchlorate, and chloroform/isoamyl alcohol (24 : 1) 2 ml are added. After further rotation in an extraction bench for 10 min, the samples are centrifuged at 2100*g* and 4° for 10 min and the separated supernatant (nonorganic phase) is transferred to a 15 ml vial. The chloroform extraction step may be repeated before 6 ml of ice-cold 96% (v/v) ethanol is slowly added to the nonorganic phase.

[51] R. A. Floyd, M. S. West, K. L. Eneff, J. E. Schneider, P. K. Wong, D. T. Tingey, and W. E. Hogsett, *Anal. Biochem.* **188,** 155 (1990).

[52] H. G. Claycamp, *Carcinogenesis* **13,** 1289 (1992).

[53] S. Adachi, M. Zeisig, and L. Moller, *Carcinogenesis* **16,** 253 (1995).

[54] R. Collins, M. Dusinska, C. M. Gedik, and R. Stetina, *Environ. Health Perspec.* **104** Suppl 3, 465 (1996).

[55] M. Nakajima, T. Takeuchi, and K. Morimoto, *Carcinogenesis* **17,** 787 (1996).

[56] P. Degan, M. K. Shigenaga, E.-M. Park, P. E. Alperin, and B. N. Ames, *Carcinogenesis* **12,** 865 (1991).

[57] B. Yin, R. M. Whyatt, F. P. Perera, M. C. Randall, T. B. Cooper, and R. M. Santella, *Free Rad. Biol. Med.* **18,** 1023 (1995).

Lymphocytes, Sperm, Bone Marrow, or Other Cells. Lymphocytes are extracted from 5–7 ml blood diluted 1:1 with NaCl 0.9%. To 5 ml blood suspension in a 10 ml vial, 4 ml Lymphoprep (Nycomed Oslo, Norway) is carefully added to the bottom of the vial. After centrifugation for 30 min at 350*g* and 21° without brake, the lymphocyte layer is transferred to a 10 ml vial and 10 ml NaCl 0.9% is added. The pellet after centrifugation at 350*g* for 10 min, or approximately 20×10^6 sperm or other cells, is suspended in 7 ml of a buffer pH 7.6 containing Triton X-100 1% (v/v), sucrose 0.32 *M*, Trisma base 5 m*M*, and magnesium chloride 10 m*M*. After centrifugation at 1000*g* for 10 min, the supernatant is discarded. The pellet containing the nuclei is resuspended in 575 μl ice-cold buffer, pH 6.5, of sodium citrate 5 m*M* and sodium chloride 20 m*M*. Pronase E (1.2 mg/ml) with dithiothreitol (2.35 m*M*), *N*-lauroylsarcosine (0.75% w/v; Sigma), EDTA (10 m*M*), Trisma base (10 m*M*), and butylated hydroxyanisole (BHA, 0.3%, w/v) in methanol (6%, v/v) are added to a final volume of 1.4 ml, the stated concentrations, and pH 7.5. The samples are incubated overnight at 40°. Ammonium acetate to a final concentration of 0.5 *M* and 650 μl of an ice-cold 10 m*M* Tris–1 m*M* EDTA buffer are added and the DNA is precipitated by addition of 5 ml ice-cold ethanol (96%).

The DNA is allowed to precipitate at −20° overnight, if necessary followed by centrifugation at 2000*g* for 5 min. The DNA precipitate is washed with ice-cold 70% (v/v) ethanol in a petri dish, and dried in a vial with a stream of nitrogen gas or under vacuum in a centrifuge. It is crucial that the DNA be just barely dry, as further drying will cause artifactual oxidation of guanine, particularly in air and if phenol has been employed.[52] In fact, blotting the lump of DNA with paper tissue to remove ethanol/water is sufficient drying before hydrolysis. The DNA is dissolved in 200 μl 20 m*M* sodium acetate (pH 4.8), and digested to nucleotide level at 37° with 20 μl (5 U/sample) of nuclease P1 for 30 min followed by addition of 20 μl (1 U/sample) of alkaline phosphatase in 1 *M* Tris buffer (pH 8) for 60 min 37°. Two hundred μl is transferred to HPLC autosampler vials, for tissue samples after filtration through a Whatman (Clifton, NJ) filter (pore size 30 kDa) by centrifugation at 2600*g* for 35 min at 4°.

Alternatively, to the enzymatic hydrolysis the DNA sample can be hydrolyzed to nucleobases in acid, provided an RNAse digestion step has been included. The partly dried DNA is dissolved in 60% formic acid in screw cap tubes filled with nitrogen and subjected to 130° for 30 min (some laboratories use up to 150° for 45–60 min) in evacuated sealed Pierce (Rockford, IL) hydrolysis tubes.[58] After hydrolysis the formic acid is removed by freeze drying in a centrifuge and the sample reconstituted in mobile phase.

[58] H. Kaur and B. Halliwell, *Biochem. J.* **318,** 21 (1996).

HPLC Procedures

The HPLC apparatus consists of Merck-Hitachi components (San Jose, CA), L-6000 pump, AS-2000 autosampler, column oven set at 30°, D-6000 integration software, a Waters 440 UV absorbance detector at 254 nm, and an ESA Coulochem II electrochemical detector (Bedford, MA) equipped with a 5011 cell set 100–120 mV at electrode 1 and 200–300 mV at electrode 2 according to the voltammograms. A 25 cm Beckman Ultrasphere 5 μm column (Fullerton, CA) is eluted with acetonitrile 3.5–4.0% in phosphate buffer, pH 4. The UV detector is placed after the electrochemical cell in the flow path as the latter may cause back pressure, ruining the UV cuvette, and identical results are obtained in either sequence. dG is quantified by UV absorbance and 8-oxodG by electrochemical reactions (Fig. 5). The system is calibrated with injection of 8-oxodG 0, 4, 40, and 400 n*M* and dG 100 n*M*. Peak areas are used for calculations. Calibration curves and a quality control sample (rat liver pool) are run before and after each batch of samples. Each sample is extracted in duplicate and the enzymatic digest is injected three times, 20, 30, and 40 μl as the exact DNA content is not necessarily determined.

For nucleobases the mobile phase is a buffer of citric acid 12.5 m*M*, sodium acetate 25 m*M*, and EDTA 25 μM at pH 2.5. The electrochemical 5011 cell is set at 0 mV at electrode 1 and 200–300 mV at electrode 2. Guanine can be quantified both by UV absorbance, along with the other unmodified bases, and by electrochemical detection, whereas 8-oxoGua is

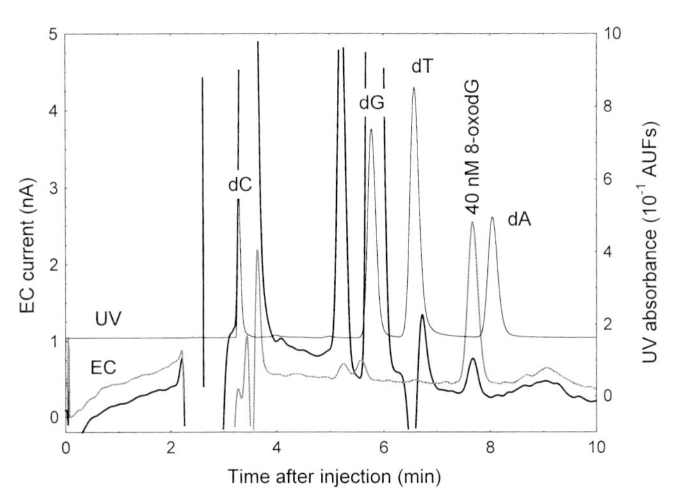

FIG. 5. Chromatogram of HPLC analysis of 8-oxodG in nuclear DNA. The 4 undamaged deoxynucleosides are shown in the UV tracing, whereas the electrochemical (EC) tracings show 8-oxodG in the hydrolyzed DNA sample (solid line) and as a standard (broken line).

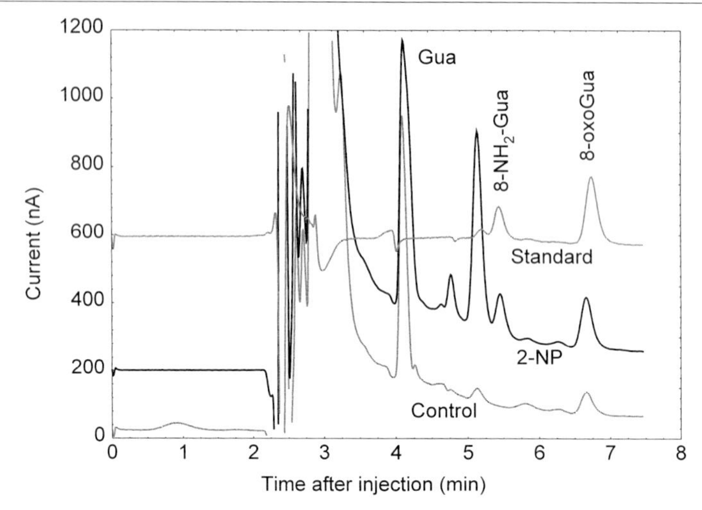

FIG. 6. Chromatogram (electrochemical tracing) of HPLC analysis of 8-oxoGua and 8-aminoGua in nuclear DNA from rats with or without pretreatment with 2-nitropropane (2-NP) and a standard solution.

quantified by electrochemical detection. If a higher pH is used, guanine elutes close to the 8-oxoGua and a guanase pretreatment is necessary to reduce the concentration.[59] Quantification is done as for 8-oxodG and dG. In addition to 8-oxoGua, the assay can measure other oxidatively modified nucleobases, 2,6-diamino-4-hydroxy-5-formamidopyrimidine (FapyGua), 5-OHmU, and 8-oxoadenine (8-oxoAde), which are electrochemically active.[58] Moreover, 8-aminoguanine, a reduction product of 8-nitroguanine that is a product of peroxynitrite attack on guanine and has received much attention, can be assayed by this method.[60,61] For the latter, calibration is achieved by subjecting 8-aminoguanosine (Sigma, St. Louis, MO) to the formic acid hydrolysis. For determination of 8-nitroguanine in DNA, the reconstituted sample, after freeze drying the formic acid hydrolyzate, is reduced with sodium hydrosulfite. Figure 6 shows electrochemical chromatograms of hydrolyzed liver DNA from rats with or without pretreatment with 2-nitropropane (100 mg/kg), which induces both 8-oxoGua and 8-aminoGua.[62]

[59] K. E. Herbert, M. D. Evans, M. T. Finnegan, S. Farooq, N. Mistry, I. D. Podmore, P. Farmer, and J. Lunec, *Free Rad. Biol. Med.* **20,** 467 (1996).

[60] V. Yermilov, J. Rubio, and H. Ohshima, *FEBS Lett.* **376,** 207 (1995).

[61] V. Yermilov, J. Rubio, M. Becchi, M. D. Friesen, B. Pignatelli, and H. Ohshima, *Carcinogenesis* **16,** 2045 (1995).

[62] R. S. Sodum, G. Nie, and E. S. Fiala, *Chem. Res. Toxicol.* **6,** 269 (1993).

Artifacts and Other Problems. The main problems with the assay relate to artifactual formation of 8-oxodG from the large amounts of dG present. This can happen at any time during sample DNA extraction and hydrolysis, as well as in the autosampler. Even in the pipes or injection ports of the autosampler, residual dG from a previous injection may be oxidized and contaminate a later injection. Accordingly, all procedures should be performed at the lowest possible temperature and we use repeated and thorough washing procedures in the autosampler, which is equipped with a cooled sample tray. Some laboratories recommend the addition of desferrioxamine for binding iron during homogenization of, e.g., liver tissue and the addition of antioxidants.[22] Some laboratories use a digestion step with both RNase and pronase before extraction with both phenol and chloroform/isoamyl alcohol.[22] However, we have tried such steps without any reduction in variation or background levels. We have tried to use the pronase digestion procedure described for cells for the organ tissues, as well as the chloroform extraction method for sperm cells. In both cases the variation and background levels increased. In our hands the most important step is the drying of the DNA after precipitation as outlined above. With our present method we have systematically passed calf thymus DNA through the individual extraction steps and subjected solutions of dG to the enzymatic and acid hydrolysis steps, and we have seen no signs of oxidation of dG in terms of increasing 8-oxodG or 8-oxoGua levels. Another problem involves variability in the EC detector response and changes in the voltammogram, particularly low responses after an analytical pause and as wear-out. The former problem can be reduced by injecting a solution containing a high concentration, e.g., 400 n*M*, 5 times or more before a batch is run, and the voltammogram should be checked frequently. Since all these sources of variation may cause considerable shifts in the measured levels, it is strongly recommended that quality control samples, e.g., from a pool of rat liver, be run with every analytical batch.

With other methods such as GC-MS SIM, ^{32}P, and immunoaffinity cleanup with slot-blot quantitation, even larger variations in obtained values from tissue and cellular DNA have been obtained in different laboratories.[63] After recent improvements in the GC-MS/SIM techniques with removal of intact bases before and/or control of temperature and oxygen during sample derivatization, the results are similar to what is achieved with the HPLC-EC methods described here.[46–48] Currently, a European interlaboratory quality comparison study is underway in order to standardize the methods and results related to 8-oxodG/-8-oxoG in tissue DNA.

[63] A. R. Collins, J. Cadet, B. Epe, and C. Gedik, *Carcinogenesis* **18,** 1833 (1997).

Antioxidant Intervention Studies Involving Biomarkers of Oxidative DNA Damage

The 24-hour urinary excretion of 8-oxodG shows a seven-fold range within the studied populations and an intersubject coefficient of variation of 30–40% (Table I). The intraindividual coefficient of variation has been 22% in 8 subjects examined twice on a controlled diet for 10 weeks[43] and 27% in groups of smokers studied two or three times.[64] For Tg and dTg excretion the interindividual coefficients of variation range from 28% to 92% (Table I).

Antioxidant supplementation could be expected to reduce the rate of oxidative DNA modification. Indeed, in a controlled smoking cessation study, the decrease in 8-oxodG excretion was a mirror of the increase in plasma vitamin C concentration.[64,65] However, so far intervention studies have not provided support for the notion of a beneficial effect of antioxidants. In smokers daily administration of β-carotene, vitamin C, vitamin E, or coenzyme Q had no effect on the excretion rate of 8-oxodG[66,67] or the RNA damage product, 8-oxoG.[38] Depletion of dietary ascorbic acid in healthy men had no effect on the level of 8-oxodG in lymphocyte DNA or the urinary excretion of 8-oxoguanosine, whereas the level of 8-oxodG in sperm DNA increased substantially.[68,69] Upon replenishment of ascorbate, the 8-oxodG level in sperm returned to the initial values.[69] In guinea pigs, the levels of 8-oxodG in liver DNA were not found to be affected by either very low or very high levels of dietary ascorbic acid or vitamin E.[70] No significant effect of vitamin E supplementation on rat liver 8-oxodG[71] and on tumor incidence and 8-oxodG in the liver of rainbow trout[72] has been found.

Ascorbic acid inhibited 8-oxodG formation in calf thymus DNA and V79 Chinese hamster cells exposed to ultraviolet radiation.[73,74] Glutathione,

[64] H. Prieme, S. Loft, M. Klarlund, K. Grønbæk, P. Tønnesen, and H. E. Poulsen, *Carcinogenesis* **19,** 347 (1998).

[65] J. Lykkesfeldt, H. Priemé, S. Loft, and H. E. Poulsen, *Br. Med. J.* **313,** 91 (1996).

[66] G. van Poppel, H. Poulsen, S. Loft, and H. Verhagen, *J. Natl. Cancer. Inst.* **87,** 310 (1995).

[67] H. Priemé, S. Loft, K. Nyyssönen, J. T. Salonen, and H. E. Poulsen, *Am. J. Clin. Nutr.* **65,** 503 (1997).

[68] R. A. Jacob, D. S. Kelley, F. S. Pianalto, M. E. Swendseid, S. M. Henning, J. Z. Zhang, B. N. Ames, C. G. Fraga, and J. H. Peters, *Am. J. Clin. Nutr.* 54 suppl. 6, 1302S (1991).

[69] G. Fraga, P. A. Motchnik, M. K. Shigenaga, H. J. Helbock, R. A. Jacob, and B. N. Ames, *Proc. Natl. Acad. Sci. USA* **88,** 11003 (1991).

[70] S. Cadenas, G. Barja, H. E. Poulsen, and S. Loft, *Carcinogenesis* **18,** 2373 (1997).

[71] K. Umegaki, S. Ikegami, and T. Ichikawa, *J. Nutritional Sci. Vitaminology* **39,** 303 (1993).

[72] J. D. Kelley, G. A. Orner, J. D. Hendricks, and D. E. Williams, *Carcinogenesis* **13,** 1639 (1992).

[73] A. Fischer-Nielsen, H. E. Poulsen, and S. Loft, *Free Rad. Biol. Med.* **13,** 121 (1992).

[74] A. Fischer-Nielsen, S. Loft, and K. G. Jensen, *Carcinogenesis* **14,** 2431 (1993).

cysteine, and vitamin C (but not vitamin E) also protected against increases in renal 8-oxodG caused by the kidney carcinogen potassium bromate in rats,[75] although tissue levels of ascorbate and glutathione, which are synthesized by rat tissues, were not measured in that study. On the other hand, vitamin E and ellagic acid, but not vitamin C, inhibited the increase in liver 8-oxodG induced by the hepatocarcinogen 2-nitropropane in rats.[76]

A potential DNA-protective effect of specific vegetable components was suggested by a 28% reduction in the rate of urinary 8-oxodG excretion after a diet with 300 g of Brussels sprouts in comparison with 300 g of noncruciferous vegetables.[77] This effect has been reproduced in rats in our laboratory (unpublished data). In a repeat experiment involving humans, however, no sign of reduced 8-oxodG excretion was seen in women and a reduction did not reach statistical significance in men after 300 g of Brussels sprouts per day for a week.[78] Nevertheless, a new method for 8-oxodG analysis was used in that study, giving extremely high values in some subjects. Indeed, cruciferous vegetables, such as Brussels sprouts and broccoli, contain certain phytochemicals which are anticarcinogenic possibly by inducing enzymes scavenging electrophiles and by mimicking the cellular protective response to oxidative stress.[79,80]

The differences between groups and effects of interventions seen in human studies of oxidized DNA bases are usually rather small, less than twofold, and for urinary excretion less than 50%. Moreover, for the nuclear steady-state levels, effects are obscured by the huge variation between and even within the various methods and laboratories. Thus, the above-mentioned negative results of antioxidant intervention trials could be due to the limited number of study subjects, i.e., a type 2 statistical error. For example, assuming that the variation is about 35%, a two-parallel-group controlled trial would need 190 persons in each group to detect a difference of 10%, 86 persons to detect a difference of 15%, and 47 persons to detect a difference of 20%, assuming $\alpha = 0.05$, $\beta = 0.20$. On the other hand, a crossover experiment with a homogeneous defined and controlled group

[75] K. Sai, T. Umemura, A. Takagi, R. Hasegawa, and Y. Kurokawa, *Jpn. J. Cancer Res.* **830,** 45 (1992).

[76] A. Takagi, K. Sai, T. Umemura, R. Hasegawa, and Y. Kurokawa, *Cancer Lett.* **91,** 139 (1995).

[77] H. Verhagen, H. E. Poulsen, S. Loft, G. van Poppel, M. I. Willems, and P. J. van Bladeren, *Carcinogenesis* **16,** 969 (1995).

[78] H. Verhagen, A. de Vries, W. A. Nijhoff, A. Schouten, G. van Poppel, W. H. M. Peters, and H. van den Berg, *Cancer Lett.* **114,** 127 (1997).

[79] T. Prestera, W. D. Holtzclaw, Y. Zhang, and P. Talalay, *Proc. Natl. Acad. Sci. USA* **90,** 2965 (1993).

[80] Y. Zhang, T. W. Kensler, C. G. Cho, G. H. Posner, and P. Talalay, *Proc. Natl. Acad. Sci. USA* **91,** 3147 (1994).

would show considerable lesser variation and differences of 30–50% could easily be detected, e.g., in a study of Brussels sprouts in 5 subjects.[66]

Urinary excretion of oxidized nucleosides and nucleobases, so far mainly 8-oxodG, represent promising biomarkers of oxidative DNA damage with a potential for establishing a relation to carcinogenesis and aging and for detecting changes in intervention trials. It should be borne in mind that even small changes, e.g., 10–15%, in oxidation rate could change the balance with DNA repair to a biologically important degree, and that detection of such a change in DNA oxidation may require large-scale controlled clinical trials.

[20] Comet Assay for Nuclear DNA Damage

By Stylianos M. Piperakis, Evangelos-E. Visvardis, and Aspasia M. Tassiou

Background

Single-cell gel electrophoresis (SCGE, comet assay) provides a very sensitive method for detecting strand breaks and measuring repair kinetics at the level of single cells. The technique was introduced in 1988 by N. P. Singh *et al.*, who modified other primary microgel electrophoresis techniques.[1,2] The unique design of the comet assay provides direct determination of the DNA damage in the responses of individual cells as well as examination of DNA damage and repair under a variety of experimental conditions. The SCGE assay is a particularly valuable technique because it allows the detection of intercellular differences in DNA damage and repair in any eukaryotic cell population. A variety of possible modifications of the assay facilitates the detection of single-stranded DNA breaks and alkali labile sites, double-stranded DNA breaks, incomplete excision repair sites, and interstrand cross-links, and increases the specificity and sensitivity of the assay. Moreover, it enables the study of different DNA repair pathways such as base excision and nucleotide excision repair.[3,4] In addition to the above, DNA fragmentation associated with cell death or related to

[1] N. P. Singh, M. T. McCoy, R. R. Tice, and E. L. Schneider, *Exp. Cell. Res.* **175,** 184 (1988).
[2] N. P. Singh, R. E. Stephens, and E. L. Schneider, *Int. J. Radiat. Biol.* **66,** 23 (1994).
[3] C. Buschfort, M. R. Muller, S. Seeber, M. F. Rajewsky, and J. Thomale, *Cancer Res.* **57,** 651 (1997).
[4] C. M. Gedik, S. W. B. Ewen, and A. R. Collins, *Int. J. Radiat. Biol.* **62,** 313 (1992).

0076-6879/99 $30.00

apoptosis can be evaluated with the comet assay.[5] There is also a great variety of DNA damaging agents that can be used in order to study DNA damage and repair with the comet assay procedure. After their treatment, the cells are embedded in agarose layers, lysed, and electrophoresed. Under fluorescence microscopy, cells with increased DNA damage display increased DNA migration from the nucleus toward the anode, thus forming the shape of a "comet" after staining with a fluorescent DNA binding dye. The comet assay utilizes a relatively small number of cells. The results can be obtained in a relatively short period of time, thus enabling the successful biomonitoring of DNA damage and repair in human cells, which makes the comet assay a very valuable tool in molecular epidemiology.

Cell Types

A great variety of cells isolated from many different sources[6] can be examined with the comet assay. Fresh or cryopreserved peripheral blood mononuclear cells are readily used from human, mouse, rat, and dog. (We have developed a method in which lymphocytes kept frozen for an indefinite period of time give the same response as fresh lymphocytes when treated with DNA damaging agents; they also show the same repair efficiency.)[8] Granulocytes from humans, epithelia (lens) from human or rat mucosal epithelia, human fibroblasts, human spermatocytes, human adenocarcinoma, lymphoma, and small cell carcinoma can be used. Other types of cells such as splenocytes, thymocytes, bone marrow cells from mouse, and brain cells from mouse and sheep have also been studied. Kidney, liver, pancreas, and testis cells from mouse and rat as well as embryos from rat have also been examined. Various cell lines can also be used in the comet assay that are derived from different types of cell culture. R_{aji} and TK6 from B-cell lines and HUT-78 from T-cell lines are currently used. From colon carcinoma cell lines, WiDr and HT-29 have been frequently used, from bladder carcinoma A1698, from cervix SiHa and HeLa, from prostate carcinoma DU-145, and from melanoma MeWo and HT-144 cell lines. Many cell lines from animal cell cultures have also been used, such as CHO and V79 from hamster, or L5178Y and SCCVII from mouse cell cultures.

Treatment

A significant number of DNA damaging factors have been studied with the comet assay and are summarized in Table I. The treatment conditions

[5] P. L. Olive, G. Frazer, and J. P. Banath, *Radiat. Res.* **136,** 130 (1993).

[6] D. W. Fairbairn, P. L. Olive, and K. L. O'Neill, *Mutation Res.* **339,** 37 (1995).

TABLE I
AGENTS USED TO INDUCE DNA DAMAGE AND EXAMINED WITH COMET ASSAY[a]

Exposure system	Dosage	Ref.	Exposure system	Dosage	Ref.
X-rays	0.1–100 Gy, 25–400 rads	*k*	Nitridazole	50–400 μM	*k*
γ-^{60}Co	1–400 Gy	*k*	Nitrogen mustard	0.5–5 μg/ml	*k*
^{137}Cs	0.1–1.75 Gy	*k*	Morphine	5–100 nM	*k*
^{125}I	160–8000 decay/cell/day	*k*	EMS	1–100 μg/ml	*k*
α-particle ^{222}Rn	0.39 Gy	*k*	Caffeine	2 mM	*k*
UV	6–24 J/m^2	*k*	Endonuclease III	1 μg/ml	*m*
UV-C (254 nm)	0.2–10 J/m^2	*k*	Superoxide dismutase	20–200 U/ml	*e, f*
UV-B (312 nm)	5–2000 J/m^2	*k*	Xanthine oxidase	2.42 μU/ml	*k*
Laser (660 nm)	1.2–7.2 J/m^2	*k*	MNNG	0.375–10 μg/ml	*b, d, g*
H_2O_2	100 nm–25 mm	*c*	DNC	0.008 mM	*b*
$CdSO_4$	0.5–5 mM	*k*	MCC	0.5–4 μg/ml	*k*
$MnCl_2$	1.5–4.5 mM	*k*	1-CMP	0.2–2 μg/ml	*k*
$KMnO_4$	1.8 mM	*k*	Lindane	0.125–1 μM	*k*
$NaCr_2O_7$	0.5–1.5 mM	*k*	Metronidazole	58.4–175.2–292.1	*k*
$NaAsO_2$	0.2–1.5 mM	*k*	BPL	100 μg/ml	*d*
TPA	1 μM	*k*	GSNO	100–500 μM	*e*
Staurosporine	5 nm	*k*	RBS	100–300 μM	*e*
Benzo[*a*]pyrene	0.3–3 mM	*b*	4-NQO	0–1 μg/ml	*g*
PhIP	0.5 mM–1.5 mM	*b*	Doxorubicin	0–5 μg/ml	*k*
IQ	1 mM–1.5 mM	*b*	RSU 1069	0–100 μg/ml	*l*
Lithoholic acid	20 μg/ml–40 μg/ml	*b*	Styrene 7,8-oxide	0.05–0.6 mM	*j*
ACNU	10–60 μg/ml	*k*	Acrylamide	10–100 μg/ml	*k*
NMMA	1 mM	*k*	NNK (*in vivo*)	50–250 mg/kg	*k*
DMM	5–40 μg/ml	*k*	NNN (*in vivo*)	500 mg/kg	*k*
SIN-1	100 μM	*e*	NDMA (*in vivo*)	1–2 mg/kg	*k*
SNOG	300 μM	*k*	Benzene (*in vivo*)	100–900 ppm	*k*
IL-1β	0.1 nm	*k*	Cisplatin (*in vivo*)	5 mg/kg	*k*
Adriamycin	1–4 μM	*k*	Streptozotocin (*in vivo*)	12.5–150 mg/kg	*k*
Bleomycin	1–300 mU/ml	*k*	Vitamin C	1.25–100 μM	*f, h*
Cyclophosphamide	25–250 μg/ml	*k*	Acetaldehyde	0–100 mM	*k*
Etoposide	0.5–50 μg/ml	*g, i*	Smoking	—	*c*
Dimetridazole	70.9–212.6–354.3 μM	*k*	Mitomycin C	10–100 μM	*k*

[a] TPA, (12-O)-tetradecanoylphorbol 13-acetate; PhiP, 2-amino-1-methyl-6-phenylimidazo[4,5-*b*]pyridine; IQ, 2-amino-3-methyl-3*H*-imidazo[4,5-*f*]quinoline; ACNU, 1-(4-amino-2-methyl-5-pyrimidinyl)methyl-3-(2-chloroethyl)-3-nitrosourea hydrochloride; NMMA, nitromonomethylarginine; DMM, dimethylmercury; SIN, 1,3-morpholinosydnonimine; SNOG, *S*-nitrosoglutathionine; IL-1β, interleukin-1β; EMS, ethylmethane sulfonate; MNNG, *N*-methyl-*N*-nitro-*N*-nitrosoguanidine; DNC, dinitrosocaffeidine; MCC, methylmercury chloride; 1-CMP, 1-chloromethylpyrene; BPL, β-propiolactone; GSNO, *S*-nitrosoglutathione; RBS, Roussin's second black salt (heptanitrosyltri-μ_3-thioxotetraferrate(1-)); 4-NQO, 4-nitroquinoline; RSU 1069, [1(2-nitro-1-imidazolyl)-3-aziridino-2-propanol]; NNK, 4-(*N*-methyl-*N*-nitrosamino)-1-(3-pyridyl)-1-butanone; NNN, *N*-nitrosonornicotine; NDMA, *N*-nitrosodimethylamine.

[b] B. L. Pool-Zobel and U. Leucht, *Mutation Res.* **375,** 105 (1997).

[c] S. M. Piperakis, E. E. Visvardis, M. Sagnou, and A. M. Tassiou, *Carcinogenesis* **19,** 695 (1998).

are dependent on the agent and the cell type under examination. In general in freshly isolated cells such as the peripheral blood mononuclear cells [lymphocytes, monocytes, NK (natural killer) cells] treatment is done in PBS or RPMI 1640 cell suspensions with or without fetal calf serum, depending on the agent. After treatment, cells are centrifuged and processed immediately for the comet assay, or they are resuspended in appropriate culture media, depending on the cell type, for the evaluation of their repair efficiency. Treated and repaired cells are pelleted and resuspended in low melting point agarose. A general outline of the comet assay protocol is given in Scheme 1.

Slide Preparation

Fully frosted microscope slides are commonly used. There are variations in the methodology reported in the literature concerning slide preparation for single-layer,[7] double-layer,[8] and "sandwich"[9] types. In the single-layer preparation, the cell pellet is mixed with 0.75% low melting point (LMP) agarose in phosphate-buffered saline (PBS), transferred to the slide (approximately 10^4–10^5 cells per slide are recommended), covered with a microscope coverslip, and refrigerated for 5 min to solidify. To anchor a single layer the slide can also be precoated with 1% LMP agarose in distilled water. Spread the LMP agarose with the pipetter tip to make a thin film and allow it to dry at room temperature. The other two variations are more commonly used. First, 1% standard agarose in PBS is layered on the slide.

[7] H. H. Evans, M. Ricanati, M.-F. Horng, Q. Jiang, J. Mencl, and P. L. Olive, *Radiat. Res.* **134,** 307 (1993).

[8] E.-E. Visvardis, A. M. Tassiou, and S. M. Piperakis, *Mutation Res.* **383,** 71 (1997).

[9] U. Plappert, E. Barthell, K. Raddatz, and H. J. Seidel, *Arch. Toxicol.* **68,** 284 (1994).

---

[d] J. E. Yendle, H. Tinwell, B. M. Elliot, and J. Ashby, *Mutation Res.* **375,** 125 (1997).

[e] C. A. Delaney, I. C. Green, J. E. Lowe, J. M. Cunningham, A. R. Butler, L. Renton, I. D'Costa, and M. H. L. Green, *Mutation Res.* **375,** 137 (1997).

[f] J. L. Re, M. P. de Meo, M. Laget, H. Guiraud, M. Castergnaro, P. Vanelle, and G. Dumenil, *Mutation Res.* **375,** 147 (1997).

[g] P. L. Olive, J. P. Banath, and R. E. Durand, *Mutation Res.* **375,** 157 (1997).

[h] N. P. Singh, *Mutation Res.* **375,** 195 (1997).

[i] P. Lebailly, C. Vigreux, T. Godard, F. Sichel, E. Bar, J. Y. LeTalaer, M. Henry-Amar, and P. Gauduchon, *Mutation Res.* **375,** 205 (1997).

[j] T. Bastlova, P. Vodicka, K. Peterkova, K. Hemminki, and B. Lambert, *Carcinogenesis* **16,** 2357 (1995).

[k] D. W. Fairbairn, P. L. Olive, and K. L. O'Neill, *Mutation Res.* **339,** 37 (1995).

[l] P. L. Olive, *Br. J. Cancer* **71,** 537 (1995).

[m] A. R. Collins, M. Ai-guo, and S. J. Duthie, *Mutation Res.* **336,** 69 (1995).

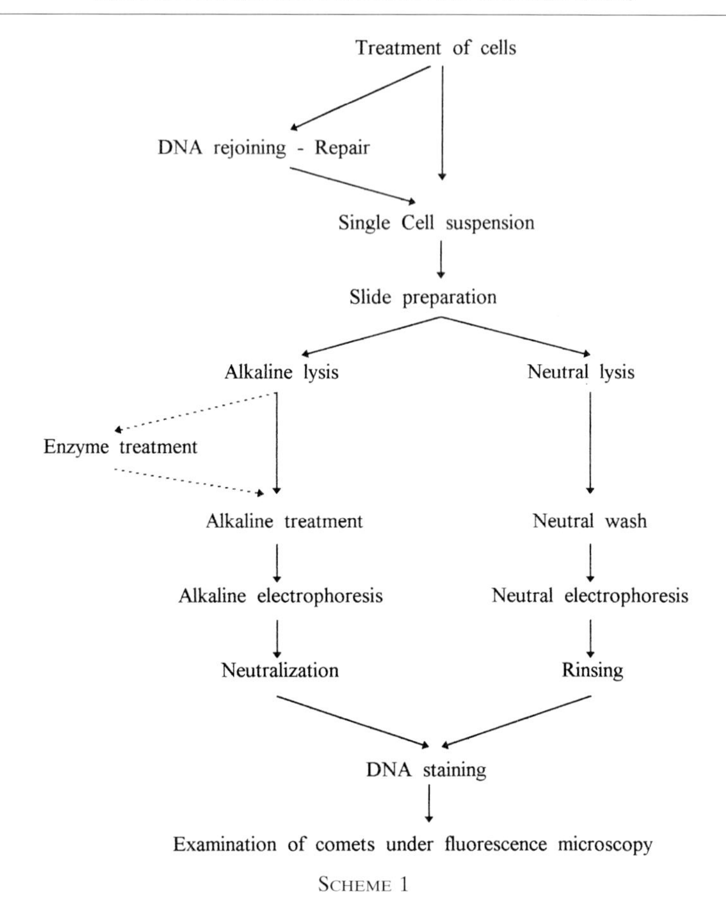

SCHEME 1

The slide is covered with a coverslip and refrigerated for 5 min to solidify. Then, the coverslip is removed and a second layer of 0.5–1% LMP agarose in PBS mixed with the cell pellet is transferred onto the slide. Cover with a coverslip and refrigerate as above (double-layer preparation). In the sandwich preparation, a third layer of 0.5–1% LMP agarose is placed on top of the other two layers and left to solidify for a few minutes as above. The fully frosted microscope slides can be used only a few times before the agarose layers begin to detach. In order to reuse them, we have developed a procedure of thorough cleaning, followed by brushing the slide with hot water. For better anchorage after the formation of the last agarose layer, we coat the slides on the periphery of the microgels with more LMP agarose. Ordinary clear slides can also be used if they are precoated with 1% standard agarose in H_2O. Spread the agarose with the pipetter tip, make a thin film on the slide, and let it dry in a warm oven or at room temperature. The cell pellet mixed with 0.5–1% LMP agarose in PBS is transferred onto the precoated slide and allowed to solidify as described above. With the latter preparation, slides can be kept for storage and later examination and the comets, which are now very close to the glass surface, do not suffer from the background fluorescence of the frosted surface.[10]

Lysis of Cells: Electrophoresis

After microgel preparations, the embedded cells are lysed and electrophoresed in order to examine for DNA breaks. Lysis can be performed under different conditions: (1) alkaline conditions for the detection of single-strand breaks and alkaline labile lesions, and (2) neutral conditions for the detection of double-strand breaks. The electrophoretical conditions depend on the lysis conditions used. An overview of the variations in the comet assay procedure reported in the literature is given in Table II.

Alkaline Lysis. The cells are lysed by gently immersing the slides in a lysis buffer (usually a high salt solution containing detergents at pH 10 to pH >12) for at least 1 hr. Usually 2.5 *M* NaCl, 100 m*M* Na_2 EDTA, 10 m*M* Tris, NaOH to pH 10.0, and 1% Triton X-100 with the possible addition of 1% sodium lauryl sulphate (SLS) and/or 10% dimethyl sulfoxide (DMSO) are used at 4° for 1 hr to remove cellular proteins. Lysis buffers such as 0.03 *M* NaOH, 1.0 *M* NaCl at about pH 12.5 with the possible addition of 0.1–0.5% SLS have also been used. The usual duration of the lysis period is 1 hr; however, extending the time up to 3 hr does not significantly alter the expressed DNA damage of the cells.[11]

Alkaline Unwinding and Electrophoresis. After lysis the slides are placed on a horizontal gel electrophoresis unit which is filled with fresh electrophoretic buffer (usually 0.3 *M* NaOH and 1 m*M* Na_2EDTA, pH > 12), at 4°. The cells are allowed to unwind in this alkaline buffer for 20 to 60 min so that the DNA single-strand breaks and alkaline labile sites are exposed prior to electrophoresis. The duration of alkali exposure seems to significantly enhance the expression of DNA damage and requires optimization for each tissue/cell type studied.[12] Electrophoresis generally has a duration from 10 to 40 min at 25 V and 300 mA, at an ambient temperature of 4°. The electrophoresis conditions depend on the levels of DNA damage expressed in the cells and the salt concentration of the running buffer. After electro-

[10] H. Vaghef, A. C. Wisen, and B. Hellman, *Pharmacol. Toxicol.* **78,** 37 (1996).
[11] Vijayalaxmi, G. S. H. Strauss, and R. R. Tice, *Mutation Res.* **292,** 123 (1993).
[12] Vijayalaxmi, R. R. Tice, and G. S. H. Strauss, *Mutation Res.* **271,** 243 (1992).

TABLE II
COMET ASSAY VARIATIONS IN TECHNIQUE PROCEDURE

Assay	Conditions	Ref.
Alkaline		
Lysis	A. 0.03 *M* NaOH, 1 *M* NaCl (pH ~ 12.5)	*a*
	A_1. 0.03 *M* NaOH, 1 *M* NaCl, 0.1–0.5% SLS	*b*
	B. 2.5 *M* NaCl, 100 m*M* EDTA, 10 m*M* Tris, 1% Triton X-100 (pH ~ 10)	*d*
	B_1. B + 1% SLS and/or 10% DMSO	*e*
DNA unwinding	A. 0.03 *M* NaOH, 1–2 m*M* EDTA (pH ~ 12.5)	*d*
	B. 0.3 *M* NaOH, 1 m*M* EDTA (pH ~ 13)	*c*
Electrophoresis	A. 0.5–0.67 V/cm	*f, a*
	A_1. 1 V/cm	
	B. 25 V, 300 mA	*b*
Neutral		
Lysis	A. 0.5% SDS, 0.03 *M* EDTA	*f, b*
	B. 2% SDS, 0.03 *M* EDTA	*g*
	C. 2.5% SDS, 0.025 *M* EDTA	*h*
Wash	A. 90 m*M* Tris, 2 m*M* EDTA, 90 m*M* boric acid	*f*
	B. Distilled water	*i*
Electrophoresis	A. 0.55–0.67 V/cm (90 m*M* Tris, 2 m*M* EDTA, 90 m*M* borate), 20–25 min	*g, f, b*
	B. 2.5 V/cm (117 m*M* Tris, 3 m*M* EDTA, 91 m*M* borate), 5 min	*i*
	C. 9 V/cm (40 m*M* Tris, 10 m*M* EDTA, 20 m*M* GAA), 3 min	*j*

[a] P. L. Olive, J. P. Banath, and H. H. Evans, *Br. J. Cancer* **67,** 522 (1993).
[b] P. L. Olive, D. Wlodek, R. E. Durand, and J. P. Banath, *Exp. Cell Res.* **198,** 259 (1992).
[c] N. P. Singh, R. R. Tice, R. E. Stephens, and E. L. Schneider, *Mutation Res.* **252,** 289 (1991).
[d] E.-E. Visvardis, A. M. Tassiou, and S. M. Piperakis, *Mutation Res.* **383,** 71 (1997).
[e] M. H. L. Green, J. E. Lowe, S. A. Harcourt, P. Akinluyi, T. Rowe, J. Cole, A. V. Anstey, and C. F. Arlett, *Mutation Res.* **273,** 137 (1992).
[f] P. L. Olive and J. P. Banath, *Int. J. Radiat. Biol.* **64,** 349 (1993).
[g] H. H. Evans, M. Ricanati, M.-F. Horng, Q. Jiang, J. Mencl, and P. L. Olive, *Radiat. Res.* **134,** 307 (1993).
[h] O. Oslting and K. J. Jochanson, *Int. J. Radiat. Biol.* **52,** 683 (1987).
[i] W. U. Müller, T. Bauch, C. Streffer, F. Niedercichholz, and W. Böcker, *Int. J. Radiat. Biol.* **65,** 315 (1994).
[j] A. Uzawa, G. Suzuki, Y. Nakata, M. Akashi, H. Ohyama, and A. Akanuma, *Radiat. Res.* **137,** 95 (1994).

phoresis, Tris buffer (0.4 *M* Tris-HCl, pH 7.5) is added gently to neutralize the excess alkali. The slides are washed 3 times, 5 min each.

Staining. The slides are stained with a fluorescent DNA binding dye for visual scoring and/or image analysis of the comets. Measurement of DNA damage with the comet assay is dependent on the binding characteristics of the fluorescent stain, although the sensitivity for detecting DNA damage is largely unaffected by the choice of stain. Various stains can be used in the comet assay procedure such as DAPI (5 μg/ml), propidium iodide (2.5–5 μg/ml), ethidium bromide (20 μg/ml), acridine orange (2 μg/ml), and Hoechst 33342 (10 μg/ml).[13–17]

Enzyme Treatment. For a more specific characterization of the origin of the expressed DNA breaks and the examination of their respective DNA repair pathways, two DNA repair enzymes, endonuclease III and formamidopyrimidine-DNA glycosylase, have been used to uncover oxidized pyrimidines and purine lesions, respectively.[18–19] The redoxyendonuclease, endonuclease III, introduces DNA breaks specifically at sites of oxidized pyrimidines, whereas the activity of formamidopyrimidine-DNA glycosylase (Fpg) is directed toward fragmented purines and those modified on the imidazole moiety such as 8-oxoguanine. In this way induced and/or unrepaired DNA damage of such type by the agent treatment can be expressed by the use of these enzymes with the comet assay. Immediately after lysis slides are washed in the enzyme reaction buffer (40 m*M* HEPES, 0.1 *M* KCl, 0.5 m*M* EDTA, 0.2 mg/ml BSA, pH 8) 3 times, 5 min each, at 4°. After the last wash excess liquid is removed using a tissue. Fifty μl of enzyme solution, 1 mg/ml endonuclease III in enzyme buffer or 22 pg/μl Fpg protein in enzyme buffer, is placed on the gel. The slides are covered with coverslips, placed into a moist box (to prevent desiccation), and incubated at 37° for 45 min for the endonuclease III treatment or for 30 min for the Fpg treatment. Subsequent steps—alkaline treatment, electrophoresis, and neutralization—are carried out as described above.

Neutral Assay

After the addition of the agarose layers, the slides are submerged in a lysis buffer which consists of 80 m*M* EDTA, 0.5–2% SDS, pH 8, at 50° for 4 hr. Other solutions have also been used (see Table II). After lysis, the slides are carefully placed in TBE buffer (usually 90 m*M* Tris, 90 m*M* boric acid, 2 m*M* EDTA) and are rinsed there for 2–16 hr at room

[13] A. Uzawa, G. Susuki, Y. Nakata, M. Akashi, H. Ohyama, and A. Akanuma, *Radiat. Res.* **137,** 25 (1994).

[14] P. L. Olive, R. E. Durant, J. Le Riche, I. A. Olivotto, and S. M. Jackson, *Cancer Res.* **53,** 737 (1993).

[15] N. P. Singh, R. R. Tice, R. E. Stephens, and E. L. Schneider, *Mutation Res.* **252,** 289 (1991).

[16] S. M. Paranjape, R. T. Kamakada, and J. T. Kadonaga, *Annu. Rev. Biochem.* **63,** 265 (1994).

[17] P. L. Olive, D. Wlodek, R. E. Durand, and J. P. Banath, *Exp. Cell. Res.* **198,** 259 (1992).

[18] A. R. Collins, S. J. Duthie, and V. L. Dobson, *Carcinogenesis* **14,** 1733 (1993).

[19] M. D. Evans, I. D. Podmore, G. J. Daly, D. Perrett, J. Lunec, and K. E. Herbert, *Biochem. Soc. Trans.* **23,** 434 (1995).

temperature. Distilled water has also been used as a rinsing solution. After rinsing, slides are placed in a horizontal gel electrophoresis chamber in a fresh TBE solution and are electrophoresed at 0.55–0.67 V/cm for 20–25 min, with the current adjusted at about 7 mA. A few variations in the electrophoretic conditions have been reported and are listed in Table II. After electrophoresis, slides are rinsed in distilled water for a few minutes and are stained (usually for 20 min in 2.5 μg/ml propidium iodide dissolved in 0.1 *M* NaCl) to examine the expressed DNA double-strand breaks of the cells under a fluorescence microscope.

Evaluation and Quantification of Comet Formation Patterns

The comets are evaluated by fluorescence microscopy. The simplest way of evaluating is to score the comets empirically on the extent of damage based on one of five predefined classes.[20] The comet formation consists of the extension of the DNA from the nucleus toward the anode, thus forming the 'tail' of the comet. The classes are defined according to tail intensity and having a value of 0, 1, 2, 3, or 4 (from undamaged, 0, to maximally damaged, 4). Representative micrographs of the five classes are shown in Fig. 1. The total score for 100 comets could range from 0 (all undamaged) to 400 (all maximally damaged). Although visual scoring many times is subjective, it shows a clear relationship to the percentage of DNA appearing in the tail as measured by image system analysis.[21] Comet length is also a commonly used parameter for the evaluation of the DNA migration and can be measured visually with a suitable eyepiece micrometer.[8,22] For the image system analysis, an epifluorescent microscope, with suitable excitation filters and with a suitable CCD (charge-coupled device) camera, is connected to a computer equipped with an image analysis software package for the comet evaluation. Many parameters are introduced in order to describe the comet formation patterns. The tail moment has been regarded as one of the most reliable parameters calculated by computerized image analysis. This parameter describes the amount of DNA damage since it refers both to the distance the DNA has migrated and to the amount of DNA that has migrated from the head region.[23] A major advantage of using the tail moment as an index of DNA damage is that both the amount of damage and the distance of its migration in the tail are repre-

[20] D. Anderson, T.-W. Yu, B. J. Phillips, and P. Schmezer, *Mutation Res.* **307,** 261 (1994).

[21] R. Collins, M. Ai-guo, and S. J. Duthie, *Mutation. Res.* **336,** 69 (1995).

[22] N. P. Singh, M. M. Graham, V. Singh, and A. Khan, *Int. J. Radiat. Biol.* **68,** 563 (1995).

[23] P. L. Olive and R. E. Durand, *J. Natl. Cancer Inst.* **85,** 707 (1992).

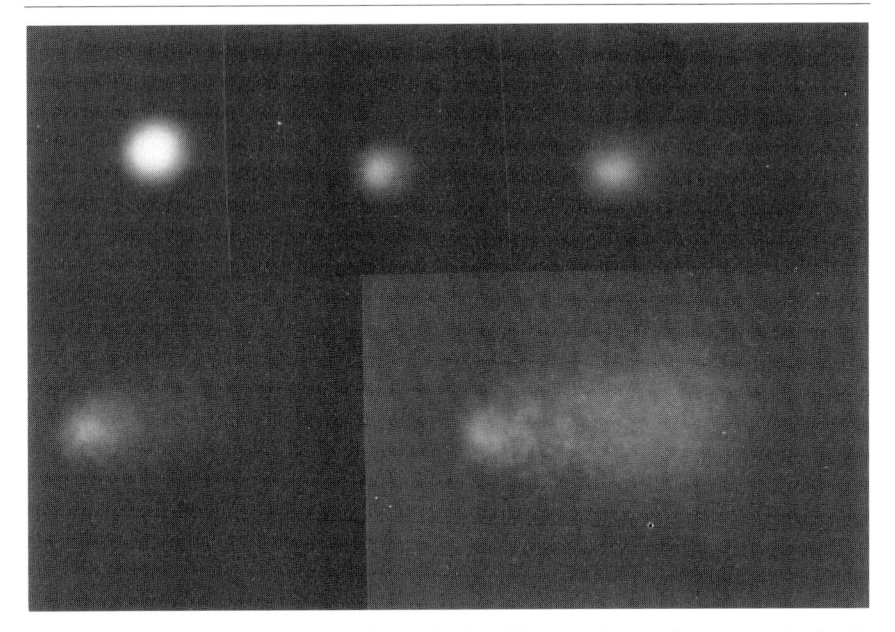

FIG. 1. Representative micrographs of the five different classes of comets stained with DAPI. First row classes: 0, 1, and 2. Second row: 3 and 4.

sented by a single number. The parameter of comet moment is used to analyze the comet formation patterns by providing a specific determination of the extent of DNA damage over a wider dose range, as it covers the cases of lower dose comets where the separate tail region is difficult to identify.[24]

Application to Study of Apoptosis

The most clearly defined biochemical event in apoptotic cells is nucleosomal fragmentation, which makes this event an obvious end point for evaluation using the comet assay. Because apoptotic DNA fragmentation is characterized by the generation of double-stranded breaks, both the neutral and alkaline assays can be used with equal efficiency for detecting breaks.[25,26] Apoptotic DNA is extensively degraded allowing most

[24] R. H. Kent, J. J. Eady, G. M. Ross, and G. G. Steel, *Int. J. Radiat. Biol.* **67,** 655 (1995).

of the comet head to migrate when subjected to electrophoresis. Under standard alkaline assay conditions, the most obvious aspect of apoptotic comets is the movement of most of the DNA from the head into the tail of the comet, which cannot be accounted for by comet length alone. Nondamaged cells have an average tail moment of approximately 2.0 or less, whereas apoptotic cell tail moment is generally greater than 30 and often higher than 60. Visual scoring of apoptosis provides a rapid screening method for determining the apoptotic fraction. Finally, of considerable significance is the fact that the highest fraction of apoptotic cells can be detected sooner using the comet assay treatment than with standard flow cytometric methods.

Acknowledgments

This work was partially supported by the EU grant No. ERBICISCT960300.

[25] O. Oslting and K. J. Jochanson, *Int. J. Radiat. Biol.* **52,** 683 (1987).
[26] D. W. Fairbairn, K. G. Carnahan, R. N. Thwaits, R. V. Grigsby, G. R. Holyoak, and K. L. O'Neill, *FEMS Microbiol. Lett.* **115,** 341 (1994).

[21] Generation of Hydroxyl Radical by Photolysis of Mercaptopyridine *N*-Oxides: Application to Redox Chemistry of Purines

By ABEL J. S. C. VIEIRA, JOÃO P. TELO, and RUI M. B. DIAS

Introduction

The reaction of the hydroxyl radical, $OH^{\cdot}$, with purines has been extensively studied owing to the major role this oxidizing radical plays in chemical damage to the DNA molecule in living tissues.[1] To provide protection against this kind of damage, a number of natural antioxidants are present in the plasma of aerobic organisms. These protective substances act either by scavenging the primary radicals or by repairing the chemical lesions. Purines such as uric acid,[2] xanthine,[3] and some of xanthine's methylated

[1] C. von Sonntag, "The Chemical Basis of Radiation Biology." Taylor and Francis, London, 1987.
[2] B. N. Ames, R. Cathcart, E. Schwiers, and P. Hochstein, *Proc. Natl. Acad. Sci. USA* **78**(11), 6858 (1981).
[3] A. J. S. C. Vieira, J. P. Telo, and R. M. B. Dias, *J. Chim. Phys.* **94,** 318 (1997).

0076-6879/99 $30.00

derivatives present in the human diet, such as caffeine[4,5] and theobromide,[6] have been shown to act as scavengers of oxygen free radicals.

The classical method to generate the OH˙ radical *in vitro* is the radiolysis of water using, for instance, γ rays or high energy electrons. However, the use of nonradiolytic methods has shown to be a convenient way to produce OH˙ and simulate the indirect effect of ionizing radiation on living systems. These so-called "radiomimetic" methods include Fenton-type chemical reactions and photolysis of hydrogen peroxide[7] and *N*-oxides of mercaptopyridines.[8–12]

Within this latter class of compounds, 4-mercaptopyridine *N*-oxide has proven to be an especially effective generator of OH˙. In this work, we illustrate this method by studying the radical redox chemistry of some biologically relevant purines, by characterizing the transient radicals by EPR spectroscopy and identifying the final stable products by HPLC.

Generation of OH˙ by Photolysis of Mercaptopyridine *N*-Oxides

2-Mercaptopyridine *N*-oxide (**1**) has been used as an antifungal and antibacterial agent and exhibits *in vitro* cytotoxic activity.[10] The biological activity of **1** is most likely due to the formation of hydroxyl radicals, since it is now well known that this compound generates OH˙ on irradiation with UV or visible light.[11,12] The 4-mercapto analog (**2**) is even more efficient as a source of OH˙ radicals on photolysis at near UV (about 350 nm), and has therefore been used to study OH˙-induced damage to DNA.[9]

4-Mercaptopyridine *N*-oxide (**2**) is conveniently prepared[8,13] from 4-chloropyridine *N*-oxide[14] by reacting this compound with thiourea and treating the resulting isothiuronium salt with sodium carbonate and sodium sulfide, followed by acidification with hydrochloric acid.

Compounds **1** and **2** exist as a thiol–thione tautomeric equilibrium [Eq.

[4] X. Shi, N. D. Dalal, and A. C. Jain, *Food Chem. Toxicol.* **29,** 1 (1991).
[5] J. P. Telo and A. J. S. C. Vieira, *J. Chem. Soc., Perkin Trans. 2,* 1755 (1997).
[6] S. Zbaida, R. Kariv, P. Fischer, and D. Gilhar, *Xenobiotica* **17,** 617 (1987).
[7] C. Hochanadel, *Rad. Res.* **17,** 286 (1962).
[8] B. Epe, D. Ballmaier, W. Adam, G. N. Grimm, and C. R. Saha-Möller, *Nucl. Acid Res.* **24**(9), 1625 (1996).
[9] W. Adam, D. Ballmaier, B. Epe, G. N. Grimm, and C. R. Saha-Möller, *Angew. Chem. Int. Ed. Engl.* **34,** 2156 (1995).
[10] J. Blatt, S. R. Taylor, and G. J. Kontoghiorghes, *Cancer Res.* **49,** 2925 (1989).
[11] J. Boivin, E. Crépon, and S. Z. Zard, *Tetrahedron Lett.* **31**(47), 6869 (1990).
[12] J. Boivin, E. Crépon, and S. Z. Zard, *Bull. Soc. Chim. Fr.* **129,** 145 (1992).
[13] D. H. R. Barton, D. Crich, and G. Kretzschmar, *J. Chem. Soc. Perkin Trans. 1,* 39 (1986).
[14] E. Ochiai, *J. Org. Chem.* **18,** 534 (1953).

(1)]. It was shown[15] that, in aqueous solution, the thione form is predominant over the respective thiol tautomers by a factor of about 54 for **1** and 3.6 for **2**. Both compounds are moderately acidic, with pK_a values[15] of 4.67 for **1** and 3.82 for **2**. The UV spectrum of the neutral form of **2** (aqueous solution at pH 2.7) presents three maxima at $\lambda_{max} = 214$ nm (log $\varepsilon = 3.9$), 285 nm (3.9), and 326 nm (4.0). At pH 7, the spectrum shows maxima at $\lambda_{max} = 225$ nm (3.8) and 322 nm (4.3).[8,15]

The UV photolysis of compound **2** in aqueous solution results in the homolysis of the N–O bond, yielding $OH^{\cdot}$ and a sulfur-centered radical, as shown in Eq. (1). The formation of both radicals was confirmed by electron paramagnetic resonance (EPR) spectroscopy using the spin-trapping technique.[16,17] This photolytic method is a clean way to produce $OH^{\cdot}$ and study its reactions. The sulfur-centered radicals $\mathbf{2}^{\cdot}$ are far less reactive than $OH^{\cdot}$ and do not interfere with its subsequent reactions. Radicals $\mathbf{2}^{\cdot}$ undergo bimolecular coupling, yielding the stable disulfide compounds **3–5**.

SH 4 5 3 6 2 N O⁻ **2** S N OH hν HO• + S• N **2•** (1)

3 **4** **5**

The photolysis of **2** to produce $OH^{\cdot}$ presents several advantages over the other nonradiolytic methods. As compared to H_2O_2, **2** is more efficiently photolyzed, allowing its use in lower concentrations. Under the same photolytic conditions,[3] **2** is quantitatively photolyzed while the extent of depletion of H_2O_2 is only 10%. The stronger absorption of **2** toward higher

[15] R. A. Jones and A. R. Katritzky, *J. Chem. Soc.*, 2937 (1960).
[16] K. J. Reszka and C. F. Chignell, *Photochem. Photobiol.* **60,** 442 (1994).
[17] K. J. Reszka and C. F. Chignell, *Photochem. Photobiol.* **61,** 269 (1995).

wavelengths allows the use of radiation of lower energies. This is experimentally more convenient and decreases the possibility of photochemically induced reactions of the substrates.

Another advantage of this method is that the OH˙ precursor is not an oxidizing species such as H_2O_2, or a reducing one such as Fe^{II}EDTA in the Fenton system. For this reason, the occurrence of undesirable redox reactions of the intermediates to be studied is greatly reduced. Moreover, a wider choice of oxidizing or reducing conditions is available, without the interference of the source of OH˙. This is especially important when it is envisaged to study the redox properties of transients formed upon reaction of OH˙ with the substrate.

The stability of **2** at high pH values allows work under basic conditions. This is not possible with the methods that use hydrogen peroxide because of its base-catalyzed disproportionation.[3,18] An additional advantage is the possibility to study the reactions of $O^{\cdot-}$, the conjugate base of OH˙ (pK_a of 11.9).[19] This radical is a weaker oxidant than OH˙, reacting by hydrogen abstraction rather than by addition.[1]

Reaction of OH˙ with Xanthine; Detection of Transient Radicals by EPR Spectroscopy

The radicals formed after the reaction of OH˙, generated by photolysis of **2**, with xanthine were studied by EPR spectroscopy. The solutions contained typically 2 m*M* xanthine, 0.5–3 m*M* 4-mercaptopyridine *N*-oxide (**2**), and 0.05 *M* sodium dibasic phosphate. The pH adjustments were made with perchloric acid and sodium hydroxide. The solutions were deaerated with argon and allowed to flow through a flat quartz cell within the spectrometer cavity, where *in situ* photolysis was performed using an optically focused high pressure Hg–Xe 1000 W UV lamp. The standard flow rate was ~0.1 ml sec^{-1}.

The reaction of the xanthine anion with OH˙ radical at pH values between 7 and 11 resulted in the observation of an EPR spectrum attributed to the xanthine radical anion, $\mathbf{6}^{\cdot-}$ (Fig. 1). At pH values higher than 12.6, the EPR spectra of the radical dianion, $\mathbf{6}^{\cdot 2-}$, is observed. The pK_a of $\mathbf{6}^{\cdot-}$ was determined as being 12.0. The same radicals were obtained after one-electron oxidation of xanthine by $SO^{\cdot -}_4$ in aqueous solution.[20]

[18] M. Kessi-Rabia, M. Gardès-Albert, R. Julien, and C. Ferradini, *J. Chim. Phys.* **92,** 1104 (1995).

[19] J. Rabani and M. S. Matheson, *J. Am. Chem. Soc.* **86,** 3175 (1964).

[20] S. R. Langman, M. C. Shohoji, J. P. Telo, A. J. S. C. Vieira, and H. M. Novais, *J. Chem. Soc. Perkin Trans. 2,* 1461 (1996).

At very high pH values $OH^{\cdot}$ deprotonates to give $O^{\cdot-}$. Under these conditions, radical $\mathbf{6}^{\cdot 2-}$ is most likely produced by hydrogen abstraction from the xanthine dianion. At moderately basic media, radical $\mathbf{6}^{\cdot -}$ may be formed either by direct oxidation of xanthine followed by deprotonation, or by addition of $OH^{\cdot}$ followed by fast dehydration [Eq. (2)].

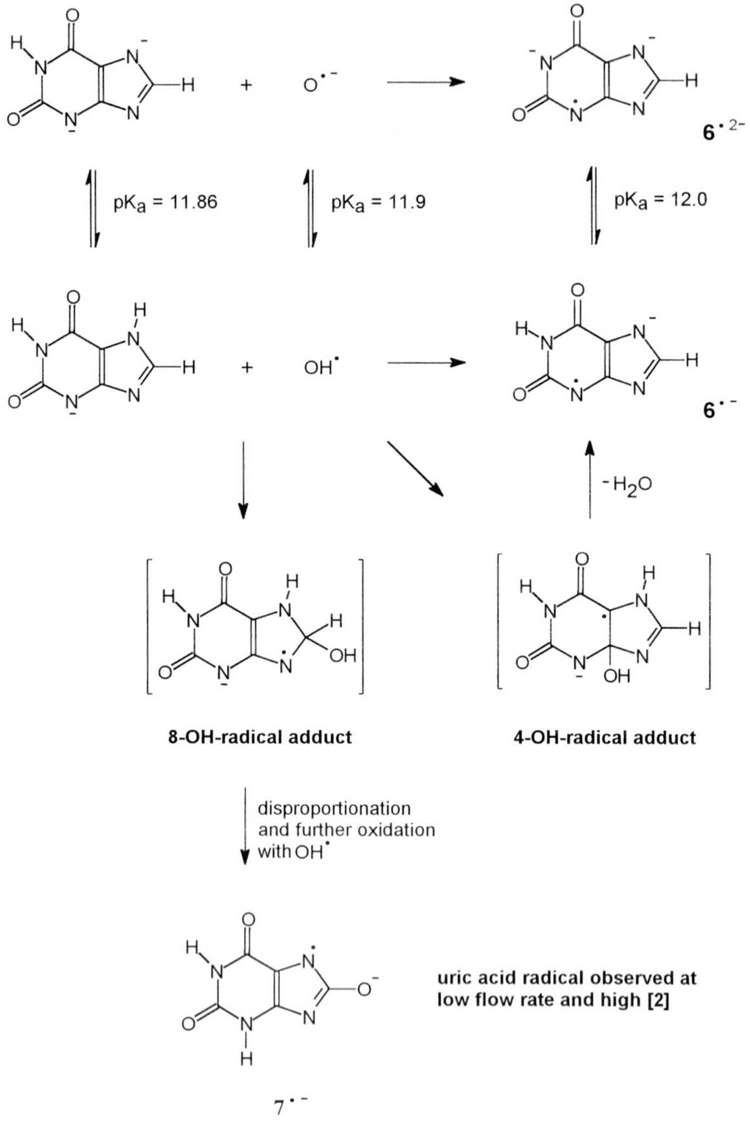

pKa = 11.86 pKa = 11.9 pKa = 12.0

$\mathbf{6}^{\cdot 2-}$

$\mathbf{6}^{\cdot -}$

$-H_2O$

(2)

8-OH-radical adduct

4-OH-radical adduct

disproportionation and further oxidation with $OH^{\cdot}$

uric acid radical observed at low flow rate and high [2]

$7^{\cdot -}$

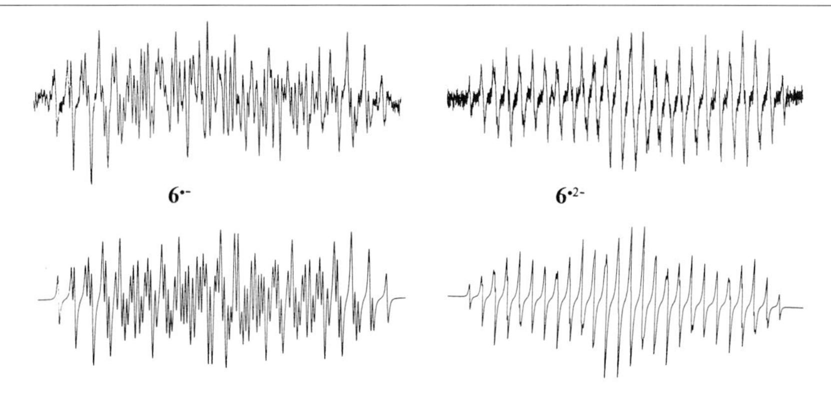

FIG. 1. EPR spectra of the radicals produced by the reaction of $OH^{\cdot}/O^{\cdot -}$ with 2 m*M* xanthine in aqueous solution at 280 K (up), together with the corresponding simulations (down). Left: Xanthine radical anion $\mathbf{6}^{\cdot -}$ at pH 10.4; right: xanthine radical dianion $\mathbf{6}^{\cdot 2-}$ at pH 12.6. [From A. J. S. C. Vieira, J. P. Telo, and R. M. B. Dias, *J. Chim. Phys.* **94,** 318 (1997).]

It is known that $OH^{\cdot}$ adds to purines[1,21–23] mainly at C-4 and C-8. In the case of xanthine, no hydroxyl adducts were detected by EPR. The 4-OH radical adduct, if formed, dehydrates rapidly to produce $\mathbf{6}^{\cdot -}$. When the concentration of **2** used was higher than that of xanthine, another EPR spectrum was observed at low flow rates. This spectrum was identical to the one obtained after oxidation of uric acid under the same conditions.[3] Previous pulse radiolysis results have shown that the $OH^{\cdot}$ adduct of xanthine at C-8 disproportionates to give uric acid and regenerate xanthine.[24] The results herein described show that the uric acid formed is further oxidized by $OH^{\cdot}$ to give the uric acid radical, $\mathbf{7}^{\cdot -}$ [Eq. (2)].

Reaction of $OH^{\cdot}$ with Adenine; Detection of Stable Final Products by HPLC

The HPLC analysis of the final products formed after the reaction of $OH^{\cdot}$ with a particular substrate is influenced by the formation of the disulfide compounds **3–5**. In the reversed-phase systems used for this analysis, the disulfides and some of the final products derived from purines have similar retention times. For this reason, care must be taken in the choice of the separation conditions.

[21] A. J. S. C. Vieira and S. Steenken, *J. Am. Chem. Soc.* **112,** 6986 (1990).

[22] A. J. S. C. Vieira and S. Steenken, *J. Chim. Phys.* **93,** 235 (1996).

[23] A. J. S. C. Vieira, L. P. Candeias, and S. Steenken, *J. Chim. Phys.* **90,** 881 (1993).

[24] J. Santamaria, C. Pasquier, C. Ferradini, and J. Pucheault, *Adv. Exper. Med. Biol.* **165A,** 185 (1984).

When OH$^{\cdot}$ reacts with the DNA base adenine, two major products are formed: 5-formamido-4,6-diaminopyrimidine (Fapy) and 8-hydroxyadenine (8-OH-Ade).[21,25] These products are easily detected by reversed-phase HPLC. In typical experiments, the solutions contained 1 m*M* adenine and 0.1 m*M* 4-mercaptopyridine *N*-oxide (**2**), and the pH was adjusted with phosphate buffers. The concentration of oxygen was varied from 0, by bubbling with argon, to saturation, by bubbling with O_2. Ferricyanide was also used as an oxidant, in concentrations in the range 1–10 μM. The solutions were left for 10 min in quartz cells at 5 cm from a 150 W medium pressure Hg lamp. After irradiation, the solutions were analyzed by reversed-phase HPLC. The eluent was a 5% (v/v) methanol aqueous solution buffered at pH 7 with 10 m*M* mono- and dibasic phosphates, flowing at 1 ml min^{-1}. The separation was performed through methylsilane (to separate Fapy) or octadecylsilane (to separate 8-OH-Ade) 250 × 4 mm columns. Detection of the products was achieved using an electrochemical or an optical diode array detector.

Both products are detected either in deaerated or aerated solutions (Fig. 2); however, the yields depend on the oxygen concentration. As expected from the mechanism of its formation [Eq. (3)], the yield of Fapy decreases with increasing oxygen concentration. This is due to the competition between ring opening followed by reduction, of the 8-OH adduct—leading to the formation of Fapy—and its oxidation, resulting in 8-OH-Ade formation.[21] Using water radiolysis as the source of OH$^{\cdot}$, the yield of Fapy is considerably lower[21,26]; this compound is not even detected if O_2 concentration is higher than 0.2 m*M*.

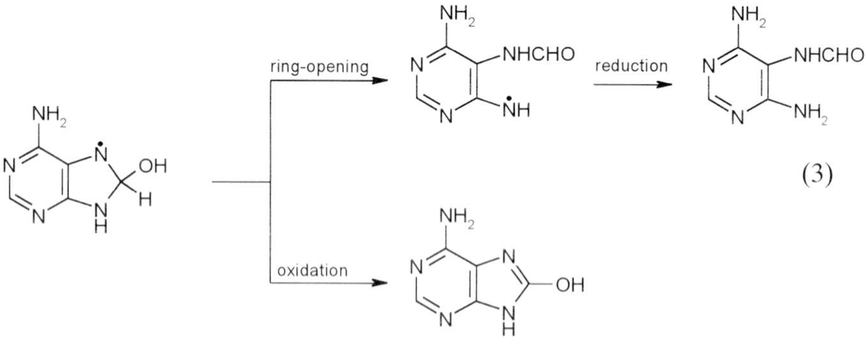

(3)

[25] A. Bonicel, N. Mariaggi, E. Hughes, and R. Téoule, *Rad. Res.* **83,** 19 (1980).

[26] R. M. B. Dias and A. J. S. C. Vieira, *J. Photochem. Photobiol.* **109**(2), 133 (1997).

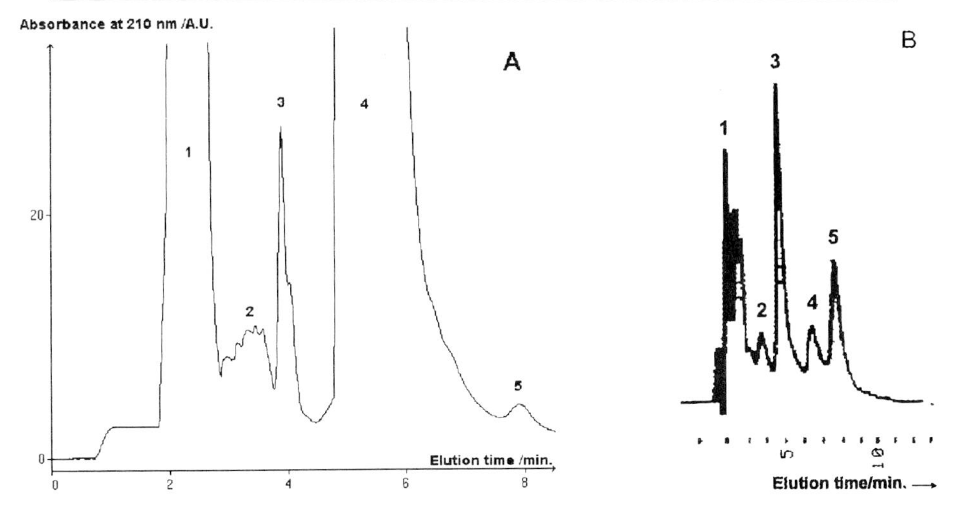

FIG. 2. (A) Optical chromatogram, using a methylsilane column and detection at 210 nm, of a solution containing initially 1.0 m*M* adenine and 0.1 m*M* **2**, aerated and at pH 7, after complete photolysis of **2**. Peak identification: 1, unretained solutes; 2, disulfide compounds; 3, 8-OH-Ade; 4, adenine; 5, Fapy. (B) Electrochemical chromatogram of an identical solution, using an octadecilsilane column and electrochemical detection at +650 mV vs NHE. Peak identification: 1, unretained solutes; 2, disulfide compounds; 3 and 4, unidentified; 5, 8-OH-Ade. Fapy is electrochemically inactive at this potential.

The yield of 8-OH-Ade also decreases with increasing oxygen concentration, although it requires the oxidation of the OH˙ adduct of adenine at C-8. When oxygen is replaced by ferricyanide as the oxidant for the Ade-8-OH˙ adduct, the yield of 8-OH-Ade increases up to three times. From these results it can be concluded that oxygen does not oxidize the C-8 hydroxyl radical-adduct by electron transfer, as ferricyanide does. Instead, oxygen adds to the position C-4 of this adduct and the subsequent elimination of superoxide from the peroxyl intermediate, necessary for the formation of 8-OH-Ade, occurs to an extent of about 30%.[26]

The addition of OH˙ radical at position C-8 of the purine system is a typical reaction pattern of a wide range of purinic compounds. Another biologically relevant example is caffeine, a naturally occurring xanthine derivative. After reaction of caffeine with OH˙ generated by UV photolysis of **2** in aqueous solution, the corresponding 8-hydroxy derivative—1,3,7-trimethyluric acid—was observed as a final stable product.[5] This is consistent with the formation of a hydroxyl radical adduct at C-8 followed by oxidation. Such a mechanism is probably responsible for the antioxidant properties attributed to this substance.[5,6]

[22] Trace Determination of Hydroxyl Radical using Fluorescence Detection

By BEIBEI LI, PETER L. GUTIERREZ, and NEIL V. BLOUGH

Introduction

Because of its importance in initiating oxidative damage, numerous methods have been developed to detect the hydroxyl radical (·OH) in biological systems.[1-4] However, in many instances these methods are not easily implemented, are subject to artifacts, are not sufficiently sensitive, or suffer from other problems.[5-7]

Here we summarize a new, highly sensitive approach for detecting and quantifying radicals, particularly ·OH, within biological systems. This approach is predicated on two well-known properties of the stable di-*tert*-alkyl nitroxides: first, their rapid reaction with carbon-centered radicals (10^7–10^9 M^{-1} sec^{-1}) to form stable *O*-alkylhydroxylamines[5-11] (radical adducts), and second, their efficient quenching of excited singlet states.[12-14] By covalently coupling a nitroxide at a short distance to a chromophore, fluorescence emission from the chromophore can be largely quenched. However, on reaction of radicals with the nitroxide moiety to form the diamagnetic (spin-paired) adducts, the intramolecular quenching pathway is eliminated and fluorescence emission increases, thus allowing radicals to

[1] G. M. Rosen, S. Pou, B. E. Britigan, and M. S. Cohen, *Methods Enzymol.* **233,** 105 (1994).
[2] H. Kaur and B. Halliwell, *Methods Enzymol.* **233,** 67 (1994).
[3] O. I. Aruoma, *Methods Enzymol.* **233,** 57 (1994).
[4] C. F. Babbs and M. G. Steiner, *Methods Enzymol.* **186,** 137 (1990).
[5] D. J. Kieber and N. V. Blough, *Anal. Chem.* **62,** 2275 (1990).
[6] C. G. Johnson, S. Caron, and N. V. Blough, *Anal. Chem.* **68,** 867 (1996).
[7] B. Li, P. L. Gutierrez, and N. V. Blough, *Anal. Chem.* **69,** 4295 (1997).
[8] J. Chateauneuf, J. Lusztyk, and K. U. Ingold, *J. Org. Chem.* **53,** 1629 (1988).
[9] A. L. J. Beckwith, V. W. Bowry, and K. U. Ingold, *J. Am. Chem. Soc.* **114,** 4983 (1992).
[10] V. W. Bowry and K. U. Ingold, *J. Am. Chem. Soc.* **114,** 4992 (1992).
[11] K. U. Ingold, "Numerical Data and Functional Relationships in Science and Technology," Vol. 13, Subvolume C, "Radical Reaction Rates in Liquids," H. Fischer, Ed., p. 166. Springer-Verlag, New York, 1983.
[12] N. V. Blough and D. J. Simpson, *J. Am. Chem. Soc.* **110,** 1915 (1988).
[13] S. A. Green, D. J. Simpson, G. Zhou, P. S. Ho, and N. V. Blough, *J. Am. Chem. Soc.* **112,** 7337 (1990).
[14] S. Herbelin and N. V. Blough, *J. Phys. Chem.* in press (1998).

0076-6879/99 $30.00

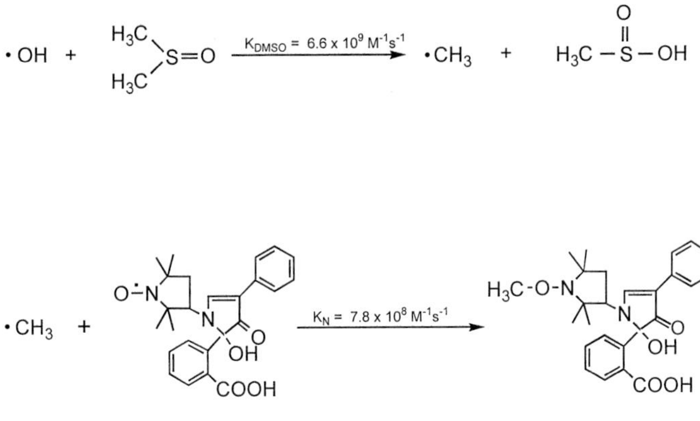

$$\cdot OH + (H_3C)_2S{=}O \xrightarrow{K_{DMSO} = 6.6 \times 10^9\ M^{-1}s^{-1}} \cdot CH_3 + H_3C{-}S({=}O){-}OH$$

$$\cdot CH_3 + \mathbf{I} \xrightarrow{K_N = 7.8 \times 10^8\ M^{-1}s^{-1}} \mathbf{II}$$

SCHEME 1

be detected optically.[12,15,16] The highly fluorescent radical adducts can then be separated by high-performance liquid chromatography (HPLC) and quantified fluorometrically.[5,7,17] In most cases, mass spectrometry can be used to help identify or confirm the structure of the radical adduct(s).[6,18]

Because the reaction between ·OH and the nitroxides does not lead to a stable product, an organic compound that reacts with ·OH to generate a carbon-centered radical must be added. This carbon-centered radical can then react with the nitroxide to form a stable *O*-alkylhydroxylamine (Scheme 1). Of the possible organic compounds that might be employed, dimethyl sulfoxide (DMSO) offers a number of advantages, including high water solubility, low toxicity to cells and tissues,[4] rapid reaction with ·OH ($6.6 \times 10^9\ M^{-1}\ sec^{-1}$)[19] and the generation of a single carbon-centered radical (the methyl radical; Scheme 1).[20] Because of the first three of these properties, the quantitative reaction of ·OH with DMSO can often be attained at DMSO concentrations compatible with cells and tissues.[4]

To trap the methyl radical, a fluorescamine-derivatized, 3-amino-2,2,5,5-tetramethyl-1-pyrrolidinyloxy free radical (3-ap) is employed (**I;** Scheme

[15] J. L. Gerlock, P. J. Zacmanidis, D. R. Bauer, D. J. Simpson, N. V. Blough, and I. T. Salmeen, *Free Rad. Res. Commun.* **10,** 119 (1990).

[16] S. Pou, Y. Huang, A. Bhan, V. S. Bhadti, R. S. Hosmane, S. Y. Wu, G. Cao, and G. M. Rosen, *Anal. Biochem.* **212,** 85 (1993).

[17] D. Kieber and N. V. Blough, *Free Rad. Res. Commun.* **10,** 109 (1990).

[18] D. J. Kieber, C. G. Johnson, and N. V. Blough, *Free Rad. Res. Commun.* **16,** 35 (1992).

[19] G. V. Buxton, C. L. Greenstock, W. P. Helman, and A. B. Ross, *J. Phys. Chem. Ref. Data* **17**(2), 513 (1988).

[20] M. K. Eberhardt and R. Colina, *J. Org. Chem.* **53,** 1071 (1988).

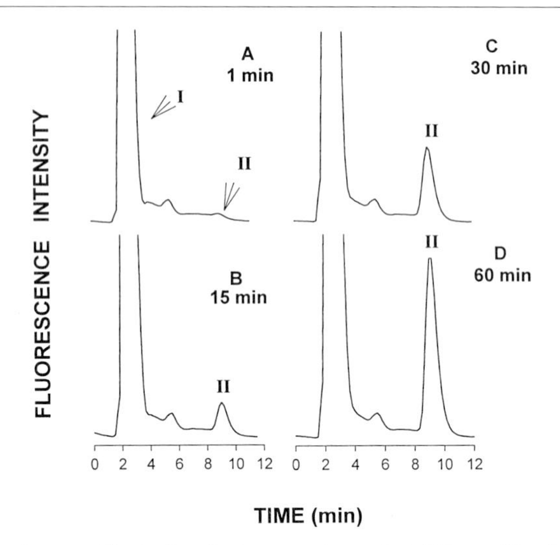

FIG. 1. Chromatograms illustrating the increase in product (**II**) resulting from a superoxide driven Fenton reaction (Scheme 2). The reaction mixture contained 500 μM **I,** 705 mM DMSO (5%), 400 μM xanthine (X), and 500 nM Fe(III)-EDTA in 100 mM, pH 7.5 phosphate (99.999%) buffer. The reaction was initiated by adding sufficient xanthine oxidase to produce superoxide at a rate of 7.5 nM sec^{-1} and terminated by HPLC injection. Product concentrations are 9.2 nM at 1 min, 126 nM at 15 min, 335 nM at 30 min, and 818 nM at 60 min. Reprinted with permission from ref. 7. Copyright 1997, American Chemical Society.

1). The choice of this particular compound was based on its ease of synthesis, the higher purity of the commercial preparations of the 3-ap precursor, and the very large difference in fluorescence quantum yields between the unreacted **(I)** and reacted **(II)** forms[5,14] (Scheme 1). The pyrrolidinyl nitroxide was chosen in an attempt to lessen or obviate possible interferences associated with the bioreduction of **I** to the hydroxylamine,[21] its reaction with superoxide,[22,23] or its reaction with metal ions[24] (see below). The product, **II,** is readily separated from **I** by reversed-phase HPLC[7] (Fig. 1) and is quantified fluorometrically with a detection limit of ~250 fmol (5 nM using a 50 μl injection loop).

[21] N. Kocherginsky and H. M. Swartz, *in* "Nitroxide Spin Labels—Reactions in Biology and Chemistry," p. 95. CRC Press, Boca Raton, FL, 1995.

[22] M. C. Krishna, D. A. Grahame, A. Samuni, J. B. Mitchell, and A. Russo, *Proc. Natl. Acad. Sci. USA* **89,** 5537 (1992).

[23] M. C. Krishna and A. Samuni, *Methods Enzymol.* **234,** 580 (1994).

[24] J. B. Mitchell, A. Samuni, M. C. Krishna, W. G. DeGraff, M. S. Ahn, U. Samuni, and A. Russo, *Biochemistry* **29,** 2802 (1990).

To establish the conditions under which the rate of ·OH formation can be determined quantitatively by this method, the dependence of the initial rate of formation of **II** on the concentration of DMSO and **I** must first be obtained.[7,25] Hydroxyl radical will react with other solution constituents (S_i) in competition with its reaction with DMSO [reactions (1) and (2)]. Similarly, in the presence of air, the methyl radical will react with dioxygen in competition with its reaction with **I** [reactions (3) and (4)].

$$\xrightarrow{F} \cdot OH + S_i \xrightarrow{k_i} \longrightarrow \text{Products} \quad (1)$$

$$\cdot OH + DMSO \xrightarrow{k_{DMSO}} \cdot CH_3 + CH_3SO_2H \quad (2)$$

$$\cdot CH_3 + O_2 \xrightarrow{k_O} CH_3O_2 \quad (3)$$

$$\cdot CH_3 + \mathbf{I} \xrightarrow{k_N} \mathbf{II} \quad (4)$$

At steady state, the initial rate of product formation is then given by

$$R = \left(\frac{d[\mathbf{II}]}{dt}\right)_o = \frac{Fk_N[\mathbf{I}]k_{DMSO}[DMSO]}{(k_o[O_2] + k_N[\mathbf{I}])\left(\sum_i k_i[S_i] + k_{DMSO}[DMSO]\right)} \quad (5)$$

To determine the concentration of DMSO needed for quantitative reaction with ·OH, the concentrations of **I** and O_2 are held constant and the dependence of R on DMSO concentration is acquired (Fig. 2). Under these conditions, Eq. (5) reduces to

$$R = \frac{k_{DMSO}[DMSO]CF}{\sum_i k_i[S_i] + k_{DMSO}[DMSO]} \quad (6)$$

where C [$= k_N[\mathbf{I}]/(k_N[\mathbf{I}] + k_O[O_2])$] is a constant. Equation (6) predicts a hyperbolic dependence of R on [DMSO]; at sufficiently high [DMSO] such that $k_{DMSO}[DMSO] \gg \Sigma_i k_i[S_i]$, quantitative reaction will be achieved (Fig. 2). A linear form of this relation can be obtained by taking the reciprocal of Eq. (6),

$$1/R = \frac{1}{CF} + \frac{k_{obs}}{k_{DMSO}CF}\left(\frac{1}{[DMSO]}\right) \quad (7)$$

[25] N. V. Blough and R. G. Zepp, *in* "Active Oxygen in Chemistry," C. S. Foote, J. S. Valentine, A. Greenberg, and J. F. Liebman, Eds., p. 280. Chapman and Hall, New York, 1995.

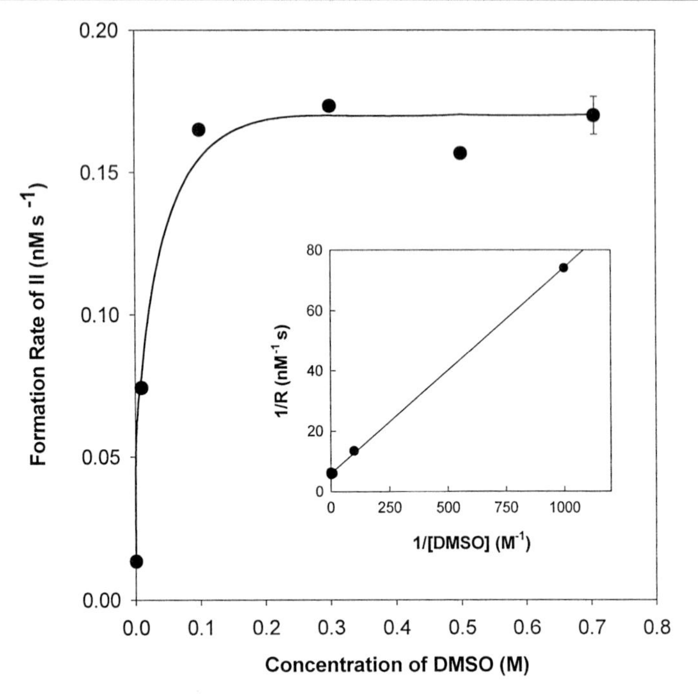

FIG. 2. Dependence of the formation rate of **II** on DMSO concentration for a suspension of JB6 cells (2×10^6 cells/ml) in PBS (pH 7.4), containing 500 μM **I** and in the presence of 100 μM diaziquone [3,6-diaziridinyl-2,5-bis(carboethoxyamino)-1,4-benzoquinone; AZQ]. The solid line represents a nonlinear least square fit of the data to a hyperbolic curve. *Inset:* Linearized form of the data (see text). Error bar represents ± one standard deviation of three measurements. Adapted from ref. 7.

where $k_{obs} = \Sigma_i k_i[S_i]$. From the values of the slope and intercept, along with the known rate constant for the reaction of ·OH with DMSO, the lifetime of ·OH ($= 1/k_{obs}$) can also be obtained.

Similarly, to determine the concentration **I** needed for quantitative reaction with the methyl radical, the concentration of DMSO is first set such that $k_{DMSO}[DMSO] \gg k_{obs}$, and the dependence of the formation rate of **II** on the concentration of **I** is determined (Fig. 3). Under these conditions, Eq. (5) reduces to

$$R = \frac{Fk_N[\mathbf{I}]}{k_O[O_2] + k_N[\mathbf{I}]} \tag{8}$$

which can be linearized by taking the reciprocal,

$$1/R = \frac{1}{F} + \frac{k_O[O_2]}{k_N F}\left(\frac{1}{[\mathbf{I}]}\right) \tag{9}$$

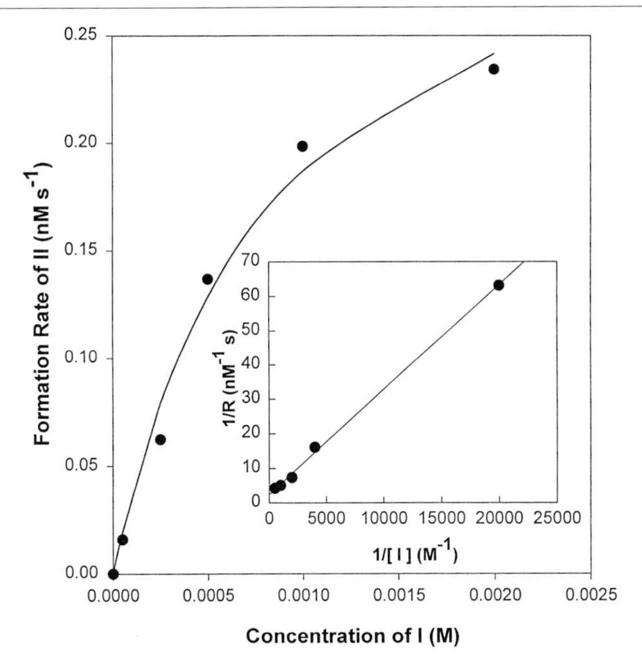

FIG. 3. Dependence of product (**II**) formation rate on the concentration of **I** in JB6 cells (2×10^6 cells/ml) in PBS, containing 100 μM AZQ and 5% DMSO. The solid line represents the hyperbolic curve generated from the values of the slope and intercept obtained from a linear least square fit to a double reciprocal plot of the data (*Inset*).

As above, when [**I**] is sufficiently high such that $k_N[\mathbf{I}] \gg k_o[O_2]$, the initial rate of **II** formation becomes equal to the rate of ·OH formation, $R = F$. From the values of the slope and intercept [Eq. (9); Fig. 3, inset], along with the concentration of O_2, the rate constant ratio for the reaction of the intermediate (methyl) radical with O_2 and **I**, respectively, can also be obtained. This ratio can be used as an independent check of the identity of the radical, if the rate constants for its reaction with O_2 and the nitroxide are known.

In principle, sufficient **I** could be added to ensure quantitative reaction with the methyl radical. In practice, lower concentrations of **I** are usually employed in order to avoid possible problems with solubility and toxicity. In this case, a rearranged form of Eq. (8) can be used to obtain F under conditions in which **I** does not react quantitatively with the methyl radical,

$$F = R\left(1 + \frac{K_O[O_2]}{k_N[\mathbf{I}]}\right) = R\left(1 + \frac{S}{In[\mathbf{I}]}\right) \quad (10)$$

where R is the measured formation rate of **II** acquired at a known concentration of **I,** and S and In are the values of the slope and intercept, respectively, obtained from the reciprocal plot [Eq. (9); Fig. 3]. A similar correction can be employed in those cases where high concentrations of DMSO are not compatible with the biological system of interest. Competition between DMSO and other hydroxyl scavengers (e.g., ethanol) can also be employed to test whether ·OH is indeed the oxidant being detected.[7]

In the following sections, the experimental procedures for applying this method to a biologically relevant model system and to cells are described.

Methods

HPLC

The HPLC that we employ for isocratic separations consists of an Eldex Model B-100-S single piston pump (Eldex, Napa, CA) followed by a 0–5000 psi pressure gauge, a Valco Model C10W injection valve (Valco, Houston, TX), and a RCM 8 × 10 cm Waters radial compression module containing a 0.5 × 10 cm Nova-PAK column with 4-μm reversed-phase (C_{18}) packing; 0.5-μm filters are placed after the pump and injector. A Spectroflow Model 757 absorbance detector (ABI Analytical, Ramsey, NJ), set to the absorption maximum of **I** and **II** (390 nm), is used in series with a Hitachi Model L-7480 fluorescence detector set to 390 nm (excitation) and 490 nm (emission maximum of **I** and **II**). Separations are performed at room temperature and a flow rate of 1 mL/min, using a mobile phase composition of 35% sodium acetate buffer (50 m*M*, pH 4.0)/65% methanol (v/v) and a 50 μl sample loop.

Synthesis and Characterization of I

Initial attempts to synthesize **I** using the approach of Pou *et al.*[16] yielded significant amounts of the nonfluorescent lactone (~50%)[14,26] as determined by HPLC. Thus, we recommend that **I** be synthesized by the one-step reaction of 3-ap with fluorescamine in acetonitrile following the procedures of Bernardo *et al.*[27]

Fluorescamine (25–100 mg) is dissolved in 3–5 ml acetonitrile. An equimolar amount of 3-ap is dissolved in 3–5 ml acetonitrile and then added dropwise with stirring to the solution of fluorescamine. Following the

[26] M. Weigele, J. F. Blount, J. P. Tengi, R. C. Czajkowski, and W. J. Leimgruber, *J. Am. Chem. Soc.* **94,** 4052 (1972).

[27] S. De Bernardo, M. Weigele, V. Toome, K. Manhart, W. Leimgruber, P. Bohlen, S. Stein, and S. Udenfriend, *Arch. Biochem. Biophys.* **163,** 390 (1974).

complete addition of 3-ap, the total volume of the solution is reduced to ~0.5 ml by flushing with dry nitrogen. Following storage at −20° overnight, a bright yellow precipitate is produced. The supernatant is decanted, and the precipitate rinsed 3 times with ethyl ether and then dried under nitrogen. The yellow product (percent yield, ~50%) is stored in the dark at −20° until needed. HPLC analysis using absorption detection shows a major component corresponding to **I** (≥95%) eluting at 3.5 min and a minor component (≤5%) eluting at 4.5 min corresponding to the lactone of **I**. Additional information on the identity and properties of **I** can be obtained through the following procedures: (1) low[6,18] and high[7,18] resolution mass spectrometry employing a desorption chemical ionization probe operating in the electron impact mode [molecular ion m/z = 417 $(M - H_2O)^+$, calculated for $C_{25}H_{25}N_2O_4$ m/z 417.18143, observed 417.18156]; (2) UV/visible spectrophotometry in 100 mM sodium phosphate buffer, pH 7.5[7] (λ_{max} at 386 nm and 268 nm, ε_{386} = 5225 M^{-1} cm^{-1} and ε_{268} = 13,900 M^{-1} cm^{-1}); (3) steady-state and time-resolved fluorescence spectroscopy in 0.2 M borate buffer, pH 8.1[14] [λ_{em} = 492 nm, quantum yield ~0.00046, fluorescence lifetime ≤20 ps]; and (4) electron paramagnetic resonance (EPR) spectroscopy in 100 mM sodium phosphate buffer, pH 7.5 (strong triplet with A_N = 16.0 G).

Synthesis and Characterization of II

Although other sources can also be employed,[5,6] the reaction of DMSO with ·OH produced via the Fenton reaction represents a convenient way of producing methyl radicals to synthesize **II** (reaction a in Scheme 2).

I (1–3 mM), 5% DMSO, and 1–3 mM H_2O_2 are combined in pH 7.5 sodium phosphate buffer to a final volume of 1–4 ml. The reaction is initiated by adding 1–3 mM Fe(II)-EDTA and allowed to proceed in the dark at room temperature for 30 min. The solution is then filtered through

$$\text{Xanthine (X)} + 2O_2 \xrightarrow{\text{Xanthine Oxidase (XO)}} 2O_2^{\cdot-} + \text{Uric Acid} \qquad (a)$$

$$2O_2^{\cdot-} + 2H^+ \longrightarrow H_2O_2 + O_2 \qquad (b)$$

$$[\text{Fe(III)-EDTA}]^- + O_2^{\cdot-} \rightleftharpoons [\text{Fe(II)-EDTA}]^{2-} + O_2 \qquad (c)$$

$$[\text{Fe(II)-EDTA}]^{2-} + O_2^{\cdot-} + 2H^+ \longrightarrow [\text{Fe(III)-EDTA}]^- + H_2O_2 \qquad (d)$$

$$[\text{Fe(II)-EDTA}]^{2-} + H_2O_2 \longrightarrow [\text{Fe(III)-EDTA}]^- + OH^- + \cdot OH \qquad (e)$$

SCHEME 2

a 0.2 μm filter and loaded onto the HPLC injection loop, consisting of three guard columns (Upchurch, 2 mm i.d. × 2 cm) connected in series with PEEK tubing (0.02 in. i.d.). These guard columns are individually packed with material obtained from a Waters C_{18} Sep-Pak and activated before sample loading by flushing with 15 ml of methanol and then 15 ml of Milli-Q water. Following injection, the peak corresponding to the product (elution time, ~9 min) is collected directly from the effluent of the HPLC column into an acid-washed vial and then extracted into chloroform. After chloroform evaporation, the purified adduct is stored at −20° in the dark until needed. High resolution mass spectrometry (EI)[7] can be employed to confirm the structure of **II** $(M - H_2O)^+$: calculated *m/z,* 432.20490 ($C_{26}H_{28}N_2O_4$), observed *m/z,* 432.20308. The fluorescence quantum yield and lifetime of **II** in pH 8.1, 0.2 *M* borate buffer are 0.097 and 4.6 nsec, respectively.[14]

HPLC Calibration

Purified **II** is first dissolved in a small volume of DMSO and then diluted into sodium phosphate buffer, pH 7.5. The concentration of **II** in this stock solution is determined spectrophotometrically at 386 nm (ε_{386} = 5225 M^{-1} cm^{-1}). A standard curve of fluorescence peak area vs the concentration of **II** injected into the HPLC is then generated by the dilution of this stock.

Detection of Hydroxyl Radical in Model Systems and in JB6 Cells

Superoxide Driven Fenton Reaction

Xanthine (X)/xanthine oxidase (XO)[28] in the presence of low concentrations of Fe(III)-EDTA (Scheme 2) provides a convenient way of producing low levels of ·OH which can be used to test the performance of an ·OH detection method.

In this system,[7,28] the rate of superoxide (and H_2O_2) production is controlled by the concentration of XO, whereas ·OH production rates are constrained by the concentration of Fe(III)-EDTA. Reactions are carried out in pH 7.5, 100 m*M* sodium phosphate buffer (99.999%), containing 400 μM X and appropriate concentrations of **I,** DMSO, XO, and Fe(III)-EDTA. Reactions are initiated by adding XO, allowed to proceed for appropriate times in the dark at 25°, and then terminated by injection onto the HPLC. XO and Fe(III)-EDTA concentrations can be varied systematically to produce ·OH at well-controlled rates.

[28] B. Halliwell, *FEBS Lett.* **92,** 321 (1978).

Rates of superoxide formation are determined spectrophotometrically at 550 nm using the SOD-dependent reduction of ferricytochrome *c*.[7,16,29] **I** and DMSO are varied systematically to establish the conditions for quantitative ·OH determination (Introduction, Figs. 2 and 3).[7] If necessary, the rates of product formation (**II**) can be corrected to ·OH formation rates using Eq. (10).

Determination of Hydroxyl Radical in Cells

The effect of the anticancer quinone, diaziquone [3,6-diaziridinyl-2,5-bis(carboethoxyamino)-1,4-benzoquinone; AZQ], on ·OH production by mouse epidermal cells (JB6) represents one example of the application of this method to a biological system.[30,31]

JB6 cells are maintained in continuous culture in Earle's modified essential medium (EMEM) supplemented with 10% fetal calf serum (FCS). Cells are incubated at 37° in a humidified atmosphere of 5% (v/v) CO_2. Cell viability is determined by trypan blue exclusion following each experiment and cell numbers are determined using a hemacytometer.

Cells in exponential growth are harvested by trypsinization. Appropriate concentrations of **I** and DMSO are added to the cell suspensions in PBS (phosphate-buffered saline, pH 7.4) to a final volume of 1 ml. Reactions are initiated by adding AZQ. The cells are then incubated in 37° in a humidified atmosphere of 5% CO_2 for appropriate periods of time. After a given time, the cells are centrifuged for 2 min at 2000–3000*g* (5415C Eppendorf centrifuge), and the supernatant analyzed by HPLC. Incubation times are measured from the addition of AZQ to the injection onto the HPLC. At the low concentrations of AZQ added, the cells are consistently 86% viable after 1.5 hr incubation.

The concentration of DMSO and **I** is varied systematically to establish the appropriate conditions for determining the rate of ·OH production quantitatively, and if necessary the rates of product formation are then corrected using Eq. (10).

Following addition of AZQ to JB6 cells, the formation of ·OH exhibits a lag over the first 40 min, followed by a linear increase (Fig. 4). Rates of ·OH production determined from the linear portion of the curves increase linearly with increasing AZQ concentration. Although this trend is observed consistently under a given set of experimental conditions, the absolute magnitude of the ·OH formation rates can change significantly depending on the batches of FCS or medium employed in the cell culture.

[29] B. F. Van Gelder and E. C. Slater, *Biochim. Biophys. Acta* **58,** 593 (1962).

[30] L. M. Weiner, *Methods Enzymol.* **233,** 92 (1994).

[31] P. L. Gutierrez, *Free Rad. Biol. Med.* **6,** 405 (1989).

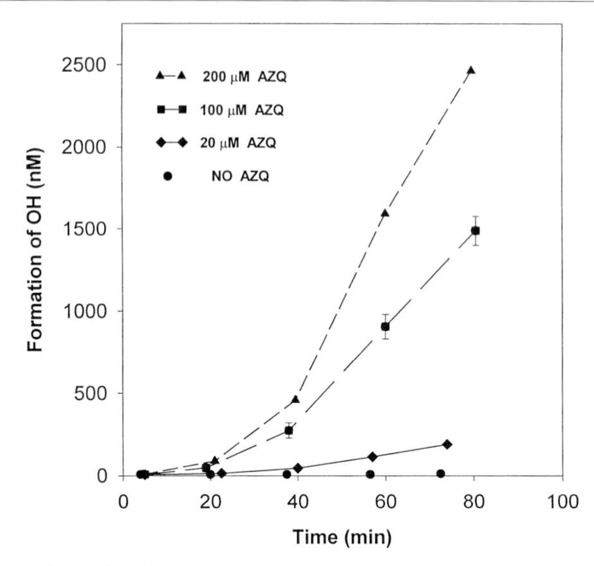

FIG. 4. The effect of AZQ concentration on the time courses of ·OH formation in a suspension of JB6 cells (2×10^6 cells/ml) in PBS (pH 7.4), containing 500 μM **I** and 705 mM DMSO (5%). Lines are drawn through the data points. Error bars represent ± one standard deviation of three measurements. Reprinted with permission from ref. 7. Copyright 1997, American Chemical Society.

We attribute this result to variable levels of trace metal contamination introduced by the different batches of serum and/or culture media. Addition of trace concentrations (e.g., 50 nM) of Fe(III)-EDTA substantially increases the rates of ·OH production, whereas addition of 100 μM diethylenetriaminepentaacetic acid (DTPA), a strong metal chelator, substantially depresses these rates. These results highlight the need to control the trace metal content in these cell systems very carefully.

A significant remaining question is whether this method measures intracellular as well as extracellular ·OH production. Detection of intracellular ·OH requires that the probe partition into the cell. Discrimination between intra- and extracellular localization of **I** is achievable using a paramagnetic broadening agent, such as potassium chromium(III) oxalate (CrOx), which does not enter the cell and thus broadens only the lines of extracellular **I**.[23] Within 2 min of its addition, **I** is partially protected from exchange broadening in the presence of the cells (Fig. 5), thus implying that it partitions into the interior of JB6 cells rapidly (in the presence of DMSO). This result argues that the probe is probably also measuring intracellular ·OH production at least in part.

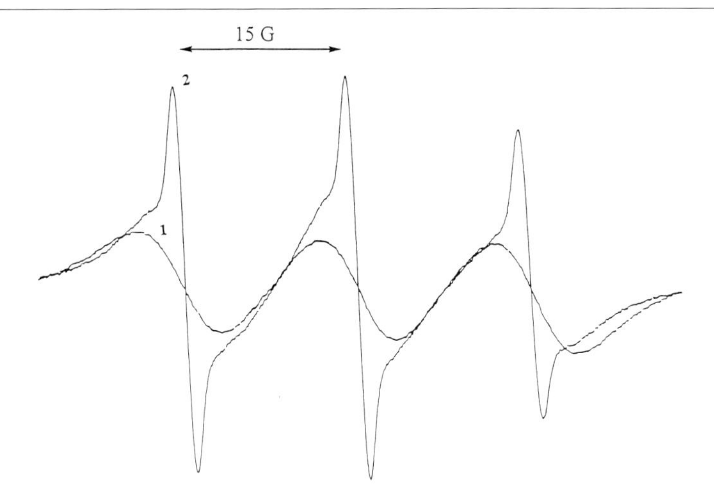

FIG. 5. EPR spectrum of 500 μM **I** in the presence of JB6 cells (1×10^8 cells/ml) suspended in PBS containing 705 mM DMSO (5%) and 80 mM of the broadening agent CrOx (2). EPR spectra were recorded 2 min after addition of **I** to the cells. Spectrum 1 is the control containing only 500 μM **I,** 705 mM DMSO (5%), and 80 mM CrOx in PBS.

Tests for Possible Interferences

Kinetics of Probe Bioreduction

Nitroxides are known to be reduced to their corresponding hydroxylamines by cells and tissues at rates dependent on the structure of the nitroxide.[21] If **I** is reduced rapidly by cells during the course of an experiment, quantitative estimates of ·OH production will not be obtained (although qualitative evidence for the formation of ·OH may still be acquired). Thus, it is essential to examine the kinetics of probe bioreduction in the system of interest to establish the appropriate conditions for the ·OH measurements.

To measure bioreduction rates, cells ($1.1–80 \times 10^6$ cells/ml) are suspended in PBS containing 500 μM **I** and 5% DMSO and then incubated at 37°. After appropriate incubation periods, aliquots are drawn into 50 μl capillaries and placed within standard 3 mm i.d. quartz EPR tubes. EPR spectra are then recorded on an EPR spectrometer with the standard instrument settings: frequency, 9.77 GHz; microwave power, 10 mW; modulation amplitude, 1.0 G; sweep width, 60 G; time constant, 10 msec; and scan time, 50 sec. Incubation times are measured from the addition of **I** to the time of EPR data collection.

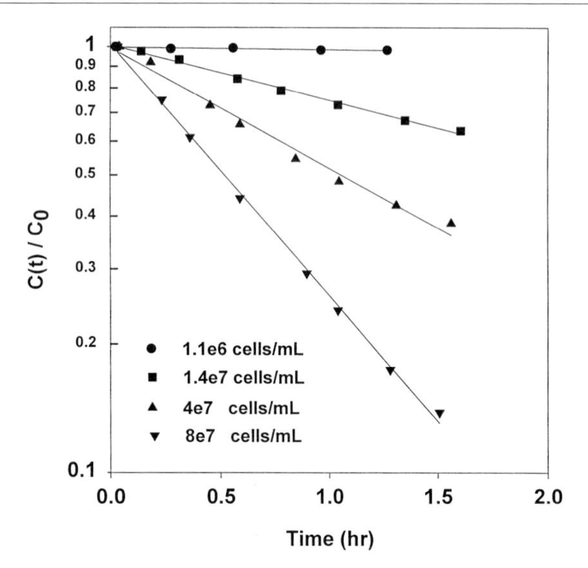

FIG. 6. Kinetics of **I** loss in JB6 cells by bioreduction. Cells ($1.1–80 \times 10^6$ cells/ml) were suspended in PBS containing 500 μM **I** and 705 mM DMSO (5%) and incubated at 37°. Aliquots were removed after appropriate times and EPR spectra recorded. $C(t)/C_0$ represents the fraction of the original EPR signal. The solid line represents a linear least square fit of the data points: (●) 1.6×10^6 cells/ml, (■) 1.4×10^7 cells/ml, (▲) 4×10^7 cells/ml, and (▼) 8×10^7 cells/ml.

In JB6 cells, the bioreduction of **I** exhibits first-order decay kinetics (Fig. 6) and is very slow at the cell densities employed in the ·OH experiments ($2–4.5 \times 10^6$ cells/ml). For example, at a cell density of 5×10^6 cells/ml, only ~10% of **I** is consumed by bioreduction over the course of 1 hr. This amount of probe loss will not interfere significantly with the ·OH assay.

Superoxide Reaction

Depending on their structure, certain nitroxides are known to react with superoxide in the presence of reductants such as NADPH and thiols.[16,22,32] Some nitroxides are also thought to act as SOD mimics.[22,23] If **I** were to undergo these reactions, its use as a quantitative probe of ·OH production could be limited because of its perturbation of the radical system (e.g., see Scheme 2). To test whether superoxide reacts with **I** (or other probes of this type), the effect of **I** on the rate of the SOD-inhibitable reduction of ferricytochrome *c* by superoxide can be examined.[7] Here

[32] E. Finkelstein, G. M. Rosen, and E. J. Rauckman, *Biochim. Biophys. Acta* **802,** 90 (1984).

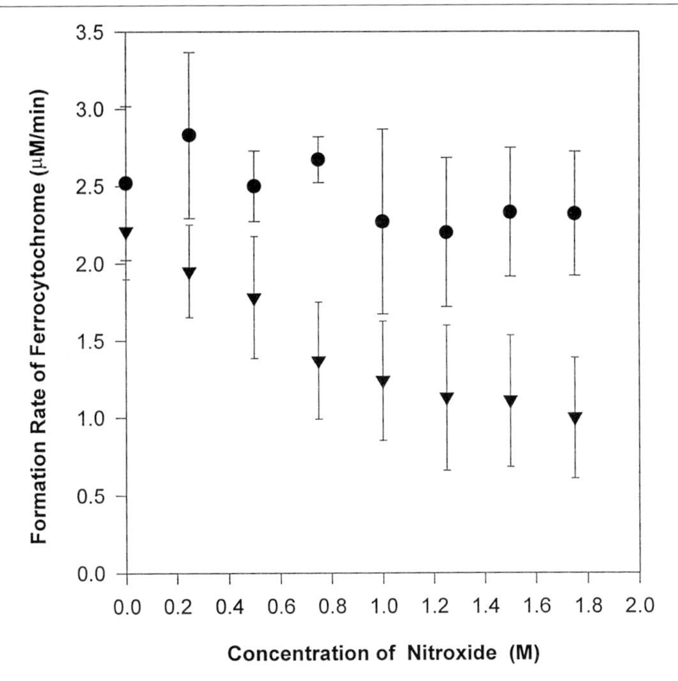

FIG. 7. Effect of **I** (●), and fluorescamine-derivatized, 4-amino-2,2,6,6-tetramethyl-1-piperidinyloxy (▼) on the rate of reduction of ferricytochrome *c* by superoxide in the presence of 1 m*M* NADPH. Solutions contain 20 μ*M* ferricytochrome *c*, 400 μ*M* X, 25 m*M* DMSO, and sufficient XO to produce superoxide at a rate of 2.5 μ*M*/min. Superoxide production rates were obtained from the difference in rates measured in the absence and presence of 30 units/ml SOD. Error bars represent ± one standard deviation of three measurements.

superoxide is generated at a well-known rate using X/XO (see above). The rate of probe loss in the presence of superoxide and reductant can also be followed by EPR (see above).[16,22]

In the presence of 1 m*M* NADPH, the rate of ferricytochrome *c* reduction by superoxide is unaffected by concentrations of **I** up to 1.8 m*M* (Fig. 7). In contrast, the fluorescamine-derivatized, 4-amino-2,2,6,6-tetramethyl-1-piperidinyloxy free radical inhibits ferricytochrome *c* reduction by superoxide over this concentration range, indicating that this compound does react with superoxide in the presence of reductant.[22,23] Further, when **I** (50 μ*M*) is subjected to a superoxide flux of 10 μ*M*/min in the presence of 200 μ*M* NADPH, no significant loss of EPR signal is observed. These results indicate that **I** neither reacts rapidly with superoxide in the presence of NADPH nor strongly catalyzes superoxide dismutation. This conclusion is further supported by the fact that the formation rate of **II** exhibits a hyperbolic dependence on the concentration of **I**. If **I** reacted rapidly with superoxide, acted as an efficient SOD mimic, or rapidly oxidized the transition metals involved in ·OH formation (Scheme 2), this type of dependence would not be observed.[7]

Acknowledgments

This work was supported by the National Institutes of Health (1R55GM44966; NVB), the Office of Naval Research (N00014-95-10201; NVB), the National Cancer Institute (CA53491; PLG), and the State of Maryland (NVB). We thank Dr. Paul Amstad for the use of the JB6 cell line.

Section II

Assays in Cells, Body Fluids, and Tissues

Articles 23 through 33

[23] Isoforms of Mammalian Peroxiredoxin That Reduce Peroxides in Presence of Thioredoxin

By HO ZOON CHAE, SANG WON KANG, and SUE GOO RHEE

Stepwise reduction of O_2 during normal metabolic processes produces reactive oxygen species (ROS) such as $O_2^{\cdot -}$, H_2O_2, and ·OH. To guard against these cytotoxic oxygen species, organisms have developed a number of antioxidant enzymes, including superoxide dismutase (SOD), catalase, and glutathione peroxidase. Cellular processes also generate reactive sulfur species (RSS), such as thiyl radicals (RS·), disulfide radical anions ($RSSR^{\cdot -}$), and peroxysulfenyl radicals (RSOO·). In the presence of O_2 and an electron donor such as ascorbate or a thiol compound (RSH), iron generates ROS. When a thiol is used as an electron donor, RSS are generated in addition to ROS. These two metal-catalyzed oxidation systems, the ascorbate oxidation system (Fe^{3+}, O_2, and ascorbate) and the thiol oxidation system (Fe^{3+}, O_2, and RSH), both cause damage to a variety of cellular components that can be prevented by addition of catalase or glutathione peroxidase, both of which remove H_2O_2.

We have previously purified a 25-kDa enzyme from yeast that prevents damage induced by the thiol oxidation system but not that induced by the ascorbate oxidation system, despite the fact that the degree of oxidative stress is similar for the two systems as judged by the comparable extent of induced inactivation of glutamine synthetase.[1,2] We postulated that the 25-kDa enzyme eliminates RSS, restricting its activity to the thiol-containing system, and named this protein thiol-specific antioxidant (TSA). Although the exact nature of the oxidant eliminated by TSA was not known at that time, the importance of TSA as an antioxidant was readily apparent as the application of oxidative pressure to yeast resulted in an increase in the synthesis of TSA, and TSA protein constituted 0.7% of total soluble protein from yeast grown aerobically.[3] Yeast TSA gene was cloned and sequenced.[4] It shows no significant homology to any known catalase, SOD, or peroxidase enzymes. This lack of homology is consistent with the observation that

[1] S. G. Rhee, K. Kim, I.-H. Kim, and E. R. Stadtman, Methods in Enzymology, Vol. 186, p. 478 (1990).
[2] K. Kim, I.-H. Kim, K. Y. Lee, S. G. Rhee, and E. R. Stadtman, *J. Biol. Chem.* **263,** 4704 (1988).
[3] I.-H. Kim, K. Kim, and S. G. Rhee, *Proc. Natl. Acad. Sci. USA* **86,** 6018 (1989).
[4] H. Z. Chae, I. H. Kim, K. Kim, and S. G. Rhee, *J. Biol. Chem.* **268,** 16815 (1993).

0076-6879/99 $30.00

TSA does not possess the catalytic activity characteristic of conventional antioxidant enzymes.

A database search revealed a number of proteins from a variety of organisms that show similarity to TSA. These homologous proteins were named the peroxiredoxin (Prx) family.[5,6] The TSA molecule is not associated with any obvious redox cofactors. Rather, a majority of the Prx proteins contain two conserved cysteine residues, which correspond to Cys-47 and Cys-170 in yeast TSA. The two conserved cysteines do not form an intramolecular disulfide. Instead, TSA exists predominantly as a dimer linked by two identical disulfide bonds between Cys-47 and Cys-170 as evidenced by direct isolation of dimeric tryptic peptides in which one monomer contains Cys-47 and the other contains Cys-170.[7] A TSA mutant with Cys-47 replaced by Ser (C47S mutant) failed to protect glutamine synthetase from the thiol oxidation system, whereas wild type and C170S mutant were equally protective. Thus, Cys-47, but not Cys-170, appears to be the site of oxidation.

We have demonstrated that the antioxidant activity of yeast TSA is attributable to its ability to reduce H_2O_2.[8] The apparent specific requirement for a thiol for antioxidant function was due to the fact that an intermolecular disulfide linkage of oxidized TSA can be reduced by a thiol but not by ascorbate.[8] We have shown that thioredoxin is the physiological electron donor for the reduction of Prx.[8] TSA was thus the first peroxidase to be identified for which Trx is the immediate electron donor, and it was therefore renamed Trx peroxidase (TPx). Despite this finding, the TSA (TPx) homologs were termed the Prx family but not the TPx family because not all members use Trx as the hydrogen donor. For example, the 21-kDa AhpC protein of *Salmonella typhimurium,* which shares 40% amino acid sequence identity with yeast TPx,[5] reduces alkyl peroxides through the thiol group of a redox-sensitive Cys residue; it is subsequently rereduced by the 57-kDa AhpF protein, which, in turn, accepts hydrogen from either NADH or NADPH.[9,10] Both AhpC and AhpF proteins copurify with each other and their genes are contiguous in both *S. typhimurium* and *Escherichia coli.*[9,10] Another TPx homolog from *Crithidia fasciculata,* Cf21

[5] H. Z. Chae, K. Robison, L. B. Poole, G. Church, G. Strotz, and S. G. Rhee, *Proc. Natl. Acad. Sci. USA* **91,** 7017 (1994).

[6] S. G. Rhee and H. Z. Chae, *Mol. Cells* **4,** 137 (1994).

[7] H. Z. Chae, T. B. Uhm, and S. G. Rhee, *Proc. Natl. Acad. Sci. USA* **91,** 7022 (1994).

[8] H. Z. Chae, S. J. Chung, and S. G. Rhee, *J. Biol. Chem.* **269,** 27670 (1994).

[9] F. S. Jacobson, R. W. Morgan, M. F. Christman, and B. N. Ames, *J. Biol. Chem.* **264,** 1488 (1989).

[10] G. Storz, F. S. Jacobson, L. A. Tartaglia, R. W. Norgan, L. A. Silveira, and B. N. Ames, *J. Bacteriol.* **171,** 2049 (1989).

protein, accepts hydrogen from NADPH via trypanothione and Cf16 protein.[11]

Mammalian Prx was purified from rat brain on the basis of its ability to protect glutamine synthetase from the oxidation system comprising dithiothreitol (DTT), Fe^{3+}, and O_2. The deduced amino acid sequence of rat brain Prx was 60% identical to that of yeast TPx. Twelve complete amino acid sequences have been reported for the mammalian Prx family members: five (PAG, NKEFA, NKEFB, TSA, and MER5) from human, four (MSP23, OSF3, TSA, and MER5) from mouse, two (TSA and HBP23) from rat, and one (SP22) from cow. With the exception of TSA, all mammalian Prx proteins were initially characterized without reference to antioxidant function: PAG was identified as the 22-kDa product of a gene whose expression increases during proliferation of human epithelial cells[12]; NKEFA and NKEFB as cytosolic factors from human red blood cells that enhance the activity of natural killer cells[13]; MER5 as the product of a housekeeping gene that is preferentially expressed in murine erythroleukemic cells during the early period of differentiation[14]; MSP23 as a 23-kDa stress-induced protein in mouse peritoneal macrophages[15]; OSF3 as the product of a gene that is specifically expressed in mouse osteoblastic cells[16]; SP22 as a substrate of a mitochondrial ATP-dependent protease in bovine adrenal cortex[17]; and HBP23 as a 23-kDa heme-binding protein purified from rat liver cytosol.[18]

Among the five reported human Prx sequences (PAG, NKEFA, TSA, NKEFB, and MER5), TSA is identical to NKEFB, and NKEFA is identical to PAG, respectively.[19] Thus, there are three distinct human Prx proteins: PAG/NKEFA, TSA/NKEFB, and MER5. Similarly, because MSP23 is identical to OSF3, the four reported mouse sequences actually correspond to three distinct proteins: MSP23/OSF3, TSA, and MER5. The three human

[11] E. Nogoceke, D. U. Gommel, M. Kiess, H. M. Kalisz, and L. Flohè, *Biol. Chem.* **378,** 827 (1997).

[12] M.-T. Prospéri, D. Fervus, I. Karczinski, and G. Goubin, *J. Biol. Chem.* **268,** 11050 (1993).

[13] M. Shau, L. H. Butterfield, R. Chiu, and A. Kim, *Immunogenetics* **40,** 129 (1994).

[14] T. Yamamoto, Y. Matsui, S. Natori, and M. Obinata, *Gene* **80,** 337 (1989).

[15] T. Ishii, M. Yamada, H. Sato, M. Matsue, S. Taketani, K. Nakayama, Y. Sugita, and S. Bannai, *J. Biol. Chem.* **268,** 18633 (1993).

[16] S. Kawai, S. Takeshita, M. Okazaki, R. Kikuno, A. Kudo, and E. Amann, *J. Biochem.* **115,** 641 (1994).

[17] S. Watabe, H. Kohno, H. Kouyama, T. Hiroi, N. Yago, and T. Nakazawa, *J. Biochem.* **115,** 648 (1994).

[18] S.-I. Iwahara, H. Satoh, D.-X. Song, J. Webb, A. L. Burlingame, Y. Nagae, and U. Muller-Eberhard, *Biochemistry* **34,** 13398 (1995).

[19] S. W. Kang, H. Z. Chae, M. S. Seo, K. Kim, I. C. Baines, and S. G. Rhee, *J. Biol. Chem.* **273,** 6297 (1998).

Prx proteins show only 60 to 80% sequence identity to each other, but each shares >90% identity with a corresponding mouse homolog. Each of the two rat proteins (HBP23 and TSA) and bovine SP22 show >92% sequence identity to one of the three human or mouse proteins. Therefore, the nine mammalian Prx proteins can be divided into three types (Prx I, II, and III).

Prx I proteins (human PAG/NKEFA, mouse MSP23/OSF3, and rat HBP23) consist of 199 residues and share >95% sequence identity. Rat HBP23 differs by only five and six amino acids, respectively, from human PAG/NKEFA and mouse MSP23/OSF3. The few differences among the three proteins constitute conserved amino acid substitutions. Prx II proteins comprise rat TSA, human TSA/NKEFB, and mouse TSA, all of which consist of 198 residues and share 92 to 98% sequence identity, with the differences mainly attributable to conserved substitutions. Among the three types of Prx proteins, Prx II sequences are most similar, showing 62% identity, to yeast TPx. Prx III proteins include mouse and human MER5 and bovine SP22. The 257-amino acid sequence of MER5[14] deduced from the cDNA sequence is substantially larger than the 195- or 194-residue sequence of SP22 determined directly by peptide sequencing.[17] Amino-terminal sequence analysis showed that SP22 purified from mitochondria of bovine adrenal cortex was a roughly equal mixture of the 195-residue protein with the NH_2-terminal sequence APAVTQH and the 194-residue protein with the NH_2-terminal sequence PAVTQH.[17] The SP22 sequence aligns with 93% identity with the COOH-terminal 195 (or 194) residues of MER5. These results suggest that maturation of Prx III in mitochondria requires truncation of the NH_2-terminal 63 (or 62) residues.

All three mammalian Prx proteins were shown to exhibit peroxidase activity toward H_2O_2 in the presence of the thioredoxin system (thioredoxin, thioredoxin reductase, and NADPH) or nonphysiological hydrogen donor DTT.[19] Here, we describe two different assay procedures for the detection of Prx enzymes and purification procedures of mammalian Prx I, II, and III.

Reagents

- Glutamine synthetase from *E. coli:* Purified as described or purchased from Sigma.
- Glutamine synthetase assay solution: γ-Glutamyltransferase assay mixture contains 100 m*M* HEPES–NaOH, 0.4 m*M* ADP, 150 m*M* glutamine, 10 m*M* potassium arsenate, 20 m*M* hydroxylamine, and 0.4 m*M* $MnCl_2$.
- Glutamine synthetase assay stop mixture: One liter of solution containing 55 g of $FeCl_3 \cdot 6H_2O$, 20 g of trichloroacetic acid, and 21 ml of concentrated HCl.

Thioredoxin: Rat thioredoxin cDNA is obtained by polymerase chain reaction amplification and expressed in *E. coli*. Recombinant thioredoxin is purified by heat treatment at 65° for 15 min and sequential chromatographies on Sephacryl S-100 HR gel filtration and HPLC DEAE-5PW ion-exchange column.

Thioredoxin reductase: Purified from rat liver as described.[19]

Assay Procedure

Assay Based on Protection of Glutamine Synthetase

This assay procedure is based on the fact that glutamine synthetase is inactivated by oxygen radicals generated from the thiol oxidation system (Fe^{3+}, O_2, and DTT) and that Prx protein can prevent this oxidative inactivation by removing H_2O_2, the precursor of radical species. Assays are performed in a 50-μl reaction mixture containing 0.5 μg of glutamine synthetase, 10 m*M* DTT, 3 μM $FeCl_3$, 50 m*M* HEPES–NaOH (pH 7.0), and a source of Prx. The inactivation reaction is initiated by adding 10 μl of a freshly prepared mixture of $FeCl_3$ and DTT to 40 μl of solution containing glutamine synthetase and Prx. After 10 min of inactivation reaction at 30°, the remaining activity of glutamine synthetase is measured by adding 2 ml of γ-glutamyltransferase assay mixture. The inactivation reaction time should be adjusted in order to achieve optimal degree of glutamine synthetase inactivation. The glutamine synthetase reaction is incubated at 30° for 5 min, then terminated by the addition of 1 ml of stop mixture. The absorbance resulting from the γ-glutamyl hydroxamate-Fe^{3+} complex is measured at 540 nm.[21] Because the remaining activity of glutamine synthetase can be measured enzymatically, this assay is sensitive. But because the extent of glutamine synthetase inactivation is not proportional to the concentration of H_2O_2, this assay is not quantitative.

Assay Coupled to NADPH Oxidation

This assay is based on the fact that the reduction of H_2O_2 by a Prx is coupled to the oxidation of NADPH via thioredoxin and thioredoxin reductase. A reaction mixture containing 50 m*M* HEPES–NaOH (pH 7.0), 250 μM NADPH, 50 n*M* thioredoxin reductase, 5 μM thioredoxin, and 0.5 m*M* H_2O_2 is placed in a cuvette and equilibrated at 30°. The reaction volume can be varied depending on the availability of protein components. The reaction is initiated by adding a source of Prx and monitored with

[20] M. Luthman and A. Holmgren, *Biochemistry* **21,** 6628 (1982).

[21] B. M. Shapiro and E. R. Stadtman, *Methods Enzymol.* **XVIIA,** 910 (1970).

time as A_{340}. Because thioredoxin reductase and thioredoxin, which contain redox-sensitive cysteines, can, albeit at a low rate, catalyze the H_2O_2-dependent oxidation of NADPH, background NADPH oxidation in the absence of Prx is subtracted. The assay is linear with time. However, at a higher concentration of H_2O_2, the rate might decrease with time because of irreversible oxidation of the active site cysteine of Prx to sulfinic or sulfonic acids.[8,18] For Prx I, II, and III, the K_m for thioredoxin is 3 to 6 μM and the K_m for H_2O_2 is <20 μM.

Purification Procedure

Glutamine synthetase protection is mainly measured during purification. Mammalian tissue extracts nevertheless contain a variety of proteins (for example, catalase, glutathione peroxidase, and iron-chelating protein) that also can protect glutamine synthetase from inactivation by DTT/Fe^{3+}/O_2. Therefore, it is recommended that fractions from the first column chromatographic step during Prx purification also be assayed by the procedure following NADPH oxidation or by immunoblot analysis. Antibodies specific to each Prx isoform have been prepared.[19]

Purification of Prx I

Purification of Prx I (NKEFA) from human erythroid cells has been reported.[22]

Purification of Prx II

Five bovine brains are freshly obtained from a local slaughterhouse and homogenized with a Polytron (Brinkman Instrument Inc., Westbury, NY) in 3 liters of buffer containing 20 m*M* Tris-HCl (pH 7.5) and 2 m*M* phenylmethylsulfonyl fluoride (PMSF). The homogenate is centrifuged for 30 min at 13,000*g* at 4°, and the resulting supernatant is adjusted to pH 4.8 with 1 *M* acetic acid. After 30 min at 4° with gentle stirring, the precipitate is removed by centrifugation and the supernatant is adjusted to pH 7.5 with Tris base.

The resulting supernatant is applied to a DEAE-Sephacel column (5 × 40 cm) that has been equilibrated with 50 m*M* Tris-HCl (pH 7.5). Proteins are eluted with a linear NaCl gradient from 0 to 400 m*M* in 6 liters of equilibration buffer. Fractions of 20 ml are collected and assayed for Prx. Peak fractions (fractions 130–170) are pooled, and which proteins that are soluble in a 30% saturated ammonium sulfate solution but precipitate in a 50% saturated ammonium sulfate solution are collected by centrifu-

[22] H. Shau, R. K. Gupta, and S. H. Golub, *Cell. Immunol.* **147,** 1 (1993).

gation. The 30–50% ammonium sulfate precipitate is dissolved in 50 ml of a solution containing 20 m*M* HEPES–NaOH (pH 7.2) and 200 m*M* ammonium sulfate, and insoluble materials are removed by centrifugation.

One-quarter of the clear supernatant (13 ml) is applied at a flow rate of 5 ml/min to an HPLC preparative TSK phenyl-5PW column (21.5 × 150 cm) that has been equilibrated with 20 m*M* HEPES (pH 7.2) and 1 *M* ammonium sulfate. Proteins are eluted at a flow rate of 5 ml/min by a decreasing ammonium sulfate gradient from 1.0 to 0 *M* for 35 min, and elution is continued for 15 min with the same HEPES buffer without ammonium sulfate. Prx elutes at the end of the gradient as a single peak. This chromatographic step is repeated with the remaining three portions and peak fractions (fractions 47–49) from all four portions are combined and concentrated in an Amicon concentrator. Concentrated Prx fractions are washed with a solution containing 20 m*M* sodium phosphate buffer (pH 7.0) and 100 m*M* NaCl, and finally concentrated to 3 ml.

One-sixth (0.5 ml) of the concentrated sample from the above step is applied to a TSK-G3000SW column (7.5 × 600 mm) that has been equilibrated with a solution containing 20 m*M* sodium phosphate (pH 7.0) and 100 m*M* NaCl. Proteins are eluted at a flow rate of 0.5 ml/min with the same buffer. A broad peak of Prx activity is apparent. This gel-filtration step is repeated with the remaining five portions of the sample. Peak fractions (fractions 13–16) from each portion are combined, concentrated, washed in Centricon-10 microconcentrators (Amicon, Danvers, MA) with 20 m*M* HEPES (pH 7.2), and concentrated finally to 3.5 ml.

The concentrated sample (3.5 ml) from the above step is applied at a flow rate of 1 ml/min to a TSK heparin-5PW HPLC column (7.5 × 75 mm) that has been equilibrated with 20 m*M* HEPES (pH 7.2). The column is first washed with equilibration buffer for 10 min, and then eluted with a linear NaCl gradient from 0 to 1 *M* for 40 min. Prx activity is detected in fractions that emerged during loading of the sample and washing of the column. An additional peak of Prx activity is seen in fractions eluted with the NaCl gradient. Because the profile of activity corresponding to unbound material closely resembles the protein profile monitored by absorbance at 280 nm, whereas the profile of activity corresponding to bound material does not, the unbound material fractions are pooled, concentrated, washed with a solution containing 20 m*M* sodium phosphate (pH 7.0) and 100 m*M* NaCl, divided into portions, and stored at −70°.

Purification of Prx III

Prx III protein is purified from rat liver mitochondria. Mitochondria are prepared from 30 g of rat liver by differential centrifugation with isotonic sucrose as described[23] and are sonicated in 50 mM HEPES–NaOH buffer (pH 7.0). The resulting lysate is centrifuged for 30 min at 300,000g for 30 min, and each half of the resulting supernatant is applied separately to a TSK phenyl-5PW column (21.5 × 150 cm) that has been equilibrated with 20 mM HEPES–NaOH (pH 7.0) containing 1 M ammonium sulfate. Proteins are eluted at a flow rate of 5 ml/min by a decreasing ammonium sulfate gradient from 1.0 to 0 M for 40 min, and elution is then continued for 10 min with same buffer without ammonium sulfate. Fractions of 5 ml are collected and assayed. Peak fractions (fraction 47–50) are pooled and concentrated in a Centriprep-30 (Amicon, Denvers, MA). The concentrated sample (4 ml, 28 mg protein) is applied to a Mono Q HR 10/10 column that has been equilibrated with 20 mM Tris-HCl (pH 7.5). Proteins are eluted at a flow rate of 2 ml/min with a linear NaCl gradient from 0 to 400 mM for 40 min. Fractions of 2 ml are collected. Peak fractions (fractions 20–24) are pooled, concentrated, and washed with 5 mM sodium phosphate buffer (pH 7.0). The washed sample (4 ml, 2 mg protein) is applied to an HPLC hydroxyapatite HCA column (4 × 75 mm), and proteins are eluted at a flow rate of 1 ml/min with an increasing sodium phosphate gradient from 0 to 200 mM for 30 min. Fractions of 1 ml are collected. Peak fractions (fractions 16 and 17) are pooled and concentrated. Purity estimated on a sodium dodecyl sulfate–polyacrylamide gel is >95%.

[23] E. Weinbach, *Anal. Biochem.* **2,** 335 (1961).

[24] Preparation and Assay of Mammalian Thioredoxin and Thioredoxin Reductase

By Elias S. J. Arnér, Liangwei Zhong, and Arne Holmgren

Introduction

Thioredoxin reductase (TrxR) catalyzes the NADPH-dependent reduction of the active site disulfide in oxidized thioredoxin (Trx-S_2) to a vicinal dithiol in reduced thioredoxin [Trx-$(SH)_2$] [reaction (1)]:

$$\text{NADPH} + \text{H}^+ + \text{Trx-S}_2 \rightleftarrows \text{NADP}^+ + \text{Trx-(SH)}_2 \quad (1)$$

$$\text{Trx-(SH)}_2 + \text{protein-S}_2 \rightleftarrows \text{Trx-S}_2 + \text{protein-(SH)}_2 \quad (2)$$

0076-6879/99 $30.00

A thioredoxin of about 12 kDa exists in the cytosol of all living cells, having a large number of activities.[1–6] Reduced thioredoxin is a powerful protein-disulfide reductase [reaction (2)] required to regenerate an active site dithiol after each catalytic cycle in, e.g., ribonucleotide reductase or methionine sulfoxide reductases. Disulfide reduction by thioredoxin involves redox regulation of enzymes and transcription factors. Secreted thioredoxin has cytokine-like effects on certain mammalian cells probably also involving thiol–disulfide exchange.[2–7] Thioredoxin, NADPH, and thioredoxin reductase (the thioredoxin system) also play an important role in signal transduction and defense against oxidative stress.

Thioredoxin reductase from *Escherichia coli* has been crystallized[1] and a high resolution X-ray structure shows surprisingly large differences from other members of the pyridine nucleotide-disulfide oxidoreductase family, notably glutathione reductase which is the best characterized.[8,9] Thus, the two subunits of 35 kDa are smaller than the 50 kDa subunits of glutathione reductase from all species. Furthermore, the active site cysteine residues of *E. coli* TrxR are located in the central NADPH domain and separated by two amino acids (Cys-Ala-Thr-Cys), in contrast to the active site in glutathione reductase which is Cys-Val-Asn-Val-Gly-Cys and located in the N-terminal FAD domain, suggesting convergent evolution of the two enzymes.[9] The structural features of TrxR from *E. coli* and a high specificity for its homologous Trx are typical also for TrxR from prokaryotes, lower eukaryotes such as yeast, or cytosol from plants.[1,8–10]

Procedures for preparation and assay of thioredoxin and its reductase from *E. coli* as well as mammalian sources were previously described in this series.[3] Since then, novel results have demonstrated the intriguing property of human, rat, and bovine cytosolic thioredoxin reductase as being a selenoprotein. Cloning of the genes coding for TrxR from humans and rat as well as extensive peptide sequencing of the bovine enzyme has

[1] C. H. J. Williams, "Chemistry and Biochemistry of Flavoenzymes" (F. Müller, Eds.), p. 121. CRC Press, Boca Raton, FL, 1992.

[2] A. Holmgren, *Annu. Rev. Biochem.* **54,** 237 (1985).

[3] A. Holmgren and M. Björnstedt, Methods in Enzymology, Vol. 252, p. 199 (1995).

[4] A. Holmgren, *J. Biol. Chem.* **264,** 13963 (1989).

[5] J. Yodoi and T. Tursz, *Adv. Cancer Res.* **57,** 381 (1991).

[6] A. Holmgren, *Structure* **3,** 239 (1995).

[7] G. Powis, M. Briehl, and J. Oblong, *Pharmac. Ther.* **68,** 149 (1995).

[8] G. Waksman, T. S. Krishna, C. H. Williams, Jr., and J. Kuriyan, *J. Mol. Biol.* **236,** 800 (1994).

[9] J. Kuriyan, T. S. R. Krishna, L. Wong, B. Guenther, A. Pahler, C. H. J. Williams, and P. Model, *Nature* **352,** 172 (1991).

[10] S. Dai, M. Saarinen, S. Ramaswamy, Y. Meyer, J. P. Jacquot, and H. Eklund, *J. Mol. Biol.* **264,** 1044 (1996).

shed new light on structure–function mechanisms, including inactivation by inhibitors and the role of thioredoxin reductase in oxidative stress.

Genes and Proteins

It has long been known that mammalian TrxR has strikingly different properties compared to the enzyme from *E. coli* and lower organisms.[2–4] The enzymes from calf liver and thymus and rat liver were first purified to homogeneity and showed subunits with a molecular mass of 58 kDa.[11,12] The mammalian thioredoxin reductases including that of human placenta reported to be 60–65 kDa[13] are thereby larger than the *E. coli* enzyme.

In 1995 a putative human placental cDNA clone for TrxR was reported, identified by classic hybridization of a cDNA library with degenerate oligonucleotides based on peptide sequences from the purified protein.[14] An open reading frame of the 3.8 kb cDNA clone encoded a protein of either 54 or 60 kDa, depending on which ATG was used as a translational start site—the lower M_r being suggested by the authors due to agreement with the N-terminal sequence Gly-Pro-Glu-Asp-Leu-.[14] Expression of the cDNA either as a glutathione *S*-transferase fusion protein or with a His_6 tag yielded a recovered protein product that comigrated on SDS–PAGE with—and was immunologically indistinguishable from—the native protein, but was not incorporating FAD and lacked activity.[14] Using this cDNA clone, the chromosomal localization was designated to position 12q23-q24.1 by *in situ* hybridization.[15] A second human cDNA clone has been sequenced, being nearly identical to the previously described clone but not initially identified as TrxR. This cDNA was suggested to be translated from the first ATG in the open reading frame, thereby giving a protein of 60 kDa.[16] The identity of these cDNA clones as truly coding for TrxR has been confirmed by identity or near identity with peptides from enzymatically active enzyme purified from other sources, such as T-lymphoblastic cells[17] or calf thymus.[18,19] A rat cDNA clone has also been characterized, conclusively establishing the primary structure of mammalian thioredoxin reductases.[18,19]

[11] A. Holmgren, *J. Biol. Chem.* **252,** 4600 (1977).
[12] M. Luthman and A. Holmgren, *Biochemistry* **21,** 6628 (1982).
[13] J. E. Oblong, P. Y. Gasdaska, K. Sherrill, and G. Powis, *Biochemistry* **32,** 7271 (1993).
[14] P. Y. Gasdaska, J. R. Gasdaska, S. Cochran, and G. Powis, *FEBS Lett.* **373,** 5 (1995).
[15] J. R. Gasdaska, P. Y. Gasdaska, A. Gallegos, and G. Powis, *Genomics* **37,** 257 (1996).
[16] R. Koishi, I. Kawashima, C. Yoshimura, M. Sugawara, and N. Serizawa, *J. Biol. Chem.* **272,** 2570 (1997).
[17] V. N. Gladyshev, K.-T. Jeang, and T. C. Stadtman, *Proc. Natl. Acad. Sci. USA* **93,** 6146 (1996).
[18] A. Holmgren, E. Arnér, F. Åslund, M. Björnstedt, Z. Liangwei, J. Ljung, H. Nakamura, and D. Nikitovic, "Oxidative Stress, Cancer, AIDS and Neurodegenerative Diseases" (L. Montagnier, R. Olivier, and C. Pasquier, Eds.), p. 229. Marcel Dekker, New York, 1998.
[19] L. Zhong, E. S. J. Arnér, J. Ljung, F. Åslund, and A. Holmgren, *J. Biol. Chem.* **273,** 8581 (1998).

TrxR from *Plasmodium falciparum* has been cloned, sequenced, and expressed as enzymatically active enzyme and displayed high sequence homology to the mammalian enzyme,[20,21] although it is not a selenoprotein like the latter. In the *Caenorhabditis elegans* genome there are two potential TrxR genes, one possibly a selenoprotein and the other not.[19]

The amino acid sequences of mammalian TrxR revealed a strikingly high homology to glutathione reductase.[14,19] The conserved features of all the structural components of glutathione reductase are preserved in mammalian TrxR, including a redox active disulfide motif in the N-terminal FAD region, the NADPH binding region, and the carboxyterminal interface region that governs the association of the two subunits in the homodimeric holoenzyme. The principle and overall structure and mechanism of action of mammalian TrxR should resemble that of glutathione reductase, lipoamide dehydrogenase, and mercuric reductase.[1] Reported spectral features of oxidized and NADPH-reduced human TrxR support this notion.[22]

Mammalian Thioredoxin Reductase: A Selenoprotein

Stadtman and co-workers reported in 1996 the isolation from human lung adenocarcinoma cells of a selenocysteine containing homodimeric flavoprotein of 57-kDa subunits, which catalyzed the NADPH-dependent reduction of DTNB and thioredoxin.[23] These features were all identical to those of classically purified TrxR.[11,12] The selenoenzyme was, however, not immunoreactive with anti-rat liver TrxR polyclonal antibodies. The same group also reported the presence of another selenocysteine-containing TrxR in human T-lymphoblasts, which was a homodimer of 55-kDa subunits and, in contrast to the enzyme from lung adenocarcinoma cells, reacted with anti-rat TrxR antibodies.[17] In this report,[17] it was shown that the selenocysteine residue corresponded to the position of TGA in the codon triplet TGA GGT TAA, thought to be a stop codon in the previously reported putative human placental TrxR cDNA clone,[14] indicating the presence of two additional amino acids at the C-terminal of the enzyme: -Sec-Gly. We found that the same carboxyterminal motif was present in bovine TrxR and in a rat cDNA clone.[18,19] We also found consensus eukaryotic selenocysteine insertion sequences (SECIS elements) in the 3′-untranslated region of both the human and the rat cDNA clones, which are required

[20] S. Müller, T. W. Gilberger, P. M. Färber, K. Becker, R. H. Schirmer, and R. D. Walter, *Mol. Biochem. Parasitol.* **80,** 215 (1996).

[21] K. Becker, P. M. Färber, C.-M. von der Lieth, and S. Müller, "Flavins and Flavoproteins" (K. J. Stevenson, V. Massey, and C. H. Williams, Jr., Eds.). University Press, Calgary, 1996.

[22] L. D. Arscott, S. Gromer, R. H. Schirmer, K. Becker, and C. H. J. Williams, *Proc. Natl. Acad. Sci. USA* **94,** 3621 (1997).

[23] T. Tamura and T. C. Stadtman, *Proc. Natl. Acad. Sci. USA* **93,** 1006 (1996).

FIG. 1. Schematic drawing of the mammalian TrxR structure. Shown are the FAD, NADPH, central and interface domains, based on the high homology to glutathione reductase and other enzymes of the pyridine disulfide oxidoreductase family.[1,19] The sequences of the glutathione reductase-like redox active motif in the N-terminal part of the protein and the carboxy-terminal motif including selenocysteine are shown by one-letter amino acid encoding. Key residues with redox activity are shown in bold. U, Selenocysteine.

in mammalian selenoprotein genes for TGA to be encoded as selenocysteine.[24,25] We also showed that removal of the selenocysteine residue by carboxypeptidase digestion of NADPH reduced enzyme resulted in loss of catalytic activity.[18,19] This explains why the recombinant protein is not expressed as a catalytically active enzyme in *E. coli,* since the machinery for selenoprotein synthesis differs significantly between prokaryotes and eukaryotes, making selenocysteine insertion using direct heterologous expression of a mammalian selenoprotein gene in prokaryotes impossible.[24–28]

Our results have demonstrated that the carboxy-terminal selenocysteine participates in a redox active selenyl-sulfide bridge with its neighboring cysteine residue.[29] The principal structure and the redox active residues of mammalian TrxR are shown in Fig. 1. The carboxy-terminal redox active Sec-Cys motif of mammalian TrxR should play a major role in explaining the extraordinary wide substrate specificity and presence of many inhibitors of this enzyme (see below).

Purification of Mammalian Thioredoxin Reductase

Source

Mammalian TrxR has been purified to homogeneity from calf thymus and liver,[11] rat liver,[12] human placenta,[13,22] human lung adenocarcinoma cells,[23] a human Jurkat T-cell line,[17] and mouse ascites cells.[30] There have been several modifications of the basic purification scheme.[12] Given here is a protocol using calf thymus as starting material but also applicable to rat liver or human placenta.

[24] S. C. Low and M. J. Berry, *TIBS* **21,** 203 (1996).
[25] R. Walczak, E. Westhof, P. Carbon, and A. Krol, *RNA* **2,** 367 (1996).
[26] A. Böck, K. Forchhammer, J. Heider, W. Leinfelder, G. Sawers, B. Veprek, and F. Zinoni, *Mol. Microbiol.* **5,** 515 (1991).
[27] T. C. Stadtman, *Annu. Rev. Biochemistry* **65,** 83 (1996).
[28] P. Tormay and A. Böck, *J. Bacteriol.* **179,** 576 (1997).
[29] L. Zhong, E. S. J. Arnér, and A. Holmgren, manuscript (1998).
[30] S. Gromer, R. H. Schirmer, and K. Becker, *FEBS Lett.* **412,** 318 (1997).

Purification Procedure

The tissue (calf thymus), which was immediately frozen when obtained, is allowed to slowly thaw and all subsequent purification steps are performed at 4°C unless stated otherwise. The tissue is homogenized in 50 m*M* Tris-HCl, 1 m*M* EDTA, pH 7.5 (TE buffer) with 1 m*M* phenylmethylsulfonyl fluoride as protease inhibitor (1 kg of tissue is homogenized in about 1 liter of TE buffer and then the homogenate is diluted with TE buffer to 5 liter). One drop of octanol can be added to reduce foam formation. On centrifugation for 30 min (10,000 rpm in a GSA rotor, Sorvall, CT, about 14,000*g*) the supernatant is recovered (since no detergent is added, mitochondria, membrane fractions, nuclei, and cell debris are left in the pellet and are discarded). The recovered supernatant is adjusted to pH 5.0 by dropwise addition of 1 *M* acetic acid, with continuously stirring. Precipitate is removed by a second centrifugation (30 min, 10,000 rpm in a Sorvall GSA rotor) after which the supernatant is recovered and adjusted to pH 7.5 with 1 *M* NH_4OH under continuous stirring. Earlier protocols included heating of this supernatant, but this step has now been excluded in order to increase yield of active enzyme.[13] To the supernatant, ammonium sulfate is added to 40% saturation (243 g/liter) and pH is held at pH 7.5 (otherwise there is a risk of loss of the TrxR-bound FAD). The sample is again centrifuged (30 min, 10,000 rpm in a Sorvall GSA rotor) and the recovered supernatant is saturated with 80% ammonium sulfate (285 g/liter) under stirring. Care is taken to hold pH at 7.5, whereafter the sample is left at stirring for more than 2 hr (usually over night). On another centrifugation (30 min, 10,000 rpm in a Sorvall GSA rotor) the supernatant is discarded and the pellet is redissolved in TE buffer (about 200 ml). This sample is either dialyzed against TE buffer or desalted with a 2 liter Sephadex G-50 column equilibrated with TE buffer. Fractions containing NADPH-dependent DTNB reduction (see below) are pooled and then incubated with 2 m*M* (final concentration) dithiothreitol (DTT) for 30 min at room temperature. The reduced sample is then applied to a column of DEAE cellulose (500 ml), equilibrated with TE buffer. The column is rinsed with two volumes of TE buffer or until a low absorbance at 280 nm is reached. Then TrxR is eluted with a linear gradient from 0 to 300 m*M* NaCl in TE buffer, 4000 ml total volume (4 ml/min). TrxR activity is determined using the DTNB assay[11] as described below. The enzyme is eluted at 120 m*M* NaCl. TrxR-containing fractions are combined and dialyzed extensively against TE buffer. The dialyzed sample is incubated with 2 m*M* (final concentration) DTT for 15 min at room temperature and then applied to a 25 ml 2′,5′-ADP-Sepharose column. The column is washed with three volumes of TE buffer and protein is eluted with a linear gradient from 0 to 600 m*M* NaCl in TE, 400 ml total volume (5 ml/min). Protein is measured in collected fractions as well as DTNB reduction to detect TrxR activity. Fractions with TrxR are combined and dialyzed against TE buffer and applied to a 45 ml Q-Sepharose column equilibrated with TE. Enzyme is eluted with a linear gradient from 0 to 400 mM NaCl in TE buffer, total volume 400 ml (1 ml/min). Fractions containing TrxR are combined, dialyzed against TE buffer, and applied to a 5 ml ω-(aminohexyl)agarose column. Enzyme is eluted using a linear gradient from 0 to 700 mM NaCl in TE buffer, total volume 100 ml (0.3 ml/min). The enzyme purity at this stage is often better than 90% and the enzyme can be used for activity assays. Fractions with DTNB reduction are analyzed on SDS–PAGE and pure fractions of TrxR (M_r 57,000) are combined, concentrated, and changed to fresh TE buffer using a Filtront (Fitron Technology Corporation, Northborough, MA) (10 kDa cutoff) centrifugation. This purification scheme gives a total of about 3–5 mg of pure TrxR from 1 kg of tissue with a yield of about 20%. Aliquots (3–10 mg/ml) are stored at −70°C until use.

The visible absorption spectrum of TrxR is typical for flavoproteins. Oxidized mammalian TrxR has two absorption maxima, at 380 and 465 nm, with the ratio between A_{280} and A_{460} being 8.0.[12] Concentration of pure TrxR is determined by the difference in absorption between 280 nm and 310 nm ($A_{280} - A_{310} = 0.7$ mg/ml)[12] or by absorbance at 460 nm using an extinction coefficient of 22.6 mM^{-1}cm^{-1} for the dimeric holoenzyme, containing two FAD.

Calf thymus TrxR showed a K_m value of 3.0 μM for wild-type human Trx and a K_{cat} of 3500 min^{-1}. The corresponding K_m value of human placenta TrxR was 2.0 μM and a K_{cat} of 2800 min^{-1}.[31] Rat liver TrxR has a K_m value of 2.5 μM for rat liver Trx or calf liver Trx with a K_{cat} of 3000 min^{-1}.[12] The K_m value of rat liver TrxR for *E. coli* Trx is 35 μM and the K_{cat} is 3000 min^{-1}.[12] In general there is thus full cross-reactivity between mammalian TrxR and Trx from different species. The K_m value of DTNB was 660 μM and the K_{cat} 4000 min^{-1}.[12]

Standard Assays of Mammalian Thioredoxin Reductase and Definition of Units

Spectrophotometric Assay using DTNB as Substrate

5,5′-Dithiobis(2-nitrobenzoic acid) (DTNB, Ellman's reagent) is a model substrate for TrxR,[11] based on reaction (3)

$$\text{DTNB} + \text{NADPH} + \text{H}^+ \underset{}{\overset{\text{TrxR}}{\rightleftharpoons}} 2\,\text{TNB} + \text{NADP}^+ \qquad (3)$$

[31] X. Ren, M. Björnstedt, B. Shen, M. L. Ericson, and A. Holmgren, *Biochemistry* **32,** 9701 (1993).

where TNB(5′-thionitrobenzoic acid) is yellow and has an absorbance maximum at 412 nm. The standard assay, used during the purification steps, is performed as follows: A reaction mixture is made containing 200 μM NADPH, 5 mM DTNB, 10 mM EDTA, and 0.2 mg/ml bovine serum albumin in 100 mM potassium phosphate buffer, pH 7.0. Reaction mixture (500 μl) is added to two semimicrocuvettes, to one of which is added protein sample and to the other, used as a reference cuvette, the same volume of buffer is added. TrxR activity is then determined by the increase of absorbance at 412 nm, utilizing the extinction coefficient of 13,600 $M^{-1}cm^{-1}$ for TNB.[3,11,12]

Spectrophotometric Assay of Insulin Reduction

The thioredoxin system catalyzes a very efficient reduction of the two interchain disulfides of insulin [reactions (1) and (2)][11,12,32] which can be utilized as a specific thioredoxin-coupled assay of TrxR.

First, make a stock solution of insulin by suspending 50 mg insulin in 2.5 ml 50 mM Tris-HCl, pH 7.5, and adjust pH to 3 with 1 M HCl to dissolve the protein completely. The solution is then titrated back to pH 7 with 1 M NaOH and the final volume is adjusted to 5.0 ml. This clear stock solution of insulin (10 mg/ml) can be kept at −20°C. Note that all preparations of insulin are made without EDTA, which will give hexameric zinc insulin.

In two semimicrocuvettes make a reaction mixture with final concentrations of 150 μM NADPH, 160 μM (1 mg/ml) insulin, 1 mM EDTA, and 3 μM *E. coli* Trx in 50 mM phosphate buffer, pH 7.0. *E. coli* Trx is prepared as described previously[3] or is commercially available from IMCO Ltd. (Stockholm, Sweden, Fax: +46-8-728 4776). To one cuvette protein is added, whereas only buffer is added to the other (reference) cuvette, and oxidation of NADPH is followed as the initial decrease of absorbance at 340 nm (extinction coefficient 6200 $M^{-1}cm^{-1}$).

Definition of Enzyme Activity and Possible Correlation to Selenium Content

One unit of enzyme is based upon the DTNB assay described above and is defined as 2 μmol TNB formed per minute (i.e., 1 μmol DTNB reduced and 1 μmol NADPH oxidized per min). The specific activity of pure TrxR following the scheme given above is usually around 22 U/mg. However, specific activities over 30 U/mg may also be obtained; why is yet not clear. These variations in specific activity may be due to variations in selenium content or oxidation state. Possibly, selenocysteine insertion in

[32] A. Holmgren, *J. Biol. Chem.* **254,** 9113 (1979).

TABLE I
SUBSTRATES OF MAMMALIAN THIOREDOXIN REDUCTASE

Substrate	Reference
Disulfide proteins[a]	
Thioredoxins	11, 12
Protein disulfide isomerase, CaBP1 and CaBP2 (ERp72)	45, 46
NK-lysin	47
Low molecular weight disulfides[a]	
DTNB	11
Lipoic acid	48
L-Cystine	12
Selenium-containing substrates	
Glutathione peroxidase	49
Selenodiglutathione	50
Selenocystine	51
Selenite	42
Peroxides	
15-HPETE	51
Hydrogen peroxide	51
Other substrates	
Alloxan	52
Menadione (Vitamin K)	12
GSNO	53
Dehydroascorbate	54

[a] Compounds which are no direct substrates of mammalian TrxR include insulin and GSSG[11,12]; their reduction requires Trx as electron donor.

TrxR may be variable *in vivo*, since selenium-deprived animals have reduced TrxR activities to different degrees, depending on which tissue is examined.[33] Also, it cannot be excluded that the purification procedure might reduce the selenium content and thereby the specific activity of TrxR, as in the case of glutathione peroxidase.[34]

Assay of Reduction of Other Substrates by Mammalian Thioredoxin Reductase

Mammalian TrxR will not only reduce thioredoxins from different species,[11] but also several other substrates as listed in Table I. It is likely that further substrates will be added to this list.

[33] K. E. Hill, G. W. McCollum, M. E. Boeglin, and R. F. Burk, *Biochem. Biophys. Res. Commun.* **234,** 293 (1997).

[34] J. W. Forstrom, J. J. Zakowski, and A. L. Tappel, *Biochemistry* **17,** 2639 (1978).

Principle of Assay

As a general principle, TrxR activity can be assessed spectrophotometrically following the oxidation of NADPH at 340 nm. Care must be taken, however, to avoid interference from substrates or products with potential absorbance at 340 nm (such as many aromatic compounds) and the inherent NADPH oxidase activity of mammalian TrxR, albeit low,[35] has to be taken into account, especially if for some reason higher (>50 n*M*) enzyme concentrations are utilized. It is necessary to include EDTA in the reaction buffer since mammalian TrxR is inhibited by heavy metal ions.[36]

Reagents

50 m*M* Tris-HCl containing 1 m*M* EDTA, pH 7.5 (TE buffer)
β-NADPH, 10 or 40 m*M*
Mammalian TrxR (freshly diluted to 5 μ*M* TrxR in TE buffer)
Substrate to be analyzed

Procedure

For a standard assay of direct reduction of substrates by mammalian TrxR, a reaction mixture is prepared in semimicrocuvettes containing TE, 150 μ*M* NADPH, and mammalian TrxR (between 5 and 50 n*M* final concentration). To the reference cuvette, add only TE and NADPH. Add substrate (in a small volume) to both cuvettes and follow decrease of absorbance at 340 nm. The activity can be calculated from the extinction coefficient of 6200 $cm^{-1}M^{-1}$ of NADPH. Assays are usually performed at 20°C.

When the effect of Trx is to be determined in reduction of a particular substrate, 20 n*M* TrxR and increasing concentrations of Trx (typically 500 n*M* to 5 μM) can be used in the assay, included from the preincubation step. If wild-type mammalian Trx is used, there is a risk of aggregation with nonlinearity or inactivation in the assay, due to the additional structural half-cystines. This can be overcome utilizing serine mutants lacking these residues or by the use of Trx from *E. coli,* which lack additional sulfhydryl groups.[11,31]

Inhibitors of Mammalian Thioredoxin Reductase

The unusual structure–function properties of mammalian TrxR are also reflected by several different inhibitors. These include several drugs in

[35] E. S. J. Arnér, M. Björnstedt, and A. Holmgren, *J. Biol. Chem.* **270,** 3479 (1995).
[36] E. Martínez-Galisteo, C. García-Alfonso, C. A. Padilla, J. A. Bárcena, and J. López-Barea, *Mol. Cellul. Biochem.* **109,** 61 (1992).

clinical use, such as antitumor quinones,[37,38] nitrosoureas,[39] or 13-*cis*-retinoic acid,[40] with the common denominator that they are electrophilic compounds. The most obvious target for inhibition of mammalian TrxR by these compounds is alkylation of the penultimate carboxyterminal selenocysteine residue (see Fig. 1).

We identified 1-chloro-2,4-dinitrobenzene (DNCB) as an irreversible inhibitor of mammalian TrxR, which induced an NADPH oxidase activity of the enzyme.[35] The enzyme had to be preincubated with NADPH for inhibition to occur, and then TrxR was rapidly inactivated for reduction of all types of substrates, tested with thioredoxin, DTNB, and selenite.[35] We have identified the alkylated residues by endoproteinase cleavage of the alkylated enzyme, supplemented by analysis of alkylated peptides using Edman degradation and mass spectrometry. Interestingly, both the penultimate selenocysteine residue and its adjacent cysteine residue had been alkylated.[41] The basic mechanism of inhibition regarding the other inhibitors of mammalian TrxR has not been elucidated in detail, but a common denominator is that the enzyme has to be reduced by NADPH for inhibition to occur. This is in agreement with the fact that the carboxyterminal selenocysteine residue makes a covalent bond with the adjacent cysteine[29] and that this bond has to be reduced by NADPH for alkylating agents to react with the enzyme.

Principle of Assay for Analysis of Inhibition

With a carboxyterminal motif compromised by an inhibitor, it is still possible that the glutathione reductase-like redox active disulfide motif in the N-terminal part of the enzyme remains nonmodified and therefore the enzyme, possibly, may still exhibit redox activity catalyzed solely by this motif. We have unpublished observations showing that this N-terminal motif may catalyze NADPH-dependent reduction of DTNB without involvement of the carboxy-terminal motif, while the latter is essential for reduction of most, if not all, of the other substrates of the enzyme. Therefore, a careful analysis of inhibition of this enzyme is warranted and the inhibition of at least three different reactions should be analyzed (i.e., reduction of the principle substrate Trx, that of DTNB, and that of a low molecular weight nondisulfide substrate, such as selenite).

[37] B.-L. Mau and G. Powis, *Biochem. Pharm.* **43,** 1613 (1992).
[38] B.-L. Mau and G. Powis, *Biochem. Pharm.* **43,** 1621 (1992).
[39] K. U. Schallreuter, F. K. Gleason, and J. M. Wood, *Biochim. Biophys. Acta* **1054,** 14 (1990).
[40] K. U. Schallreuter and J. M. Wood, *Biochem. Biophys. Res. Commun.* **160,** 573 (1989).
[41] J. Nordberg, L. Zhong, A. Holmgren, and E. S. J. Arnér, *J. Biol. Chem.* **273,** 10835 (1998).

Reagents

50 mM Tris-HCl containing 1 mM EDTA, pH 7.5 (TE buffer)
β-NADPH, 10 or 40 mM
Mammalian TrxR (freshly diluted to 5 μM TrxR in TE buffer)
E. coli Trx (see insulin reduction assay above)
Insulin (10 mg/ml, see insulin reduction assay above)
DTNB [63 mM in 99.5% (v/v) ethanol]
Sodium selenite
Inhibitor to be analyzed

Procedure

For initial analysis of a potential inhibitor, a reaction mixture is prepared in semimicrocuvettes (500 μl) containing 200 μM NADPH and purified mammalian TrxR (between 5 and 50 nM final concentration) in TE. To the reference cuvette, add only TE and NADPH. Preincubate 10 min with NADPH and then add inhibitor. After different times, either of the following substrates of TrxR are added to both reference and sample cuvettes: (1) 5 mM DTNB, (2) 160 μM (1 mg/ml) insulin and 10 μM *E. coli* thioredoxin, or (3) 100 μM selenite. When DTNB is used as a substrate (see above), increase of absorbance at 412 nm is followed[11] while consumption of NADPH is followed at 340 nm when insulin and thioredoxin[11] or selenite[42] are used as substrates. In all cases, activity is compared to a control without inhibitor. For inhibition of TrxR in cells, inhibitors can be added to cell cultures and the TrxR activity can thereafter be measured in cell lysates as described earlier[3] and summarized as follows.

Determination of TrxR Activity in Cell Lysates

Principle

The assay of TrxR activity in biological material is based on the Trx-coupled insulin reduction assay described above, but is instead an endpoint assay measuring the number of SH groups formed in the added insulin, since decrease of absorbance at 340 nm is difficult to measure in cell lysates due to interfering enzyme activities. Cell lysates are prepared by conventional method of choice. Note, however, that if detergent is included and/or mitochondrial content is recovered a mitochondrial isoenzyme might be analyzed in addition to the cytosolic TrxR, the properties of the former not yet being fully characterized (see below).

[42] S. Kumar, M. Björnstedt, and A. Holmgren, *Eur. J. Biochem.* **207,** 435 (1992).

Reagents

HEPES buffer, 1.0 *M*, pH 7.6
EDTA, 0.2 *M*
β-NADPH (40 m*M*)
Mammalian TrxR (freshly diluted to 20 n*M* TrxR in TE buffer containing 0.1 mg/ml bovine serum albumin)
Insulin (10 mg/ml, see insulin reduction assay above)
DTNB (25 m*M* in 99.5% ethanol)
6 *M* Guanidine hydrochloride in 0.2 *M* Tris-HCl, pH 8.0
E. coli Trx 60 μ*M*

Procedure

Test tubes are prepared with 40 μl of a reaction mixture of 200 μl HEPES buffer, 40 μl EDTA, 40 μl NADPH, and 500 μl insulin and are held on ice. Mammalian TrxR is added to some test tubes (0, 10, 20, 30, 40, 50, and 70 μl) to make a standard curve and different amounts of cell extracts (usually 20 μg of protein is sufficient) are added to other test tubes. Thioredoxin (10 μl) is added to two tubes for each sample, whereas 10 μl of water is added to two other tubes for each sample. Then all are then adjusted to a total volume of 120 μl using water. The two tubes for each sample without Trx will provide a background absorbance to be subtracted from that determined in the presence of Trx. All tubes are transferred to 37°C and incubated for 20 min. The reaction is stopped by addition of 500 μl 1 m*M* DTNB/6*M* guanidine hydrochloride in 0.2 *M* Tris-HCl, pH 8.0, and absorbance at 412 nm is determined. As a control for total SH groups, protein from each sample is kept on ice during the incubation, and immediately prior to addition of DTNB/guanidine-hydrochloride, Trx and reaction mixture are added to the cell extract. The absorbance of this control is substrated from the value of the sample determined above.

Isoenzymes

As mentioned, TrxR prepared from T lymphoblasts or lung adenocarcinoma cells differed in immunoreactivity and molecular weight.[17,23] Two isoenzymes with these properties were shown to differ also in heparin binding,[43] the basis of which is not yet clarified. There is also evidence of a mitochondrial 54 kDa isoenzyme.[44] In our purification scheme, mitochon-

[43] S. Y. Liu and T. C. Stadtman, *Proc. Natl. Acad. Sci. USA* **94,** 6138 (1997).
[44] M. P. Rigobello, M. T. Callegaro, E. Barzon, M. Benetti, and A. Bindoli, *Free Radic. Biol. Med.* **24,** 370 (1998).

dria are discarded at an early stage and the mitochondrial isoenzyme has therefore not been present in the TrxR preparations used by us. The different isoenzyme forms of TrxR await further characterization.

Concluding Remarks

Mammalian thioredoxin reductase has gained increased interest due to its wide reductive capacity, the discovery of selenium in the enzyme, and its lipid hydroperoxide reductase activity. The role of the enzyme in defense against oxidative stress or in cell signaling is inadequately characterized as yet. In addition, since thioredoxin shows a growing number of new roles in redox regulation of cellular processes and as an extracellular cytokine, the interest in TrxR in these contexts naturally follows. Future studies of mammalian thioredoxin systems should be an exciting area of research, yielding results required for a deeper understanding of thiol redox control and mechanisms protecting against oxidative stress.

[45] J. Lundstrom-Ljung, U. Birnbach, K. Rupp, H. D. Soling, and A. Holmgren, *FEBS Lett.* **357,** 305 (1995).
[46] J. Lundstrom and A. Holmgren, *Biochemistry* **32,** 6649 (1993).
[47] M. Andersson, A. Holmgren, and G. Spyrou, *J. Biol. Chem.* **271,** 10116 (1996).
[48] E. S. J. Arnér, J. Nordberg, and A. Holmgren, *Biochem. Biophys. Res. Commun.* **225,** 268 (1996).
[49] M. Björnstedt, J. Xue, W. Huang, B. Åkesson, and A. Holmgren, *J. Biol. Chem.* **269,** 29382 (1994).
[50] M. Björnstedt, S. Kumar, and A. Holmgren, *J. Biol. Chem.* **267,** 8030 (1992).
[51] M. Björnstedt, M. Hamberg, S. Kumar, J. Xue, and A. Holmgren, *J. Biol. Chem.* **270,** 11761 (1995).
[52] A. Holmgren and C. Lyckeborg, *Proc. Natl. Acad. Sci. USA* **77,** 5149 (1980).
[53] D. Nikitovic and A. Holmgren, *J. Biol. Chem.* **271,** 19180 (1996).
[54] J. M. May, S. Mendiratta, K. E. Hill, and R. F. Burk, *J. Biol. Chem.* **272,** 22607 (1997).

[25] Methionine Sulfoxide Reductase in Antioxidant Defense

By JACKOB MOSKOVITZ, BARBARA S. BERLETT, J. MICHAEL POSTON, and EARL R. STADTMAN

Introduction

Free methionine (Met) and surface-exposed methionine residues in proteins are readily oxidized to methionine sulfoxide [Met(O)] by any one of a large number of different reactive oxygen species (ROS), including H_2O_2, HOCl, ozone, peroxynitrite, alkylperoxides, and hydroxyl radical. However, most, if not all, cells contain methionine sulfoxide reductases (MsrA) that catalyze the thioredoxin [$T(SH)_2$]-dependent reduction of Met(O) back to Met. The oxidation of methionine residues of some proteins may lead either to activation or to inactivation of their biological activities,[1] whereas the oxidation of one or more methionine residues in other proteins may have little or no effect on biological function.[2] This has led to the suggestion that MsrA may have multiple biological functions. (a) It may serve to repair oxidative protein damage of some proteins. (b) It may play an important role in the regulation of enzyme (protein) activities by facilitating the interconversion of specific methionine residues of these proteins between oxidized and reduced forms, as occurs, for example, in regulating the ability of bacteria to adhere to prokaryotic cells,[3] in the regulation of various plasma proteinase activities and hormones (reviewed by Vogt,[1] Brot and Weissbach,[4] and Swaim and Pizzo[5]), in the calmodulin-dependent activation of plasma membrane Ca^{2+}-ATPase,[6] and in the modulation of potassium channel function.[7] (c) MsrA might also serve as an antioxidant enzyme to protect some enzymes from oxidative damage by various ROS.[2]

This antioxidant function is derived from the coupling of methionine oxidation with reactions catalyzed by MsrA and thioredoxin reductase (TR). For example, Met(O) formed by the oxidation of methionine by H_2O_2 [reaction (1)] can be reduced back to Met by the [$T(SH)_2$]-dependent reaction (2), catalyzed by MsrA, and the $T(S_2)$ so formed can be reduced back to $Tt(SH)_2$ by the action of TR [reaction (3)]. The sum of these reactions [reaction (4)] describes the scavenging of H_2O_2 by NADPH, i.e., an NADPH peroxidase activity. It is evident from this example that, if H_2O_2 in reaction (1) is replaced by another form of ROS, the overall reaction (4) would describe the NADPH-dependent reduction (scavenging) of that form of ROS. In view of the fact that Met residues of proteins are by far the most sensitive to oxidation by almost all forms of ROS, it is evident that reactions (1)–(4) provide a mechanism for the first line of

[1] W. Vogt, *Free Radical Biol. Med.* **18,** 93 (1995).

[2] R. L. Levine, L. Mosoni, B. S. Berlett, and E. R. Stadtman, *Proc. Natl. Acad. Sci. USA* **93,** 15036 (1996).

[3] T. M. Wizemann, J. Moskovitz, B. J. Pearce, D. Cundell, C. G. Arvidson, M. So, H. Weissbach, N. Brot, and H. R. Masure, *Proc. Natl. Acad. Sci. USA* **93,** 7985 (1996).

[4] N. Brot and H. Weissbach, *Arch. Biochem. Biophys.* **223,** 271 (1983).

[5] M. W. Swaim and S. V. Pizzo, *J. Leukocyte Biol.* **43,** 365 (1988).

[6] Y. Yao, D. Yin, G. S. Jas, K. Kuczera, T. D. Williams, C. Schoneich, and T. C. Squier, *Biochemistry* **35,** 2767 (1996).

[7] M. A. Ciorba, S. H. Heinemann, H. Weissbach, N. Brot, and T. Hoshi, *Proc. Natl. Acad. Sci. USA* **94,** 9932 (1997).

antioxidant defense to protect a given enzyme from more serious oxidative damage of other, more critical, amino acid residues. Moreover, in a broader sense, the cyclic oxidation–reduction of free methionine and surface-exposed methionine residues of all proteins in the cell may collectively constitute an important mechanism for the intracellular scavenging of almost any form of ROS.

$$H_2O_2 + Met \longrightarrow Met(O) + H_2O \tag{1}$$

$$Met(O) + T(SH)_2 \xrightarrow{MsrA} Met + T(S_2) \tag{2}$$

$$T(S_2) + NADPH + H^+ \xrightarrow{TR} T(SH)_2 + NADP^+ \tag{3}$$

$$H_2O_2 + NADPH + H^+ \longrightarrow 2H_2O + NADP^+ \tag{4}$$

A role of MsrA as a repair enzyme of oxidative damage to methionine is indicated by the fact that a null *msrA* mutant of *Escherichia coli* is more sensitive to H_2O_2 than its wild-type parent strain.[8] In addition, the MsrA protein is highly expressed in neutrophils,[9] alveolar macrophages, retinal epithelial pigment cells, and cells of the renal medulla.[10] In all of these cells, except for the kidney cells, high levels of H_2O_2 are known to be produced under certain physiological conditions. It has been shown that a null mutant of *msrA* in yeast is more sensitive to H_2O_2 and accumulates higher amounts of free and protein-bound methionine sulfoxide, in comparison to its wild-type parental strain, under oxidative stress conditions.[11] These findings show for the first time that MsrA is responsible for the reduction of methionine sulfoxide *in vivo* as well as *in vitro,* in eukaryotic cells. Also, the results support the proposition that MsrA possesses an antioxidant function. The ability of MsrA to repair oxidative damage *in vivo* may be of singular importance if methionine residues serve as antioxidants as has been proposed by Levine *et al.*[2]

Expression and Purification of Recombinant Eukaryotic MsrA

Yeast MsrA

The GenBank Accession No. U18796 cosmid 9379, denoted as a homolog of *pilB,* has been shown to be the *Saccharomyces cerevisiae msrA.* This open reading frame is amplified by PCR using *S. cerevisiae* DNA as

[8] J. Moskovitz, A. M. Rahman, J. Strassman, S. O. Yancey, S. R. Kushner, N. Brot, and H. Weissbach, *J. Bacteriol.* **177,** 502 (1995).

[9] N. Brot, H. Fliss, T. Coleman, and H. Weissbach, Methods in Enzymology, Vol. 107, p. 352 (1984).

[10] J. Moskovitz, N. A. Jenkins, D. J. Gilbert, N. G. Copeland, J. F. Jursky, H. Weissbach, and N. Brot, *Proc. Natl. Acad. Sci. USA* **93,** 3205 (1996).

[11] J. Moskovitz, B. S. Berlett, J. M. Poston, and E. R. Stadtman, *Proc. Natl. Acad. Sci. USA* **94,** 9585 (1997).

template, a 5′-sense primer (H1) containing a *Bam*HI site (5′-CTGGAG GGATCCATGGCTGTCGCTGCCAAC), and a 3′-reverse complement primer (H2) with *Hind*III site (H2: 5′-AGGGCAAAGCTTCTAAA AAAGCTACATTTC). PCR (polymerase chain reaction) was performed for one cycle of 5 min at 94°, followed by 30 cycles of 30 sec at 94°, 60 sec at 50°, and 90 sec at 72°. Both the amplified product and pQE-30 (Qiagen, Valencia, CA) are digested with *Bam*HI and *Hind*III, and the PCR fragment encompassing the complete yeast *msrA* coding region is ligated into the restricted pQE-30 using T4 DNA ligase (Boehringer Mannheim, Indianapolis, IN). *E. coli* cells (M15) are transformed with an aliquot of the ligation mixture and grown in LB medium containing ampicillin (100 μg/ml) and kanamycin (25 μg/ml). When cells reach an A_{600} of 0.8, isopropyl-β-D-thiogalactoside (IPTG) is added to a final concentration of 1 m*M*, and the growth continued for an additional 5 hr. The cells are harvested by centrifugation and resuspended in 50 m*M* sodium phosphate (pH 8.0) and 300 m*M* NaCl (buffer A) and sonicated. The lysate is centrifuged, the supernatant is applied to Ni-NTA resin (Qiagen, Valencia, CA), and following extensive washing with buffer A, the protein is eluted with buffer A containing 400 m*M* imidazole. The purity of the protein is analyzed by SDS–PAGE.

Bovine MsrA

The *msrA* open reading frame is amplified by PCR using pJM200 as template,[12] a 5′ sense primer containing a *Bam*HI site (5′-CGGCGGATCC AT-GCTCTCGATCACCAGG), and a 3′ reverse complement primer with an *Eco*RI site (5′GGGGAGAATTCACTTTTTAATACCCAGGGGAC). Reaction conditions are as decribed above. Both the amplified product and pGEX-2T are digested with *Bam*HI and *Eco*RI and the PCR fragment encompassing the complete bovine *msrA* coding region is ligated into the restricted pGEX-2T using T4 DNA ligase (Boehringer Mannheim, Indianapolis, IN). *E. coli* mutant cells lacking the *msrA* gene (SK8779, *msrA1*::*kan*)[8] are transformed with an aliquot of the ligation mixture and grown in LB medium containing ampicillin (50 μg/ml). When cells reach an A_{600} of 0.4, IPTG is added to a final concentration of 1 m*M* and the growth continued for an additional 4 hr. The cells are collected by centrifugation, suspended in buffer B (50 m*M* Tris-HCl, pH 7.4, and 150 m*M* NaCl), and sonicated. The lysate is centrifuged and the supernatant is applied to a glutathione-agarose resin (Sigma, St. Louis, MO). The resin is preequilibrated in PBS and incubated with the cell supernatant for 16 hr at 4°. The slurry is centrifuged at 1000*g* for 10 min and the supernatant removed. The

[12] J. Moskovitz, H. Weissbach, and N. Brot, *Proc. Natl. Acad. Sci. USA* **93,** 2095 (1996).

slurry is washed extensively with buffer B followed by buffer C (buffer B plus 5 mM $CaCl_2$). After resuspending in 10 ml of buffer C, 200 units of thrombin are added to the slurry and incubated at room temperature for 2 hr. After centrifugation at 1000g for 10 min, the supernatant is removed and dialyzed against 25 mM Tris-HCl (pH 7.4). Minor contaminating proteins are removed from the preparation by chromatography on a DE-52 column which is eluted with 100 mM NaCl in 25 mM Tris-HCl (pH 7.4).

Methionine Sulfoxide Reductase Assays

The ability of methionine sulfoxide reductase to reduce *free* methionine sulfoxide is assayed by using [^{3}H]Met(O) as substrate, prepared as described by Brot *et al.*[13] The reaction mixture (30 μl) contains 15 mM dithiotreitol, 25 mM Tris-HCl (pH 7.5), 16.7 μM [^{3}H]Met(O), and yeast extract or pure MsrA. Following incubation at 37° the reaction is stopped by adding 0.33 mM of Met(O), and conversion of [^{3}H]Met(O) to [^{3}H]Met is analyzed by thin-layer chromatography on a silica gel plate using the solvent n-butanol : acetic acid : water (v/v) (60 : 12 : 25). After ninhydrin treatment of the plate, the spot that corresponds to the migration of Met is extracted by water and the radioactivity is measured. The reduction of *protein-bound* methionine sulfoxide by MsrA is assayed using either N-acetyl[^{3}H]Met(O)[13] or dabsyl-Met(O)[14] as substrate. In the latter assay the amino group of the methionine sulfoxide has been derivatized with dabsyl chloride (4-N,N-dimethylaminoazobenzene-4′-sulfonyl chloride). Reduction of the dabsylmethionine sulfoxide to dabsylmethionine is determined by means of an HPLC technique. Reaction mixtures (100 μl) containing 20 mM Tris-HCl (pH 7.5), 10 mM $MgCl_2$, 30 mM KCl, 20 mM dithiothreitol (DTT), 20 μM dabsylmethionine sulfoxide, and an aliquot of purified MsrA or cell extract are incubated at 37° for 30 min and stopped by the addition of 200 μl acetonitrile. After centrifugation for 5 min (top speed in a Beckman Microfuge 12 centrifuge), 10 μl of the clear supernatant solution is injected onto a 10 cm C_{18} column (Apex, Jones Chromatography, Denver, CO) equilibrated at 50° with buffer (0.14 M sodium acetate, 0.5 ml/liter triethylamine (pH 6.1)) containing 30% acetonitrile. Using a linear gradient from 30 to 70% acetonitrile over 11 min, the dabsylmethionine sulfoxide is eluted at 2.6 min, whereas the dabsylmethionine is eluted at 5.0 min, as monitored by peak integration at 436 nm (one pmol of dabsylmethionine gives 340 area units). The column is washed for 5 min with the solvent containing 70% acetonitrile and reequilibrated at 30% before the next injection. The dabsylmethionine sulfoxide is prepared by reacting an aqueous solution of

[13] N. Brot, J. Werth, K. Koster, and H. Weissbach, *Anal. Biochem.* **122,** 291 (1982).

[14] G. Minetti, C. Balduini, and A. Brovelli, *Ital. J. Biochem.* **43,** 273 (1994).

methionine sulfoxide in 100 m*M* bicarbonate buffer (pH 9.0) with dabsyl chloride in acetonitrile. After 10 min at 70°, the dabsylmethionine sulfoxide is purified from the reaction mixture by passage through a silica gel column and stepwise elution with aqueous solutions of methanol and methanol–acetonitrile. Purity of the eluates is checked by thin-layer chromatography on silica gel plates using the solvent *n*-butanol : acetic acid : water (v/v) (60 : 12 : 25). The dabsylmethionine sulfoxide is collected by evaporating the solvent and is stored at room temperature.

Determination of Methionine Sulfoxide in Yeast Extracts

Yeast cells are grown aerobically in synthetic complete medium at 30° with or without H_2O_2 (1 m*M*) or 2,2-azobis(2-amidinopropane) dihydrochloride (AAPH) (6 m*M*). When cell density reaches 300 Klett units (turbidity measured in a Klett–Summerson colorimeter at 540 nm), cells are spun down and washed five times with phosphate-buffered saline (PBS) prior to their disruption in French pressure cell in buffer D [6 *M* guanidine chloride and 500 m*M* potassium phosphate (pH 2.5)]. Following centrifugation at 20,000*g* for 20 min, the supernatant solutions are passed through microconcentrators (Microcon 3, Amicon, Beverly, MA). In each case, the flowthrough is collected for free amino acid analysis, whereas the retained material is kept for protein-bound amino acid analysis, after extensive washing with buffer B. The protein moiety of each preparation is subjected to CNBr cleavage as follows (Levine *et al.*[2]); in samples where protein-bound Met(O) is determined, 10 μg of protein from yeast extract is subjected to 100 m*M* CNBr in 70% formic acid for 1 hr at 70°, and thereafter samples are dried by vacuum centrifugation. The assay procedure is based on the fact that on treatment with CNBr, Met residues are converted to homoserine[15] but Met(O) residues are unaffected, whereas during acid hydrolysis in the presence of 1 m*M* DTT, Met(O) residues are converted to Met and Met residues are unaffected. It follows, therefore, that the amount of Met present in acid hydrolyzates of proteins that have not been subjected to CNBr treatment is a measure of Met + Met(O), and the amount of Met in acid hydrolyzate of CNBr-treated protein is a specific measure of Met(O); the difference in Met measurement, with and without CNBr treatment, is therefore a measure of Met only. Hydrolysis of samples is performed in the presence of 6 *N* hydrogen chloride and 1 m*M* DTT at 155° for 45 min, and amino acid analyses are carried out on samples with and without CNBr treatment, as previously described.[16]

[15] H. Fliss, H. Weissbach, and N. Brot, *Proc. Natl. Acad. Sci. USA* **80,** 7160 (1983).

[16] V. Y. Reddy, P. E. Desrochers, S. V. Pizzo, S. L. Gonias, J. A. Sahakian, R. L. Levine, and S. J. Weiss, *J. Biol. Chem.* **269,** 4683 (1994).

[26] Determination of Tissue Susceptibility to Oxidative Stress by Enhanced Luminescence Technique

By P. VENDITTI, T. DE LEO, and S. DI MEO

Introduction

A considerable body of evidence now exists attesting that free radical-initiated oxidative stress is a major cause of tissue damage in several pathological conditions. Because antioxidant capacity of tissues may play an important role in influencing their susceptibility to oxidative challenge, much research has focused on the assessment of the tissue antioxidant defense systems.[1–4] The determinations of single enzyme activity and single scavenger concentration supply only a limited evaluation of the ability of tissue to prevent generation of reactive oxygen species or to scavenge them. Therefore, interest has been focused on the measurement of total antioxidant capacity of tissues. We have developed an enhanced luminescence method[5,6] which supplies the level of tissue antioxidants expressed as equivalent concentration of desferrioxamine. This method is advantageous compared with other methods which have been set up only for biological fluids.[7,8]

Tissue susceptibility to oxidation is also dependent on the activity of metal ions able to interact with oxygen species, such as the superoxide radical and hydrogen peroxide, and to produce highly reactive oxygen species such as hydroxyl radicals. Here we present an enhanced luminescence method which allows the evaluation of tissue susceptibility to oxidative stress and its modification in a variety of physiopathological conditions.

[1] G. W. Burton and K. U. Ingold, *in* "CRC Handbook of Free Radicals and Antioxidants in Biomedicine (II)" (J. Miquel, A. T. Quintanilha and H. Weber, Eds.), p. 29. CRC Press, Boca Raton, FL, 1989.
[2] K. J. A. Davies, *in* "Oxidative Damage and Repair: Chemical, Biological, and Medical Aspects" (K. J. A. Davies, Ed.), p. 17. Pergamon Press, Elmsford, NY, 1991.
[3] E. Niki, *in* "Oxidative Damage and Repair: Chemical, Biological and Medical Aspects" (K. J. A. Davies, Ed.), p. 57. Pergamon Press, Elmsford, NY, 1991.
[4] B. P. Yu, *Physiol. Rev.* **74,** 139 (1994).
[5] P. Venditti, S. Di Meo, P. de Martino Rosaroll, and T. De Leo, *Arch. Physiol. Biochem.* **103,** 484 (1995).
[6] S. Di Meo, P. Venditti, and T. De Leo, *Experientia* **52,** 786 (1996).
[7] T. P. Whitehead, G. H. G. Thorpe, and S. R. J. Maxwell, *Anal. Chim. Acta* **266,** 265 (1992).
[8] A. Ghiselli, M. Serafini, G. Maiani, E. Azzini, and A. Ferro-Luzzi, *Free Radic. Biol. Med.* **18,** 29 (1995).

0076-6879/99 $30.00

The method has been tested on blood, liver, heart, and muscle,[6] as well as on mitochondrial preparations from rat tissues.[6]

Procedures

Preparation of Samples

Tissue Preparation. Blood samples are collected, and other tissues are rapidly excised and placed into petri dishes containing ice-cold isolation medium (IM) consisting of 125 m*M* KCl, 2 m*M* EDTA, 15 m*M* Tris, pH 7.4. Heart great vessels and valves are trimmed away and ventricles and atria are cut open and rinsed free of blood. Muscles and livers are freed from connective tissue. After the tissues are weighed, 20% (w/v) homogenates are prepared with a Potter–Elvehjem homogenizer set at a standard velocity (500 rpm) for 2 min in IM.

Aliquots of the homogenates are diluted with an equal volume of 15 m*M* Tris containing 0.2% Lubrol at pH 8.5 (final concentration 10% w/v), while blood samples are added with nine volumes of 16.7 m*M* Tris containing 0.11% Lubrol at pH 8.5 (final concentration 10% v/v).

Further dilutions of samples up to a final concentration of 0.002% are prepared with 15 m*M* Tris, pH 8.5.

Preparation of Mitochondrial Fractions. The homogenates are freed from debris and nuclei by centrifugation at 500*g* for 10 min. Crude mitochondrial fractions are isolated by centrifugation of the resulting supernatants at 12,000*g* for 10 min. The mitochondrial pellets are resuspended in the respective homogenization media and centrifuged at the same sedimentation velocity. Mitochondrial preparations are washed in this manner three times before final resuspension in the homogenization medium.

The protein content is determined and several dilutions of the mitochondrial suspensions are prepared with 15 m*M* Tris, pH 8.5, in a range of protein concentrations from 20 to 0.005 mg/ml.

Enhanced Luminescence Assay

The sample response to oxidative stress is determined by using reagents and instrumentation of the commercially available Amerlite system (Johnson & Johnson Clinical Diagnostics Ltd., Amersham, UK), widely used in immunoassays for diagnostic purposes in endocrinology and oncology. A reaction mixture is obtained by dissolving a tablet containing substrate in excess (sodium perborate) and signal generating reagents (sodium benzoate, indophenol, and luminol) (Amerlite Signal Reagent Tablets) in buffer at pH 8.6 (Amerlite Signal Reagent Buffer). The assays are performed in microtiter plates (96 wells). Enhanced chemiluminescence reactions are initiated by the addition of 250 μl of the reaction mixture to (i) 25 μl of 22 or 44 ng/ml horseradish peroxidase (Sigma Chemical CO., St. Louis, MO) in 15 m*M* Tris, pH 8.5, placed in the first two microwells; (ii) 25 μl of samples.

Plates are incubated at 37°C for 30 sec with shaking and then transferred to a luminescence analyzer (Amerlite Analiser), where the light intensity is measured at 2 min from the beginning of the reaction. Emission values are supplied as percentages of the emission of the peroxidase standard. In all tissues, the perborate-induced light emission exhibits a peculiar dependence on sample concentration (Fig. 1). A simple equation describing the relationship between emission (E) and concentration (C) is the following:

$$E = aC/\exp(bC)$$

where a and b are constants. The function analysis shows that b is the inverse of the homogenate concentration related to the emission maximum (E_{max}), which in turn is determined by a and b values ($E_{max} = a/eb$). The best fitting of data to the theoretical equation is obtained by the method of multiple regression, using Fig. P Program (Biosoft, Cambridge, UK).

Biological Meaning of a and b Parameters

Determinations performed by supplementing liver homogenates with either substances containing iron or antioxidants[9] have shown that (i) the a value is related to the concentration of ligand-bound iron ions able to react with H_2O_2, released by perborate, producing $\cdot OH$ radicals; (ii) the b value is related to the concentration of antioxidants. The above relationships have been supported by results obtained with the tissues containing different concentrations of hemoproteins and antioxidants.[6]

In analyzing the results, however, the possibility of reciprocal interactions between the two types of substances must be considered. Homogenates supplemented by a preventive antioxidant such as desferrioxamine exhibit lower values of both a and b.[9] This can be explained by considering that a preventive antioxidant can bind metal ions in forms that will not generate reactive species, with a consequent decrease in concentration of the species catalyzing radical production. Thus, EDTA-Fe^{2+} and cytochrome *c* exert a stimulatory effect on light emission only when added to the homogenates at low concentrations, the decreased light emission at high concentrations being due to a decrease of the a value and an increase of the b value.[9] The effect on the b parameter of the above iron ligands is

[9] S. Di Meo, P. Venditti, M. C. Piro, and T. De Leo, *Arch. Physiol. Biochem.* **103,** 187 (1995).

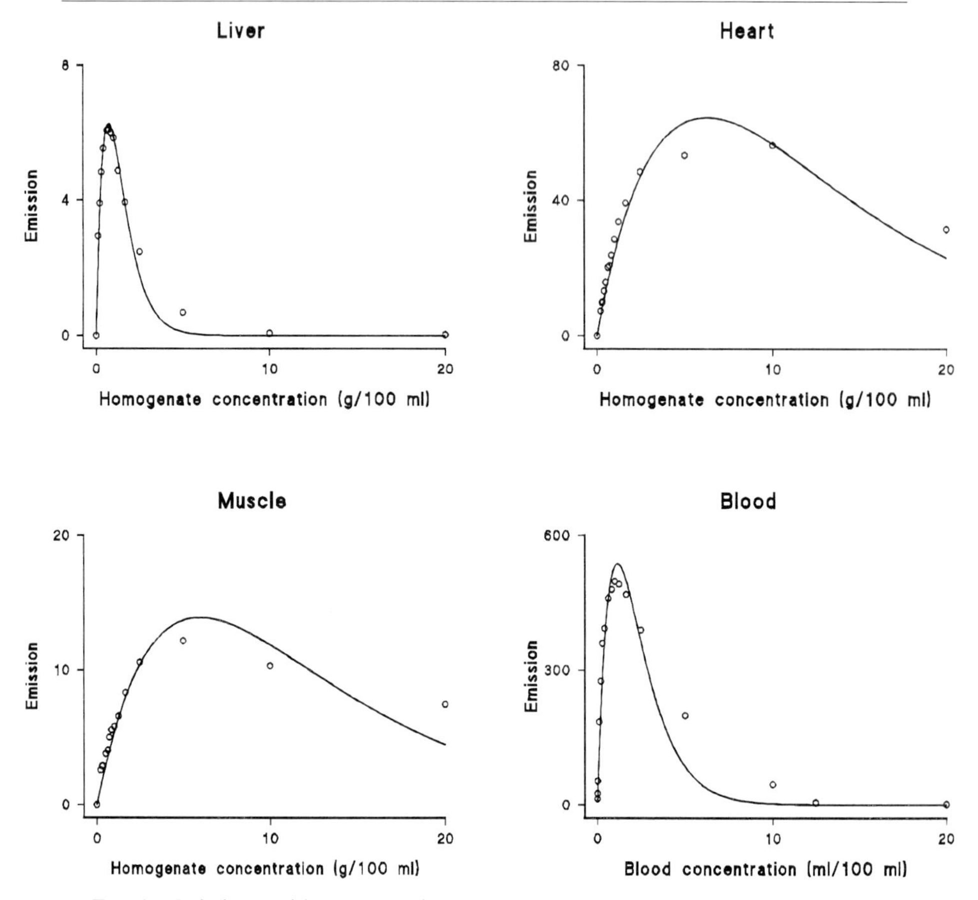

FIG. 1. Variations, with concentrations of rat tissue homogenates and blood, of light emission from a luminescent reaction. Emission values are given as percentage of an arbitrary standard (44 ng/ml peroxidase). The curves are computed from experimental data using the equation $E = aC/\exp(bC)$. Modified by Di Meo *et al.*,[6] with permission.

not surprising, since they can interact with hydroxyl radical formed at the site of metal binding and such a "scavenger" action becomes prevalent at high concentrations.[10,11] The effect on the a parameter is due to the competition for H_2O_2 with other cellular systems which produce radicals, but do not trap them efficiently.

The above statement indicates that, although a and b are not strongly correlated to the concentrations of catalysts of radical reactions and antioxi-

[10] E. Cadenas, A. Boveris, and B. Chance, *Biochem. J.* **187,** 131 (1980).
[11] J. M. C. Gutteridge, R. Richmond, and B. Halliwell, *Biochem. J.* **184,** 469 (1979).

TABLE I
PARAMETERS CHARACTERIZING LIGHT EMISSION FROM HOMOGENATES OF RAT TISSUES STRESSED WITH SODIUM PERBORATE[a]

Tissue	Parameters[b]		
	a	b	E_{max}
Liver	25.2 ± 4.0	1.35 ± 0.10	6.8 ± 0.7
Heart	34.9 ± 4.1	0.18 ± 0.02[c]	71.3 ± 0.8[c]
Muscle	8.2 ± 1.3[c,d]	0.16 ± 0.01[c]	18.8 ± 2.1[c,d]
Blood	1046 ± 161[c,d,e]	0.66 ± 0.05[c,d,e]	582 ± 60[c,d,e]

[a] Modified from S. Di Meo *et al.*,[6] with permission.
[b] Mean ± SE of eight experiments.
[c] Significant ($p < 0.05$) vs liver by Student–Newman–Keuls test.
[d] Significant ($p < 0.05$) vs heart by Student–Newman–Keuls test.
[e] Significant ($p < 0.05$) vs muscle by Student–Newman–Keuls test.

dants, they can be considered as indicators of the tissue ability to produce ·OH radicals and to trap or prevent generation of such reactive species, respectively.

In our system the ·OH radicals released into the solution and escaped to scavengers are available for detector molecules and originate luminescent reaction. Thus, light emission, in particular the E_{max} value, supplies an evaluation of susceptibility of tissue to oxidative stress.

Results

Tissue Susceptibility to Oxidative Stress

Using the present method we have demonstrated that different rat tissues exhibit different susceptibility to oxidative stress (Fig. 1). The order of susceptibility to oxidants drawn from E_{max} (Table I) is blood > heart > muscle > liver. The great susceptibility of blood depends on its high hemoglobin content (high a value), whereas the poor susceptibility of liver depends on both the relatively low content of cytochromes[6] (low a value) and the high antioxidant capacity[6] (high b value).

Mitochondrial Susceptibility to Oxidative Stress

By analyzing the response of the mitochondrial suspensions from different tissues to oxidative stress, a dependence of light emission on protein concentration has been found which can be described by the equation $E = aC/\exp(bC)$ (Fig. 2). The order of susceptibility of the mitochondria is muscle > heart > liver. Also, in this case the a and b values which affect

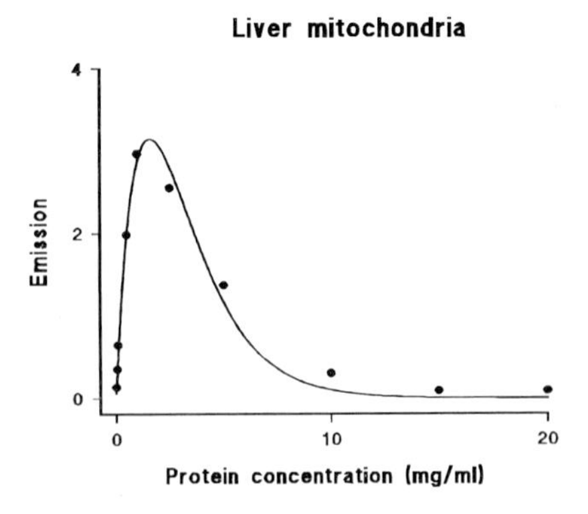

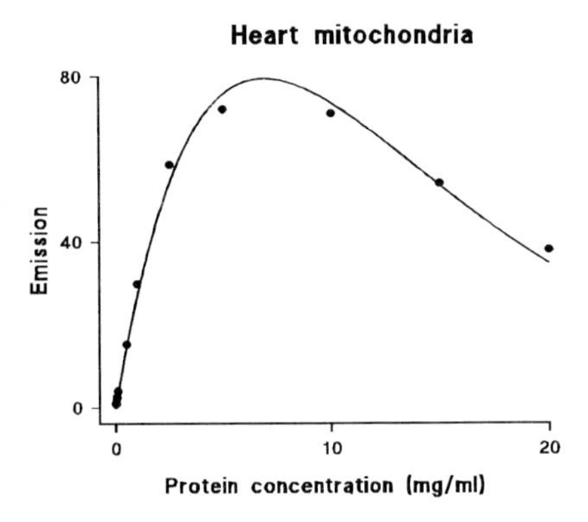

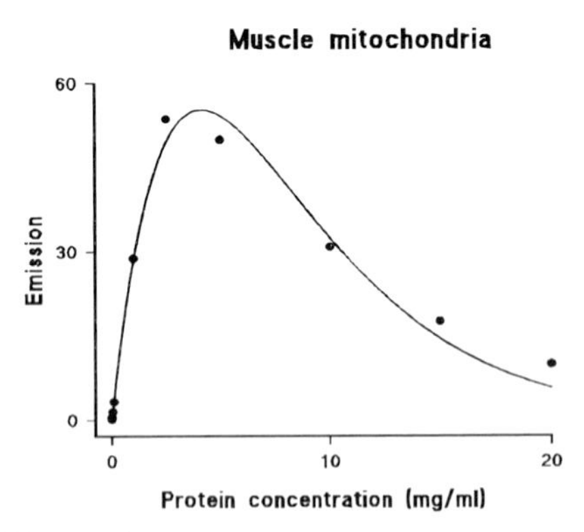

FIG. 2. Variations, with protein concentration of mitochondrial preparations from rat tissues, of light emission from a luminescent reaction. Emission values are given as percentage of an arbitrary standard (44 ng/ml peroxidase). The curves are computed from experimental data using the equation $E = aC/\exp(bC)$. Modified by Di Meo *et al.*,[6] with permission.

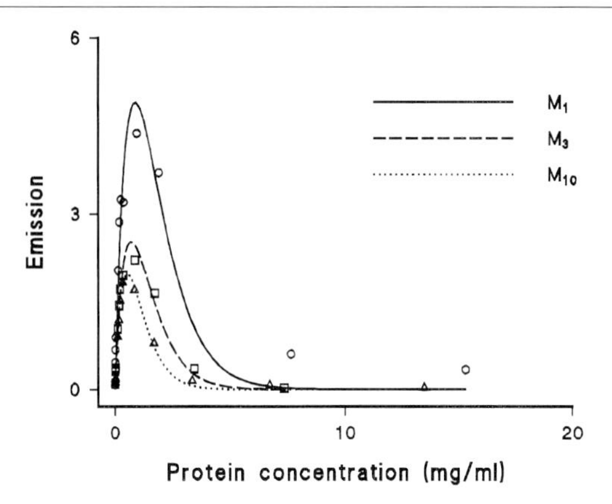

FIG. 3. Variations, with protein concentrations of mitochondrial fractions from rat liver, of light emission from a luminescent reaction. Emission values are given as percentage of an arbitrary standard (44 ng/ml peroxidase). The curves are computed from experimental data using the equation $E = aC/\exp(bC)$. M_1, M_3, and M_{10} are mitochondrial fractions obtained by series of sequential centrifugation steps lasting 10 min at 1000, 3000, and 10,000*g*, respectively. Modified by Venditti *et al.*,[12] with permission of Karger, Basel.

the light emission are related to mitochondrial content of cytochromes and antioxidants, respectively.[6]

Susceptibility to Oxidative Stress of Mitochondrial Fractions from Rat Liver

The method has been also used to determine the susceptibility to oxidative stress of three mitochondrial fractions resolved by differential centrifugation from rat liver in different thyroid states.[12]

The fraction obtained at 1000*g* (M_1) exhibited a greater oxidative capacity and smaller capacity for opposing oxidative stress than the fractions obtained at 3000*g* (M_3) and 10,000*g* (M_{10}) (Fig. 3). Susceptibility to antioxidants increased in all the mitochondrial fractions in the hypothyroid state and decreased in the hyperthyroid state. These results have been explained by the modulation exerted by thyroid hormone, during the normal maturation process, on cytochrome content of mitochondria, which in turn determines both respiratory characteristics and sensitivity to prooxidants. This sensitivity and the decrease of antioxidant capacity, likely due to free radical production associated with the increased oxygen flux, make the mitochondria at higher density less able to respond to an oxidative challenge.

[12] P. Venditti, S. di Meo, and T. De Leo, *Cell. Physiol. Biochem.* **6,** 283 (1996).

Comments

By the enhanced chemiluminescence technique a rapid and sensitive evaluation of the susceptibility of tissues and cellular organelles to oxidative stress is possible. Actually, a similar result could be obtained by hydroperoxide-initiated chemiluminescence. Such a method has allowed the determination of free radical production in organs,[13] tissue homogenate,[14] isolated mitochondria and microsomes,[15] and submitochondrial particles.[16] It also has the advantages of being noninvasive and providing continuous monitoring. However, it exhibits a low signal-to-background ratio and low intensity light emission. Moreover, our method is also able to supply information on characteristics of analyzed systems determining the susceptibility to oxidative stress.

[13] A. Boveris, E. Cadenas, R. Reiter, M. Filipkowski, Y. Nakase, and B. Chance, *Proc. Natl. Acad. Sci. USA* **77,** 347 (1980).

[14] E. Cadenas, A. Varsavsky, A. Boveris, and B. Chance, *Biochem. J.* **198,** 645 (1981).

[15] A. Boveris, B. Chance, M. Filipkowski, Y. Nakase and K. G. Paul, *in* "Frontiers of Biological Energetics: From Electron to Tissues" (A. Scarpa, P. L. Dutton, and J. S. Leigh, Jr., Eds.), Vol. 2, p. 974. Academic Press, New York, 1981.

[16] E. Cadenas, A. Boveris, and B. Chance, *Biochem. J.* **186,** 659 (1980).

[27] Measurement of Hydroxyl Radical by Salicylate in Striatum of Intact Brain

By MIDORI HIRAMATSU and MAKIKO KOMATSU

Introduction

Active oxygen species are associated with neurological disorders and it is therefore desirable to detect hydroxyl radicals in the brain of animal models *in vivo*. Salicylate reacts easily with hydroxyl radicals to form 2,3- and 2,5-dihydroxybenzoate, and these substances can be analyzed by high-performance liquid chromatography (HPLC).[1] Extracellular hydroxyl radicals reacted with salicylate in the brain have been evaluated using a microdialysis method in animal models of ischemia[2,3] and neurodegenera-

[1] B. Halliwell, H. Kaur, and M. Ingelman-Sundberg, *Free Rad. Biol. Med.* **10,** 439 (1991).

[2] C. Yang, N. Lin, P. Tsai, L. Liu, and J. Kuo, *Free Rad. Biol. Med.* **20,** 245 (1996).

[3] M. Y. Globus, R. Busto, B. Lin, H. Schnippering, and M. D. Ginsberg, *J. Neurochem.* **65,** 1250 (1995).

0076-6879/99 $30.00

tive diseases.[4] We have detected extracellular hydroxyl radicals with salicylate in the rat striatum using the microdialysis method and the HPLC system.

Materials and Methods

Animal Preparation

Male Wistar rats of 200 g were used for the experiments. The rat is anesthetized with sodium pentobarbital, and the head of the rat is mounted on a stereotaxic apparatus. With a midline incision, the skull is then exposed, and a burr hole is drilled in the skull for insertion of a dialysis probe. The guide cannula (AG-8, Eicom Corp., Tokyo) is implanted into the striatum (3.3 mm anterior and 1.6 mm left lateral to the bregma, and 3 mm from the brain surface) for a microdialysis probe and into the cerebral ventricle (0.92 mm anterior, 1.5 mm left lateral to the bregma and 3.5 mm from the brain surface) for administration of guanidino compounds, and is left for 2 days. Then the microdialysis probe (A-1-8-02, 2 mm length, Eicom Corp., Tokyo) is inserted into the guide cannula, and the brain is perfused with Ringer's solution (147 m*M* NaCl, 4 m*M* KCl, 2.3 m*M* $CaCl_2$, 100 μM EDTA, pH 7.4) for 2–4 hr at a flow rate of 3 μl/min. Then the sample is perfused with Ringer's solution containing 5 m*M* salicylate for 1 hr. Perfused dialyzate samples are collected every 10 min using a fraction collector (EF-80C, Eicom Corp., Tokyo) at a rate of 3 μl/min regulated by a microsyringe pump (Model 100, Neuroscience Inc., Tokyo). The average dihydroxybenzoate concentration of three fractions is expressed as a basal level.

HPLC System for Determination of 2,3- and 2,5-Dihydroxybenzoate

2,3- and 2,5-Dihydroxybenzoate in the 10 μl sample loop are analyzed by automated sample injector (AS-950, JASCO, Tokyo) connected to a high-performance liquid chromatography system (Chromatocorder 21, System Instruments Co., Ltd.; PU-980 for pump, JASCO; Amperometric Detector, Model ICA-3060, TOA Electrics Ltd., Japan). Conditions for analysis of 2,3- and 2,5-dihydroxybenzoate are as follows: mobile phase solution consists of water, acetonitrile, and tetrahydrofuran (75 : 15 : 10) including 0.142% chloroacetate and is delivered by a pump at rate of 1.0 ml/min. The column is J'sphere ODS-H80 (150 $\times$ 4.6 mm, YMC Co. Ltd., Japan) and the column temperature is 30°.

[4] T. S. Smith and J. P. Bennett, Jr., *Brain Res.* **765,** 183 (1997).

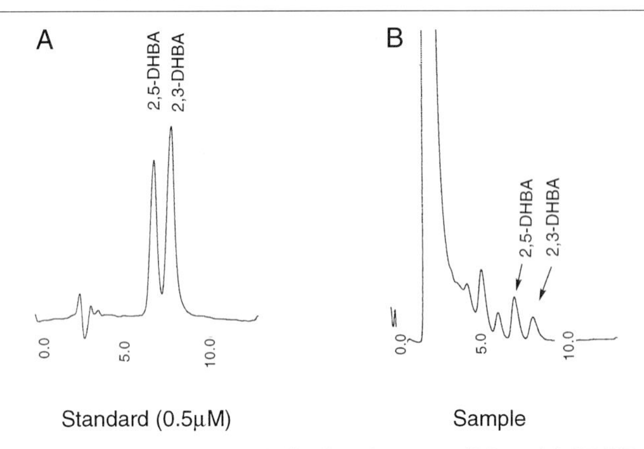

FIG. 1. Chromatogram of 2,3- and 2,5-dihydroxybenzoate (2,3- and 2,5-DHBA). (A) Standard, (B) sample (perfusate).

Effect of Guanidino Compounds on Extracellular Hydroxyl Radicals

Methylguanidine[5] and α-guanidinoglutaric acid[6] induce seizures and convulsions in animals and they also generate hydroxyl radicals and other

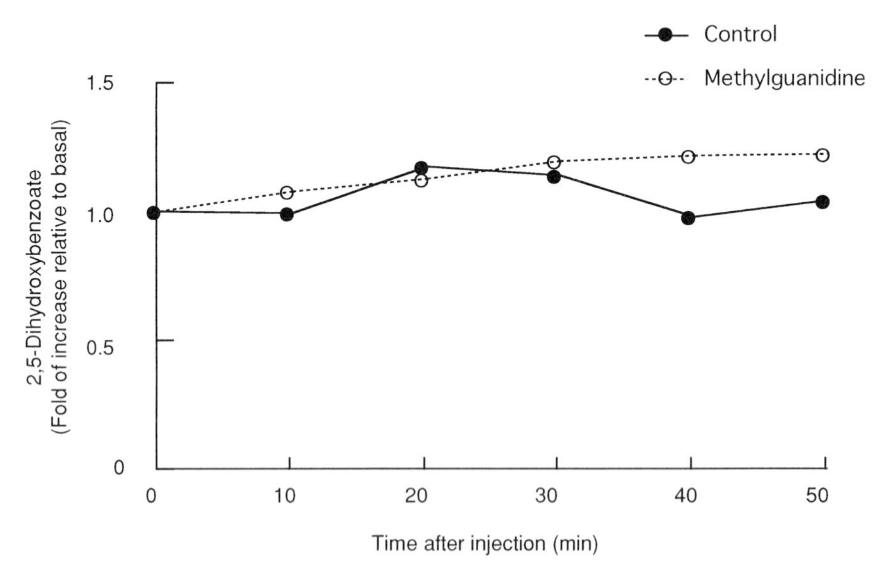

FIG. 2. Effect of methylguanidine on extracellular concentration of 2,5-dihydroxybenzoate.

[5] M. Matsumoto, K. Kobayashi, H. Kishikawa, and A. Mori, *IRCS Med. Sci.* **4,** 65 (1976).
[6] H. Shiraga, M. Hiramatsu, and A. Mori, *J. Neurochem.* **47,** 1832 (1986).

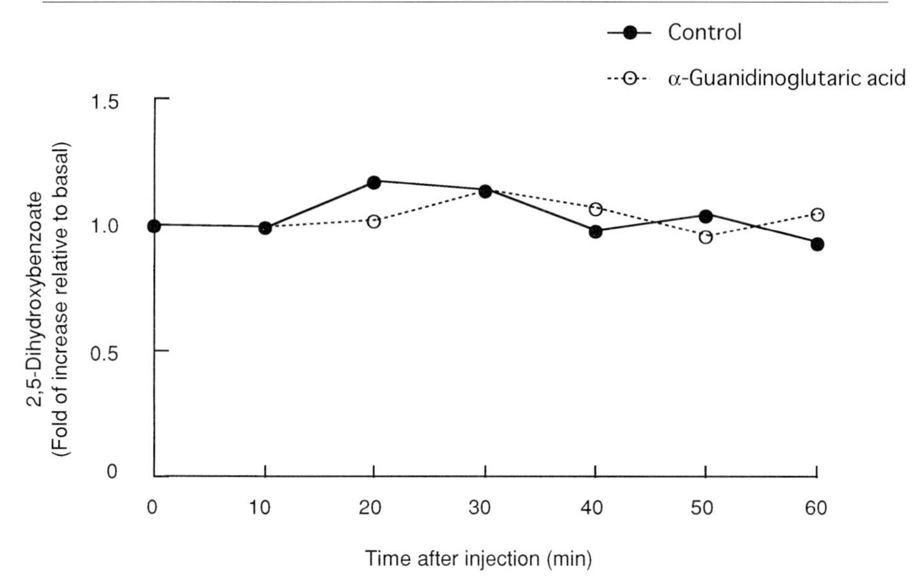

FIG. 3. Effect of α-guanidinoglutaric acid on extracellular concentration of 2,5-dihydroxybenzoate.

free radicals in solutions.[7] After intraventricular administration of methylguanidine and α-guanidinoglutaric acid extracellular hydroxyl radicals were analyzed in the striatum of rats. The chromatogram of 2,3- and 2,5-dihydroxybenzoate is shown in Fig. 1. Methylguanidine generated hydroxyl radicals from 40 to 50 min after administration of methylguanidine. In the case of α-guanidinoglutaric acid, no change in the concentration of hydroxyl radicals was found after administration (Figs. 2 and 3). We detected a slight change in the concentration of hydroxyl radicals in the extracellular striatum of rats after administration of methylguanidine in a free moving state and found that hydroxyl radicals were involved in seizures. This method is valuable for *in vivo* study to follow the generation of extracellular hydroxyl radicals in the brain of animal models with neurological disorders.

[7] A. Mori, M. Kohno, T. Masumizu, Y. Noda, and L. Packer, *Biochem. Mol. Biol. Int.* **40,** 135 (1996).

[28] Analytical and Numerical Techniques for Evaluation of Free Radical Damage in Cultured Cells using Imaging Cytometry and Fluorescent Indicators

By STEPHEN E. BUXSER, GERI SAWADA, and THOMAS J. RAUB

Introduction

To understand oxidative stress-related damage in biological systems at the cellular level, ideally, one needs to determine the type and location of reactive oxygen species (ROS) occurring in the cells under consideration and to observe the temporal relationship between ROS generation and the onset of pathophysiological events. Approaches have included the use of spin trap molecules, whose behavior in response to ROS reactivity may be measured by electron spin resonance (ESR). Additionally, homogenization of cells or tissues followed by extraction and measurement of ROS products, such as protein carbonyl adducts or lipid peroxides, has been used. Both of these methods have limitations. Obtaining a clear signal from homogeneous species in ESR is difficult in complex systems and, furthermore, cannot be used to discern events in subcellular sites. Use of tissue disaggregation and cell homogenates can result in artifacts, including redistribution of ROS and their targets and introduction of transition metals which may catalyze oxidation that is not physiologically relevant. Also, alteration of oxidative events or ROS targets by proteolysis may occur. To circumvent some of these difficulties, noninvasive and nondestructive methods may be applied in living cells and tissues. For example, flow cytometry and fluorescent probes have been applied to living cell suspensions. In addition, imaging cytometry using fluorescent probes in cells in culture attached to substrate, which is more sensitive and of higher resolution than flow cytometry,[1] has gained popularity.

To advance the study of ROS research, the development of novel methods to monitor oxidative reactions associated with particular cells and specific subcellular sites is needed. One technique gaining considerable use within the last few years is confocal laser scanning microscopy (CLSM) using fluorescent molecules sensitive to oxidation. This is a noninvasive method that provides a high spatial and temporal or real-time resolution

[1] A. S. Waggoner, *in* "Applications of Fluorescence in the Biomedical Sciences" (D. Lansing-Taylor, A. S. Waggoner, R. F. Murphy, F. Lanni, and R. R. Birge, Eds.), p. 3. Alan R. Liss, New York, 1986.

0076-6879/99 $30.00

TABLE I
IDEAL PROPERTIES FOR FLUORESCENT INDICATORS OF CELLULAR OXIDATIVE STRESS[a]

Large extinction coefficient where $\varepsilon \geq 60{,}000$ liter/mol cm
High quantum yield in local environment where $\tau > 0.3$
Insensitive to microenvironment pH, solvation, etc.
Optimal excitation wavelength avoiding inner filter effect and autofluorescence of cellular material such as flavins, NADH, etc.
Good photostability; low phototoxicity
Conversion from nonfluorescent to fluorescent form that involves specific ROS
Suitable rate of reaction with ROS
Low spontaneous oxidation or high signal-to-noise ratio
Cell membrane permeability where significant intracellular concentrations are achieved in a relatively short time
Efficient intracellular trapping of oxidized species
Appropriate cellular distribution relative to location of ROS
Minimal perturbation of cellular functions; negligible cytotoxicity

[a] Portions from Waggoner[1] and Royall and Ischirpoulos.[9]

of ROS reactions within single, living cells or small clusters of cells.[2,3] Some properties important in the application of such fluorescent probes in this technique are listed in Table I. Some of the most commonly used indicators are 2′,7′-dichlorofluorescin diacetate (DCFH-DA, 2′,7′-dichlorodihydrofluorescein diacetate), dihydrorhodamine 123 (DHR), and dihydroethidium or dihydroethidine (DHE) (Table II). Other measures of oxidative stress that have been used, but will not be discussed further here, include oxidant-induced decreases in autofluorescence of mitochondrial pyridine nucleotides[4] and oxidative increases in fluorescence of 4-carboxyldihydrotetramethylrosamine succinimidyl ester covalently attached to cell surface proteins.[5]

Indicator or probe molecules, originally employed predominantly to measure respiratory burst activity in phagocytes using flow cytometry,[6] are nonfluorescent until they react with certain ROS yielding a fluorescent product whose appearance is quantifiable. Chemical structures and oxidation reactions are given in Fig. 1. The nonfluorescent substrate readily penetrates through the cell plasma membrane, and the oxidized fluorophore is trapped intracellularly (Table II and Fig. 2). DCFH-DA enters the cell passively and is deacetylated by intracellular esterase activity generating

[2] L. M. Henderson and J. B. Chappell, *Eur. J. Biochem.* **217,** 973 (1993).
[3] N. Sarvazyan, *Am. J. Physiol.* **271** (*Heart & Circ. Physiol.* **40**), H2079 (1996).
[4] A.-L. Nieminen, A. M. Byrne, B. Herman, and J. J. Lemasters, *Am. J. Physiol.* **272** (*Cell Physiol.* **41**), C1286 (1997).
[5] M. A. Model, M. A. KuKuruga, and R. F. Todd, *J. Immunol. Methods* **202,** 105 (1997).
[6] G. Rothe and G. Valet, *Methods Enzymol.* **233,** 539 (1994).

TABLE II
COMMONLY USED FLUORESCENT INDICATORS OF CELLULAR OXIDATIVE STRESS

Nonfluorescent probe	Molecular weight	λ_{EX} [a] (nm)	λ_{EM} [a] (nm)	ε ($\times 10^{-3}$)	ROS reactivity	Cellular location
2′,7′-Dichlorofluorescin diacetate	487	504	529	90	H_2O_2 + peroxidase	Cytosol
Dihydrorhodamine 123	346	507	529	101	H_2O_2 + peroxidase	Mitochondria
Dihydroethidium [b]	315	480	620	5.6	$O_2^- > H_2O_2$ + peroxidase	Nucleus

[a] Excitation and emission wavelengths obtained from Haugland.[10]
[b] Also called hydroethidine; fluorescence increases 10-fold on binding DNA.

nonfluorescent 2′,7′-dichlorofluorescin (DCFH). In turn, the DCFH reacts with ROS to generate 2′,7′-dichlorofluorescein (DCF)[7,8] (Fig. 1). Spontaneous deacetylation occurs, but it is a slow process.[9] DCFH is located diffusely throughout the cytoplasm of a cell and is not irreversibly trapped there. Approximately 90% of either DCFH or DCF leaked from endothelial cells within 1 hr after labeling.[9] The rate of loss is dependent on the type of cells used and can be influenced by the effects of oxidative stress on membrane integrity. Chemical modifications of DCFH-DA, such as carboxylation with and without esterification (Molecular Probes, Inc., Eugene, OR), have been suggested to improve performance over longer time intervals by decreasing leakage[10]; unfortunately, there are no studies demonstrating their comparative performance. Studies by Trayner *et al.*[11] and G. Sawada (unpublished, 1996) suggest that the signal-to-noise ratio for the carboxylated DCFH-DA is low and may be attributed to decreased cellular accumulation.[12] A chloromethylated derivative of DCFH-DA (Molecular Probes, Inc.), believed to be trapped intracellularly by formation of a conjugate with glutathione,[10] has been used in an experimental design involving hours of exposure,[13] but we know of no direct comparisons to DCFH-DA. Most protocols counteract the loss of DCFH/DCF from cells by maintaining a concentration of DCFH-DA in the extracellular medium during the experiment. In contrast to DCFH-DA, only approximately 15% of oxidized DHR or Rhodamine

[7] D. A. Bass, J. W. Parce, L. R. Dechatelet, P. Szejda, M. C. Seeds, and M. Thomas, *J. Immunol.* **130,** 1910 (1983).
[8] R. Cathcart, E. Schweirs, and B. N. Ames, *Anal. Biochem.* **134,** 111 (1983).
[9] J. A. Royall and H. Ischiropoulos, *Arch. Biochem. Biophys.* **302,** 348 (1993).
[10] R. P. Haugland, "Molecular Probes Handbook of Fluorescent Probes and Research Chemicals," 6th Ed., Chapter 21, Section 21.4, p. 491 (1996).
[11] I. D. Trayner, A. P. Rayner, G. E. Freeman, and F. Farzaneh, *J. Immunol. Methods* **186,** 275 (1995).
[12] S. J. Vowells, S. Sekhsaria, H. L. Malech, M. Shalit, and T. A. Fleisher, *J. Immunol. Methods* **178,** 98 (1995).
[13] K. Polyak, Y. Xia, J. L. Zweier, K. W. Kinzler, and B. Vogelstein, *Nature* **389,** 300 (1997).

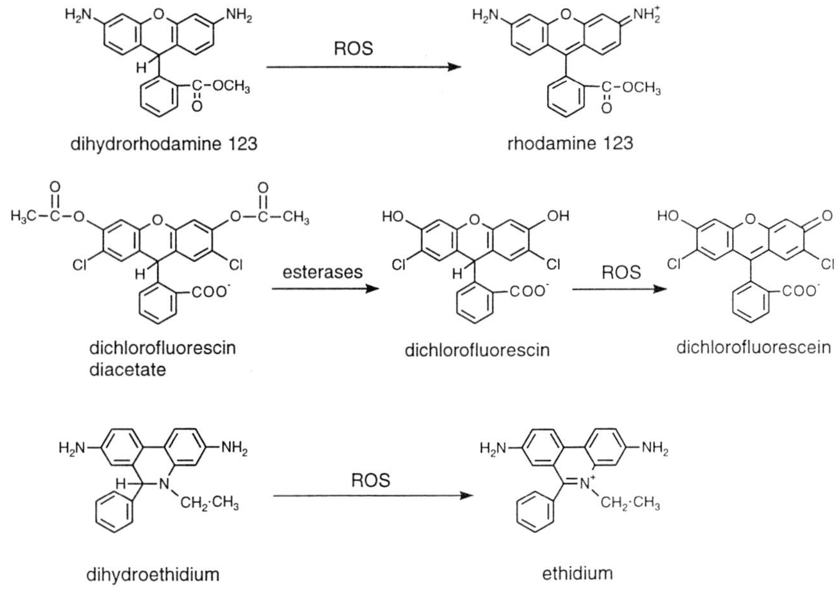

dihydrorhodamine 123 rhodamine 123

dichlorofluorescin diacetate dichlorofluorescin dichlorofluorescein

dihydroethidium ethidium

FIG. 1. Chemical structures and oxidative reactions for the three most commonly used fluorescent indicators of cellular oxidative stress. Dichlorofluorescin diacetate is a membrane-permeable compound deacetylated by intracellular esterases forming a substrate that is a less permeable and more readily oxidized than the *pro* form.

123 is lost from endothelial cells.[9] This is attributed to low affinity, high capacity binding to mitochondria[14] (Fig. 2). A similar intracellular sink exists for oxidized DHE or ethidium, which partitions into the nucleus by intercalating into DNA.[15] In contrast to DCF, neither of the oxidized forms of DHR or DHE appear to be membrane permeable.[2,16] In addition, Rhodamine 123 and ethidium are substrates for multidrug resistance efflux pumps, which reduce cellular concentrations of substrates by active transport out of the cells.[16] Therefore, expression of the pumps in the cell types under study and the potential response of the pumps to oxidative stress should be considered, since they may influence interpretation of the results.

Multiple reactive oxygen products and metabolites are generated within a cell following extracellular addition of a single ROS initiator. However, the probes measure general oxidant concentrations in cells, since none of the indicators listed react with a high degree of selectivity with specific

[14] L. B. Chen, *Methods Cell Biol.* **29,** 103 (1989).

[15] C. D. Bucana, I. Saiki, and R. Nayer, *J. Histochem. Cytochem.* **34,** 1109 (1986).

[16] C. D. Bucana, R. Giavazzi, R. Nayar, C. A. O'Brian, C. Seid, L. E. Earnest, and D. Fan, *Exp. Cell Res.* **190,** 69 (1990).

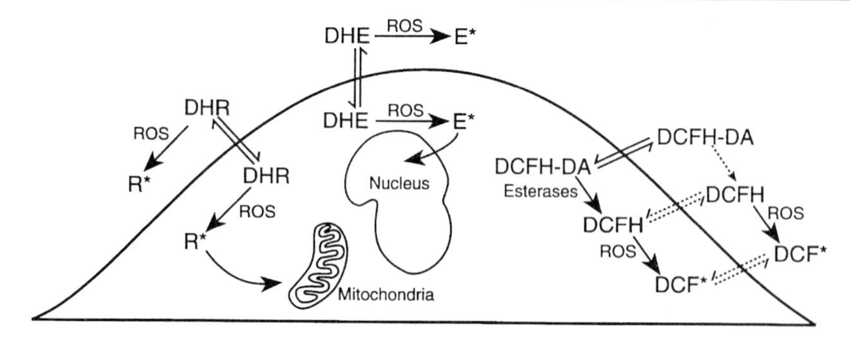

FIG. 2. Subcellular distribution of the three most commonly used indicators (and their oxidized fluorescent species) of cellular oxidative stress. *Left:* Dihydrorhodamine 123 (DHR) reacts with reactive oxygen species (ROS) forming the fluorescent product rhodamine 123 (R*) which binds mitochondria. *Center:* Dihydroethidium (DHE) reacts with ROS forming the fluorescent product ethidium (E*) which partitions into the nucleus upon binding DNA. *Right:* 2′,7′-Dichlorofluorescin diacetate (DCFH-DA) is deacetylated by cellular esterases yielding 2′,7′-dichlorofluorescin (DCFH), which reacts with ROS forming the fluorescent product 2′,7′-dichlorofluorescein (DCF*). The dotted reactions indicate slower processes.

ROS.[6,9] This makes it difficult to identify the role of specific ROS in cellular responses to oxidative stress. Nevertheless, there are some general reaction preferences among the probes, and these may influence practical aspects of their use. For example, DCFH and DHR do not appear to react readily with O_2^-.[17,18] In contrast, DHE, while reacting more specifically with O_2^-, is spontaneously oxidized 10-fold more rapidly than DCFH and DHR, contributing to higher background fluorescence and a lower signal-to-noise ratio.[19,20]

There are also differences in sensitivity among the various probes for reaction under particular experimental conditions. For example, in testing the cellular response to less potent physiological stimuli such as *N*-formyl-methonylleucylphenylalanine or the cytokine tumor necrosis factor-α, DHR was approximately three times more sensitive than DCFH or DHE.[6,12,20–22] Additionally, a weak or absent fluorescent signal may be due to differences in the distribution of the probe molecule and the major ROS within cells,[20]

[17] G. Rothe, A. Emmendörffer, A. Oser, J. Roesler, and G. Valet, *J. Immunol. Methods* **138,** 133 (1991).
[18] W. O. Carter, P. K. Narayanan, and J. P. Robinson, *J. Leukocyte Biol.* **55,** 253 (1994).
[19] G. Rothe and G. Valet, *J. Leukocyte Biol.* **47,** 440 (1990).
[20] G. Packham, R. A. Ashmum, and J. L. Cleveland, *J. Immunol.* **156,** 2792 (1996).
[21] G. Rothe, A. Oser, and G. Valet, *Naturwissenschaften* **75,** 354 (1988).
[22] A. Emmendörffer, M. Hecht, M.-L. Lohrmann-Matthes, and J. Roesler, *J. Immunol. Methods* **131,** 269 (1990).

thereby reducing reactivity, because the probe and ROS are not in close proximity to each other on the molecular scale. Thus, both the degree of reactivity and localization are important considerations in choosing the elements of an experimental system.

A number of processes may interfere with simple interpretation of a fluorescence signal observed in cells. For example, measurement of ROS may be influenced by intracellular peroxidase levels, as well as by the degree of an oxidative insult. Both DCFH and DHR react very slowly with H_2O_2, and the reaction rate is increased markedly in the presence of peroxidase. Therefore, only cells expressing sufficient levels of peroxidase may be detected.[7,2,23] Additionally, conversion of the indicator to a fluorescent form may not always be attributable to ROS. Thus, DCFH can be oxidized by ferrous ion (Fe^{2+}) or by cellular oxidases or peroxidases that are released from membrane compartments in response to certain stimuli.[24,25] DCFH, for example, but not DHE, is oxidized by release of myeloperoxidase from lysosomes of phagocytes degranulating during sepsis.[6] Finally, ROS, especially H_2O_2, generated by a cell can rapidly diffuse significant distances and react with an indicator in a neighboring cell that is not generating ROS.[2]

Some General Methodological Considerations

Cell culture models have proven to be useful for studying oxidative stress in systems less complex than tissues, but still provide the opportunity to examine events in living cells. Thus, cell-based assays may be used to understand the reactions and events of ROS-mediated cytotoxicity, the endogenous protective antioxidant mechanisms within cells, and cellular responses to oxidative stress. Some of the complications in cell model systems can be reduced through selection of experimental conditions. For example, using simple physiological buffers instead of complex cell culture medium can eliminate the potential role of medium components, such as amino acids and vitamins, as ROS scavengers. Likewise, serum use may be minimized for short-term experiments, such as those done on the order of hours, to reduce or eliminate multiple endogenous antioxidants in serum which would complicate the cellular system.[26]

[23] L. J. van Pelt, R. van Zwieten, R. S. Weenig, D. Roos, A. J. Verhoeven, and B. G. J. M. Bolscher, *J. Immunol. Methods* **191,** 187 (1996).

[24] C. P. LeBel, H. Ischiropoulos, and S. C. Bondy, *Chem. Res. Toxicol.* **5,** 227 (1992).

[25] H. Zhu, G. L. Bannenberg, P. Moldéus, and H. G. Shertzer, *Arch. Toxicol.* **68,** 582 (1994).

[26] M. J. Richard and P. Guirard, *in* "Analysis of Free Radical in Biological Systems" (A. E. Favier, J. Cadet, B. Kalyanaraman, M. Fontecave, and J.-L. Pierre, Eds.), p. 261. Birhäuser Verlag, Basel, 1995.

Cytometry, microscopy, and fluorimetry have been used with cells in culture and fluorescent ROS-sensitive probes. It is necessary to optimize cell loading and fluorescence detection for the specific cells, culture conditions, and instrumentation used in a particular experiment. Although the specific conditions used to study ROS-sensitive probes with cells have varied widely, below we offer some general guidelines regarding how to prepare the probes for use. The indicator molecules mentioned here are available from Molecular Probes, Inc. (Eugene, OR) although dihydroethidium also is sold as hydroethidine by Sigma Chemical Co. (St. Louis, MO) and Polysciences, Inc. (Warrington, PA). 2′,7′-Dichlorofluorescin diacetate may be prepared as a concentrated (up to 60 m*M*) stock solution in dimethyl sulfoxide (DMSO) or ethanol. Dihydroethidium and dihydrorhodamine 123 are prepared in DMSO or *N*,*N*-dimethylformamide. Stock solutions may be stored in the dark at −20° for months under nitrogen or argon, and can be repeatedly frozen and thawed. In order to load the ROS probe into suspended cells or cells adhered to a substrate, the cells may be incubated for periods of time ranging from 5 min up to several hours. Final concentrations used for cell loading routinely consist of 10–20 μM DCFH-DA, 1–5 μM DHR, or 10–150 μM DHE in phosphate-buffered saline (PBS) or phenol-free culture medium. Fetal calf serum is sometimes included, if the cells require it for survival or for minimizing spontaneous damage. Loading solutions of the probes are best made up fresh, but can be kept chilled and protected from light, if used during the course of 1 day. After loading, it is best to initiate ROS generation without delay, especially if a rinse step is used to remove extracellular indicator. It is important to test incubation and washing steps carefully, since these may lead to substantial leakage of the indicator molecules from inside the cells. Additionally, the probes may be phototoxic to living cells, if excitation is excessively intense or prolonged. For example, DHR ceases to be a supravital dye when exposed to intense blue or green light, such as when using a rhodamine filter set, causing photobleaching and release of free radicals. Also, prolonged exposure (>16 hr) of cells to the dye may be toxic depending on the amount of dye initially accumulated and maintained.[14] All cells do not take up the fluorescent probes to the same degree. For example, some tumor cell lines accumulate 5- to 20-fold more DHR than nontransformed cells, probably because of elevated mitochondrial and plasma membrane potentials.[27] Especially in such extreme cases, the potential cytotoxicity due to the probe itself and the necessary excitation conditions to observe the probe should be tested during optimization of the experimental design.

[27] B. M. Kinsey, A. I. Kassis, F. Fayad, W. W. Layne, and S. J. Adelstein, *J. Med. Chem.* **30,** 1757 (1987).

Application of Mathematics to Fundamental Chemistry of Free Radical Reactions

Nonlinear Least Squares Analysis

Fully understanding ROS action in cells requires the ability to measure the fundamental kinetic properties of the reactions. This may be best accomplished by applying an understanding of the underlying chemistry of such reactions to cellular systems. Curve-fitting techniques utilizing nonlinear least squares regression may now be routinely applied, considering the nearly universal adaptation of high speed personal computers. Moreover, many laboratory instruments now make it simple to incorporate complicated data with many readings into files that may be readily analyzed using such techniques. A major advantage of the proper use of curve fitting techniques is the derivation of characteristic parameters from an entire time course or dose–response experiment. Such parameters may be considerably more robust than simple comparisons between individual data points observed at single time intervals. With readily available computer power and appropriate nonlinear least squares analysis software, the failure to apply appropriate mathematical analyses to the data is unnecessary and even constitutes a waste of effort by failing to fully exploit data gathered in what may be expensive and time consuming experiments.

In order to quantify ROS reactions in cells, especially kinetic parameters, we strongly recommend the use of nonlinear least squares analysis of the data. A simple but efficient and highly versatile program was written a number of years ago by Yamaoka *et al.*[28] Although the program was originally written for use in pharmacokinetic analysis, it is readily adapted to more general use and has the advantage of being written for the DOS environment, making it available even for older computers. However, its use requires some knowledge of the computer programming language BASIC. A number of commercial programs are available for nonlinear least squares analysis, including Sigmaplot (Windows version 2.01 and above, Jandel Scientific, San Rafael, CA) and Scientist (MicroMath Scientific Software, Salt Lake City, UT). Numerous other programs probably also include methods for inputting data to a nonlinear least squares analysis. A general description of some of the uses and pitfalls of nonlinear least squares analysis, including a nontechnical description of the mathematics of the approach, is available.[29] Statistical analyses used in comparing the quality of the fit of equations to data are also described by Motulsky and

[28] K. Yamaoka, Y. Tanagawara, T. Nakagawa, and T. Uno, *J. Pharm. Dyn.* **4,** 879 (1981).
[29] H. J. Motulsky and L. A. Ransnas, *FASEB J.* **1,** 365 (1987).

Ransnas.[29] A more detailed description of nonlinear least squares analysis, including advantages and limitations as well as technical and mathematical considerations, is found in Johnson and Faunt.[30]

Equations Describing Kinetics of ROS Activity

Free radical reactions may be broken down into three fundamental steps: initiation, chain propagation, and termination.[31] These steps may be represented in schematic form as:

$$\mathrm{G} \xrightarrow{k_i} \mathrm{R}\cdot + \mathrm{R}\cdot \xrightarrow{k_t} \text{recombination products}$$

where G is the concentration of the free radical generating species, $\mathrm{R}\cdot$ is the concentration of free radicals generated from G, k_i is the characteristic rate constant for the initiation of free radical generation, and k_t is the characteristic rate constant for the termination of the free radical equation. It is useful to note that, although the free radicals are represented identically as $\mathrm{R}\cdot$ in this schematic representation, they need not be identical but may be chemically distinct products of the free radical generator, G. The important characteristic is that two free radicals are generated from each generator molecule. Even though the cascade of free radicals ultimately resulting from the generator molecules may be complex and consist of multiple chemical species, the process can be represented in simple equation form.[32–37] This is due to the fundamental property of one generator molecule producing two free radicals. Additionally, an important practical point is that the equations are simple, if free radical generation is at steady state. The presteady-state conditions are more complex[32,33] and, consequently, experimental parameters for the presteady-state condition are more difficult to quantify precisely.

The differential equation describing free radical generation for the schematic shown above is

$$\frac{d\mathrm{R}\cdot}{dt} = k_i\mathrm{G}_0 - k_t\mathrm{R}\cdot^2 \tag{1}$$

[30] M. L. Johnson and L. M. Faunt, *Methods Enzymol.* **210,** 1 (1992).

[31] G. M. Barrow, "Physical Chemistry," 2nd ed., p. 458. McGraw-Hill, New York, 1966.

[32] R. McKenna, F. J. Kezdy, and D. E. Epps, *Anal. Biochem.* **196,** 444 (1991).

[33] D. E. Epps and J. M. McCall, *in* "Handbook of Synthetic Antioxidants" (L. Packer and E. Cadenas, Eds.), p. 95. Marcel Dekker, New York, 1997.

[34] D. E. Decker, S. M. Vroegop, and S. E. Buxser, *Biochem. Pharmacol.* **50,** 1063 (1995).

[35] L. D. Horwitz, J. S. Wallner, D. E. Decker, and S. E. Buxser, *Free Rad. Med. Biol.* **21,** 743 (1996).

[36] S. M. Vroegop, D. E. Decker, and S. E. Buxser, *Free Rad. Biol. Med.* **18,** 141 (1995).

[37] G. A. Sawada, T. J. Raub, D. E. Decker, and S. E. Buxser, *Cytometry* **25,** 254 (1996).

At steady state, free radical generation is quadratic and may be solved for the free radical concentration:

$$R\cdot = \sqrt{G_0 \frac{k_i}{k_t}} \tag{2}$$

Thus, the steady-state concentration of free radicals is directly proportional to the square root of the initial concentration of the free radical generator. This is a fundamental equation that may be used to show that a reaction conforms to properties consistent with the chemistry of free radical generation.

Precise analysis of data generated through the use of fluorescent ROS markers requires consideration of equations describing the reaction of the indicator molecule with free radicals. If indicator molecules are inactivated by the reaction, such as those occurring if the reaction yields a nonfluorescent product from molecules that were fluorescent before reacting, then a schematic of the reaction is

$$I \xrightarrow{k_d R\cdot} X, \tag{3}$$

where I is the concentration of the indicator molecules, X is the concentration of the inactivated, i.e., nonfluorescent, indicator molecules, and k_d is the rate constant for the reaction between the indicator molecules and free radicals. The rate of the reaction may be described as

$$\frac{dI}{dt} = -R\cdot k_d I \tag{4}$$

and in integrated form the concentration of the indicator at a given time, t, relative to the initial concentration of the indicator, I_0, is given by

$$I = I_0 e^{-R\cdot k_d t} \tag{5}$$

Incorporating the concentration of free radicals at steady state as a function of the concentration of the free radical generating species, i.e., Eq. (2), yields

$$I = I_0 e^{-k_d t \sqrt{\frac{k_i}{k_t} G_0}} \tag{6}$$

The loss of the fluorescent form of the indicator molecule is proportional to the square root of the concentration of the free radical generator species, G_0. One may observe the rate of loss of the indicator and characterize the results by an experimentally determined rate constant, k_{exp}:

$$k_{exp} = k_d \sqrt{\frac{k_i}{k_t} G_0} \propto \sqrt{G_0} \tag{7}$$

Thus, using an instrument capable of gathering multiple measurements over short time intervals, the experimental observations consist of observing the rate of disappearance of the fluorescent signal as a result of reaction with ROS.

Conversely, if the reaction of reactive oxygen species results in the generation of fluorescence, then the applicable equation based on the same principles as described above is

$$I = X_0\left(1 - e^{-k_d t\sqrt{\frac{k_i}{k_t}G_0}}\right) \tag{8}$$

In Eq. (8), I is the concentration of fluorescent indicator molecules formed by reaction with the free radicals generated from the free radical generator, G_0. X_0 is the initial concentration of the nonfluorescent indicator species before reaction with free radicals.

As already mentioned, the equations outlined above depend on the assumption of steady-state free radical generation. Equations without this assumption are substantially more complex[32,33] and render the data analysis more difficult. In experiments carried out in a cell culture system using cumene hydroperoxide or hydrogen peroxide as free radical generator species,[37] we determined that steady-state free radical concentrations were achieved after 300 sec or less. Of course, other culture conditions, different cells, or a different free radical generator may result in substantially different times required for reaching steady-state conditions. A useful test for steady-state conditions is provided by agreement of an equation for the first-order process with the experimental data, following a lag period during which the reaction approaches the steady-state free radical concentration. Additionally, a series of concentrations of the generator, G, may be used to obtain rate constants for the changes in fluorescence at each concentration of G. As shown previously[37] and described further below, the rate constants may be used to determine the agreement between the data and the free radical-dependent model delineated above in Eq. (7). This approach may be used to provide evidence that the free radical generation is essentially steady state and that the changes in fluorescence are due to free radicals. Results from a useful starting experiment for the characterization of a cellular system are shown graphically in Figs. 3 and 4. A range of concentrations of the free radical generator, G, may be used to generate a series of curves showing the fluorescence of the marker, I. From the fitted curves, the parameters of the first-order process are calculated and used to check the agreement between the data and the expected square root relationship between the rate constants for fluorescence generation and the concentration of the generating species.

If a marker that becomes fluorescent on reaction with reactive oxygen

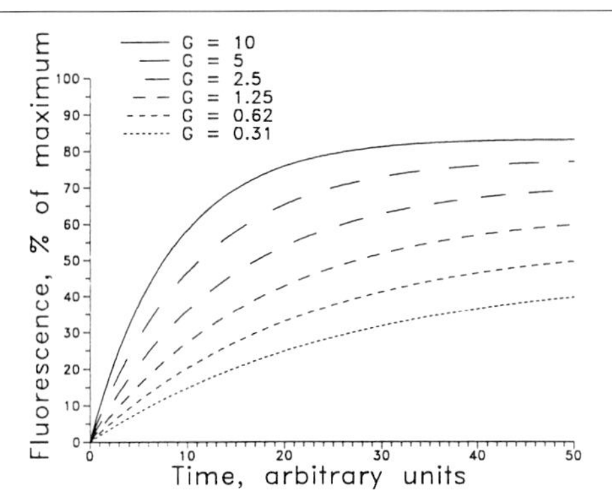

FIG. 3. Rate of fluorescence generation for a range of concentrations of a free radical generator species, G.

species (ROS) is used to detect ROS activity, a first-order increase in fluorescence may be recorded when the free radical generation is at steady state. Figure 3 shows the shape of time course curves resulting from such an experiment. The rate of fluorescence generation is directly proportional to the concentration of the free radical generator species. If the rate constants for the fluorescence generated in the presence of the various concentrations of G are calculated and plotted vs the concentration of G, the pattern shown in Fig. 4 may be observed. For this example, the k_{exp} calculated from Fig. 3 are shown as open circles. The line shown is generated using Eq. (7) showing the relationship between the rate constants and the square root of the concentration of the free radical generator, G.

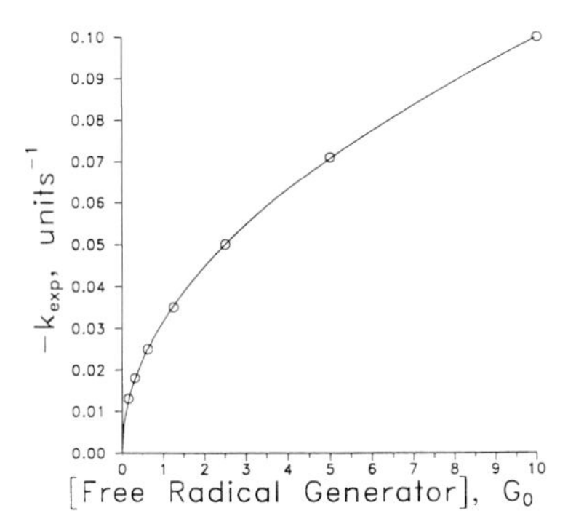

FIG. 4. Effect of the concentration of the free radical generator on the experimental rate constants, k_{exp}, calculated from experiments.

The situation becomes somewhat more complicated, if in addition to the generator, G, and free radical species, R·, a free radical scavenging species, S, is also present. In the simplest case where the species S is a simple scavenger that reacts with a single free radical to form a stable intermediate radical species, R·S, the defining differential equation is:

$$\frac{d\mathrm{R}\cdot}{dt} = k_\mathrm{i}\mathrm{G}_0 - k_\mathrm{t}\mathrm{R}\cdot - k_\mathrm{s}\mathrm{R}\cdot\mathrm{S} \tag{9}$$

where k_s is the rate constant for free radical scavenging by the scavenger, S. At steady state Eq. (9) is equal to zero. If the primary means by which free radicals are removed from the reaction is via reaction with S, then k_t is small relative to k_s, the $k_\mathrm{t}\mathrm{R}\cdot^2$ term is inconsequential, and the equation for steady-state free radical generation simplifies to

$$\mathrm{R}\cdot = \frac{k_\mathrm{i}}{k_\mathrm{s}\mathrm{S}}\mathrm{G}_0 \tag{10}$$

This shows a direct and linear proportionality between the concentration of free radicals and the concentration of the generating species, G_0, in contrast to the proportionality between the square root of G_0 and R· described above [Eq. (2)]. Comparing the fit of data to the two equations, especially in the presence or absence of the scavenging species, S, allows us to determine some of the details of the radical reaction. Of course, if the scavenger is not the predominant means by which free radicals are removed from the reaction, then Eq. (2) will conform to the data better than does Eq. (10). In contrast, if both the rate constant for the scavenger, S, and the normal termination characterized by the rate constant k_t, described above, are within severalfold of each other in magnitude, the $k_\mathrm{t}\mathrm{R}\cdot^2$ term does not drop out of the equation. At steady state, the equation is quadratic in $\mathrm{R}\cdot^2$ and may be solved for R·:

$$\mathrm{R}\cdot = \frac{-k_\mathrm{s}\mathrm{S} + \sqrt{(k_\mathrm{s}\mathrm{S})^2 + 4k_\mathrm{t}k_\mathrm{i}\mathrm{G}_0}}{2k_\mathrm{t}} \tag{11}$$

Substituting Eq. (11) into the value for R· from Eq. (8) yields

$$\mathrm{I} = \mathrm{X}_0\left(1 - e^{-k_\mathrm{d}t\frac{-k_\mathrm{s}\mathrm{S} + \sqrt{(k_\mathrm{s}\mathrm{S})^2 + 4k_\mathrm{t}k_\mathrm{i}\mathrm{G}_0}}{2k_\mathrm{t}}}\right) \tag{12}$$

As may be seen in Eq. (12) if $k_\mathrm{s}\mathrm{S}$ is small relative to $k_\mathrm{d}t$ and $k_\mathrm{t}k_\mathrm{i}$, then the effect of the scavenger is small. That is, if the concentration of S is low, its effect is likely to be undetectable. Although the former result is not surprising, the equation also shows that a small rate constant for the reaction between the scavenger, S, and the ROS will have a similar effect. One must match the detection system to the scavenging species in order to be able to detect effects readily using the fluorescent probe. Thus, a highly reactive fluorescence probe will not be useful in detecting protection by a substantially less reactive scavenger, since the ROS will react more readily with the marker species and be used up by the time reaction with the scavenger becomes significant. If the reactivity of the scavenger is within severalfold of the reactivity of the marker species, then the relationship between the concentration of the generating species, G, added and the k_{exp} of the reactions at the various concentrations of G will be reduced relative to the G vs k_{exp} curve observed in the absence of the scavenger species. If the rate of fluorescence is observed, as described above for the uninhibited reaction, the curves for the time courses may appear, as shown in Fig. 5. Thus, the slope of a curve changes either as the concentration of S increases or, for a group of compounds, as k_s changes. Specifically, the experimental rate constants change in accord with Eq. (12). The effect of the scavenger, S, may also be observed by plotting the initial concentration of the free radical generator, G_0, vs the experimentally derived rate constants, k_{exp}, at a range of concentrations of S. As shown in Fig. 6, the expected square root relationship between the k_{exp} and G_0 is altered by the presence of S. As the concentration of S increases, the fluorescence is reduced relative to the uninhibited reaction, and the curve flattens.

Additional Experimental Considerations

Effect of Dye Leakage on Calculated Rates of Reaction with Free Radicals

An experimental complication results from the propensity of the indicator molecules to leak from cells. If the measured species is initially fluores-

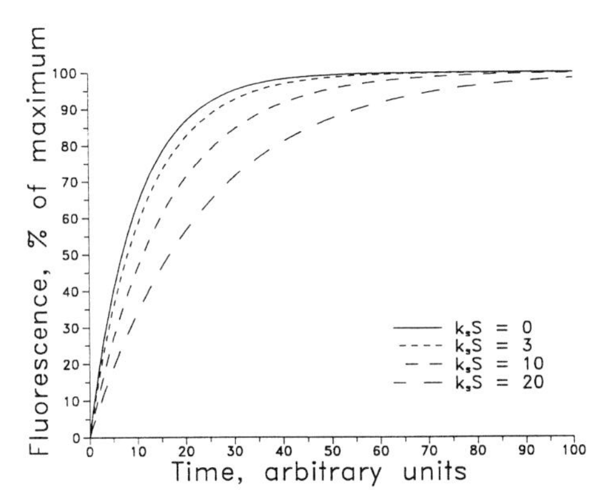

FIG. 5. Effect of changes in the scavenger concentration, S, or the rate constant for free radical scavenging, k_s, on the rate of fluorescence generation.

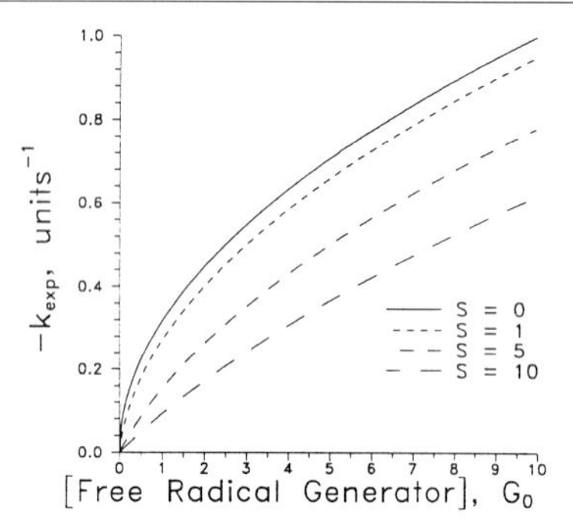

FIG. 6. Effect of the presence of a free radical scavenger, S, on the experimental rate constants observed over a range of free radical generator concentrations, G_0, where

$$k_{exp} = -k_d t \frac{-k_s S + \sqrt{(k_s S)^2 + 4k_t k_i G_0}}{2k_t}$$

cent and the experimental parameter used is the decrease in fluorescence, then reaction of the marker molecules with free radicals will contribute to the loss of signal inside cells, but so will loss due to leakage. The loss of fluorescence due to leakage will not be readily distinguished from a loss of fluorescence due to reaction with free radicals inside the cells. Conversely, in systems where the marker species becomes fluorescent on reaction with free radicals, changes in fluorescence will be reduced by leakage of the nonfluorescent species before reaction with free radicals can take place. Loss of the marker species will not be easily distinguished from decreased reaction with free radicals. One needs to account for the effects due to leakage and be able to distinguish it from the effect of an ROS reaction. Schematically, leakage may be described as

$$\begin{array}{l} X \xrightarrow{k_d R\cdot} I \\ \downarrow k_l \\ E \end{array}$$

where X, I, and $k_d R\cdot$ have meanings as described above, E is the concentration of X that leaks to the extracellular compartment before reaction with ROS, and k_l is the rate constant for the leakage of X. In this case, the following rate equations [Eq. (13)] apply:

$$\begin{aligned} \frac{dI}{dt} &= k_d R \cdot X \\ \frac{dX}{dt} &= -k_d R \cdot X - k_l X \end{aligned} \tag{13}$$

Solved for the fluorescent product, I, the equations yield

$$I = \frac{k_d R \cdot X_0}{k_d R \cdot + k_l}(1 - e^{-(k_d R \cdot + k_l)t}) \tag{14}$$

Note that the form of Eq. (14) is unchanged relative to Eq. (10), but the meaning of the parameters is changed. In Eq. (12), the magnitude of the reaction, i.e., the maximum fluorescence, is simply the initial concentration of $X = X_0$. However, under conditions where X leaks from the cells appreciably during the time course of the experiments, the maximum fluorescence is reduced proportionally to the relative rate constants of the reaction with free radicals and the rate constant for leakage, as

$$\frac{k_d R \cdot X_0}{k_d R \cdot + k_l}$$

Similarly, the rate constant for the reaction changes when leakage is significant as

$$k_{exp} = k_d R \cdot + k_l \tag{15}$$

That is, the rate constant observed in the experiment will be the sum of the rate constant for the free radical reaction and the rate constant for leakage. Thus, both the maximum fluorescence and the rate of fluorescence generation will be different than if leakage did not occur. Experimentally, this will result in the effects on the fluorescence data shown in Fig. 7.

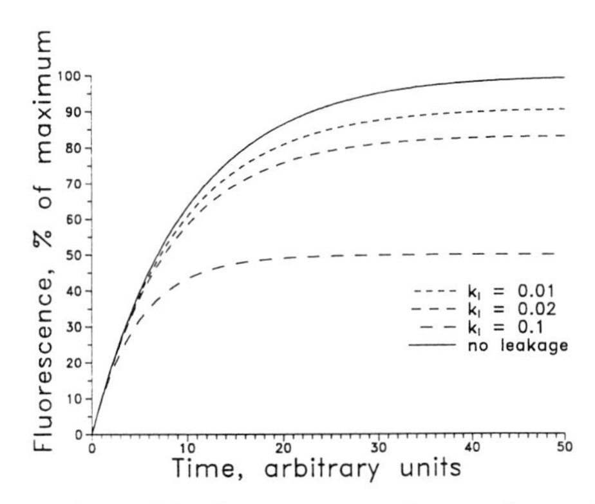

FIG. 7. Effect of the leakage of the fluorescence marker species on the rate and maximum fluorescence in cells.

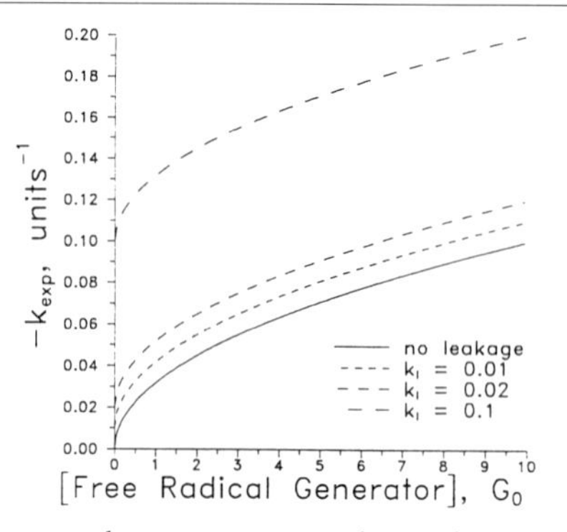

FIG. 8. Effect of leakage on the rate constants observed over a range of concentrations of the free radical generator, G.

In the presence of leakage, the relationship between k_{exp} and the concentration of the generator species, G, is altered, as shown in Fig. 8. In this case, the plots are shifted up so that the intercept at $G_0 = 0$ is no longer equal to zero. For marker species with various spontaneous leakage rates, the intercept will increase linearly in direct proportion to the rate constant for leakage of the marker, as indicated in Eq. (15).

If the experiment is designed using a fluorescent marker that loses fluorescence when reacted with free radicals, then the applicable equation is

$$I = I_0 e^{-(k_d R\cdot + k_l)t} \tag{16}$$

where the rate constant for the loss of fluorescence parameter is altered from Eq. (7) by the additon of the term k_l. In this case, there is no loss of maximum fluorescence, since maximum fluorescence is defined as the total amount of signal detected from the initiation of fluorescence readings. Leakage will increase the rate of loss of the fluorescence in proportion to the size of k_l relative to the product of $k_d R\cdot$.

Not only may leakage of the prereacted nonfluorescent species take place, but also the species fluorescent after reaction with free radicals may leak. In this case, the scheme and the resulting equations are somewhat more complex. Schematically, the situation may be outlined as

$$\begin{array}{ccc} I & \xrightarrow{k_d R\cdot} & X \\ \downarrow k_E & & \downarrow k_X \\ E & & X' \end{array}$$

where E is the concentration of the marker leaked before reaction with free radicals, k_E is the rate constant for the process, X′ is the concentration of marker leaked after reaction with free radicals, and k_X is the rate constant for the latter process. The differential equations [Eq. (17)] describing the process are

$$\frac{dX}{dt} = k_d R\cdot - k_x X$$
$$\frac{dI}{dt} = -k_d R\cdot I - k_E I \tag{17}$$

Solved for the concentration of the fluorescent marker molecule, X, the equations yield

$$X = I_0 \frac{k_d R\cdot}{k_d R\cdot + k_l - k_x} (e^{-(k_d R\cdot + k_l)t}) \tag{18}$$

Equation (18) is appropriate if the rate constants for leakage, k_l and k_X, are different. If $k_l = k_X$, Eq. (18) is slightly simpler and can be readily derived from Eq. (15). Data from an experiment where leakage of both forms of the marker molecule occurs may resemble the pattern shown in Fig. 9.

For compounds with a range of k_x values, the shape of the curves may resemble those shown in Fig. 10, as the value of the k_x rate constant changes. In other words, as the rate of leakage of the fluorescent marker molecules increases, the signal inside the cells will decay more rapidly. Obviously, this will complicate interpretation of the results, if the leakage rate is rapid relative to the time course of the experiment.

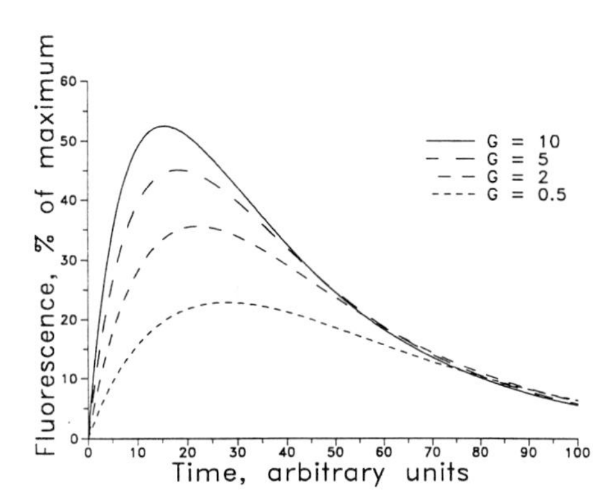

FIG. 9. Time course of fluorescence when the marker species leaks both before and after reaction with free radicals.

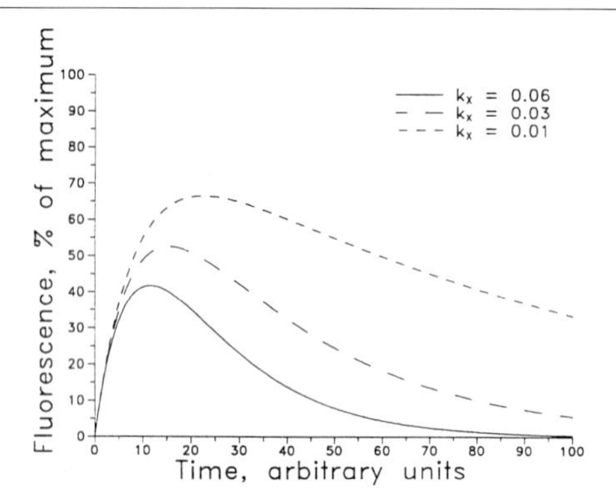

FIG. 10. Effect of changes in the rate constant for leakage of the fluorescent product from cells.

Data Pooling for Optimal Derivation of Parameters Important for Analyzing Digitized Image Analysis Data from Cells

Fluorescence image analysis using an instrument such as the ACAS 570 Interactive Laser Cytometer (Meridian Instruments, Inc., Okemos, MI) gathers data over a defined region on a surface with living cells. Regions containing clusters of cells are chosen visually, and then the regions are scanned repeatedly at regular time intervals for the generation of a time course of fluorescence activation or inactivation. Inevitably, cells growing in clusters will not be distributed evenly over the growth surface. Also, the size of individual clusters may vary considerably. Choice of particular cell clusters on a surface containing less than confluent cells can influence the data analysis. Even if considerable effort is expended in order to get the cells distributed as evenly as possible on the original growth surface, the number of cells surviving and growing in individual clusters will vary. Since the uptake of a fluorescent dye will depend on the number and size of cells in a given cluster, we must expect that the amount of dye absorbed will result in variance among cell clusters with regard to the brightness of fluorescence. How data are pooled and analyzed from the various sizes of clusters found within a given treatment protocol will have to be considered carefully in order to prevent introducing artifacts due simply to clustering and the choice of particular cell clusters. For example, in a study reported previously,[37] we compared the effect of pooling data from individual cell clusters before and after fitting to the expected first-order time courses; if the data were pooled before curve fitting, then the apparent standard deviations were large. This would have resulted in very imprecise calculations in the determination of the rate constants for progress of the free radical reaction with the fluorescent probe. However, if the data from each individual cell cluster were analyzed separately, the calculation of individual rate constants was precise. The major difference was found to be due to differences in the maximum fluorescence of each cell cluster, probably due to differences in the number of cells in each cluster and/or the size of cells making up each cluster. Once this adjustment was accounted for, the agreement among rate constants calculated from individual data sets was remarkably high. Thus, we were able to draw the conclusion that the rate of reaction in individual cell clusters was very uniform among the samples, even though the maximum fluorescence varied broadly. The resulting parameters calculated from the pooled data are thus the result of replicate calculations of the rate constants and/or maximum fluorescence.

[29] *In Vivo* Measurement of Hydrogen Peroxide by Microelectrode

By HIDEKATSU YOKOYAMA

Introduction

Hydrogen peroxide is one of the active oxygen species, but hydrogen peroxide itself is very stable in aqueous media (its half-life is 10^8 centuries at a temperature of 37°), so it does not exert a direct cytotoxic effect. However, the hydroxyl radical that has been transformed from hydrogen peroxide by the catalytic action of transition metals appears to be the most toxic active oxygen species, even in a small amount. It has been hypothesized that hydrogen abstraction from polyunsaturated lipid acids, which make up the cellular membrane, by hydroxyl radical plays a causative role in cell injuries. Because the hydroxyl radical is highly reactive, it is difficult to detect it *in vivo.*[1-3] Therefore, to investigate the cellular toxicity of active oxygen, *in vivo* measurement of hydrogen peroxide as a precursor of the hydroxyl radical is important.

Hydrogen peroxide is produced via many enzymatic reactions. Xanthine oxidase, aldehyde oxygenase, NAD(P)H oxygenase, NADPH–cytochrome-*c* reductase, and a number of other enzymes generate superoxide first, then hydrogen peroxide is produced from superoxide by spontaneous disproportionation. Amino acid oxidase, amine oxidase, glucose oxidase, and others

[1] B. Halliwell and J. M. C. Gutteridge, *Arch. Biochem. Biophys.* **246,** 501 (1986).
[2] B. Halliwell, *Acta. Neurol. Scand.* **126,** 23 (1989).
[3] E. Niki and H. Shimazaki, "Active Oxygen." Ishiyaku Shuppan, Tokyo, 1987.

0076-6879/99 $30.00

generate hydrogen peroxide directly. Hydrogen peroxide is readily scavenged *in vivo* by a scavenging system, such as glutathione peroxidases.[1-3]

For conventional determinations of hydrogen peroxide, absorptiometry (e.g., horseradish peroxidase, cytochrome *c* peroxidase, alcohol dehydrogenase, and titanium methods) or fluorometry (e.g., scopoletin and fluorescein methods) is employed. These methods have been applied in *in vitro* experiments but are not suitable for *in vivo* studies.[4-9] Electrochemical techniques using microelectrodes have been employed for quantitative analysis of a target substance existing in the microregion and has provided good results *in vivo*.[10-13] The classical electrochemical detection of hydrogen peroxide was represented by simple amperometry.[14] With this method, one can obtain a hydrogen peroxide concentration from oxidative current when electrodes are charged at the oxidative potential for hydrogen peroxide. However, because all substances having oxidative potentials that are lower than that of hydrogen peroxide are oxidized together with hydrogen peroxide, this method would have no selectivity *in vivo.*

Differential double-pulse amperometry (DDPA) is a highly sensitive and selective electrochemical technique. The DDPA technique uses a stepped pulse where the first pulse is lower than the characteristic oxidation potential of a target substance, while the second pulse is higher than the oxidative potential. The concentration of the target substance can be determined by the difference between current values derived from these pulses. The response is emphasized for an analyte with an onset oxidation potential in the region between the first and second pulses. In addition, the capacitive and residual currents are suppressed. These characteristics contribute to sensitivity and selectivity of the technique. For hydrogen peroxide measurement, the first and second steps of the pulses are set, respectively, at 750 mV

[4] A. Boveris, N. Oshino, and B. Chance, *Biochem. J.* **128,** 617 (1972).

[5] T. Yonetani, *J. Biol. Chem.* **240,** 4509 (1972).

[6] M. Nakano, Y. Tsustumi, and T. S. Danowski, *Proc. Soc. Exp. Biol. Med.* **159,** 960 (1968).

[7] H. Pobiner, *Anal. Chem.* **33,** 1423 (1961).

[8] W. A. Andreae, *Nature* **175,** 859 (1955).

[9] A. S. Kenston and R. Brandt, *Anal. Biochem.* **11,** 1 (1965).

[10] T. Nakazato and A. Akiyama, *J. Neurochem.* **51,** 1007 (1988).

[11] T. Malinski, F. Bailey, Z. G. Zhang, and M. Chopp, *J. Cereb. Blood Flow Metab.* **13,** 355 (1993).

[12] H. Yokoyama, N. Mori, K. Osonoe, N. Kasai, M. Hiramatsu, T. Yoshimura, T. Matsue, I. Uchida, N. Kobayashi, N. Tsuchihashi, and S. Niwa, *Physchiatry and Clinical Neurosciences* **49,** S277 (1995).

[13] H. Yokoyama, N. Mori, N. Kasai, T. Matsue, I. Uchida, N. Kobayashi, N. Tsuchihashi, T. Yoshimura, M. Hiramatsu, and S. Niwa, *Denki Kagaku* **63,** 1167 (1995).

[14] A. J. Bard and L. R. Faulkner, "Electrochemical Methods." John Wiley & Sons Inc., New York, 1980.

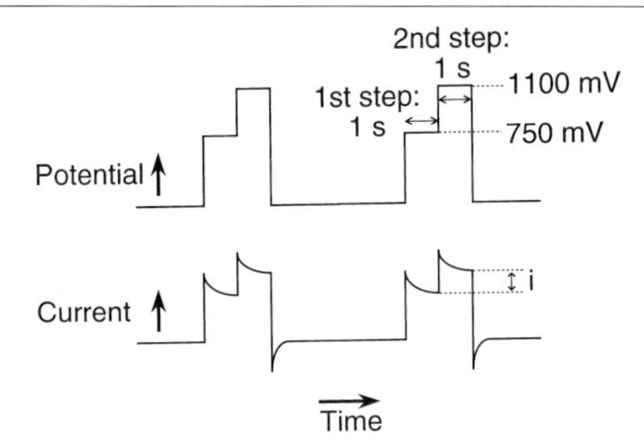

FIG. 1. Schematic diagram of DDPA for hydrogen peroxide detection (from ref. 16). (Reproduced from "Biosensors & Bioelectronics" with kind permission from Elsevier Science.)

and 1100 mV vs reference level (=Ag|AgCl). The electrode reaction of hydrogen peroxide,

$$H_2O_2 \rightarrow O_2 + 2H^+ + 2e^-$$

shows that the hydrogen peroxide decomposition does not produce the electrochemical signals under the above-stated conditions. Furthermore, the rate of this oxidation reaction is not high, so 1 sec is suitable for the duration of each step of the pulse for hydrogen peroxide detection (Fig. 1).[15,16] Tamiya *et al.*[15] demonstrated the measurement of the hydrogen peroxide concentration with high sensitivity and selectivity by DDPA, using a cylinder microelectrode of carbon fiber *in vitro*. This technique is considered to be suitable for *in vivo* measurement.

We have developed a Pt-disk microelectrode to conduct DDPA to detect hydrogen peroxide in the brain of a freely moving rat.[16] In this chapter, the fabrication, electrochemical response, and *in vivo* application of this electrode are described.

Fabrication of Platinum Disk Microelectrode

Working Electrode

A Pt-disk microelectrode serves as a working electrode. A Pt wire (30 μm in diameter) is connected to a copper wire (0.3 mm in diameter) by using an electric welder (ME-35A & MT-510A, Miyachi Technos, Ja-

[15] E. Tamiya, Y. Sugiura, A. Akiyama, and I. Karube, *Ann. NY Acad. Sci.* **613,** 396 (1990).

[16] H. Yokoyama, H. Tsuchihashi, N. Kasai, T. Matsue, I. Uchida, N. Mori, H. Ohya-Nishiguchi, and H. Kamada, *Biosens. Bioelectron.* **12,** 1037 (1997).

pan). The other end of the copper wire is connected to a gold-plated pin connector. The Pt wire is insulated with melted lead glass, which is melted by a heater coil (5 mm in outer diameter, constructed from 4 turns of nichrome wire). This coil is connected to a volt slider (V-130-5, Yamabishi, Japan) via enamel resistors (0.1 Ω). The isolated Pt wire is enclosed in a glass capillary tube (1 mm in outer diameter, 15 mm in length, Narishige, Japan) by using epoxy resin (Araldite, Ciba-Geigy, Switzerland). One day later, the tip of the electrode is polished by using alumina powder (Alpha micropolish alumina B & C, Buehler, Lake Bluff) to make a flat disk surface (Fig. 2). The rigidity of the electrode is reinforced by the glass capillary filled with epoxy resin, which is especially important when one is working with the freely moving animals. A glass capillary of this size is acceptable for implantation into the brain of a rat.[13,16]

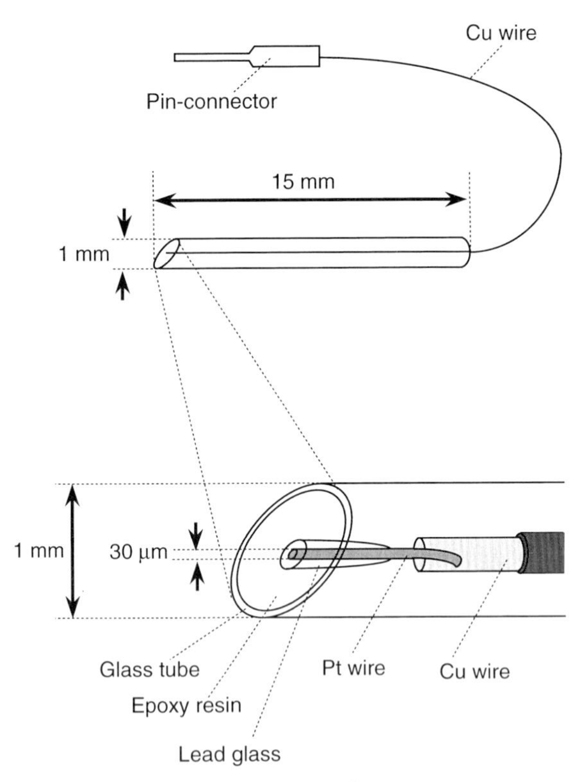

FIG. 2. Structure of the Pt-disk microelectrode (from ref. 16). (Reproduced with slight modification from "Biosensors & Bioelectronics" with kind permission from Elsevier Science.)

Reference/Counter Electrode

An Ag|AgCl electrode is used as a reference/counter electrode. Ag|AgCl electrodes of two different sizes are fabricated. The larger one is used to test the electrochemical response of platinum microelectrodes, while the smaller one serves for *in vivo* experiments.

The larger electrode is constructed as follows. Silver wire (0.5 mm in diameter) and Pt wire (0.5 mm in diameter) are coiled to make 8 turns (3 mm in outer diameter, 15 mm in length). Both coils are dipped in a 1 *M* HCl solution, and a constant bias current (0.3 mA) is sent through them by a galvanostat (HA-502, Hokuto Denko Ltd., Japan) (Ag is the cathode, Pt the anode) for 60 min. The coiled Ag|AgCl wire thus obtained is inserted in a sample tube (0.5 ml in volume), which is filled with a saturated KCl solution. The sample tube is connected to a salt bridge, which consists of a glass capillary tube (1 mm in outer diameter, 50 mm in length, Narishige, Japan) filled with agar that contains saturated KCl (Fig. 3a).

The smaller electrode is constructed as follows. Silver wire (0.5 mm in diameter, 15 mm in length) is soldered to a copper wire (0.3 mm in diameter). The copper wire is then connected to a gold-plated pin connector. The Ag wire and the above-described coiled Pt wire are dipped in a 1 *M* HCl solution, and a constant bias current (0.3 mA) is applied for 3 min. The Ag|AgCl wire thus obtained is inserted into a glass capillary tube (1 mm in outer diameter, 15 mm in length, Narishige, Japan) which is filled with agar that contains saturated KCl and is fixed with dental cement (Fig. 3b).

One day after fabrication, the difference between the potential of the Ag|AgCl electrode thus obtained and that of a commercially available Ag|AgCl electrode (RE-1B, BAS, Lafayette) in a saturated KCl solution is measured; then the electrode with a potential difference of less than 1 mV (larger electrode) or 10 mV (smaller electrode) can be used in the following experiments. The reference electrodes are stored in a saturated KCl solution.

Computer-Controlled Electrochemical System

A computer-controlled electrochemical system is used for analysis. It consists of a current amplifier, personal computer, and A/D and D/A converters (Fig. 4). A patch/whole cell clamp amplifier (CEZ-2300, Nihon Kouden Co., Ltd., Japan) is used as a current amplifier. The amplifier has been modified. As a result, the ratio of external input potential and bias potential at the electrodes is 1:1, and the current-detecting resistance is 50 MΩ. The external potential (i.e., the bias potential at the electrodes) is controlled by a personal computer (PC9801BA, NEC, Japan) via a D/A

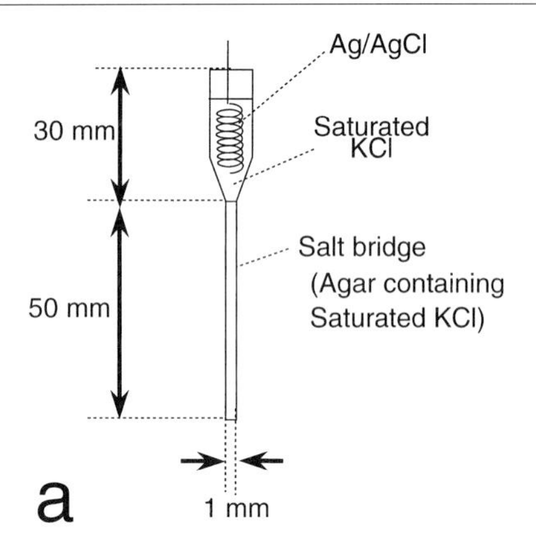

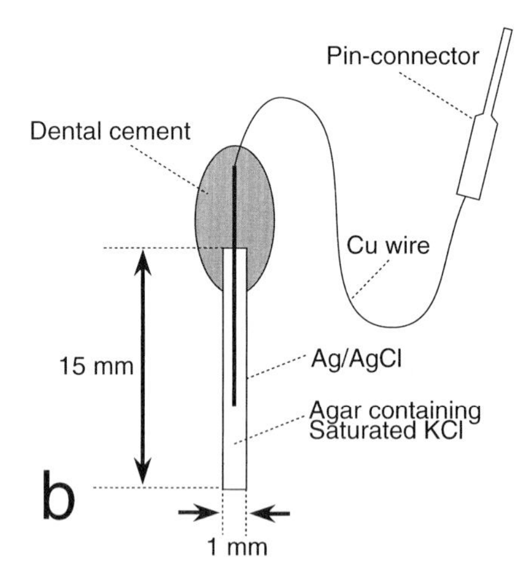

FIG. 3. Structure of two sizes: (a) larger electrode; (b) smaller electrode of the Ag|AgCl microelectrode.

converter (DAJ98, Canopus Co., Ltd., Japan). One can perform cyclic potential scans and DDPA under computer control. Currents through the electrodes are amplified and are collected via an A/D converter (ADJ98, Canopus Co., Ltd., Japan), using the personal computer.

Shielded Box

Electrochemical experiments are performed inside a shielded box to avoid system noise. It is an aluminum case measuring 40 × 40 × 30 mm. The ground level of the box is common to the reference level and flame

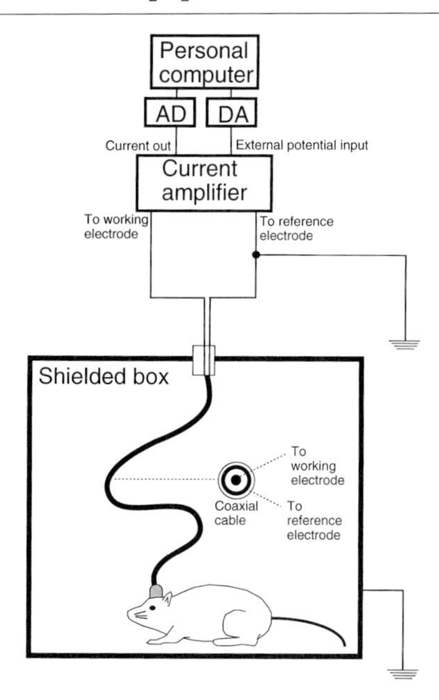

FIG. 4. Schema of the computer-controlled electrochemical system and shielded box.

ground of the amplifier. The signal ground is floating. In the box, the inner conductor of the coaxial cable is connected to the working electrode, and the outer conductor to the reference electrode via pin connectors (Fig. 4).

Electrochemical Response of Platinum Disk Microelectrode

Area of Platinum Disk Microelectrode

The Pt-disk microelectrode and Ag|AgCl electrode (the larger electrode) are immersed in a 10 m*M* ferrocyanide (Wako Pure Chemical Industries Ltd., Japan) aqueous solution in an electrolytic cell. Subsequently, a cyclic potential scan (0–600 mV vs Ag|AgCl; scan speed, 10 mV/sec) is performed. The area of the electrode is estimated by employing the following formula:[17]

$$r = i/(4nFDC)$$

where r, radius (cm); i, oxidative current (A); n, valence electron number; F, Faraday constant; D, diffusion coefficient; and C, concentation (mol/

[17] K. Aoki, K. Akimoto, K. Tokuda, and H. Masuda, *J. Electroanal. Chem.* **171,** 219 (1984).

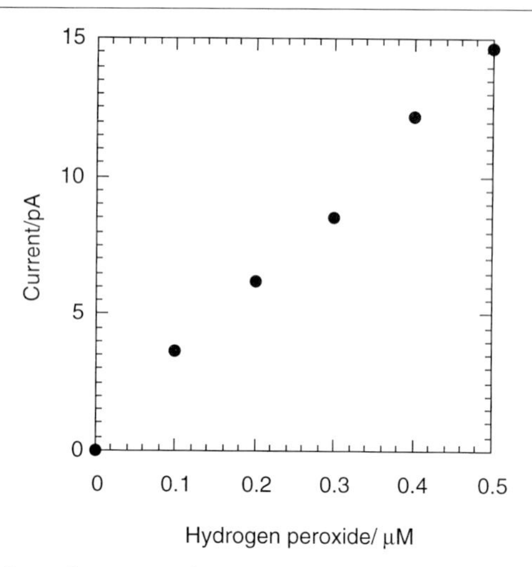

FIG. 5. Typical plots of current values at the Pt-disk microelectrode vs the concentrations of hydrogen peroxide (from ref. 16). The electrode responded linearly (correlation coefficient, 0.998). The detection limit was estimated to be about 0.03 μM at a signal-to-noise ratio of 3. (Reproduced from "Biosensors & Bioelectrons" with kind permission from Elsevier Science.)

cm^3). When the difference between the areas derived from this experiment and from the predicted values is less than 10%, the electrode can be applied in the following experiments.

Calibration of Platinum Disk Microelectrode

Solutions of hydrogen peroxide at various concentrations are prepared by injecting 30% (w/w) of a hydrogen peroxide solution (Wako Pure Chemical Industries, Ltd., Japan) into an electrolytic cell that holds the Pt-disk microelectrode and Ag | AgCl electrode (larger electrode) and contains a phosphate buffer solution (pH 7.4). We investigated the performance of the electrodes by DDPA (first step: 750 mV, 1 sec; second step: 1100 mV, 1 sec; see Fig. 1) for hydrogen peroxide at various concentrations. After adding hydrogen peroxide to the solution in steps, the current increased linearly with the concentration (Fig. 5). The detection limit was 0.03 μM at a signal-to-noise ratio of 3.[16]

For each fabricated electrode, the current values are plotted against hydrogen peroxide concentrations to obtain a calibration curve. Using this calibration curve, an electrode with a correlation coefficient greater than 0.95 can be applied to the following experiments.

With repeated calibrations for the same electrode, the reproducibility of the current response was confirmed. The performance of the electrode had not changed 1 week after fabrication.[16]

To examine the responses to other oxidizable species, DDPA under the same operating conditions was conducted for dopamine, uric acid, and ascorbic acid. No incremental responses were observed.[16] These results were consistent with those given in a previous report where DDPA with identical parameters showed high sensitivity and selectivity in detecting hydrogen peroxide and did not respond to nonspecific oxidizable species such as ascorbic acid.[15]

In Vivo Application of Platinum Disk Microelectrode

Implantation of Platinum Disk Microelectrode

Here, I describe *in vivo* measurement of hydrogen peroxide of the brain of a rat by a Pt-disk microelectrode. Male Wistar rats (200 g) are anesthetized with pentobarbital (50 mg/kg, i.p.) and the working electrode is stereotaxically implanted into the target region. The reference electrode (the smaller Ag | AgCl electrode) is placed on the dura over the contralateral parietal cortex. The electrodes are then firmly anchored to the skull with miniature screws and dental cement. The pin connectors of the electrodes are also fixed by dental cement. Via these connectors, the electrodes are connected to the coaxial cable in the shielded box.

In Vivo Measurement of Hydrogen Peroxide

One day after implantation, the rat is placed in the shielded box and connected to the coaxial cable in the box (Fig. 4). To estimate the offset current, a cyclic potential scan (0–300 mV vs Ag | AgCl; scan speed, 50 mV/sec) is performed. The offset current is estimated from the difference between oxidation and reduction currents. The rats which show an offset current smaller than 20 nA are deemed suitable for use in the following experiments.

To measure hydrogen peroxide *in vivo,* DDPA (first step: 750 mV, 1 sec; second step: 1100 mV, 1 sec; see Fig. 1) for hydrogen peroxide is performed. Following a 2-hr DDPA as a stabilization period, oxidation currents of hydrogen peroxide by DDPA are monitored continuously while observing the behavior of the animals that are allowed to move about freely. The currents recorded for each animal are converted to hydrogen peroxide concentrations by using the calibration curve.

Example

In the presented method, one can record temporal changes in the hydrogen peroxide concentration but not basal concentration. Thus, this method is suitable for measuring *in vivo* changes in the hydrogen peroxide content after administration of an agent that presumably augments hydrogen peroxide generation.

In theory, hydrogen peroxide is produced in a process of dopamine metabolism by monoamine oxidase. It is believed, therefore, that an increase in dopamine release induces an overproduction of hydrogen peroxide.[18] Because amphetamine stimulates dopamine release in striatum,[19] amphetamine injection will cause an increase in the intrastriatal hydrogen peroxide content. As an example of *in vivo* measurements of hydrogen peroxide, the Pt-disk microelectrode was implanted into the right striatum of a rat. After intraperitoneal injection of amphetamine, hydrogen peroxide concentrations were measured while the behavioral changes were monitored for 40 min. Amphetamine injection led to significant augmentation of hydrogen peroxide, the elevation of which depended on the dose of amphetamine.[16] This is consistent with a previous report on the increase in dopamine release caused by amphetamines and indirect evidence of the production of hydrogen peroxide via dopamine metabolism.[18]

Conclusion

In this chapter, the fabrication, electrochemical response, and *in vivo* application of the Pt-disk microelectrode for hydrogen peroxide detection were described. This electrode has a potential capability of measuring hydrogen peroxide generation in any small area where the electrode can be implanted. Hydrogen peroxide analysis by the Pt-disk microelectrode will be useful for clarifying the pathogenesis of many diseases derived from excessive production of active oxygen.

Acknowledgment

This work was supported by Grants-in-Aid for Scientific Research (08770803 and 09770768) from the Ministry of Education, Science, Sport and Culture, Japan.

[18] G. Cohen and M. B. Spina, *in* "Progress in Parkinson Research" (F. Hefti and W. J. Weiner, Eds.), p. 119. Plenum Press, New York, 1988.

[19] T. Zetterstrom, T. Sharp, C. A. Marsden, and U. Ungerstedt, *J. Neurochem.* **41,** 1769 (1983).

[30] Overall Low Molecular Weight Antioxidant Activity of Biological Fluids and Tissues by Cyclic Voltammetry

By RON KOHEN, ELIE BEIT-YANNAI, ELLIOT M. BERRY, and OREN TIROSH

Introduction

The need for a method to quantify overall antioxidant activity is obvious. Antioxidants can be classified into two major groups, antioxidant enzymes and low molecular weight antioxidants (LMWA).[1] The latter is a large group which contains a few hundred compounds, while the enzymatic antioxidants comprise a small group of a limited number of proteins. Much is known concerning the antioxidant enzyme group, their mechanism of activity, their role following oxidative stress, and their regulation mechanisms. There are well-established methods for evaluating these enzymes, and the relationship between the various proteins can be easily elucidated. In contrast, much less is known about the LMWA. These compounds have several biological advantages over the enzyme group. They are small molecules of a hydrophilic or lipophilic nature and can therefore easily penetrate into cells and be concentrated at high concentrations at specific locations where oxidative damage might occur. They exhibit a wide spectrum of activity toward a large variety of reactive oxygen species (ROS) and can be regenerated by the cell. These compounds are derived from various sources, including endogenous biosynthesis and waste products of the living cell. The majority of these compounds originate from dietary sources. A review of the literature reveals numerous methods for evaluating specific LMWA. Technically, however, it is impossible to evaluate all of these compounds in biological environments. There is a lack of knowledge concerning the distribution of LMWA in different biological sites and thus, incorrect conclusions concerning the role of specific antioxidants in specific tissues may be reached. In order to overcome these difficulties several methods have been developed for evaluating total LMWA activity.[2–5] However, these methods are not sufficient to give a complete picture of the overall antioxidant profile of biological fluids and tissues. The method described here is designed to

[1] R. Kohen, D. Fanberstein, and O. Tirosh, *Arch. Gerontol. Geriatr.* **24,** 103 (1997).
[2] C. Rice-Evans and N. J. Miller, Methods in Enzymology, Vol. 234, p. 279 (1994).
[3] D. D. M. Wayner, G. W. Burton, K. U. Ingold, and S. Locke, *FEBS Lett.* **187,** 33 (1985).
[4] R. J. DeLange and A. N. Glaser, *Anal. Biochem.* **177,** 300 (1989).
[5] T. P. Whitehead, G. H. G. Thorpe, and S. R. J. Maxwell, *Anal. Chim. Acta* **266,** 265 (1992).

METHODS IN ENZYMOLOGY, VOL. 300
0076-6879/99 $30.00

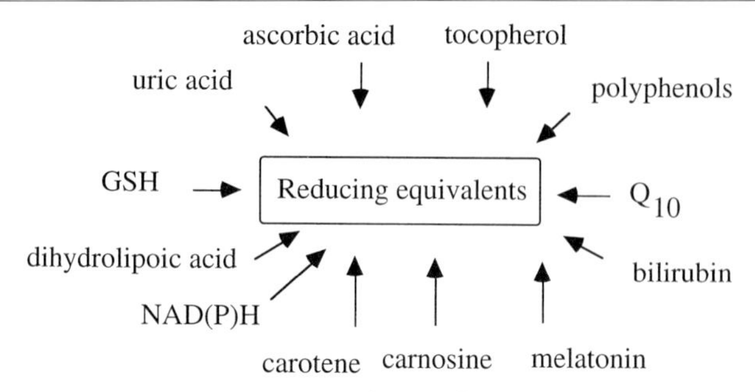

FIG. 1. Low molecular weight antioxidants (LMWA) which act directly with reactive oxygen species and possess reducing properties.

evaluate the total antioxidant activity of lipophilic and hydrophilic LMWA without specific determination of the various compounds.

Principle of Antioxidant Quantification

A common chemical structure of LMWA which act directly (scavengers) was the basis for their overall quantification. Examination of the chemical nature of the scavengers revealed that these molecules possess reducing properties.[1] They can donate electrons to the oxidizing species and by doing so, neutralize it (Fig. 1). Therefore, it was assumed that evaluation of the overall reducing power of a biological fluid or tissue homogenate would reflect its antioxidant activity, which is derived from the various reducing antioxidants.

Principle and Procedure of Cyclic Voltammetry Measurements

We used the cyclic voltammeter (CV) instrument (BAS, West Lafayette, IN, Model CV-1B cyclic voltammeter apparatus) to evaluate reducing power. This methodology allowed us to conduct rapid screening of the tested sample and to measure the ability of the tested sample to donate electron(s) (oxidation potential). It also allowed us to calculate the overall concentration of the reducing agents present in the sample tested without measuring specific compounds.[1,6]

Following preparation, the sample is introduced into a well in which three electrodes are placed: the working electrode (WE) (e.g., glassy carbon), which is 3.3 mm in diameter, the reference electrode (Ag | AgCl), and the auxiliary electrode (platinum wire). The potential is applied to the

[6] S. Chevion, E. M. Berry, N. Kitrossky, and R. Kohen, *Free Rad. Biol. Med.* **22,** 411 (1997).

working electrode at a constant rate (100 mV/sec) either toward the positive potential (evaluation of reducing equivalents) or toward the negative potential (evaluation of oxidizing species). During operation of the CV, a potential current curve is recorded (cyclic voltammogram). Figure 2A shows a typical cyclic voltammogram composed of ascorbic acid and β-NADPH. As can be seen, the phosphate-buffered saline (PBS) used for the cyclic voltammetry preparations does not, by itself, possess any reducing power. The reducing power is determined from the cyclic voltammogram and is composed of two parameters: the peak potential [$E_{p(a)}$] and the anodic current (AC). The ability of the reducing compound to donate its electron (to the working electrode) is evaluated by its $E_{p(a)}$ and is calculated from the voltage applied to the working electrode (x axis). The $E_{p(a)}$ is measured, in our experiments, at the half increase of the current at each anodic wave (AW) and is referred as $E_{1/2}$ (Fig. 2).[1,6] Each compound has its own $E_{1/2}$, and the lower the potential, the higher the ability of the compound to act as a reducing agent. If the two compounds present in the sample possess close or similar oxidation potentials, the two waves may overlie each other, as shown in Fig. 2B for ascorbic acid and uric acid. In biological samples such as rat brain two anodic waves were detected.[7] Each one of them may be related to a specific compound or to a variety of compounds possessing a similar (close) $E_{p(a)}$. The evaluation of the $E_{p(a)}$ of the detected AWs may indicate the type of the different reducing antioxidants present in the sample tested (their peak potentials) and therefore, their overall capacity to react with ROS. The lower the peak potential, the higher the ability of the compounds composing the wave to donate their electrons, indicating stronger reducing power. Besides measuring the $E_{p(a)}$ which correlates with the type of antioxidant, the overall concentration of the reducing antioxidants can be calculated.[1] This is done by measuring the AC from the y axis of each wave, as shown in Fig. 2. Changes in the concentration of one or more LMWA in each group of molecules would result in a change in the anodic current of the AW of this group, as shown in Eq. (1). A decrease in the AC may indicate a reduction in the levels of the compound, while an increase may indicate an increase in its concentration.

$$i_t = nFAD_0\,(\partial C_0/\partial_x), \qquad x = 0, t \tag{1}$$

where i_t is current at time t; n, number of electrons; F, Faraday's constant; A, electrode area; C_0, concentration of an oxidizable species (mol/cm^3); D_0, diffusion coefficient (cm^2/sec); t, time (sec); and x, distance from the electrode.

[7] E. Beit-Yannai, R. Kohen, M. Horowitz, V. Trembovler, and S. Shohami, *J. Cereb. Blood. Flow Metab.* **17,** 273 (1997).

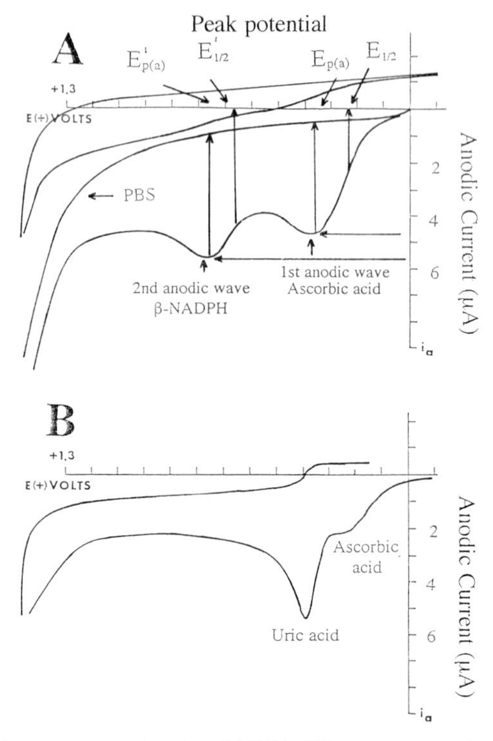

FIG. 2. Cyclic voltammogram of various LMWA. The measurement was carried out using a BAS cyclic voltammeter at a scan rate of 100 mV/sec and using glassy carbon as the working electrode, Ag | AgCl as the reference electrode, and a platinum wire as the auxiliary electrode. The measurement was initiated toward the positive potential. (A) Cyclic voltammogram of ascorbic acid (500 μM) and β-NADPH (500 μM). (B) Cyclic voltammogram of ascorbic acid (500 μM) and uric acid (500 μM).

The CV by itself cannot provide specific information on the exact nature of the LMWA. It can, however, supply information concerning the type of the various antioxidants in the sample and their total concentration.

Method for Water-Soluble LMWA

Procedure for Biological Fluids

Various biological fluids, including plasma, serum, cerebrospinal fluids, seminal fluids, gastric juice, and saliva,[1,6,8] have been tested. The samples are collected from animals or human volunteers. The samples are analyzed

[8] R. Kohen, O. Tirosh, and K. Kopolovich, *Exp. Gerontol.* **27,** 161 (1992).

as collected or following dilution with phosphate-buffered saline, pH 7.2, at a ratio of 1:1. Samples are analyzed fresh or following storage at −70° for up to 3 months.

Some biological fluids required special treatment in their collection, as follows: Blood samples should be collected in heparin, citrate, or siliconized tubes (not EDTA) to prevent coagulation. Plasma should be separated from the blood components extremely carefully to avoid hemolysis of red blood cells during separation. It is important that the working electrode be polished intensively following every measurement to remove biological components such as proteins that can be adsorbed onto its surface. The polishing is carried out using a polishing kit (BAS PK-1, Bioanalytical Systems, West Lafayette, IN) and alumina. All cyclic voltammograms are performed between −0 mV and 1.3 V. The volume needed for each measurement is at least 0.5 ml.

Measurements of Saliva, Plasma, and Cerebrospinal Fluids

It has been found that all the biological fluids tested demonstrate reducing power and possess one or two anodic waves. Several examples are presented. Figure 3A shows a cyclic voltammogram of saliva.[8] One anodic wave is found, indicating one group of reducing LMWA. It has been previously shown that the compounds composing the wave correlate with the antioxidant activity of saliva. Figure 3B shows a cyclic voltammogram of plasma collected from human subjects. Two anodic waves can be seen at different $E_{1/2}$, indicating two groups of reducing LMWA. While the first group is composed of uric acid and ascorbic acid, as detected by HPLC equipped with an electrochemical detector, the second is composed of compounds present in red blood cells which leak into the plasma during the separation procedure. Careful separation will prevent the occurrence of the second wave. It has been found that this wave is composed of at least 50% NADPH. Exposure of plasma to oxidative stress results in a significant decrease in the anodic wave, suggesting consumption of the antioxidants composing the wave.[6] Table I shows the effect of peroxyl radicals induced by the azo compound 2,2′-azobis(2-amidopropane) hydrochloride (AAPH) on the first anodic wave (reduction in the AC). Exposure to copper ions results in a 31% reduction in the AC of the first wave, and exposure to irradiation results in a 79% reduction in the AC. Characterization of the compounds composing the wave is carried out by incubating the samples with various enzymes such as uricase (urate oxidase, EC 1.7.3.3) and ascorbate oxidase (EC 1.10.3.3), or by HPLC with electrochemical detection.[6,9] Figure 3C shows a cyclic voltammogram of rat cerebrospinal fluid.

[9] P. A. Motchnik, B. Frei, and B. N. Ames, *Methods Enzymol.* **234,** 269 (1994).

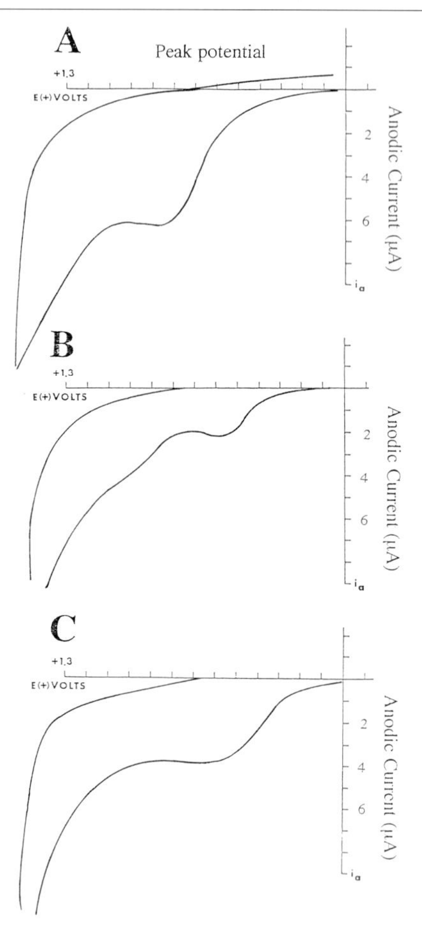

FIG. 3. Cyclic voltammograms of several biological fluids. The measurement was carried out using a BAS cyclic voltammeter at a scan rate of 100 mV/sec and using glassy carbon as the working electrode, Ag|AgCl as the reference electrode, and a platinum wire as the auxiliary electrode. The measurement was initiated toward the positive potential. (A) Cyclic voltammogram of human saliva from an individual aged 30 years. (B) Cyclic voltammogram of human plasma collected with citrate as the anticoagulant. (C) Cyclic voltammogram of cerebrospinal fluid.

Preparing Tissue Samples for Cyclic Voltammetry

In order to determine the overall reducing power of biological tissues, samples of at least 30–500 mg of the tested tissue are removed and homogenized. On removal from the animals, the samples are introduced into Eppendorf tubes and are deep frozen in liquid air for a few minutes. The

TABLE I
CHANGES IN LEVELS OF LMWA FOLLOWING OXIDATIVE STRESS AS EVALUATED BY THE ANODIC CURRENT

Oxidative stress inducer	Concentration	% Change of anodic current
AAPH	200 mM	-92 ± 2
$CuSO_4$	400 μM	-31 ± 2
γ-irradiation	1,000,000 rad	-79 ± 3

homogenization of the frozen samples is carried out either immediately or following a period of storage at $-75°$ for up to 6 months. [In brain tissue longer periods of storage (>6 months) were found to affect the results. It is therefore recommended that analysis be carried out immediately following freezing of the sample, or within a short period of time.] The homogenization is carried out in 1 ml PBS on ice using a mechanical homogenizer. Following homogenization the samples are centrifuged at 1000g at 4° for 10 min to remove large insoluble particles. The pH of all homogenates is measured and adjusted to 7.2. It is possible to perform the homogenization using other buffers at different pH values. One must take care, however, that the buffer contains electrolytes sufficient for conducting the CV measurement, and that comparison of various samples will be done at a similar pH. The electrochemical response and the various anodic waves are sensitive to pH values. Following centrifugation, samples of at least 1 ml are analyzed in the CV instrument.

Evaluation of Water-Soluble LMWA of Rat Liver, Heart, and Brain

Several cyclic voltammograms of rat tissues are presented in Fig. 4. All the tissues tested exhibited two or three anodic waves. Figure 4A shows a typical voltammogram of rat liver obtained from 4-month-old rats. Two anodic waves are shown, indicating two groups of reducing LMWA.[10] Figure 4B shows a cyclic voltammogram of rat heart. It can be seen that rat heart possesses two anodic waves, indicating two groups of LMWA. Measurements taken in animals of various ages show that the $E_{P(a)}$ is not changed in most of the tissues, indicating that the type of LMWA remains constant over the life span of the animal.[1] However, significant changes have been found in the total concentration of the LMWA at various ages as measured

[10] R. Kohen, O. Tirosh, and R. Gorodetsky, *Free Rad. Res. Commun.* **17,** 239 (1992).

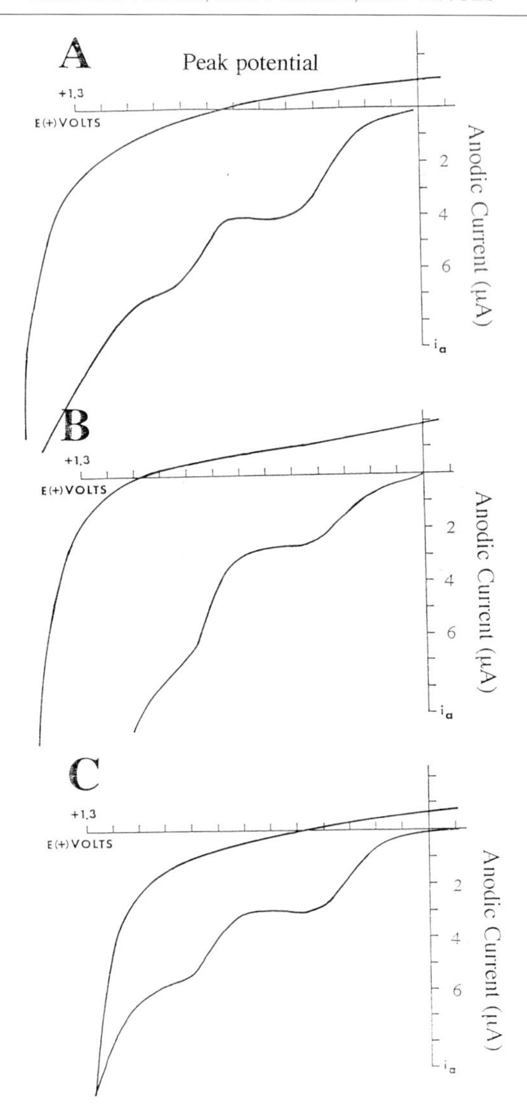

FIG. 4. Cyclic voltammograms of several rat tissues. The measurement was carried out following homogenization in PBS using a BAS cyclic voltammeter at a scan rate of 100 mV/sec and using glassy carbon as the working electrode, Ag | AgCl as the reference electrode, and a platinum wire as the auxiliary electrode. The measurement was initiated toward the positive potential. (A) Cyclic voltammogram of rat liver. (B) Cyclic voltammogram of rat heart. (C) Cyclic voltammogram of rat brain.

by the AC. Figure 4C shows a typical cyclic voltammogram of rat brain. Two anodic waves were detected at peak potentials of 400 ± 30 mV and 750 ± 40 mW.

Methods for Lipid-Soluble LMWA in Biological Tissues

General Procedure

While the procedure described above for biological fluids and tissues relates to water-soluble antioxidants (ascorbic acid, uric acid, carnosine, NADH, NADPH, etc.), the lipid-soluble LMWA require different procedures. It has been found that lipid-soluble LMWA, such as tocopherol, carotene, lipoic acid, and coenzyme Q_{10}, can be quantified in the CV instrument. For this purpose an extraction procedure of the lipophilic LMWA is used on 0.5–1 g of the tested tissue [brain, gastrointestinal mucosa, liver, lung, skin, kidney, heart, and whole blood (10 ml)]. The extraction procedure is based on a modified protocol described by Motchnik *et al.*[9] The extraction is carried out using a mixture of ethanol : hexane (1 : 5, v/v) followed by centrifugation for 5 min at 1000*g*. This procedure is repeated twice and the upper layers are combined and taken for analysis.

Evaluation of Lipid-Soluble LMWA in Rat Brain

An example of a cyclic voltammogram of the lipid-soluble LMWA extracted using the procedures described above is shown in Fig. 5. The extraction procedure is as follows: The rat is sacrificed and its brain removed immediately and weighed.[7] The brain is homogenized in a 1 : 4 (w/v) mixture of ethanol : hexane (1 : 5). The homogenization is conducted using a mechanical homogenizer for 3 min on ice. The homogenate is then centrifuged at 1000*g* for 15 min at 4°. The upper layer is separated and the lower layer is extracted again with a 1 : 4 (w/v) mixture of ethanol : hexane (1 : 5). Following 3 min of vortex, the sample is centrifuged at 1000*g* for 15 min at 4°. The two upper layers are combined and the organic solvents removed by evaporation. The residue obtained is dissolved in an acetonitrile solution containing 1% *tert*-butylammonium perchlorate as an ion-pairing agent. During the extraction process the sample must be protected from light and kept on ice. A cyclic voltammogram of rat brain lipophilic extraction reveals two anodic waves, indicating two lipid-soluble LMWA or two groups of LMWA. The various anodic waves are characterized using HPLC equipped with an electrochemical detector according to the procedure described by Motchnik *et al.*[9] It has been found that tocopherol (both alpha and gamma

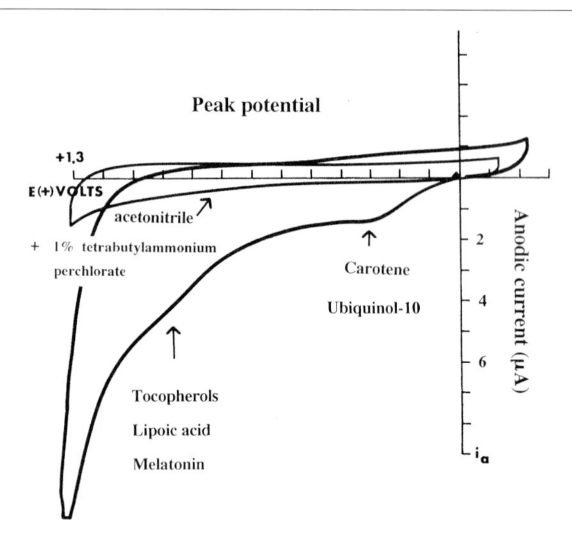

FIG. 5. Cyclic voltammogram of lipid extraction of rat brain. The sample was prepared as described using an organic extraction with a mixture of ethanol : hexane (1 : 5). The measurement was done using a BAS cyclic voltammeter at a scan rate of 100 mV/sec. The working electrode was glassy carbon and the reference electrode was Ag|AgCl. All the samples contained 1% *tert*-butylammonium perchlorate as an ion-pairing agent.

possess similar oxidation potential under the conditions used) and β-carotene are major constituents, as shown in Fig. 5. Other lipophilic LMWA can also be quantified using this methodology. By spiking various lipidic LMWA into the brain mixture, it has been found that the anodic wave at $E_{1/2}$ of 980 mV may relate to lipoic acid in its oxidized form, while coenzyme Q_{10} appears in the first anodic wave around 300–400 mV. Melatonin has been found to possess an $E_{1/2}$ of 750 mV.

Quantification of LMWA in Their Oxidized State

In order to quantify both the reduced and oxidized forms of the LMWA in a certain sample, the cyclic voltammogram measurement should be conducted twice. In the first run the quantification will be conducted toward the positive potential in order to determine the $E_{P(a)}$ of the reducing compounds in the sample. Following polishing of the glassy carbon electrode, the cyclic voltammogram measurement will be performed as a whole cycle, starting toward the negative potential, followed by a concomitant measurement toward the positive potential. The oxidized fraction of the compounds tested is determined by their reduction on the surface of the electrode (negative potentials), followed by their oxidation in the positive potential.

Such measurement can be conducted only if the reduction/oxidation products are stable and the redox process is reversible. One can evaluate the ratio between the oxidized and reduced forms of a reversible redox couple in a biological sample by such measurements. In order to conduct this analysis oxygen should be removed from the mixture containing the tested compound. This can be achieved by bubbling nitrogen (99.999%) into the solution for 10 min prior to measurement. The tested solution is flushed with nitrogen throughout the measurement.

Quantification of Oxidative Damage using Cyclic Voltammetry

The methdology described here is also suitable for evaluating oxidative damage of biological samples. It has been found that following exposure to oxidative stress, *Escherichia coli* cells lose their reducing power prior to any noticeable change in their survival curve. Similar observations have been made on enzyme activity and tissue integrity and function.[11] Following exposure to oxidative stress, the enzyme lactate dehydrogenase loses its reducing properties in parallel to a change in its activity.[11]

Comments, Accuracy of the Method, and Conclusions

Evaluation of antioxidant activity using the cyclic voltammetry approach has several advantages over other methods. Cyclic voltammetry measurements allow the evaluation of the antioxidant status of both water- and lipid-soluble LMWA. The outcome of such measurements can be used for the assessment of the overall antioxidant activity, which is derived from the reducing LMWA (scavengers) without measuring specific compounds. Examination of cyclic voltammograms indicates the types of the various compounds responsible for the antioxidant activity (the ability to donate electrons) of a biological sample and their total concentration. Preparation of samples for such evaluation is simple and does not require advanced procedures. The measurement itself is rapid and the results can be analyzed immediately. Therefore, this method is especially suitable for screening large numbers of samples. The overall reducing power profile (antioxidant profile) can be easily determined and compared among various populations. The sensitivity of this procedure is relative low. One can usually only determine reducing equivalents in biological samples using the cyclic voltammeter to as low as 1–10 μM. However, such sensitivity is sufficient for determining physiological concentrations of reducing antioxidants. Cyclic voltammetry is suitable for measuring compounds in systems at equilib-

[11] R. Kohen, *J. Pharmacol. Toxicol. Methods* **29,** 185 (1993).

rium, a state which does not normally occur in biological environments. Therefore, such measurement will not reflect the overall LMWA present in the sample, but will reflect a temporary state, that when the sample is measured. Kinetic parameters should also be taken into consideration. The use of a glassy carbon electrode is not suitable for all the LMWA, as not all the LMWA donate their electrons to the glassy carbon electrode at a sufficient rate. For example, thiol compounds such as glutathione will not be detected by the glassy carbon electrode. In order to overcome these difficulties one must use a different electrode, such as gold or platinum, to evaluate the whole spectrum of LMWA (e.g., an Au | Hg electrode is needed for glutathione determination), or use a different rate for applying the potential. Therefore, it is important to use the same conditions throughout the experiments for comparison purposes. Another approach is to use a reversible electrochemical couple which can interact with most of the LMWA in the sample, and to measure its cyclic voltammogram. Characterization of the compounds composing the waves should be done using enzymatic and chromatographic methods. The cyclic voltammetry methodology may be used in a large variety of clinical and pathological situations, including diabetes, irradiation therapy, brain degenerative diseases, head trauma, and the aging process, and as an efficient tool for evaluating the overall antioxidant status of biological samples.

Acknowledgments

This work was supported in part by the Foundation for Research and Development and the Bergman Foundation. We would like to thank Dr. D. Mandler for critical reading of the manuscript and valuable comments and suggestions.

[31] Estimation of Hydroxyl Free Radical Levels *in Vivo* Based on Liquid Chromatography with Electrochemical Detection

By IAN N. ACWORTH, MIKHAIL B. BOGDANOV, DOUGLAS R. MCCABE, and M. FLINT BEAL

Introduction

The hydroxyl free radical (HO·) is an extremely aggressive molecule and reacts at diffusion-controlled rates (typical second-order rate constants of 10^9 to 10^{10} M^{-1} sec^{-1})[1] with virtually any molecule in close proximity. Consequently, HO· has an incredibly short half-life (10^{-10} sec) and only diffuses a few angstroms from its site of production before reacting.[2,3]

The hydroxyl free radical (HO·) can be formed in a variety of ways, including the homolytic cleavage of water induced by ionizing radiation [Eq. (1)]; the reaction between the superoxide radical anion and hypochlorous acid in neutrophils [Eq. (2)]; and the reaction of hydrogen peroxide with biologically prevalent oxidizable metals such as iron and copper (the Fenton reaction) [Eq. (3)].[1,4–7] The Fenton reaction may also be promoted by the superoxide radical anion (the Haber–Weiss reaction) [Eq. (4)].

$$H_2O \rightarrow H\cdot + HO\cdot \tag{1}$$
$$O_2^{\cdot -} + HOCl \rightarrow O_2 + HO\cdot + Cl^- \tag{2}$$
$$H_2O_2 + Metal^{n+} \rightarrow HO\cdot + HO^- + Metal^{(n+1)+} \tag{3}$$
$$O_2^{\cdot -} + Metal^{(n+1)+} \rightarrow O_2 + Metal^{n+} \tag{4}$$

Peroxynitrous acid, formed from the reaction between NO· and $O_2^{\cdot -}$ [Eqs. (5)–(7)], is capable not only of nitrating a variety of substrates, but also of mimicking the oxidation reactions of HO·.[8,9]

[1] I. N. Acworth, D. R. McCabe, and T. J. Maher, *in* "Antioxidants, Oxidants and Free Radicals" (S. Baskin and H. Salem, Eds.), p. 23. Taylor and Francis, London, 1997.
[2] F. Antunes, A. Salvador, H. S. Marhino, R. Alves, and R. E. Pinto, *Free Rad. Biol. Med.* **21,** 917 (1996).
[3] W. A. Pryor, *Ann. Rev. Physiol.* **48,** 657 (1986).
[4] B. Halliwell and J. M. C. Gutteridge, *Methods Enzymol.* **186,** 1 (1990).
[5] B. Halliwell and J. M. C. Gutteridge, *FEBS Lett.* **307,** 108 (1992).
[6] L. P. Candeias, K. B. Patel, M. R. L. Stratford, and P. Wardman, *FEBS Lett.* **333,** 151 (1993).
[7] I. N. Acworth, B. Bailey, and T. J. Maher, *in* "Neurochemical Markers of Degenerative Nervous Diseases" (G. A. Qureshi, S. Parvez, and S. H. Parvez, Eds.), p. 3. HPLC-HPCE **7,** VSP The Netherlands, 1997.
[8] J. P. Crow and J. S. Beckman, *Adv. Pharmacol.* **34,** 17 (1995).
[9] W. A. Pryor and G. L. Squadrito, *Am. J. Physiol.* **268,** L699 (1995).

0076-6879/99 $30.00

$$NO\cdot + O_2^{\cdot -} \rightarrow ONO_2^- \quad (5)$$
$$ONO_2^- + H^+ \rightarrow ONO_2H \quad (6)$$
$$ONO_2H \rightarrow HO\cdot + NO_2\cdot \quad (7)$$

However, such oxidation reactions need not necessarily involve the direct formation of "free" HO·.[9,10]

When HO· is produced *in vivo,* it can readily react with many extremely important biomolecules, including DNA, lipids, proteins, carbohydrates, and a variety of low molecular weight species. This not only damages these molecules, but can also lead to the production of cytotoxic species.[1,7,11,12] Such HO·-induced damage is thought to underlie a variety of diseases (e.g., arthritis, cancer, neurodegenerative diseases and reperfusion damage).[1,7,11] Furthermore, it has been hypothesized that the accumulation of such free radical-induced lesions over time may be the primary cause of aging.[13]

As expected it is virtually impossible to measure HO· directly *in vivo* because of its extreme reactivity, short half-life, and low levels. Of all the approaches in literature that claim to measure HO· production, two are showing great promise. One approach, electron spin resonance (ESR) detection, captures HO· by reacting it with a suitable spin trap such as phenyl-*tert*-butylnitrone or 5,5-dimethylpyrroline *N*-oxide. Not only is the lifetime of HO· extended, but the HO·–spin trap adduct still remains paramagnetic and can thus be measured by ESR. Unfortunately, ESR approaches are fraught with problems, including difficulty of interpretation of spectra, artifacts, spin trap toxicity, and quenching of the spin trap by biological systems. These issues have been reviewed elsewhere.[1,7,11,14,15] Although ESR-based methods are routinely used for *in vitro* studies, with few exceptions this approach is generally regarded as unsuitable for routine *in vivo* experimentation.

A second approach, aromatic hydroxylation, is mechanistically similar to the spin-trap ESR-based methods discussed above, but it does not use "classical" spin trap agents, nor do the products formed have to be paramagnetic. Thus, referring to aromatic hydroxylation as a spin-trap method is quite erroneous. Aromatic hydroxylation uses the formation of products from the reaction between HO· and a suitable aromatic species as a marker

[10] S. Goldstein, G. L. Squadrito, W. A. Pryor, and G. Czapski, *Free Rad. Biol. Med.* **21,** 965 (1996).

[11] B. Halliwell and J. M. C. Gutteridge, *in* "Free Radicals in Biology and Medicine." Clarendon Press, Oxford, 1989.

[12] H. Esterbauer, R. J. Schaur, and H. Zollner, *Free Rad. Biol. Med.* **11,** 81 (1991).

[13] D. Harman, *J. Gerontol.* **11,** 298 (1956); *EXS.* **62,** 1 (1992).

[14] H. Kaur and B. Halliwell, *Methods Enzymol.* **233,** 67 (1994).

[15] I. Yamazaki and L. Piette, *J. Biol. Chem.* **265,** 13589 (1990).

of HO· production.[1,7,11,14] The importance and mechanisms of aromatic hydroxylation assays in biological systems were first put forward by Halliwell.[14,16] Aromatic hydroxylation reactions can be further divided into those methods which measure an endogenous marker (e.g., the production of 8-hydroxy-2′-deoxyguanosine from 2′-guanosine or tyrosine isomers from phenylalanine) and, as discussed in greater detail below, those which measure an HO· adduct formed from an exogenously administered scavenging agent [e.g., salicyclic acid (SA) and 4-hydroxybenzoic acid (4-HBA)]. The substrates and products of aromatic hydroxylation reactions generally lend themselves to analysis by high-performance liquid chromatography (HPLC). Of all the detection schemes for HPLC, electrochemical detection (ECD) is usually the detector of choice as it can measure many scavengers and products with good selectivity and high sensitivity, making it ideal for determination of low amounts of analytes even in complex biological matrices.

Reaction between HO· and Salicyclic Acid

SA readily reacts with HO·. SA's hydroxyl group is *o*- and *p*-directing (one of the *o*-positions is blocked by the carboxylate group) so that reaction with HO· results in the production of two compounds: the *o*-isomer (relative to the hydroxyl group) 2,3-dihydroxybenzoic acid (2,3-DHBA) and the *p*-isomer 2,5-dihydroxybenzoic acid (2,5-DHBA). Floyd *et al.* developed the original HPLC-based method utilizing SA as an HO· scavenging agent.[17] They used HPLC-ECD to measure HO· production as reflected by changes in DHBA levels (SA was not measured) and found that HO· production could be stimulated by several *in vitro* systems.[17] Since then many HPLC-based HO·–SA scavenging approaches have been reported in the literature and several have been used to study HO· production *in vivo* (see McCabe *et al.*[18] for current literature). However, it is now known that some of these approaches may not be suitable for *in vivo* studies. First, biological systems are much more complex than *in vitro* models and, as shown in Fig. 1, administered SA can undergo several metabolic fates, including the HO·-independent formation of 2,5-DHBA by cytochrome P450.[19] Thus, those methods based on measurement of 2,5-DHBA as a marker of HO· production are likely to overestimate production of this reactive oxygen

[16] B. Halliwell, *FEBS Lett.* **92,** 321 (1978).

[17] R. A. Floyd, J. J. Watson, and P. K. Wong, *J. Biochem. Biophys. Meth.* **10,** 221 (1984).

[18] D. R. McCabe, T. J. Maher, and I. N. Acworth, *J. Chromatogr. B* **691,** 23 (1997).

[19] M. Ingelman-Sundberg, H. Kaur, Y. Terelius, J. O. Persson, and B. Halliwell, *Biochem. J.* **276,** 753 (1991).

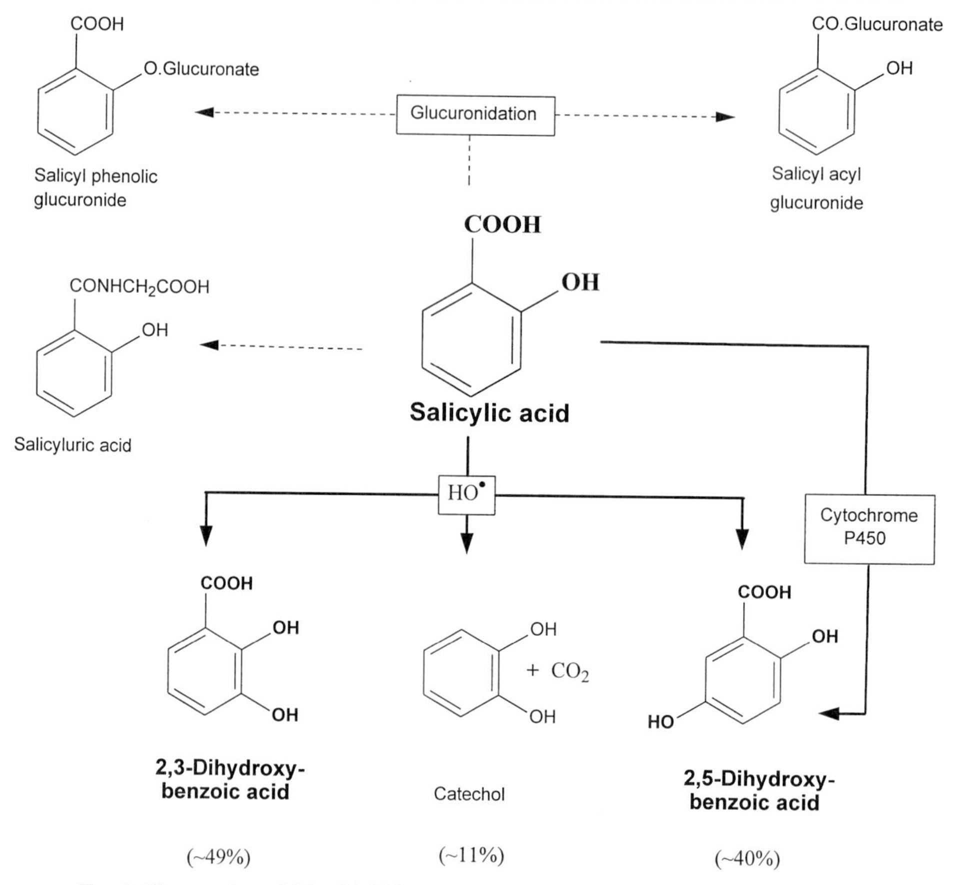

FIG. 1. The reaction of SA with HO·. The percent given in parentheses represents the *in vitro* metabolite yield when SA is exposed to conditions promoting HO· production [H. Kaur and B. Halliwell, *Methods Enzymol.* **233,** 67 (1994)]. The figure also shows the catabolism of SA *in vivo.*

species *in vivo.* Fortunately, 2,3-DHBA appears to be a better marker for HO· production and its level has been shown to be increased in body fluids under conditions of increased HO· production (e.g., in subjects consuming excess alcohol, in rheumatoid patients taking aspirin, in rats exposed to hyperoxia or ionizing radiation, or in animals receiving the redox-cycling drug adriamycin) (see Kaur and Halliwell[14] and references therein). Second, the correct dose of SA is very important. Too little SA, and not enough of it will reach the site of HO· production, so that the formation of the DHBAs may be below the detection limit of the analytical system. Too much SA, and unwanted physiological consequences may occur.[14,18] A wide range of doses can be found in the literature and, not surprisingly, the highest doses are often associated with most insensitive analytical equipment. Third, many reports ignore the importance of SA itself, and all too frequently, DHBA data are presented without normalization for SA levels (altered DHBA levels could be due to differences in SA concentrations at the site of HO· production resulting from variability in SA's distribution and metabolism, or accuracy and location of its administration).

Salicyclic Acid Method

We have developed a single-detector, HPLC–ECD-based method to accurately and simultaneously quantitiate 2,3-DHBA, 2,5-DHBA, and SA in different biological samples. The electrochemical detector uses two serially placed coulometric sensors with the upstream (lower potential) electrode specifically measuring DHBA isomers, and the downstream (higher potential) electrode specifically measuring SA. The analytical advantages of coulometric detection over amperometric detection have been extensively discussed elsewhere.[20,21] To show method feasibility, rats are decapitated 30 min after they receive saline (i.p.) or 100 mg/kg (i.p.) SA. Tissue samples (brain, kidney, liver, and blood) are immediately removed and kept on ice. The brain is then dissected into the striata, hippocampi, and cortex. The solid tissues are ultrasonicated in ice-cold 0.2 *M* perchloric acid (containing 100 μM EDTA and 100 μM sodium metabisulfite (1:5 or 1:10, w/v). Following centrifugation (12 500*g*, 5 min at 4°) the supernatant is passed through a 0.2 μm filter by centrifugation (as above). A 10 μl volume is used for analysis. Blood is centrifuged (as above) and the resulting serum is deproteinized using perchloric acid (as above) 1:5 (v/v), centrifuged and filtered as above. Stock standard solutions are prepared at a concentration of 1 mg/ml in 0.2 *M* perchloric acid containing EDTA and sodium metabisulfite (as above) and these are stable at 4° for several months.[18] The DHBAs and SA are resolved on a proprietary DHBA-250 analytical (3 × 250 mm; 5 μm) column (ESA, Inc.) using a mobile phase composed of 50 m*M* sodium acetate and 50 m*M* sodium citrate, 8% methanol, 2% 2-propanol, (v/v/v), final pH 2.5. The mobile phase is passed through the system at 0.5 ml/min and all analyses are performed at 27°. All analytes are determined electrochemically using a dual potentiostat electrochemical detector (Coulochem II, ESA, Inc., Chelmsford, MA) and dual coulometric

[20] I. N. Acworth and M. Bowers, *in* "Coulometric Electrode Array Detectors for HPLC" (I. N. Acworth, M. Naoi, S. Parvez, and S. H. Parvez, Eds.), p. 3. HPLC-HPCE **6,** VSP The Netherlands, 1996.

[21] C. N. Svendsen, *Analyst* **118,** 123 (1993).

electrode analytical cell (Model 5010, ESA, Inc). The upstream electrode is set at +250 mV and the downstream at +750 mV (vs palladium reference). Optimal potentials are determined by construction of hydrodynamic voltammograms for each analyte.[18] A third coulometric electrode is placed before the injector and is used to electrochemically clean the mobile phase (+775 mV).

Results and Discussion of Salicyclic Acid

The analysis was completed in under 20 min (Fig. 2), had a limit of detection for the DHBAs of ~1 pg on column, and was free from any known interference.[18] The tissue levels of DHBAs, SA, and the DHBA/SA ratios are presented in Table I and show that this method permits the simultaneous determination of DHBA isomers and SA in a variety of tissues. The use of a single detector (with wide dynamic range) permits the simultaneous measurement of DHBA and SA, whose concentrations typically occur with at least 3 orders of magnitude difference. It also overcomes the complications and costs associated with multiple detector systems (see, for example, refs. 22 and 23). With low picogram sensitivity for the DHBA isomers, the SA dose can be reduced (30–100 mg/kg i.p.) from the normal literature dose range, ensuring that possible physiological complications are minimized. At these lower doses there is less likely to be an inhibitory effect of SA on cyclooxygenase and less of an interference with prostaglandin production. Thus, our approach may be suitable for the study of the role of HO· in ischemia–reperfusion injury, a situation known to be associated with increased production of reactive oxygen species by cyclooxygenase and involvement of prostaglandins in its etiology.

One of the major advantages of using SA as a free radical scavenger is due to its reactivity. SA reacts with a rate constant of 5×10^9–10^{10} M^{-1} sec^{-1}, which is approximately 2.5 and 22 times more rapid than its reaction with phenylalanine and 2′-deoxyguanosine, respectively.[14,24,25] However, the SA approach is not without its problems. For example, the SA method splits the HO· level between two adducts, one of which (2,5-DHBA) is an unreliable marker. Thus, some of the signal is wasted, so that the lower limits of detection of the SA approach would not be expected to be as good as for methods which produce only one marker (see 4-hydroxybenzoic acid below). Also, the SA method is unsuitable for some experiments using microdialysis perfusion, but only if SA is to be included in the aCSF

[22] S. L. Smith, P. K. Andrus, J.-R. Zhang, and E. D. Hall, *J. Neurotrauma* **11,** 393 (1994).

[23] P. Grammas, G.-J. Liu, K. Wood, and R. A. Floyd, *Free Rad. Biol. Med.* **14,** 553 (1993).

[24] M. Grootveld and B. Halliwell, *Biochem. J.* **237,** 499 (1986).

[25] J. E. Schneider, M. M. Browning, X. Zhu, K. L. Eneff, and R. A. Floyd, *Mutation Res.* **214,** 23 (1989).

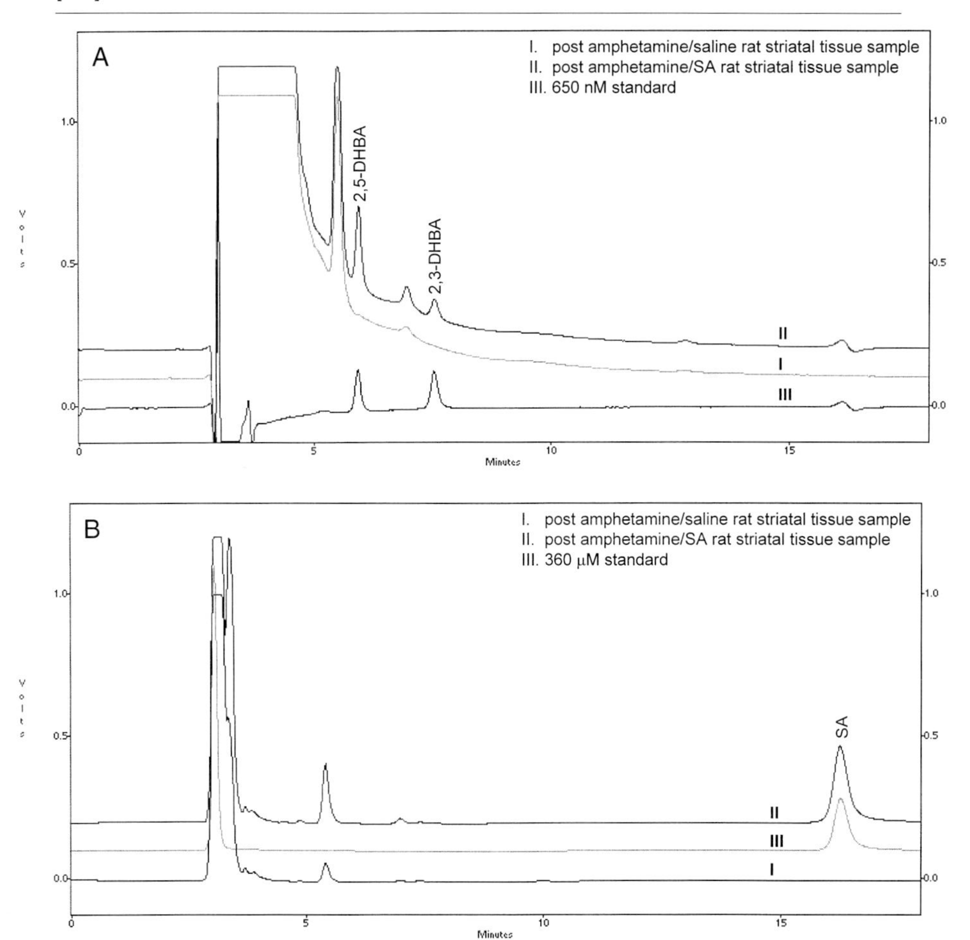

FIG. 2. (A) Production of DHBA isomers in rat striatal tissue samples 30 min after peripheral SA administration (100 mg/kg i.p.) in amphetamine (5 mg/kg i.p.) pretreated rats. Chromatograms are presented at a gain range of 100 nA full scale. Analytes were measured at +250 mV vs palladium reference (electrode 1). No DHBA isomers were found in untreated animals. The striatal concentration of 2,5-DHBA and 2,3-DHBA following SA administration were 0.72 and 0.15 nmol/g (wet weight), respectively. (B) Corresponding measurement of SA levels in samples presented in (A). SA was measured at +750 mV (electrode 2) and a gain range of 10 μA. SA was not detected in saline treated animals. The striatal SA level was 0.29 μmol/g (wet weight) in SA treated animals.

TABLE I
TISSUE PRODUCTION OF DHBA ISOMERS FOLLOWING TREATMENT WITH SALICYCLIC ACID[a]

Tissue	2,3-DHBA (pmol/g)	2,5-DHBA (pmol/g)	SA (nmol/g)	2,3-DHBA/SA ($\times 10^{-3}$)	2,5-DHBA/SA ($\times 10^{-3}$)
Kidney[b]	780	6170	298	2.6	20.7
Liver[b]	268	3344	206	1.3	16.2
Brain					
Striatum	93	48	28	3.3	1.7
Cortex	114	136	48	2.4	2.8
Serum[b,c]	753	10 130	935	0.8	10.8

[a] Animals were killed 30 min post-SA administration (100 mg/kg i.p.). Data taken from D. R. McCabe, T. J. Maher, and I. N. Acworth, *J. Chromatogr. B* **691,** 23 (1997), with permission of the authors and Elsevier Science B.V.

[b] Basal (control) DHBA levels were: 2,3-DHBA (54, 49 pmol/g and 36 pmol/ml for liver, kidney, and serum, respectively); 2,5-DHBA (190 pmol/g and 214 pmol/ml for kidney and serum, respectively).

[c] Serum levels are presented in pmol/ml.

perfusion medium (see below). A major problem with all methods utilizing SA and catechols is that they have high affinities for iron(III), which may interfere with iron-dependent HO· production.[14] Also, SA can be derived from the diet, and can thus affect data obtained with the SA method.[26]

Several papers in literature attempt to extend the SA–DHBA method to include other analytes of interest (e.g., dopamine, serotonin).[7,18] These methods tend to fall into two categories. The first group uses a single electrode to measure a variety of metabolites in conjunction with SA.[14] To do so, a high potential must be applied, leading to complicated chromatograms, a greater probability of analyte coelutions, and a poorer sensitivity due to increased noise. The second group just measures metabolites and DHBAs but without SA.[27] Consequently, a lower applied potential can be used, but without measurement of SA, the DHBA data cannot be normalized. Beal *et al.* have used gradient HPLC with coulometric electrode array detection to measure many different biologically relevant metabolites, DHBAs, and SA in different biological samples.[28] Coulometric array detection overcomes many of the sensitivity and selectivity problems associated with single electrode approaches.[20,21]

[26] P. L. Janssen, P. C. Hollman, D. P. Venema, W. A. van Staveren, and M. B. Katan, *Nutr. Rev.* **54,** 357 (1996).

[27] M. Patthy, I. Kiraly, and I. Sziraki, *J. Chromatogr. B* **664,** 247 (1995).

[28] M. F. Beal, R. J. Ferrante, R. Henshaw, R. T. Matthews, P. K. Chan, N. W. Kowall, C. J. Epstein, and J. B. Schulz, *J. Neurochem.* **65,** 919 (1995).

Microdialysis perfusion is a novel sampling technique permitting unique insights into the chemical functioning of the extracellular space. The principles of microdialysis perfusion have been detailed elsewhere.[29,30] Several researchers have used microdialysis perfusion in combination with SA-scavenging methods to examine production of HO· in living tissues (see McCabe *et al.*[18] and references therein). While developing our SA-scavenging method we found the spontaneous formation of DHBAs when SA was dissolved in the artificial cerebrospinal fluid (aCSF) used to perfuse the microdialysis probe.[18] Similarly, Montgomery *et al.* found that DHBAs were formed when either SA or 4-HBA were dissolved in aCSF.[31] They reported that spontaneous DHBA production was stimulated by metal in the fluid path (e.g., syringe Luer-lock, metal tubing of the probe), the inclusion of ascorbate in the aCSF, the use of low grade salts in the aCSF, the reuse of probes, and the duration of the experiment. Arning *et al.* (1997) also reported problems with spontaneous formation of DHBA isomers in microdialysis perfusates and found that DHBA production was promoted by perchloric acid and by inclusion of homocysteine in the perfusion medium.[32] These findings suggest that inclusion of SA (and possibly other scavengers) is going to be problematic for microdialysis approaches, even if the amount of metal in contact with the perfusion medium is minimized and ascorbate is not included. As SA readily passes through the blood–brain barrier (BBB), perhaps a better approach is to administer SA via the periphery. However, any SA dialyzing into the probe may spontaneously form DHBAs and lead to artificially elevated DHBA levels. A further complication with microdialysis is due to the stability of the DHBAs. At neutral pH (the pH of the aCSF) the DHBAs are unstable.[18] This suggests that unless the aCSF is acidified immediately after leaving the probe, interpretation of the significance of DHBA levels is difficult.

Use of Salicyclic Acid Method in Assessing Oxidative Injury

The SA method has also proven useful for assessing oxidative injury in brain tissue specimens and has been applied to both rat and mouse models.

[29] U. Ungerstedt, *in* "Measurement of Neurotransmitter Release *in Vivo*" (C. A. Marsden, Ed.), IBRO Handbook Series: Methods in the Neurosciences, Vol. 6, p. 81. John Wiley and Sons, New York, 1984.

[30] T. E. Robinson and J. B. Justice, "Microdialysis in the Neurosciences," Techniques in the Behavior and Neural Sciences, Vol. 7. Elsevier, New York, 1991.

[31] J. Montgomery, L. Ste-Marie, D. Boismenu, and L. Vachon, *Free Rad. Biol. Med.* **19,** 927 (1995).

[32] E. Arning, K. Hyland, G. F. Carl, and T. Bottiglieri, 27th Meeting of The Society for Neurosciences, New Orleans, Poster #109.21 (1997).

Typically, SA is administered intraperitoneally at a dose of 150–200 mg/kg. A mitochondrial toxin is then administered either systemically or intrastriatally. Animals are sacrificed 1 and 3 hr after administration of the toxin. Brain tissue is then dissected and homogenized in 0.1 *M* perchloric acid. SA, 2,3-, and 2,5-DHBAs are then separated and quantitated using HPLC with coulometric electrode array detection.[33] Both SA and DHBAs can be cleanly resolved using this methodology.

Prior studies have shown that systemic administration of 3-nitropropionic acid is followed by significant increases in 2,3- and 2,5-DHBAs 3 hr after administration of the toxin.[34] These increases were significantly attenuated by preadministration of a neuronal nitric oxide synthase inhibitor, 7-nitrindazole.[35] This suggests that they may be mediated by peroxynitrite, which can act in a hydroxyl radical-like manner (see below). Furthermore, significant increases in 2,3- and 2,5-DHBAs were seen following administration of 3-nitropropionic acid to mice. These increases were significantly attenuated in mice overexpressing superoxide dismutase.[36] Administration of either the mitochondrial toxin malonate or direct excitotoxins such as *N*-methyl-D-aspartate, kainic acid, or AMPA were also associated with increases in both DHBA isomers. These increases were attenuated following malonate injection by the free radical spin trap 2-sulfo-*N*-*tert*-butylphenylnitrone.[33] The increases produced by malonate and *N*-methyl-D-aspartate were also significantly attenuated by administration of 7-nitroindazole, suggesting that these increases were associated with peroxynitrite generation.[35] Studies in animals with a knockout of the neuronal nitric oxide synthase gene also showed that malonate induced increases in the 2,3- and 2,5-DHBAs were attenuated in these mice.[37] This correlated with significant neuroprotective effects. We also found that several growth factors can attenuate increases in the DHBA isomers induced by 1-methyl-4-phenylpyridinium.[38] The SA trapping assay for HO· generation therefore

[33] J. B. Schulz, D. R. Henshaw, D. Siwek, B. G. Jenkins, R. J. Ferrante, P. B. Cipolloni, N. W. Kowall, B. R. Rosen, and M. F. Beal, *J. Neurochem.* **64,** 2239 (1995).

[34] J. B. Schulz, D. R. Henshaw, U. MacGarvey, and M. F. Beal, *Neurochem. Int.* **29,** 167 (1996).

[35] J. B. Schulz, R. T. Matthews, B. G. Jenkins, R. J. Ferrante, D. Siwek, D. R. Henshaw, P. B. Cipolloni, P. Mecocci, N. W. Kowall, B. R. Rosen, and M. F. Beal, *J. Neurosci.* **15,** 8419 (1995).

[36] M. F. Beal, R. J. Ferrante, R. Henshaw, R. T. Matthews, P. H. Chan, N. W. Kowall, C. J. Epstein, and J. B. Schulz, *J. Neurochem.* **65,** 919 (1995).

[37] J. B. Schulz, P. L. Huang, R. T. Matthews, D. Passov, M. C. Fishman, and M. F. Beal, *J. Neurochem.* **67,** 430 (1996).

[38] P. B. Kirschner, B. G. Jenkins, J. B. Schulz, S. P. Finkelstein, R. T. Matthews, B. R. Rosen, and M. F. Beal, *Brain. Res.* **713,** 178 (1996).

appears to be a useful procedure for assessing free radical generation in brain tissue samples, as well as in assessing the effects of therapeutic interventions.

Reaction between HO· and 4-Hydroxybenzoic Acid

4-Hydroxybenzoic acid (4-HBA), in addition to SA, has been demonstrated to be another potentially useful HO· scavenging agent *in vivo*.[39] The reaction between 4-HBA and HO· can produce two possible DHBA isomers: the 3,4- and 2,4-DHBAs. Only 3,4-DHBA is quantitatively important. 4-HBA has an HO· trapping efficiency comparable to that of SA, and, although to date its use has been limited to animal studies, it appears to show some advantages over the SA method. First, neither the 3,4- nor 2,4-DHBA isomers result from enzymatic hydroxylation of 4-HBA by cytochrome P450. Second, as 3,4-DHBA is the major product, the signal is not being split between two isomers, as is the case with the SA method. Thus, the 4-HBA method would be expected to be a more sensitive assay. Third, 4-HBA is without any reported physiological effects, e.g., unlike SA (and acetylsalicyclic acid which is also used as a precursor of SA), 4-HBA does not inhibit cyclooxygenase. However, the 4-HBA method is not without its limitations. 4-HBA is synthesized from tyrosine by the intestinal flora in significant amounts and may enter the blood system. Since 4-HBA readily crosses the BBB, basal levels of 4-HBA, which are not related to the exogenous 4-HBA challenge, can sometimes be observed in brain ECF. This does not appear to be too much of a problem, as the levels of 4-HBA following its systemic administration are much greater than its preadministration level. It is worth noting that the level of 4-HBA can be markedly elevated in pathological conditions such as in the case of leaky gut syndrome.

4-Hydroxybenzoic Acid Method

We have developed an HPLC-ECD assay for simultaneous measurements of 3,4-DHBA and 4-HBA in brain microdialysis samples in awake mice following systemic administration of 4-HBA. The general approach, utilizing an electrochemical detector with two serially placed coulometric electrodes, is similar to that described above for the measurements of 2,3-, 2,5-DHBA and SA.

[39] L. Ste-Marie, D. Boismenu, L. Vachon, and J. Montgomery, *Anal. Biochem.* **241,** 67 (1996).

Male C57BL mice (25–30 g body weight) are anesthetized with halothane. A nonmetallic polypropylene-based concentric microdialysis probe (membrane length 2 mm; o.d. 220 μm; molecular weight cutoff 18,000) is placed into the right striatum (coordinates from bregma: AP + 0.5; ML 1.7; V −4.8 mm). The microdialysis fluid path consists of either fused silica or PEEK tubing. Any metal parts (e.g., syringe Luer lock/needle) are passivated with 6 *M* nitric acid to inhibit the spontaneous production of 3,4-DHBA from 4-HBA. The *in vitro* probe recoveries ($n = 4$) for 3,4-DHBA and 4-HBA were 17.47 ± 1.23% and 15.29 ± 0.99%, respectively. After an 18- to 24-hr recovery period animals are perfused at 1.0 μl/min with an aCSF composed of (in m*M*): 145 NaCl, 2.7 KCl, 1.2 $CaCl_2$, 1.0 $MgCl_2$, pH 7.4. A bolus dose of 4-HBA (400 mg/kg i.p. in saline) is administered following a 2 hr equilibration period. Dialysis perfusates are then collected every 20 min into 5 μl 0.5 *M* perchloric acid (to prevent autoxidation) and samples are analyzed within the next 20 min. Analytes are separated on a SuperODS (4.6 × 50 mm, 2 μm) column (TosoHaas, Montgomeryville, PA) kept at 29°. The mobile phase consists of 100 m*M* NaH_2PO_4 (pH 2.8 with phosphoric acid) and 6.5% methanol (v/v); it is passed through the system at 1.0 ml/min. Analytes are detected using a dual potentiostat electrochemical detector (Coulochem II, ESA, Inc.) and a dual coulometric electrode analytical cell (Model 5011, ESA, Inc.). The potentials applied to the first (upstream) and the second (downstream) electrodes are +150 mV (optimized for 3,4-DHBA) and +700 mV (optimized for 4-HBA), respectively. The limit of detection for 3,4-DHBA has been found to be ~1 pg on column. The analysis is completed in under 8 min (Fig. 3).

Results and Discussion of 4-Hydroxybenzoic Acid Method

In our initial experiments we attempted to administer 4-HBA (1–5 m*M*) locally through the microdialysis probe by including it in the aCSF. However, despite the lack of any metal components in the microdialysis system, we found spontaneous *in vitro* production of 3,4-DHBA in agreement with Montgomery *et al.*[31] (see above). Spontaneous 3,4-DHBA production was directly correlated to the concentration of 4-HBA, with as much as 15 n*M* 3,4-DHBA being produced from 5 m*M* 4-HBA. Interestingly, no production of 3,4-DHBA was found when aCSF containing 5 m*M* 4-HBA was incubated at room temperature for 72 hr. This suggests that 3,4-DHBA production can be influenced by components of the microdialysis system. Although as yet undetermined, these may include the formation of H_2O_2 (which can produce $HO\cdot$ with metals such as iron diffusing from the brain ECF into the probe) [Eq. (3)], permeability of PEEK tubing to oxygen, and possibly the production of reactive oxygen species by the epoxy glue used during the manufacture of the microdialysis probe.

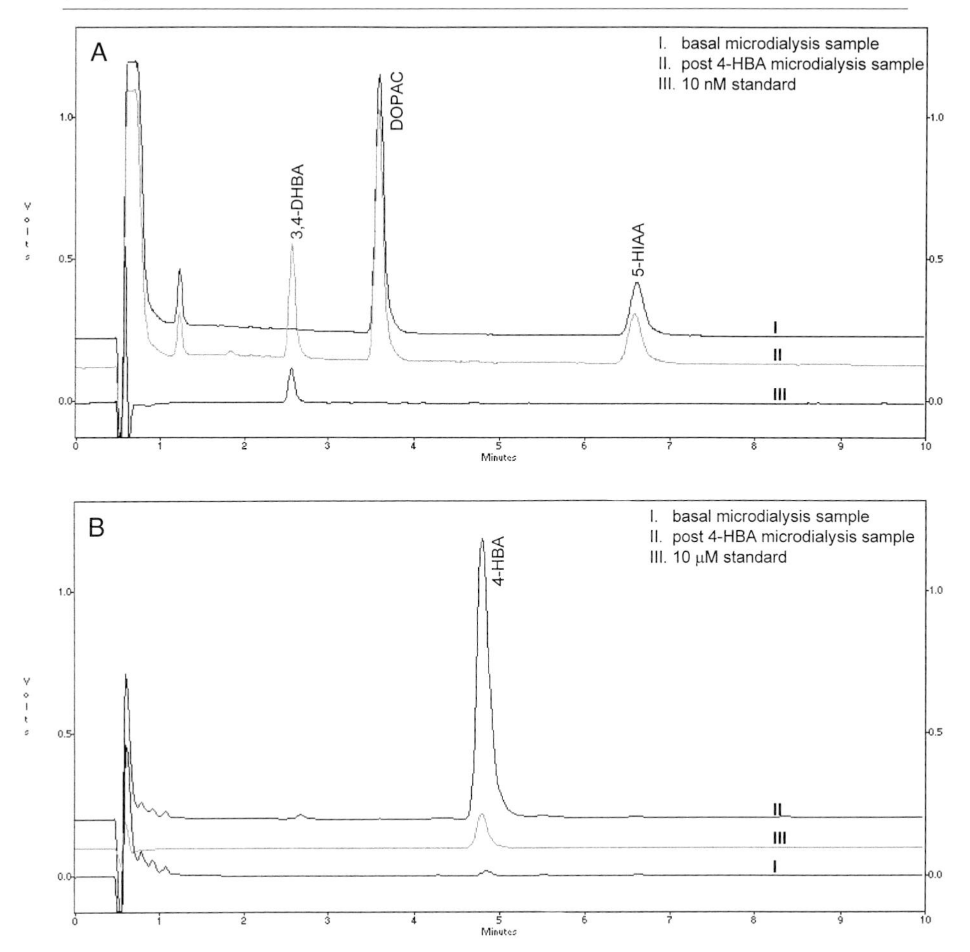

FIG. 3. Measurement of HO· production using the 4-HBA method. Mouse striatal ECF samples were obtained using microdialysis in awake animals. (A) Chromatograms showing the production of 3,4-DHBA 40 min after peripheral administration of 4-HBA (400 mg/kg i.p.). Chromatograms are presented at a gain range of 10 nA full scale. Analytes were measured at +150 mV vs palladium reference (electrode 1). No 3,4-DHBA was found in saline treated animals. The concentration of 3,4-DHBA in 4-HBA treated animals was 40 n*M* (uncorrected for *in vitro* recovery). (B) Corresponding measurement of 4-HBA levels in samples presented in (A). 4-HBA was measured at +700 mV (electrode 2) and a gain range of 10 μA. The striatal ECF level of 4-HBA was 2 and 105 μM in control and 4-HBA treated animals, respectively.

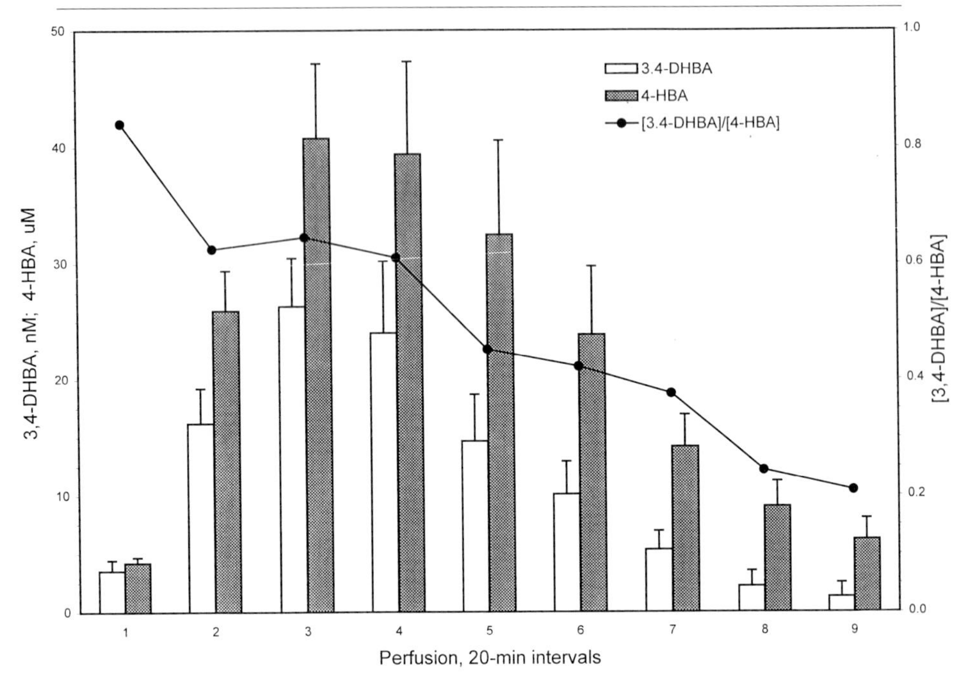

FIG. 4. Time course for 3,4-DHBA and 4-HBA levels in striatal microdialysis perfusates obtained from awake mice. 4-HBA was administered systematically (400 mg/kg i.p.) after an equilibration period (at least 2 hr after the start of perfusion). Data are presented as means ± SEM (n = 9) of the absolute 3,4-DHBA (nM) and 4-HBA (μM) concentrations or as the ratios [3,4-DHBA]/[4-HBA] in consecutive 20 min samples.

Basal 4-HBA levels were 0.870 ± 0.096 μM (n = 9). No significant amount of 3,4-DHBA was detected in basal mice striatal microdialysis perfusates (Fig. 3). The time course for the passage of 4-HBA through the BBB and striatal production of 3,4-DHBA following a systemic 4-HBA dose (400 mg/kg i.p.) is shown in Fig. 4.

DHBAs and Other Potential Markers

Other potential markers of HO· production include the formation of *o*- and *m*-tyrosine isomers from phenylalanine (*p*-tyrosine is the "normal" tyrosine isomer formed by the action of phenylalanine hydroxylase on phenylalanine).[1,7,14,40] Markers for peroxynitrous acid (and possibly other reactive nitrogen species) include 3-nitrotyrosine (3NT) and dityrosine,

[40] J.-Z. Sun, H. Kaur, B. Halliwell, X.-Y. Li, and R. Bolli, *Circul. Res.* **73,** 534 (1993).

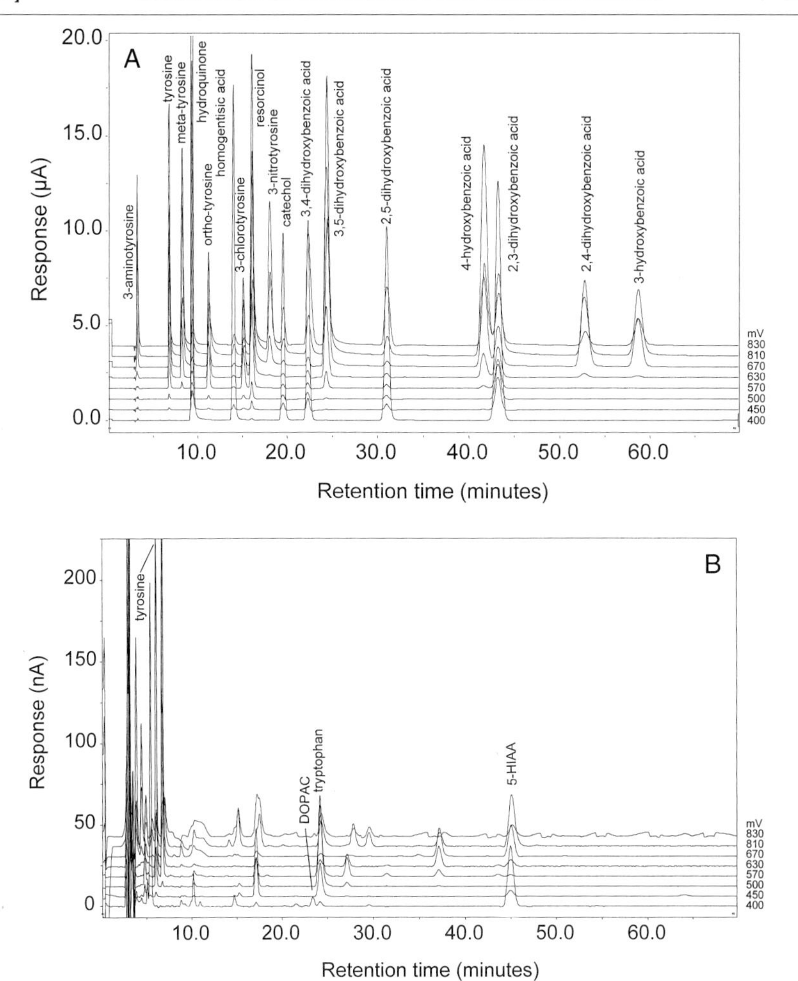

FIG. 5. (A) Isocratic coulometric array chromatogram showing the simultaneous measurement of possible ROS and RNS markers and precursors, including tyrosine isomers, 3-nitrotyrosine, and 3-chlorotyrosine (~10 μg/ml standard). Dityrosine (not shown for clarity) eluted just after tyrosine but at a lower oxidation potential. (B) Basal striatal microdialysis sample (3 mm loop probe) obtained from anesthetized rat and presented at 100 times the sensitivity of (A). Note that although markers were not measured in the basal microdialysis sample, the majority of them would elute in areas of the chromatogram devoid of endogenous analytes.

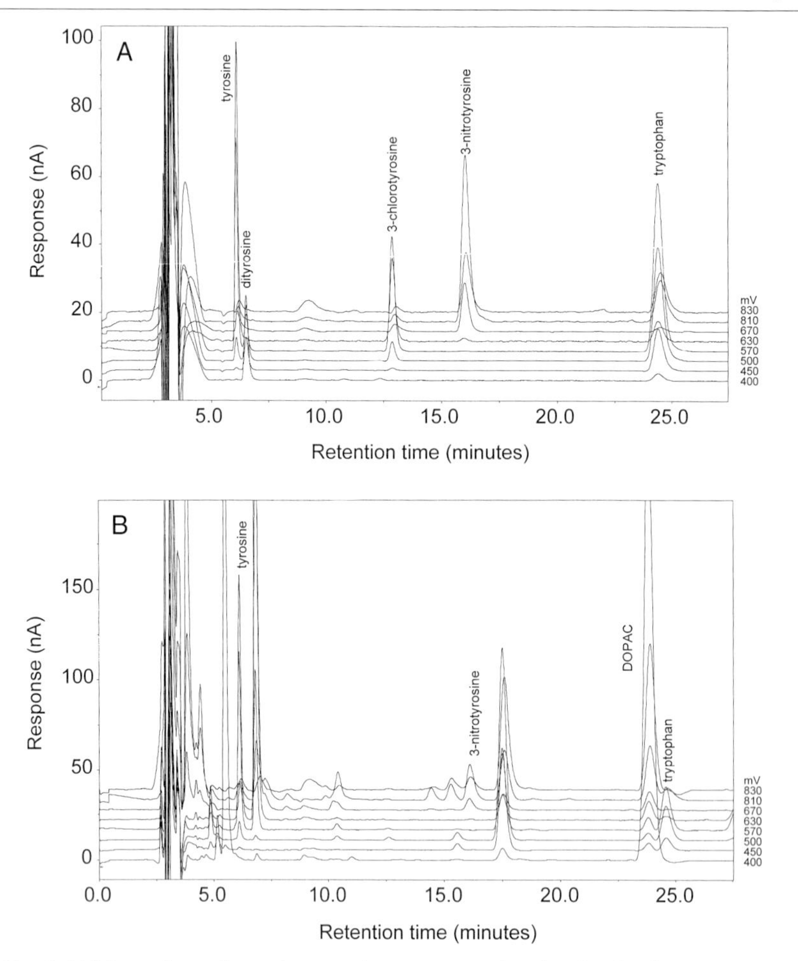

FIG. 6. (A) Isocratic coulometric array chromatogram showing the simultaneous measurement of tryptophan, tyrosine, and tyrosine markers (2 ng on column). (B) Basal striatal microdialysis sample obtained from an anesthetized rat showing the presence of 3-nitrotyrosine following its peripheral administration (10 mg/kg i.v.).

while 3-chlorotyrosine is thought to be an indicator of hypochlorous acid production.[1,7,41]

HPLC with coulometric array detection has been used previously to measure 3NT (free and protein bound), DHBAs, and 3,4-dihydroxyphenyl-

[41] S. L. Hazen and J. W. Heinecke, *J. Clin. Invest.* **99,** 2075 (1997).

alanine (protein bound) *in vitro* and *in vivo*.[36,42–46] We explored the feasibility of using such an approach to measure simultaneously several reported markers. In total 17 analytes were resolved on a TSKgel ODS-80T_M (4.6 × 250 mm; 5 μm) column (TosoHaas) using a mobile phase composed of 20 m*M* sodium phosphate, 8% (v/v), pH 3.2, at 31° (Fig. 5). Analytes were resolved voltammetrically using an array of eight serially placed coulometric electrodes (set at +400, +450, +500, +570, +630, +670, +810, and +830 mV vs palladium reference). As shown in Fig. 5, relatively few of the analytes coeluted with endogenous compounds in striatal ECF samples which were obtained using microdialysis perfusion in anesthetized animals. (Figure 6 shows the first 25 min of the chromatogram at a higher sensitivity, again showing few coelutions.) This method may have applicability to the measurement of other endogenous oxidative metabolism markers.

Conclusions

Careful use of the SA and 4-HBA HPLC-coulometric ECD methods described above permit the accurate estimation of HO · production *in vivo*. Although these methods can be used with microdialysis sampling, the route of administration of the HO ·-scavenging agent must be carefully considered. The spontaneous production of DHBA isomers when either SA or 4-HBA are included in the aCSF (and exposed to the microdialysis fluid path) can lead to overestimation of HO · production. Currently, methods are being explored to enable the simultaneous measurement of several markers of reactive oxygen and nitrogen species production.

Acknowledgment

The authors are grateful to Prof. Timothy Maher (Massachusetts College of Pharmacy) for help with obtaining tissue samples for the SA procedure and microdialysis samples for the 3-nitrotyrosine study.

[42] W. Maruyama, Y. Hashizume, K. Matsubara, and M. Naoi, *J. Chromatogr. B* **676,** 153 (1996).

[43] J. B. Schulz, R. T. Matthews, M. M. K. Muqit, S. E. Browne, and M. F. Beal, *J. Neurochem.* **64,** 936 (1995).

[44] L. I. Brujn, M. F. Beal, M. W. Becher, J. B. Schulz, P. C. Wong, D. L. Price, and D. W. Cleveland, *Proc. Natl. Acad. Sci. USA* **94,** 7606 (1997).

[45] K. A. Skinner, J. P. Crow, H. B. Skinner, R. T. Chandler, J. A. Thompson, and D. A. Parks, *Arch. Biochem. Biophys.* **342,** 282 (1997).

[46] K. Hensley, M. L. Maidt, Q. N. Pye, C. A. Stewart, M. Wack, T. Tabatabaie, and R. A. Floyd, *Anal. Biochem.* **251,** 187 (1997).

[32] Mechanism of Interaction of *in Situ* Produced Nitroimidazole Reduction Derivatives with DNA using Electrochemical DNA Biosensor

By ANA MARIA OLIVEIRA BRETT, SÍLVIA H. P. SERRANO, MAURO A. LA-SCALEA, IVANO G. R. GUTZ, and MARIA L. CRUZ

Introduction

A DNA biosensor has been developed for the evaluation of mechanisms of interaction of drugs with DNA. Damage to DNA can occur by intercalating, alkylating, or strand-breaking agents, but most drugs bind covalently to DNA by cross-linking between two bases on the same strand. The mechanism of the biological action of nitroimidazoles, commonly used to treat infection by anaerobic bacteria, depends on the reduction of the nitro group producing intermediate species which interact with DNA, oxidizing it and resulting in strand breaking and double-helix destabilization.[1,2] The most important nitroimidazole derivatives have substituents in the N-1 position of the heterocyclic ring of 5-nitroimidazole.

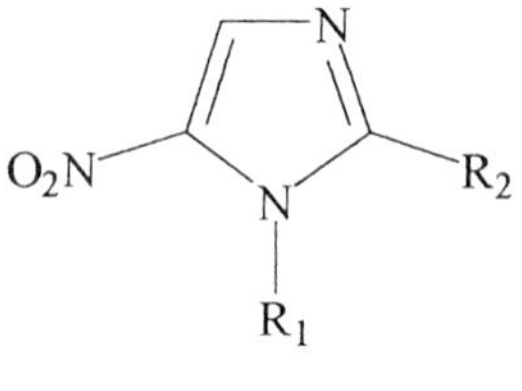

METRONIDAZOLE : R_1 = $-CH_2CH_2OH$; R_2 = $-CH_3$

TINIDAZOLE : R_1 = $-CH_2CH_2SO_2C_2H_5$; R_2 = $-CH_3$

SECNIDAZOLE : R_1 = $-CH_2CH_2(OH)CH_3$; R_2 = $-CH_3$

The electrochemical reduction of nitroimidazoles follows a complex mechanism and in theory the nitro group is able to receive up to six electrons to form the corresponding amine.[3] Under anaerobic and low oxygen pres-

[1] D. I. Edwards, *in* "DNA Binding and Nicking Agents," Vol. 2, Comprehensive Medicinal Chemistry (C. Hansch, Ed.). Pergamon Press, 1990.
[2] D. I. Edwards, *J. Antimicrob. Chemother.* **31,** 9 (1993).
[3] P. Zuman and Z. Fijalek, *J. Electroanal. Chem.* **296,** 538 (1990).

0076-6879/99 $30.00

sure conditions metronidazole (2-methyl-5-nitroimidazole-1-ethanol) follows a reduction mechanism similar to that of nitrobenzene.[3]

Electrochemical reduction of nitroimidazole derivatives[4–7] shows two reduction waves in aqueous acid media, the first involving 4 electrons and corresponding to the reduction of the nitro group to form the intermediate hydroxylamine (−NHOH) and the second involving 2 electrons and corresponding to the reduction of the hydroxylamine to amine ($-NH_2$). Using three different electrode materials, bare glassy carbon electrode, mercury thin film electrode, and DNA biosensor, we have verified[8–10] that hydroxylamine formation involves 4 electrons and is pH dependent. The results are in good agreement with previous work using the dropping mercury electrode.[11,12]

As will be shown, the DNA biosensor developed by us[8–10,13] provides a new perspective to the research and study of the mechanism of action of nitroimidazoles with DNA.

Materials and Instrumentation

Metronidazole (MTZ) and secnidazole (SCZ) are supplied by Rhodia Farma Lda Brasil, and tinidazole (TNZ) by Laboratórios Pfizer Lda Brasil. Calf thymus DNA (sodium salt, type I), is obtained from Sigma Chemical Co. and is used without further purification. Acetate buffer solutions of ionic strength 0.2 at pH 4.5 are used in all experiments and are prepared using analytical grade reagents and purified water from a Millipore (Bedford, MA) Milli-Q system (resistivity $\geq$ 18 MΩ cm). All experiments are done in deoxygenated solutions and at room temperature.

Electrochemical techniques[14] of cyclic voltammetry and differential pulse voltammetry are employed using a μAutolab potentiostat/galvanostat

[4] J. H. Tocher and D. I. Edwards, *Biochem. Pharmac.* **48,** 1089 (1994).

[5] P. J. Declerck and C. J. De Ranter, *J. Chem. Soc. Faraday Trans 1* **83,** 257 (1987).

[6] D. Dumanovic, J. Volke, and V. Vajgand, *J. Pharma. Pharmac.* **18,** 507 (1966).

[7] D. Dumanovic and J. Ciric, *Talanta* **20,** 525 (1973).

[8] A. M. Oliveira Brett, S. H. P. Serrano, I. Gutz, and M. A. La-Scalea, *Bioelectrochem. Bioenerg.* **42,** 175 (1997).

[9] A. M. Oliveira Brett, S. H. P. Serrano, I. Gutz, and M. A. La-Scalea, *Electroanaylsis* **9,** 110 (1997).

[10] A. M. Oliveira Brett, S. H. P. Serrano, I. Gutz, and M. A. La-Scalea, *Electroanalysis,* **9,** 1132 (1997).

[11] P. Zuman, Z. Fijalek, D. Dumanovic, and D. Suznjevic, *Electroanalysis* **4,** 783 (1992).

[12] C. Karakis and P. Zuman, *J. Electroanal. Chem.* **396,** 499 (1995).

[13] A. M. Oliveira Brett, S. H. P. Serrano, T. A. Macedo, D. Raimundo, M. H. Marques, and M. A. La-Scalea, *Electroanalysis* **8,** 992 (1996).

[14] C. M. A. Brett and A. M. Oliveira Brett, "Electrochemistry, Principles, Methods and Applications." Oxford University Press, Oxford, 1993.

running with GPES version 3 software (Eco-Chemie, Utrecht, Netherlands). Differential pulse voltammetry conditions are: pulse amplitude 50 mV, pulse width 70 msec, and scan rate 5 mV sec^{-1}. The working electrode is the DNA biosensor, the counter electrode is a Pt wire, and the reference electrode is a standard calomel electrode (SCE), all contained in a one-compartment cell.

DNA Biosensor

The DNA biosensor consists of a glassy carbon electrode covered by a layer of immobilized double-stranded DNA conditioned in a single-stranded solution. The structure of the DNA layer on the electrode surface after conditioning is important for electrode stability and reproducibility. Single-stranded DNA (ssDNA) is prepared by treating an accurately weighed sample of approximately 3 mg of DNA with 0.5 ml of 60% pure perchloric acid; after dissolution, 0.5 ml of 9 *M* NaOH is then added to neutralize the solution followed by addition of pH 4.5 acetate buffer until complete 10 ml ($I \cong 9$ *M*).

The DNA biosensor is prepared by covering a glassy carbon electrode (Tokai, GC20, area 0.07 cm^2) with 3 mg of DNA dissolved in 80 μl of pH 4.5 acetate buffer and leaving the electrode to dry. After drying, the electrode is immersed in acetate buffer solution and a constant potential of +1.4 V applied during 5 min. A differential pulse scan is then done from 0 V to +1.4 V vs SCE to check that no electrochemical reaction occurs on the surface of the DNA biosensor in supporting electrolyte. The electrode is then transferred to a solution containing single-stranded DNA and differential pulse voltammograms recorded in the range 0 to +1.4 V until stabilization of the peak currents that correspond to guanine (+0.776 V) and adenine (+1.065 V) electrooxidation occurs.[15,16] The electrode was then dried at room temperature and used to study the mechanism of action of nitroimidazoles with DNA.

Mechanism of Interaction of Nitroimidazoles with DNA

The DNA biosensor has been used to study the electrochemistry of metronidazole (MTZ) in detail and also that of two other nitroimidazoles: secnidazole (SCZ), and tinidazole (TNZ) which follow a similar reduction

[15] C. M. A. Brett, A. M. Oliveira Brett, and S. H. P. Serrano, *J. Electroanal. Chem.* **366,** 255 (1994).

[16] A. M. Oliveira Brett and S. H. P. Serrano, *J. Braz. Chem. Soc.* **6,** 97 (1995).

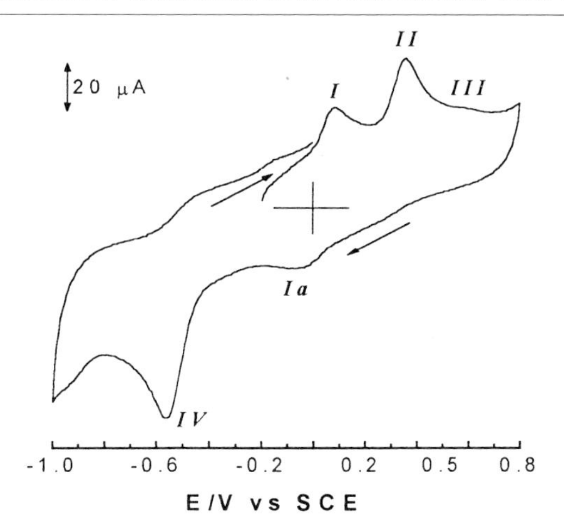

FIG. 1. Cyclic voltammograms of 1.0 mM MTZ in pH 4.5 acetate buffer at a DNA biosensor. ν, 100 mV sec^{-1}; E_{dep}, -0.6 V; t_{dep}, 2 min with stirring; and E_i, -0.2 V vs SCE.

mechanism involving a 4-electron transfer to form the corresponding hydroxylamines[1,3,11,17] as occurs with metronidazole.

The reduction of MTZ has been studied at pH 4.5 using differential pulse voltammetry with a bare glassy carbon electrode as well as with the DNA biosensor. The DNA biosensor shows the advantage of the ability to preconcentrate metronidazole on the electrode surface, and a good linear working range of 1.0 to 54.3 μmol/liter when using 2 min preconcentration at 0.0 V with stirring has been obtained.[9]

Preconcentration consists in accumulating the electroactive species onto the electrode surface at a predetermined applied potential during an optimized length of time, which will incorporate either metronidazole or its reduction products on the DNA biosensor. This enables the detection of lower concentration levels of the electroactive analyte in solution when a potential scan is applied in the determination step. In order to investigate the oxidation of the reduced nitro compound, the preconcentration potential is changed from 0.0 V to -0.6 V. Preconcentration at a potential more negative than the reduction potential of MTZ leads to preconcentration of the reduction products rather than of the starting compound.

The interaction of MTZ with DNA is shown in the cyclic voltammogram of Fig. 1. Three oxidation peaks can be seen, the first being reversible.

[17] P. J. Declerck and C. J. De Ranter, *Biochem. Pharmac.* **35,** 59 (1986).

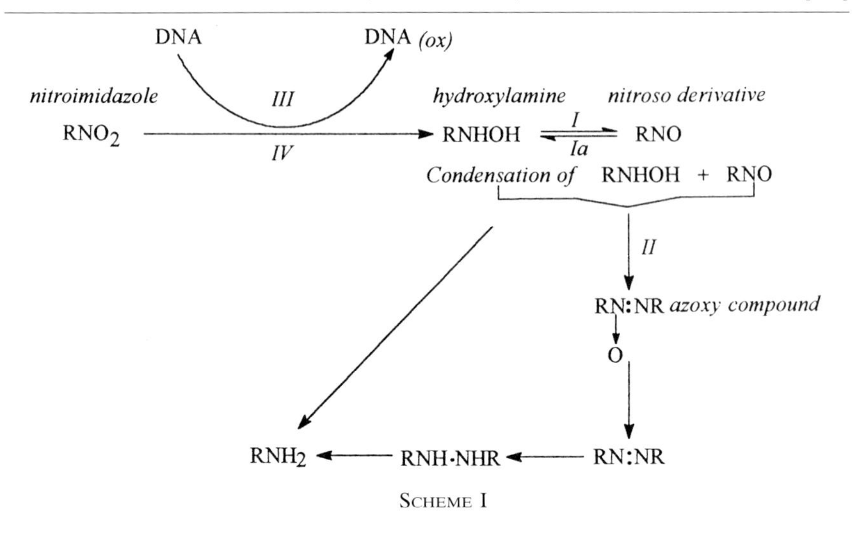

SCHEME I

Identification of the peaks in Fig. 1 is based on the electrochemical behavior of nitrobenzene.[12,18] The reduction product of MTZ at −0.6 V is the corresponding hydroxylamine which is preconcentrated on the DNA-biosensor surface at a constant applied potential for 2 min, consequently permitting its interaction with the DNA. The cyclic voltammetric scan is started in the positive direction at −0.2 V so that oxidation of the nitro compound reduction products can be detected. The first wave, *I*, corresponds to oxidation of the hydroxylamine to a nitroso derivative (E_p = 0.081 V); which can be further oxidized to an azoxycompound[19,20] peak *II* (E_p = 0.358 V), or, in the cyclic voltammetry reverse scan, reversibly reduced to hydroxylamine[12,18,21] peak *Ia* (E_p = −0.037 V), and the reduction of MTZ in solution corresponds to peak *IV* (E_p = −0.556 V).

The reduction mechanism of the nitroheterocyclic derivatives, responsible for biological activity, and the consequent formation of bimolecular azoxy, azo, and hydrazo derivative compounds can be described by Scheme I.

This shows the reversible oxidation of the hydroxylamine derivative (RNHOH) to the corresponding nitroso derivative (RNO) and a condensation reaction between the hydroxylamine and nitroso derivatives to form

[18] I. Rubinstein, *J. Electroanal. Chem.* **183,** 379 (1985).

[19] N. V. Sidgwick, "Organic Chemistry of Nitrogen." Clarendon Press, Oxford, 1937.

[20] Z. Fijalek, M. Pugia, and P. Zuman, *Electroanalysis* **5,** 65 (1993).

[21] R. Carlier, E. Raoult, A. Tallec, V. Andre, P. Gauduchon, and J.-C. Lancelot, *Electroanalysis* **9,** 79 (1997).

the azoxy compound (RNO:NR). The amount of azoxy compound formed depends on the concentration of the nitroso derivative, which rapidly reduced to hydroxylamine.[19–21]

Concerning peak *III* (E_p = 0.595 V), it is well known that hydroxylamine displays a large difference in reactivity for single-stranded vs double-stranded DNA,[22–24] which makes it a sensitive detector of H-DNA. H-DNA is the recently discovered nucleic acid triplex structure.[22] The kinetics of formation of triple-stranded structures is slower than that of double-stranded structures and is dependent on the ionic strength of the medium. This H-DNA conformation, formed by mirror-symmetric homopurine–homopyrimidine sequences, consists of an intramolecular PyPuPy triple-stranded region, i.e., the single strand of an unwound region folds back on the double strand forming Hoogsteen hydrogen bonds.

Because the DNA biosensor is prepared by covering a glassy carbon electrode with double-stranded DNA and then transferring it to a solution containing single-stranded DNA for electrochemical conditioning, it is possible that the electrode contains a small amount of H-DNA segments which could be detected by the hydroxylamine. Further evidence for this hypothesis was given by the same experiments using a glassy carbon electrode modified with either only ssDNA or only dsDNA: an irreproducible electrochemical response was obtained showing no peak *III*. The differential pulse potential of peak *III* (Fig. 2) is very similar to that of uric acid, E_p = +0.424 V vs SCE in pH 4.5 acetate buffer measured at a newly prepared DNA biosensor.[13] In fact, this comparison is not speculative and is in agreement with the literature that presents the purine bases as the principal target of the oxidative lesion of DNA.[5,25] We can therefore assume that peak *III* corresponds to the formation of deoxypurinic acid derivative[26] formed by reaction of the MTZ reduction intermediates with the DNA multilayer immobilized onto the surface of the glassy carbon electrode. Immobilization of DNA on solid substrates by means of electrostatic interactions has been shown[27] to have clear advantages compared with chemical bonding.

[22] V. N. Soyfer and V. N. Potaman, "Triple-Helical Nucleic Acids." Springer-Verlag, New York, 1995.
[23] B. Johnson, *Methods in Pharmacology* **212,** 180 (1992).
[24] E. Freese and E. B. Freese, *Biochemistry* **4,** 2419 (1965).
[25] P. J. Declerck and C. J. De Ranter, *Analysis* **15,** 148 (1987).
[26] G. Dryhurst, *J. Electrochem. Soc.* **116,** 1411 (1969).
[27] G. B. Sukhorukov, M. M. Montrel, A. I. Petrov, L. I. Shabarchina, and B. I. Sukhorukov, *Biosensors and Bioelectronics* **11,** 913 (1996).

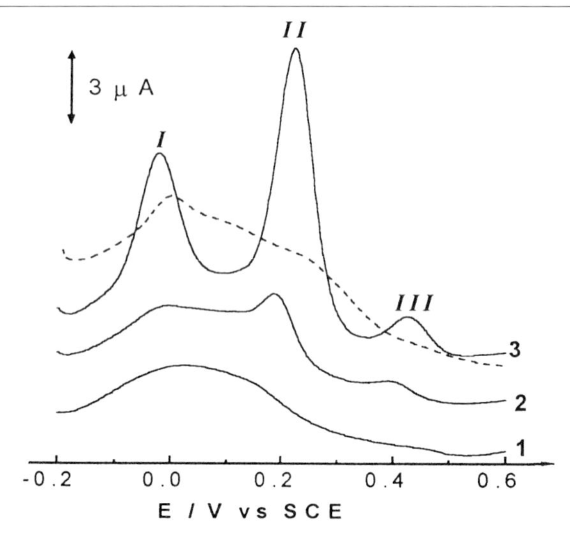

FIG. 2. Anodic stripping differential pulse voltammograms in pH 4.5 acetate buffer of MTZ: (—) (1) 0.0 m*M*; (2) 0.2 m*M*, and (3) 0.5 m*M* and E_{dep} = −0.5 V vs SCE using the DNA biosensor; (---) 1.0 m*M*; E_{dep} = −0.7 V vs SCE using a bare glassy carbon. ν, 5 mV sec^{-1}; ΔE, 50 mV; t_{dep}, 2 min with stirring.

An important parameter is the preconcentration time because of its relation with the formation of hydroxylamine that will interact with DNA and two minutes preconcentration with stirring, was found to be the best. Figure 2 also includes, for comparison, the results at a bare glassy carbon electrode; because it is not possible to preconcentrate when using this electrode, a much higher concentration, 1 m*M*, of MTZ was used—nevertheless, no peak *III* appeared. However, peak *III* can already be seen for a concentration of 0.2 m*M* when using the DNA biosensor. This confirms the assumption that peak *III* is due to interactions of the reduction intermediates with the DNA.

In order to better understand the biological activity of this type of drug, similar voltammetric experiments using the DNA biosensor were carried out for two other nitroimidazoles, secnidazole and tinidazole (Fig. 3). There is no appreciable difference in the peak potentials; the differences in the peak currents for peaks *I* and *II* can be explained, probably, by the difference in the substituent groups in the N1 position.

The reduction intermediates' interaction with DNA can be assigned to peak *III* with the same explanation as for MTZ; the nitro radical is always that principally responsible for the biochemical lesion on DNA of anaerobic microorganisms.

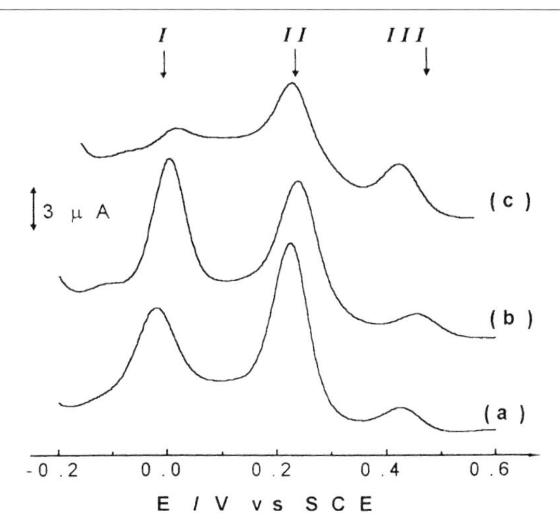

FIG. 3. Anodic stripping differential pulse voltammograms in pH 4.5 acetate buffer using a DNA biosensor of three nitroimidazoles: (a) metronidazole; (b) secnidazole; (c) tinidazole. ν, 5 mV sec^{-1}; ΔE, 50 mV; t_{dep}, 2 min with stirring, and E_{dep} = −0.5 V vs SCE; E_i, −0.2 V.

Conclusions

The newly developed DNA biosensor is a very promising tool for the investigation and study of the action of drugs specifically designed to interact with DNA. The electrochemical reduction of nitroimidazole drugs was studied in the presence of DNA immobilized onto the surface of a glassy carbon electrode. This enabled preconcentration of the drug onto the electrode surface, which was then electrochemically reduced to the corresponding hydroxylamine followed by reoxidation to give the nitroso compound and subsequently an azoxy compound. Moreover, as the target of the nitroimidazole action was the DNA, the damage caused to DNA on the electrode surface by a reduction product of this drug could be detected *in situ.*

Acknowledgments

We thank CNPq and FAPESP (Brasil) for a fellowship to M. A. La-Scalea and S. H. P. Serano, respectively. Rhodia Farma Lda and Laboratórios Pfizer Lda very kindly supplied the nitroimidazoles used in this work.

[33] Heme Oxygenase Activity Determination by High-Performance Liquid Chromatography

By STEFAN RYTER, EGIL KVAM, and REX M. TYRRELL

Introduction

Heme oxygenase, first described as a hepatic microsomal mixed function oxygenase (EC 1:14:99:3, heme-hydrogen donor:oxygen oxidoreductase), catalyzes the rate-determining step in heme catabolism.[1,2] Two structurally distinct isozymes, an inducible form, heme oxygenase-1 (HO-I), and a constitutive form, heme oxygenase-2 (HO-II), both contribute to heme oxygenase activity measured by conventional methods.[3,4] Heme oxygenase cleaves the heme substrate (ferriprotoporphyrin IX), forming biliverdin IXα, water, and carbon monoxide (CO); while releasing the heme iron.[1] NADPH:cytochrome P450 reductase assists in the HO reaction by catalyzing the reduction of the ferric heme iron as a prerequisite for the binding and activation of dioxygen.[5] Finally, NAD(P)H:biliverdin reductase completes heme metabolism by reducing biliverdin IXα, the principal product of heme oxygenase, to form bilirubin IXα.[6]

Heme oxygenase activity serves an indispensible physiological function in the regulation of hemoprotein–heme turnover, primarily in the reticuloendothelial system.[1] Additionally, HO-I has been widely studied as an integral component of stress responses, since its increased synthesis occurs in most mammalian cells following stimulation with diverse chemical and physical agents.[7,8] Many agents that induce HO-I also deplete or complex cellular reduced glutathione and alter intracellular redox potential.[9] These include thiol-reactive substances, hydrogen peroxide, nitric oxide, and ultra-

[1] R. Tenhunen, H. S. Marver, and R. Schmid, *J. Biol. Chem.* **244,** 6388 (1969).
[2] R. Tenhunen, H. S. Marver, and R. Schmid, *Proc. Natl. Acad. Sci. USA* **61,** 748 (1968).
[3] M. D. Maines, G. M. Trakshel, and R. K. Kutty, *J. Biol. Chem.* **261,** 411 (1986).
[4] G. M. Trakshel, R. K. Kutty, and M. D. Maines, *Arch. Biochem. Biophys.* **261,** 732 (1988).
[5] T. Yoshida, M. Noguchi, and G. Kikuchi, *J. Biol. Chem.* **255,** 4418 (1980).
[6] R. Tenhunen, M. E. Ross, H. S. Marver, and R. Schmid, *Biochemistry* **9,** 298 (1970).
[7] M. D. Maines, "Heme Oxygenase: Clinical Applications and Functions." CRC Press, Boca Raton, FL, 1992.
[8] S. W. Ryter and R. M. Tyrrell, *In* "Oxidative Stress and Signal Transduction (E. Cadenas, and H. J. Forman, Eds.), Chapter 15, p. 343. Chapman and Hall, New York, 1997.
[9] R. M. Tyrrell, *Methods* **11,** 313 (1997).

0076-6879/99 $30.00

violet-A (320–380 nm) radiation exposure.[10–12] Heme oxygenases may play a central role in cellular antioxidant defense mechanisms, by mediating the redistribution of heme iron to iron-storage pools, such as ferritin.[13,14] The *in vitro* antioxidant properties of biliverdin and bilirubin have also implied a hypothetical antioxidant function of heme oxygenase.[15]

Interest in heme oxygenase arises from intriguing relationships to the nitric oxide/nitric oxide synthase (NO/NOS) system.[16] In a manner analogous to NO, the carbon monoxide (CO) generated from heme oxygenase activity may function as a second messenger gas.[17] In the proposed pathway, CO binds to the heme iron of soluble guanylate cyclase (sGC), stimulating sGC to produce cyclic guanosine-3′,5′ monophosphate (cGMP). While this mechanism remains controversial, roles for heme oxygenase have been proposed in the regulation of vascular tone,[18] and in the modulation of tissue injury following ischemia–reperfusion.[19,20] Finally, the generation of CO by constitutive HO-II activity has been implicated in neural signaling processes in various systems.[21,22]

Methods for Determination of Heme Oxygenase Activity

The original method for heme oxygenase activity relies on the spectrophotometric determination of bilirubin IXα (BR). As described by Tenhunen *et al.*[1,2] and Schacter,[23] the assay measures the rates of BR production in reconstituted heme oxygenase reaction mixtures at 37°, by visible difference spectroscopy (410–500, λ_{max} 468 nm) in the absence and presence of reduced nicotinamide adenine dinucleotide phosphate (NADPH). As a source of heme oxygenase and NADPH : cytochrome P450 reductase enzyme protein,

[10] S. Keyse and R. M. Tyrrell, *Proc. Natl. Acad. Sci. USA* **86,** 99 (1989).
[11] D. Lautier, P. Luscher, and R. M. Tyrrell, *Carcinogenesis* **13,** 227 (1992).
[12] R. Foresti, J. E. Clark. C. J. Green, and R. Motterlini, *J. Biol. Chem.* **272,** 18411 (1997).
[13] G. F. Vile, S. Basu-Modak, C. Waltner, and R. M. Tyrrell, *Proc. Natl. Acad. Sci. USA* **91,** 2607 (1994).
[14] G. F. Vile and R. M. Tyrrell, *J. Biol. Chem.* **268,** 14678 (1993).
[15] R. Stocker, Y. Yamamoto, A. McDonagh, A. Glazer, and B. Ames, *Science* **235,** 1043 (1987).
[16] M. D. Maines, *Annu. Rev. Pharmacol. Toxicol.* **37,** 517 (1997).
[17] T. Dawson and S. Snyder, *J. Neurosci.* **14,** 5147 (1994).
[18] M. Suematsu, S. Kashiwagi, T. Sano, N. Goda, Y. Shinoda, and Y. Ishimura, *Biochem. Biophys. Res. Comm.* **205,** 1333 (1994).
[19] V. S. Raju and M. D. Maines, *J. Pharmacol. Exp. Ther.* **277,** 1814 (1996).
[20] N. Maulik, D. Engelman, M. Watanabe, R. Engelman, J. Rousou, J. Flack, D. Deaton, N. Gorbunov, N. Elsayed, V. Kagen, and D. Das, *Circulation* **94,** Suppl. II, II398 (1996).
[21] A. Verma, D. J. Hirsch, C. E. Glatt, G. V. Ronnett, and S. H. Snyder, *Science* **259,** 381 (1993).
[22] R. Zhakhary, K. D. Poss, S. R. Jaffery, C. D. Ferris, S. Tonegawa, and S. H. Snyder, *Proc. Natl. Acad. Sci. USA* **94,** 14848 (1997).
[23] B. A. Schacter, *Methods Enzymol.* **52,** 367 (1978).

the assay requires the 18,000 g postmitochondrial supernatant fraction from organ tissue such as rat liver or spleen, which may be further purified as the 104,000*g* microsomal pellet fraction.[1,23] The reaction mixtures typically consist of microsomal protein, NADPH, or an NADPH-regenerating system (i.e., glucose 6-phosphate, glucose-6-phosphate dehydrogenase), heme as methemalbumin, and NAD(P)H–biliverdin reductase (BVR), in a physiological buffer at pH 7.4. Because biliverdin IXα (BV), the principal heme oxygenase reaction product, is difficult to assay by spectrophotometric methods, the quantification of heme oxygenase activity depends on the complete conversion of BV to BR by BVR.[23] BVR may be supplied from crude extracts (rat liver 105,000*g* supernatant fraction), or in purified form.[6,24] The spectrophotometric method for heme oxygenase activity has been modified by Kutty and Maines[25] such that BR is measured by the difference between O.D. at 464 nm and 530 nm in a chloroform extract of the reaction mixture. This heme oxygenase assay has been widely used for the determination of heme oxygenase in organ tissue, yet criticized for insensitivity, spectral interference in complete reaction mixtures, or poor efficiency of BR extraction in chloroform.[25–27] Furthermore, the molar extinction coefficient for BR varies greatly with the protein composition of the reaction mixtures.[1,23]

Alternative radioisotopic heme oxygenase methods, which use ^{14}C-labeled heme as the heme oxygenase substrate, have been described for heme oxygenase activity, whereby the [^{14}C]bilirubin formed in the reaction is recrystallized from chloroform extracts,[2,27] or isolated by thin-layer chromatography.[28] Heme oxygenase activity may also be estimated by measuring CO evolution, using gas chromatographic methods.[29,30]

This paper describes detailed methodology for a heme oxygenase activity assay, which utilizes high-performance liquid chromatography (HPLC) for the separation and detection of heme metabolites, based on modifications of previously described methods.[26,31,32] The resolution of individual

[24] R. K. Kutty and M. D. Maines, *J. Biol. Chem.* **256,** 3956 (1981).

[25] R. K. Kutty and M. D. Maines, *J. Biol. Chem.* **257,** 9944 (1982).

[26] B. C. Lincoln, A. Mayer, and H. L. Bonkovsky, *Anal. Biochem.* **170,** 485 (1988).

[27] R. Tenhunen, *Anal. Biochem.* **45,** 600 (1972).

[28] E. E. Sierra and L. M. Nutter, *Anal. Biochem.* **200,** 27 (1992).

[29] H. Vremen and D. Stevenson, *Anal. Biochem.* **168,** 31 (1988).

[30] E. Cavallin-Stahl, G. I. Jonnson, and B. A. Lundh, *Scan. J. Clin. Lab.* **38,** 69 (1978).

[31] H. L. Bonkovsky, S. G. Wood, S. K. Howell, P. R. Sinclair, B. Lincoln, J. F. Healy, and J. F. Sinclair, *Anal. Biochem.* **155,** 56 (1986).

[32] T. C. Lee and I. C. Ho, *Cancer Res.* **54,** 1660 (1994).

bile pigments by HPLC, including biliverdin,[31,33–36] bilirubin structural or photochemical isomers,[37] and bilirubin conjugates,[38,39] has been described previously.

Assay Method

Principle

Heme oxygenase enzymatic activity is calculated from the rate of formation of bilirubin equivalents (biliverdin + bilirubin) at 37°. Biliverdin IXα (BV) and bilirubin IXα (BR) are separated by HPLC, and detected at 405 nm using visible absorbance spectroscopy. Despite the submaximal absorbance of BV and BR at this wavelength, both pigments are detected on a single chromatograph at picomole sensitivity. Thus, complete conversion of BV to BR, as required for spectrophotometric methods,[23] is not necessary since both metabolites are quantifiable. The protein extraction procedures described (Ryter *et al.*[40]) result in cosedimentation of biliverdin reductase (BVR), which may become limiting only under heme oxygenase overexpression. Therefore, the supplementation of reactions with BVR, as described previously,[1] may be included with the present protocol, but is not strictly necessary for the quantification of heme oxygenase activity. The HPLC parameters for the separation and detection of tetrapyrroles are based on the modification of protocols by Lincoln *et al.*[26] and Bonkovsky *et al.*[31] as described in Ryter *et al.*[40] We have routinely used the following protocol for the determination of HO activity in cultured adherent monolayers of mammalian cells.

Cell Culture

For the determination of heme oxygenase activity, we have routinely used human HeLa cells. The cells are seeded at a density of 3×10^6 cells per 15 cm plate in 25 ml Earle's minimum essential medium (EMEM)

[33] B. C. Lincoln, T. Y. Aw, and H. L. Bonkovsky, *Biochim. Biophys. Acta* **992,** 49 (1989).
[34] M. Noguchi, T. Yoshida, and G. Kikuchi, *J. Biochem. Tokyo* **91,** 1479 (1982).
[35] R. D. Rassmussen, W. H. Yokoyama, S. G. Blumenthal, D. E. Bergstrom, and B. H. Ruebner, *Anal. Biochem.* **101,** 66 (1980).
[36] R. K. Kutty and M. D. Maines, *J. Biol. Chem.* **257,** 9952 (1982).
[37] A. F. McDonagh, L. Palma, F. Trull, and D. Lightner, *J. Am. Chem. Soc.* **104,** 6865 (1982).
[38] W. Spivak and M. C. Carey, *Biochem. J.* **225,** 787 (1985).
[39] S. Onishi, S. Itoh, S. N. Kawade, K. Isobe, and S., Sugiyama, *Biochem. J.* **185,** 281 (1980).
[40] S. W. Ryter, E. Kvam. L. Richman, F. Hartmann, and R. M. Tyrrell, *Free Rad. Biol. Med.* **24,** 959 (1998).

[supplemented with 0.2% (w/v) sodium bicarbonate, 2m*M* L-glutamine (Gibco-BRL, Paisly, Scotland), 50 U/ml penicillin, 50 μg/ml streptomycin (Animed, Allschwil, Switzerland), and 15% (v/v) fetal calf serum (FCS) (Seromed, Biochrom KG, Berlin, Germany) (heat inactivated at 56°)]. The cells are cultured at 37° in humidified incubators containing an atmosphere of 95% air–5% CO_2 (v/v), for 3 days, at which time they reach 70–90% confluency. In principle, the starting cell number must be determined empirically for each cell type. We recommend using 3–5 confluent 15 cm petri dishes per microsomal protein extract, which should yield sufficient protein for 2–4 independent determinations. Alternatively, if crude extract is used for the determination of heme oxygenase activity, then one 15 cm plate may be sufficient. Treatment of cultures with heme oxygenase inducing agents may be performed at near confluence (60 hr) and should allow sufficient time before harvest (8–14 hr) for *de novo* synthesis of enzyme protein. It is not recommended that the medium be changed during the treatment period, since restimulation with fresh growth medium can result in the induction of endogenous HO-I activity. Therefore, conditioned growth medium should be restored to cultures following drug removal. The treatment of cultures with sodium metaarsenite can be used as a positive control for the induction of HO-I activity in culture as follows:

Treatment with Sodium Arsenite

1. Seed cells 3 days prior to assay in 15 cm culture dishes and culture to 70–90% confluency as described above.
2. After 60 hr of culture, remove culture media and retain it in a sterile bottle. Rinse the cells with 15 ml phosphate-buffered saline (PBS) at room temperature. Apply 12 ml PBS containing 10–50 μM sodium metaarsenite (Sigma, St. Louis, MO). Incubate cells at 37° for 30 min.
3. After treatment, aspirate the buffer containing the sodium arsenite, and rinse the monolayers three time with sterile PBS. Replace the original conditioned growth media, and incubate the cultures in a 37° incubator for an additional 12–14 hr prior to harvesting.

Preparation of Microsomal Protein Extracts from Cultured Cells

Materials. All centrifugation equipment for the protein extraction procedure should be precooled at 4°. The procedure requires a refrigerated centrifuge with capacity for 50 ml tubes (Beckman JB6, Beckman, Palo Alto, CA), a refrigerated ultracentrifuge (Centrikon T20-60, Kontron Instruments, Zurich, Switzerland), a fixed-angle ultracentrifuge rotor (TFT65.13, Kontron), an Eppendorf microfuge (Eppendorf Scientific, Westbury, NY), and a tabletop sonicator (sonifier B-12, Branson Sonic

Power, Danbury, CT). Additional materials include 13.5 ml polyallomer ultracentrifuge tubes (Kontron, Switzerland), polypropylene microtubes (1.5 ml) (Eppendorf), PCR microtubes (0.5 ml) (Treff AG, Degersheim, Switzerland), ice buckets, a flat ice tray (50 × 150 cm), and ice-cold sterile phosphate-buffered saline (PBS). The ethylenediaminetetraacetic acid (EDTA), trishydroxymethylaminomethane hydrochloride (Tris), sucrose (saccharose), and bovine serum albumin (BSA), are obtained from Sigma Chemical Company (St. Louis, MO). The protease inhibitors 4-(2-aminoethyl)benzylsulfonyl fluoride hydrochloride (AEBSF, pefabloc SC), leupeptin (Ac-Leu-Leu-arginal-$\frac{1}{2}H_2SO_4$), and pepstatin (isovaleryl-L-Val-L-Val-statyl-L-alanylstatin), are obtained from Boehringer Mannheim (Mannheim, Germany). The protein assay reagent kit (Coomassie Brilliant Blue G-250) is from Bio-Rad (Hercules, CA).

Solutions

Solution A: 0.25 *M* sucrose, 20 m*M* Tris-HCl, pH 7.4. Autoclave, and store at 4°.[26] Protease inhibitor solutions: The AEBSF 50 mg/ml and the leupeptin 4 mg/ml stock solutions are prepared in distilled water. The pepstatin 4 mg/ml stock solution is prepared in 100% methanol. Protease inhibitor solutions are aliquoted in 20 μl volumes, stored frozen at −20°, and added to buffers immediately before use.

Cell scraping solution: Phosphate-buffered saline (PBS), 1 m*M* EDTA, pH 8.0. (To 500 ml PBS, add 0.5 ml of a 1 *M* EDTA, pH 8.0 stock.) Store at 4°.

Procedure

1. Place tissue culture dishes on a large flat tray of wet ice. Aspirate tissue culture medium using a tabletop vacuum flask, and rinse cell monolayers twice with ice-cold PBS to remove all residual medium. To each dish, add 4 ml ice-cold PBS containing 1 m*M* EDTA, pH 8.0, and let stand for 5 min. Harvest cells on ice by mechanical scraping with a rubber policeman. Gently pipette the suspension with a 10 ml Pipetman and pool into 50 ml conical tubes (Falcon, France). Use an additional 2 ml of PBS to recover residual cells from tissue culture plates and pool.

2. Pellet the cells at 4° in a refrigerated centrifuge (J6B, Beckman, Palo Alto, CA) at 1000 rpm. Aspirate the supernatant from the cell pellets. Resuspend the cell pellets in 800 μl ice-cold homogenization buffer A (0.25 *M* sucrose, 20 m*M* Tris, pH 7.4), containing the protease inhibitors: AEBSF (50 μg/ml), leupeptin (4 μg/ml), pepstatin (4 μg/ml). The protease inhibitors should be added to the solution immediately before use. The use of AEBSF alone is not sufficient and will result in a decreased yield of

active enzyme protein. Transfer the suspension to 1.5 ml Eppendorf microcentrifuge tubes on ice.

3. Sonicate the cell suspensions by placing the sonicator tip inside the cell suspension while retaining the tube completely immersed in ice. Sonicate twice for 15 sec at low power output (20 W), using a Branson sonifier B-12 sonicator (Branson Sonic Power, Danbury, CT).

4. Centrifuge the sonicated suspensions at 4° in an Eppendorf microfuge at 15,000*g* for 20 min. If whitish pellet material is floating in the supernatant, repeat the centrifugation step until a clear supernatant is obtained.

5. Transfer the supernatant (~1 ml) directly to prechilled polyallomer ultrafuge tubes (13.5 ml) (Kontron, Zurich, Switzerland) on ice. Equilibrate the tubes with buffer A containing freshly added protease inhibitors. Ultracentrifuge (Centrikon T20-60, Kontron Instruments, Zurich, Switzerland) at 105,000*g* for 1 hr at 4°, using a precooled fixed angle rotor (TFT65.13, Kontron). The resulting pellets should be yellowish, opaque, and extremely viscous.

6. Drain the microsomal pellets at 4°. Resuspend the pellet in 200 μl buffer A containing freshly added protease inhibitors [AEBSF (50 μg/ml), leupeptin (4 μg/ml), pepstatin (4 μg/ml)], by manual pipetting with a 200 μl capacity Pipetman. Transfer the suspension to Eppendorf microcentrifuge tubes. If necessary use another 100–200 μl buffer A to remove residual particles from the ultracentrifuge tubes and pool.

7. Homogenize the crude suspensions by sonication on ice, as described in step 3, for 10 sec at low power (20 W). Centrifuge the suspensions in an Eppendorf microcentrifuge at 4° for 5 min at 15,000*g*. Resulting suspensions should be clear and free of aggregates, and the pellet should be barely visible. If aggregates are still visible or the pellet is too large, repeat the sonication and centrifugation steps. Decant the supernatant into a fresh microcentrifuge tube and store on ice until use. It is not recommended that protein extracts be frozen at −20°, since this will result in gradual loss of activity.[23]

8. Determine the protein concentration of the final supernatant by the method of Bradford,[41] using the Bio-Rad protein assay (Bio-Rad, Hercules, CA). Bovine serum albumin is used as the standard. Lysates from four confluent 15 cm tissue culture dishes should typically provide 1–2 mg microsomal protein.

Note: While the assay as described herein has been developed using microsomal protein extracts, crude extract (15,000*g* supernatant from step 4) may also be used as a source of protein.

[41] M. M. Bradford, *Anal. Biochem.* **72,** 248 (1976).

Assembly of Heme Oxygenase Reactions

Materials. The assembly of *in vitro* heme oxygenase reactions requires a circulating water bath maintained at 37° with cover, an Eppendorf microfuge (room temperature), and a vortex mixer. The nicotinamide adenine dinucleotide phosphate reduced form (β-NADPH), D-glucose 6-phosphate, glucose-6-phosphate dehydrogenase (Type XV from baker's yeast), bovine hemin, dimethyl sulfoxide (DMSO), ethanol and mesoporphyrin are obtained from Sigma Chemical Company (St. Louis, MO). Sn-Protoporphyrin IX is obtained from Porphyrin Products (Logan, UT).

Solutions

Glucose 6-phosphate: 40 m*M* glucose 6-phosphate in solution A (0.25 m*M* sucrose, 20 m*M* Tris, pH 7.4).

β-NADPH: 20 m*M* β-NADPH in solution A (0.25 m*M* sucrose, 20 m*M* Tris, pH 7.4).

Glucose-6-phosphate dehydrogenase: Dilute the commercial preparation to a final concentration of 1U/μl with solution A (0.25 m*M* sucrose, 20 m*M* Tris, pH 7.4). Aliquot and store frozen at −20°. Thaw only once.

Hemin stock solutions: Prepare a 10 m*M* bovine hemin solution in 100% DMSO, and store at −20°. Immediately before use, dilute the stock solution to a final concentration of 2.5 m*M* by three serial additions of one volume distilled water. No precipitate should occur with this dilution scheme.

Sn-protoporphyrin-IX stock solutions: Under dark ambient conditions, prepare a concentrated solution of SnPP-IX at 10 m*M* in 100% DMSO, aliquot, and store in foil wrapped tubes at −20°. Before use, gradually dilute the stock solution to a final concentration of 2.5 m*M* by three serial additions of one volume distilled water.

Mesoporphyrin stock solution: Under dark conditions, prepare a 200 μ*M* stock solution of mesoporphyrin in 100% DMSO, aliquot, and store in foil wrapped tubes at −20°.

Stop solution (ethanol : dimethyl sulfoxide 95 : 5 (v/v), 0.4 μ*M* mesoporphyrin): TO a 10 ml aliquot of ethanol: dimethyl sulfoxide 95 : 5 (v/v), add 20 μl of the mesoporphyrin stock solution to a final concentration of 0.4 μ*M*, immediately before use. Do not store.

Procedure

1. The heme oxygenase reaction mixtures are composed of the following final concentrations: 5 mg/ml fresh microsomal protein, 1 m*M* β-NADPH,

2 m*M* glucose 6-phosphate, 1 U glucose-6-phosphate dehydrogenase, 25 μ*M* hemin, 0.25 *M* sucrose, 20 m*M* Tris, pH 7.4, in a final volume of 100 μl. Assemble the following reagents in order in Eppendorf microcentrifuge tubes on ice: Add 500 μg fresh microsomal protein suspension for each sample tube in <88 μl volume. If the protein preparation is too dilute, increase starting cell number or reduce microsomal resuspension volume. Adjust volume to 88 μl with buffer A (0.25 *M* sucrose, 20 m*M* Tris, pH 7.4, without protease inhibitors). Add 5 μl glucose 6-phosphate solution (40 m*M* stock). Add 5 μl β-NADPH solution (20 m*M* stock). Add 1 μl glucose-6-phosphate dehydrogenase (1U/μl). Add 1 μl bovine hemin (2.5 m*M* stock in 25% DMSO) to reach a final hemin concentration of 25 μ*M* in 100 μl final reaction volume.

2. Prepare negative reaction controls by substituting the NADPH solution in the above reaction assembly with 5 μl buffer A. Alternatively, prepare negative controls by assembling above reactions in stated order with the additional inclusion of 1 μl Sn-protoporphyrin-IX solution (2.5 m*M*), to a final concentration of 25 μ*M*.

3. Initiate the reactions by pipetting the reaction mix up and down several times. Vortex 5 sec, and place tubes in a circulating water bath at 37°. Place a cover or aluminum foil on the water bath to ensure dark conditions. Incubate reactions for a chosen time interval, typically between 10 and 60 min. Kinetic experiments may be performed to empirically determine the linear range for measured activity. A 20 min reaction time is recommended from previous studies with various human cell types expressing basal or overexpressed heme oxygenase activity.[40] Reaction time may have to be increased up to 1 hr to detect activity from control (uninduced) cells that express low basal activity.

4. Under dark ambient conditions, remove tubes from the water bath and place them on an Eppendorf rack. Add 100 μl stop solution (95:5 ethanol:DMSO, 0.4 μ*M* mesoporphyrin)[26] to each tube and vortex vigorously for 5 sec. Centrifuge at 15,000*g* in an Eppendorf microfuge at room temperature. The resulting pellet will be colored since it contains approximately 85% of the excess heme substrate.

5. Transfer the supernatant to PCR microfuge tubes. Ensure that samples are free of particulate matter. Samples should be clear with a slight yellowish tint. If an autosampler is used, make certain that plastic tube lids are compatible with the HPLC autosampler system (i.e., Treff 0.5 ml microtubes, AG, Degersheim, Switzerland). Reinforced tubes may block or bend autosampler needles. Inject samples into the HPLC system as soon as possible. Samples may be transported in a dark box and briefly kept at room temperature until loading.

Preparation of HPLC System

Materials. The high-performance liquid chromatography system used in the development of this assay was from Kontron Instruments (Zurich, Switzerland), equipped with a Kontron UVIKON 720 LC microdetector, a Kontron MT 450 (MS-DOS) data acquisition system, and a Kontron HPLC 360 autosampler. Heme oxygenase reactions are separated on a Waters Novapak reversed phase C_{18} steel cartridge HPLC column (3.9 mm × 15 cm, Waters Chromatography, Milford, MA). We have found that improved separation can be obtained with a Waters Novapak phenyl reversed-phase column. A tank of compressed helium gas is also required for solution degassing.

Solutions

1 *M* ammonium acetate, pH 5.2: For 1 liter, dissolve 77 g ammonium acetate with 26 ml glacial acetic acid, and bring volume to 1 liter with Millipore filtered water. Adjust pH to 5.2 with glacial acetic acid. Filter through a 45 μm bottletop filter and store tightly closed at 4°.

HPLC buffer A: 100 m*M* ammonium acetate, pH 5.2, 60% methanol (v/v). Pour 100 ml 1 *M* ammonium acetate, pH 5.2, stock solution, and 600 ml methanol (HPLC grade) into a 1 liter cylinder. Fill to 1 liter with Millipore filtered water from the source. Degas 2 min with helium prior to use.

HPLC buffer B: 100% methanol, HPLC grade (Fluka Biochemika, Buchs, Switzerland). Degas 2 min with helium prior to use.

Note: The use of ammonium phosphate as mobile phase, as described previously,[26,32] is not recommend since it may precipitate in the HPLC system at low pH.

HPLC Parameters. Program a linear gradient from 100% HPLC buffer A (100 m*M* ammonium acetate, pH 5.2) to 100% buffer B over 14 min, followed by reversion to 100% buffer A at 14 min until the termination at 19 min. The total rate is 1.5 ml/min. The visible detector is set for continuous absorbance monitoring at 405 nm. The HPLC system should be purged free of air through a precolumn check valve. Prime the HPLC with 100% HPLC buffer A for 2 min prior to the first injection. Rinse the columns after each series with 100% methanol for at least 20 min.

Calibration of Assay and Quantification of Heme Oxygenase Activity

Tetrapyrroles extracted from the heme oxygenase reaction mixture will elute in the following order with typical retention times: biliverdin IXα

(5.29 min), hemin (9.8 min), bilirubin IXα (10.9 min), mesoporphyrine (12.54 min). If bilirubin preparations containing mixed isomers are used for standard calibration, such as the commercial preparation from Sigma (St. Louis, MO), the minor structural isomers BR-IIIα and BR-XIIIα will elute immediately before and after the BR-IXα peak, at 10.39 and 11.38 min.

Solutions

Biliverdin IXα standard: Prepare 200 μM in 100% DMSO and store in the dark at $-20°$.

Bilirubin IXα standard: prepare 200 μM bilirubin IXα in 100% DMSO. If mixed isomers are used (Sigma), then the bilirubin preparation is weighed according to its weight composition of bilirubin-IXα isomer. Prepare immediately before use under dark ambient conditions and keep at room temperature in a dark bottle.

100× dilution series: Using the standard stock solutions, prepare a dilution series for each pigment ranging in concentration from 20 to 200 μM in 100% DMSO.

All other reagents are prepared as described above.

Procedure

1. Prepare biliverdin and bilirubin standards in protein solution for best results. Bilirubin standard is unstable when prepared in HPLC buffer A alone. To Eppendorf microtubes at room temperature, add 500 μg microsomal protein in <100 μl volume. For this step, protein extracts that have been stored frozen at $-20°$ may be used. Adjust volume to 100 μl with solution A. Heat the protein to 37° for 10 min in a circulating water bath. Add 100 μl stop solution. Vortex, and centrifuge at 15,000*g* for 5 min at room temperature. Recover 148.5 μl of the supernatant. In the dark, add 1.5 μl BV or BR standard from 100× concentrated stock solutions to final concentrations ranging from 0.2 to 2 μM, assuming a final volume of 150 μl. Inject immediately into the HPLC system. The detector response should be linear with BV or BR concentration up to 10 μM.

2. Construct the standard curves by plotting the detector response (integrated peak area) as a function of biliverdin and bilirubin concentration of standards. If an integration program is not available, then detector response in peak height (arbitrary units) may also be used.

3. The assay protocol assumes that 100% of the BV and BR will be recovered from heme oxygenase reactions by extraction in 95 : 5 (v/v) ethanol : dimethyl sulfoxide, which is the case for heme oxygenase reactions containing 500 μg protein and less than 2 μM BV or BR.[40] If a different protein concentration is used, recovery efficiency of BV and BR can be

assayed as follows. To Eppendorf microtubes at room temperature, add the desired amount of microsomal protein in 98 μl volume and heat to 37° for 10 min. Add 2 μl BV or BR from 100× stock solutions to final concentrations of 0.2–2 μM, assuming a final volume of 200 μl. Add 100 μl stop solution. Vortex, and centrifuge at 15,000*g* for 5 min at room temperature. Recover the supernatant to microtubes and inject immediately into the HPLC system. Calculate recovery efficiency by dividing the detector response (arbitrary units) obtained from standards prepared according to step 3, with values obtained from step 1, for identical final concentrations of standard and protein.

4. Tabulate the integrated peak areas of biliverdin IXα, bilirubin IXα, and mesoporphyrin from the chromatograms of each assay sample. From the linear standard curves, calculate the bile pigment concentrations in an injected sample in micromolar (pmol/μl). Convert concentration values to total picomoles for each pigment assuming an original stopped reaction volume of 200 μl. Bilirubin equivalents (BReq) are then calculated as the sum of the total amount of BV and BR in picomoles:

$$BR_{eq} = (x \text{ pmol}/\mu\text{l BR})200\ \mu\text{l} + (y \text{ pmol}/\mu\text{l BV})200\ \mu\text{l}$$

Bilirubin equivalents are equal to the picomoles of substrate oxidized.

5. Normalize the BR_{eq} value against the relative mesoporphyrin recovery for each sample as follows: Multiply the BR_{eq} value by the ratio of average mesoporphyrin recovery (average mV detector signal) to sample mesoporphyrin recovery (mV detector signal). The mesoporphyrin recovery ratio can be calculated from arbitrary detector units and need not be converted to picomoles.

6. Specific activity is then reported as BR_{eq} (pmol BV + BR)/mg protein/hr, using a reaction time in the linear range of the assay. If BVR occurs in excess, then the heme oxygenase activity equation reduces to pmol BR/mg protein/hr.

Troubleshooting

1. The principal causes of failure of the assay are mechanical problems of the HPLC system. Such problems can be avoided by regular maintenance of the equipment, ensuring that all solutions and samples are free of particulate matter, and rinsing the columns with 100% methanol for at least 20 min immediately following the termination of the run. Even with proper rinsing, the repeated injection of protein solution is a cause of column aging. Reversing the column direction at the first sign of peak spreading can extend column lifetime.

2. A fraction of the excess injected hemin may be retained in the column

and elute during subsequent methanol rinsing. The binding of excess heme inside the column may also affect column lifetime. The bile pigments, however, do not elute in subsequent column rinses when injected at less than 2 μM.

3. In the absence of heme oxygenase activity (i.e., using boiled extract or NADPH blanks), a substantial heme peak should elute at approximately 9.8 min, and a single mesoporphyrin peak at 12.54 min following injection of the reaction mixture under the recommended solvent profile. If the baseline is flat and no peaks at all are visible, this indicates a general problem with the HPLC system. Ensure that the injection mechanism is functioning normally, that the detector is set at the proper wavelength, and that the precolumn filter is clean.

4. Most mammalian cell types should contain detectable basal heme oxygenase activity. If basal activity is undetectable or barely detectable, increase reaction time up to 1 hr. If heme and mesoporphyrin are visible, but biliverdin or bilirubin peaks are still undetectable, then the enzyme protein may have degraded. Make sure that all steps were performed at 4° in the presence of freshly prepared protease inhibitors. Ensure that the sonicator did not heat the sample during the cell lysis or microsomal homogenization steps. Ensure that the temperature in the circulating water bath did not exceed 37° during the reaction and that the samples were completely immersed in the water during the reaction. Lack of NADPH in the reaction will also result in no detectable activity. Ensure that the NADPH or NADPH regenerating system was freshly prepared.

5. If the bilirubin is degraded, additional small peaks may occur in the chromatogram preceding the heme peak. Ensure that dark conditions were strictly followed following the initiation of the reactions. Additional unidentified peaks may also occur if the mesoporphyrin stock solution used for preparing the stop solution is repeatedly freeze–thawed.

6. Best results are obtained when all steps of the assay are completed on the day of protein extraction. Freezing microsomal extracts may cause activity loss. Freezing the alcohol-extracted heme oxygenase reaction mixtures (final step) at −70° prior to assay has been attempted. This will result in bile pigment degradation and reduction in apparent activity yield, with additional mesoporphyrin degradation peaks appearing in the chromatogram

7. We have found that there is an exponential decay of BR-IXα signal when the samples are stored in the HPLC tubes in the dark. Thus, long serial repetitions of equivalent samples in an autosampler result in serial reduction in apparent activity. Therefore, runs of less than 20 successive sample tubes are recommended if an autosampler is used. Alternatively,

the rate of dark decay can be determined experimentally from the serial repetition of bilirubin standards, and correction factors applied.

Conclusions

Previously, the method of choice for assaying the stress inducible HO-I response has been the quantification of HO-I mRNA, which is an effective means for determining the relative level of HO-I gene expression. The principal drawback of this method, however, is that it does not directly estimate how much mRNA transcript is translated into functional enzyme protein.[42] This paper describes a microassay procedure to quantify heme oxygenase activity from human cultured cells based on the HPLC separation and detection of the bile pigments BV and BR. The assay is convenient since results can be obtained on the same day. Up to 20–30 samples may be processed simultaneously, rendering this assay efficient for large scale screening of samples for heme oxygenase activity. Since both metabolites in the sequence of heme degradation are quantifiable, the assay does not require supplementation with biliverdin reductase, which is not commercially available and requires extensive preparatory work.[6,24] HPLC based heme oxygenase assays are more sensitive than the previously described spectrophotometric assay, since the possibility of spectral interference from complex reaction components during the detection of the pigments is reduced.[26] Radiochemical assays, which are also more sensitive than spectrophotometric methods,[27,28] depend on inefficient chloroform extraction of bilirubin, and may take several days for autoradiographic detection of bilirubin.[28]

The heme oxygenase assay can be applied to the study of pharmacological manipulation of heme oxygenase activity in any cell type, by incorporating any test agent in the cell culture treatment protocol. Recent widespread interest in the role of heme oxygenase in CO dependent signal transduction pathways calls for a simple and accurate heme oxygenase activity method. The assay as described may be readily applicable to quantifying basal heme oxygenase activity in cells of neural and cardiovascular origin. Furthermore, the assay can be applied to the study of agents which modulate heme oxygenase activity by directly reacting with the enzyme–substrate complex. Such compounds include metal chelators such as desferrioxamine (which accelerates the dissociation of the biliverdin–iron complex[43]) and heme ligands such as CO and NO, which interfere with the oxygenase active site.[16]

[42] R. M. Tyrrell and S. Basu-Modak, *Methods Enzymol.* **234,** 224 (1994).
[43] T. Yoshida and G. Kikuchi, *J. Biol. Chem.* **253,** 4230 (1978).

Recently the method as described herein has been applied as a screening tool to construct and verify functional heme oxygenase activity in human HO-II overexpressing clones.[40]

Acknowledgments

This work was supported by core grants from the International Association for Cancer Researh (UK), and the Department of Health (UK), contract no. 121/6378, with partial additional support from the League Against Cancer of Central Switzerland, the Neuchateloise League Against Cancer, and the Swiss National Science Foundation. S. R. and E. K. received postdoctoral fellowships from the Emma Muschamp Foundation, and the European Molecular Biology Organization (EMBO), respectively.

Section III

Oxidant and Redox-Sensitive Steps in Signal Transduction and Gene Expression

Articles 34 through 40

[34] Assay for Redox-Sensitive Kinases

By ASHOK KUMAR and BHARAT B. AGGARWAL

Introduction

The physiological and pharmacological modification of cellular redox status drastically changes the expression of a number of genes, probably by modulating the activity of some of the transcription factors such as AP-1 and NF-κB (nuclear factor κB).[1] Considerable evidence has now emerged suggesting that low levels of reactive oxygen intermediates (ROI), which a cell may be able to tolerate without major damage, may mediate the change in gene expression.[1] Studies have indicated that the activities of several protein serine/threonine and tyrosine kinases involved in cell signaling change in response to oxidative stress. Among these are mitogen-activated protein kinases (MAPK),[2] p38 mitogen-activated protein kinase,[3] c-Jun N-terminal kinase (JNK),[4,5] big mitogen-activated protein kinase (BMK),[6] IκBα kinase,[7,8] and protein tyrosine kinases of the Src family such as $p56^{lck}$ and syk/zap-70.[9–11]

In the present chapter we describe the procedures that are commonly used to assay the activity of these kinases. For this we describe the assay of the well-known redox-sensitive kinase JNK. We also show that the activity of JNK is increased on treatment of the cells with H_2O_2 and an inflammatory cytokine, tumor necrosis factor (TNF), in a time- and dose-dependent manner. Similar protocols are used to assay the activity of other kinases.

[1] J. M. Muller, R. A. Rupec, and P. A. Baeuerle, *Methods* **11,** 301 (1997).
[2] L. L. Dugan, D. J. Creedon, E. M. Johnson, Jr., and D. M. Holtzman, *Proc. Natl. Acad. Sci. USA* **94,** 4086 (1997).
[3] J. Huot, F. Houle, F. Marceau, and J. Landry, *Circ. Res.* **80,** 383 (1997).
[4] Y. Y. C. Lo, J. M. S. Wong, and T. F. Cruz, *J. Biol. Chem.* **271,** 15703 (1996).
[5] X.-L. Cui and J. G. Douglas, *Proc. Natl. Acad. Sci. USA* **94,** 3771 (1997).
[6] J.-I. Abe, M. Kusuhara, R. J. Ulevitch, B. C. Berk, and J.-D. Lee, *J. Biol. Chem.* **271,** 16586 (1996).
[7] J. A. DiDonato, M. Hayakawa, D. M. Rothwarf, F. E. Zandi, and M. Karin, *Nature* **388,** 548 (1997).
[8] C. H. Regnier, H. Y. Song, X. Gao, D. V. Goeddel, Z. Cao, and M. Rothe, *Cell* **90,** 373 (1997).
[9] S. Qin, Y. Minami, M. Hibi, T. Kurosaki, and H. Yamamura, *J. Biol. Chem.* **272,** 2098 (1997).
[10] S. R. Yan and G. Berton, *J. Biol. Chem.* **271,** 23464 (1996).
[11] K. Nakamura, T. Hori, and J. Yodoi, *Mol. Immunol.* **33,** 855 (1996).

0076-6879/99 $30.00

Materials

Penicillin, streptomycin, RPMI 1640 medium, and fetal bovine serum (FBS) are obtained from GIBCO (Grand Island, NY). ^{32}P-Labeled γ-ATP with a specific activity of 7000 Ci/mmol is obtained from ICN Pharmaceutical, Inc. (Costa Mesa, CA). Bacteria-derived recombinant human TNF, purified to homogeneity with a specific activity of 5×10^7 units/mg, is kindly provided by Genentech, Inc. (South San Francisco, CA). H_2O_2 is obtained from Sigma Chemical Co. (St. Louis, MO). Anti-JNK1 rabbit polyclonal antibody is obtained from Santa Cruz Biotechnology, Inc. (Santa Cruz, CA). Bovine serum albumin (BSA) is purchased from United States Biochemical Corp. (Cleveland, OH). Protein A/G Sepharose beads are obtained from Pierce (Rockford, IL). BL21 cells transformed with a plasmid encoding the glutathione S-transferase (GST)–c-Jun (residue 1–79) protein are kindly provided by Dr. Bing Su of the Department of Immunology, University of Texas M. D. Anderson Cancer Center (Houston, Texas). U937 cells (a human histiocytic lymphoma) are obtained from the American Type Culture Collection (Rockville, MD). The cells are periodically checked for mycoplasma contamination.

Purification of GST-c-Jun Protein

Reagents

Lysis buffer: 20 m*M* Tris, pH 8.0, 200 m*M* NaCl, 10% glycerol, 0.5% Nonidet P-40 (NP-40), 1 m*M* phenylmethylsulfonyl fluoride (PMSF), 2 μg/ml aprotinin, 2 μg/ml leupeptin, 1 m*M* dithiothreitol (DTT), and 1 m*M* EDTA.

Lysozyme: 10 mg/ml in water (store at −20°).

Glutathione agarose: 50% (v/v) in phosphate-buffered saline (PBS) (store at 4°).

Wash buffer: 50 m*M* Tris (pH 8.0), 200 m*M* NaCl, 0.1% 2-mercaptoethanol.

Elution buffer: 50 m*M* Tris, pH 8.0, 200 m*M* NaCl, 0.1% 2-mercaptoethanol, 20 m*M* glutathione.

Storage buffer: 20 m*M* Tris, pH 8.0, 150 m*M* NaCl, 20% glycerol, 1 m*M* DTT, 1 m*M* EDTA.

Ampicillin: 100 mg/ml.

Procedure

Fifty ml of LB medium containing 100 μg/ml of ampicillin (LBamp) is inoculated with either glycerol stock (50 μl) or a single colony of BL21 cells expressing GST–c-Jun (1-79). The cultures are allowed to grow overnight at 37°. The overnight grown cultures are then added into 500 ml of LBamp and again allowed to grow for another 2 hr (at this stage OD_{600} should be approximately 0.8–1.0). The temperature of these cultures is reduced to 20°, and isopropyl-β-D-thiopyranoside (IPTG) is added at a final concentration of 0.5 m*M*. The bacteria are allowed to grow for a further 2 hr before centrifugation for 10 min at 4000*g*. The bacterial pellet is resuspended in lysis buffer (10–15 ml per 500 ml of culture volume) and placed in a high-speed centrifugation tube. Lysozyme (5 mg per liter of cell culture) is added to these tubes, which are kept on ice for another 30 min. The lysates are sonicated for 2 min at 4°. The suspension is then centrifuged for 45 min at 40,000 rpm at 4°. The supernatants are collected in 50-ml conical tubes and glutathione agarose added (1.2 ml per liter of cell culture). The suspension is rotated at 4° for 30 min and centrifuged for 5 min at 1500*g*. The supernatant is discarded, and the pellet is washed two times with lysis buffer containing 1 *M* NaCl. The pellet is transferred to a 1.5-ml microfuge tube and again washed three times with lysis buffer containing 1 *M* NaCl. It is incubated at 37° for 20 min with 10 m*M* $MgCl_2$ and 0.2 m*M* ATP, washed two times with lysis buffer containing 1 *M* NaCl, and washed four times with lysis buffer without NaCl. The pellet is split in Eppendorf tubes (100 μl of agarose beads per tube) and washed three times with washing buffer. The GST–c-Jun is eluted three times at room temperature by adding 700 μl of elution buffer to each tube, rotating 10 min between each elution, and the eluates are collected and dialyzed overnight at 4° against storage buffer. The protein is concentrated by ultrafiltration using Ultrafree-15 centrifugal filter device (Millipore, Bedford, MA). The amount of GST–c-Jun is measured by standard Bradford method, and purity determined by SDS–PAGE.

Cell Culture and Preparation of Cell Extracts. For the various assays, U937 cells are grown in RPMI 1640 medium supplemented with glutamine (2 m*M*), gentamicin (50 μg/ml), and 10% fetal calf serum (FCS). The cells are seeded at a density of 1×10^5 cells/ml in T75 flasks (Falcon 3013; Becton Dickinson Labware, Lincoln Park, NJ) containing 30 ml of medium and grown at 37° in an atmosphere of 95% air and 5% CO_2. Cell cultures are split every 3–4 days.

After appropriate treatments, cells (2×10^6) are washed two times with chilled phosphate-buffered saline (PBS), pH 7.4, containing 1 m*M* sodium orthovanadate. The cells are then suspended in whole cell-extraction buffer A (20 m*M* Tris-HCl, pH 7.4, 2 m*M* EDTA, 250 m*M* NaCl, 0.1% NP-40, 1 m*M* Na_3PO_4, 2 μg/ml leupeptin, 2 μg/ml aprotinin, 1 m*M* PMSF, 0.5 μg/ml benzamidine, and 1 m*M* DTT). The cell suspension is mixed on a rotary mixer at 4° for 30 min and the extracts cleared by centrifugation at

13,000*g* for 10 min. The protein content is measured by using Bio-Rad (Richmond, CA) reagent.

Immune-Complex Kinase Assay

The assay is performed as follows:

1. Three hundred μg of cytoplasmic extracts (approximately 100 μl of cell extract if the initial number of cells used for treatments is 2×10^6) is placed in a 1.5-ml microfuge tube.
2. The total volume of the extracts is made to 700 μl using whole cell-extraction buffer A and immunoprecipitated by adding 0.3 μg anti-JNK antibody per 300 μg of cytoplasmic protein.
3. The tubes are put on a rotating shaker at 4° for 60 min followed by another 30 min with 25 μl of 50% (w/v) stock of protein A/G Sepharose beads in PBS.
4. The immune complex is separated by centrifugation at 13,000 rpm for 20 sec in a refrigerated centrifuge.
5. The beads are washed four times with buffer A and two times with kinase wash buffer containing 20 m*M* HEPES, pH 7.4, 1 m*M* DTT, 25 m*M* NaCl.
6. The kinase assay is performed for 15 min at 30° with 2 μg GST–c-Jun(1–79) as a substrate in 15 μl containing 20 m*M* HEPES, pH 7.4, 10 m*M* $MgCl_2$, 1 m*M* DTT, and 10 μCi [γ-^{32}P]ATP.
7. The reaction is stopped by adding 15 μl sodium dodecyl sulfate (SDS) sample buffer, boiled for 5 min, and subjected to 9% SDS–polyacrylamide gel electrophoresis (SDS–PAGE). The gel is allowed to run until the dye front reaches the bottom of the gel.
8. The gel is fixed, stained with Coomassie blue, destained, and dried. The gel is stained to verify the equal loading of the protein.
9. The gel is analyzed either by autoradiography or by a PhosphorImager (Molecular Dynamics, Sunnyvale, CA) and quantitated by ImageQuant Software (Molecular Dynamics, Sunnyvale, CA).

Results

Typical results obtained from the immune-complex kinase assay are shown in Fig. 1. In these experiment we examined the effect of H_2O_2 (Fig. 1A) and TNF (Fig. 1B) on the activation of JNK in U937 cells. Briefly, cells (2×10^6) were treated with different concentration of H_2O_2 for 45 min and also for different time periods with 300 μM H_2O_2. Cells were also treated with different concentrations of TNF for 10 min and for different

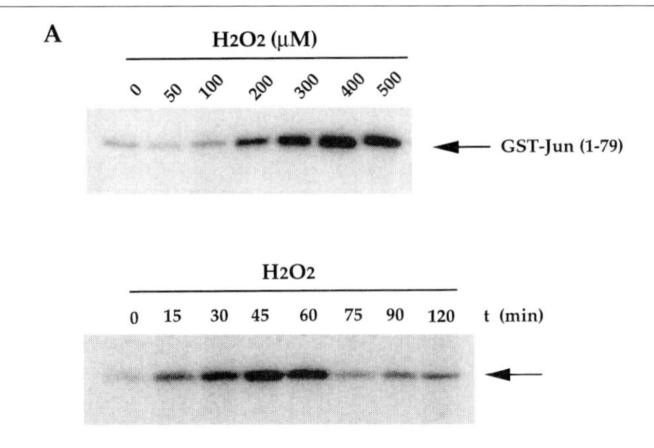

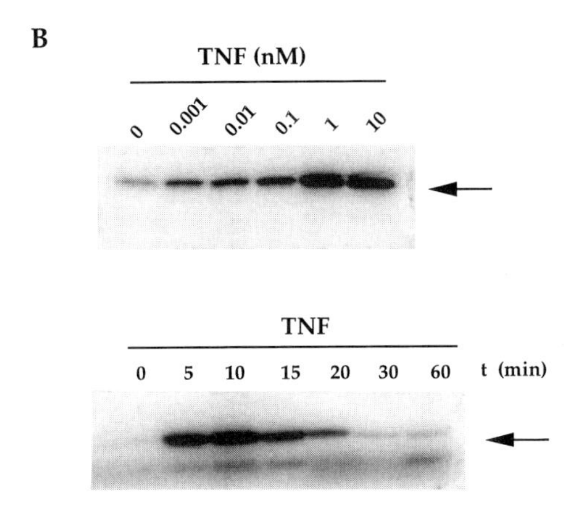

FIG. 1. Effect of H_2O_2 and TNF on the activation of JNK in U937 cells assayed by immunoprecipitation method. (A) U937 cells (2×10^6) were treated either with different concentration of H_2O_2 for 45 min (*upper*) or for different time periods with 300 μM H_2O_2 (*lower*). (B) U937 cells were treated with different concentrations of TNF for 10 min (*upper*) or for different time periods with 1 n*M* TNF (*lower*). The cells were lysed in whole cell extraction buffer, and the JNK activity was assayed using the *in vitro* immunoprecipitation method as described in the text.

time periods with 1 n*M* TNF. Thereafter, the cell extracts were prepared and assayed for JNK activity as described above. H_2O_2 dose–response indicates that maximum activation of JNK occurs at 400 μM (upper panel of Fig. 1A). The time course shows that 45–60 min is sufficient to reach the peak of activation (lower panel of Fig. 1A). The JNK activity quickly disappears at 75 min and then there is small increase again at 90 and 120 min, indicating a biphasic response. Like H_2O_2, TNF also activates JNK in a dose- and time-dependent manner with maximum activation occuring at 1 n*M* (upper panel of Fig. 1B) and within 10 min of treatment (lower panel of Fig. 1B). Unlike H_2O_2, however, biphasic activation of JNK is not seen with TNF.

Alternative Assays for JNK

In-Gel Kinase Assay

Another method commonly used to assay the JNKs is an in-gel kinase assay. This method has the advantage that it can be used to look for novel kinases. In the in-gel assay, the JNK is denatured by dissolving cell extracts in SDS and electrophoresed on an acrylamide gel polymerized in the presence of GST–c-Jun protein. In assaying the JNK activity by this method, the following general steps are involved.

1. The whole cell extracts are mixed with 10 ml of GSH-agarose suspension (Sigma) to which 10 μl of either GST or GST–c-Jun are bound. The mixture is rotated at 4° for 3 hr in a microfuge tube and pelleted by centrifugation at 14,000 rpm for 30 sec.
2. The proteins are eluted in Laemmli sample buffer and are resolved on a 9% SDS–polyacrylamide gel polymerized in the presence of 40 μg/ml of GST–c-Jun or GST.
3. The gel is washed two times for 30 min each with 20% 2-propanol and 50 m*M* Tris-HCl (pH 7.9).
4. The gel is again washed twice with 50 m*M* Tris-HCl (pH 7.9) containing 5 m*M* 2-mercaptoethanol, for 30 min each.
5. The proteins are denatured by incubating the gel twice for 15 min in 100 ml of 6 *M* urea in 50 m*M* Tris-HCl (pH 7.9) at room temperature and then renatured in several 500 ml volumes of 50 m*M* Tris-HCl (pH 8.0), 5 m*M* 2-mercaptoethanol, 0.04% (v/v) Tween 40, and 0.1% (v/v) Nonidet P-40 (NP-40) for 20 hr at 4°.
6. The gel is incubated in 500 ml kinase buffer [20 m*M* HEPES (pH 7.4), 10 m*M* $MgCl_2$, 1 m*M* DTT] for 1 hr.
7. The kinase activity is assayed by incubating the gel in 25 ml of kinase buffer containing 50 μM ATP and 5 μCi/ml of [γ-^{32}P]ATP at 30° for 1 hr.
8. Finally, the gel is washed extensively in several volumes of 5% (w/v) trichloroacetic acid, 1% (w/v) sodium pyrophosphate for 20 hr.

9. The gel is dried under vacuum at 75°, analyzed by a PhosphorImager (Molecular Dynamics, Sunnyvale, CA), and quantitated by ImageQuant Software (Molecular Dynamics, Sunnyvale, CA) for JNK activity.

Decreased Electrophoretic Mobility Assay

The phosphorylation of JNK by MEKK4 is associated with a decrease in its electrophoretic mobility on SDS–PAGE, and this can be detected in cell extracts by immunoblotting with anti-JNK antibodies. This assay is, however, only semiquantative and not suitable for detecting low levels of JNK activation. Separation of phosphorylated and dephosphorylated forms of JNK needs a 20-cm-long gel containing final concentrations of 9% (w/v) acrylamide and 0.166% (w/v) bisacrylamide. There is also a potential danger that the electrophoretic mobility may be decreased by phosphorylation at sites other than those labeled by MEKK4, rendering the assay invalid. This is of particular concern when the activation of JNK is studied using a stimulus whose effect on JNK has never been analyzed in molecular detail.

Acknowledgment

This research was supported by a grant from The Clayton Foundation for Research.

[35] Inhibition of NF-κB Activation *in Vitro* and *in Vivo*: Role of 26S Proteasome

By MATTHEW B. GRISHAM, VITO J. PALOMBELLA, PETER J. ELLIOTT, ELAINE M. CONNER, STEPHEN BRAND, HENRY L. WONG, CHRISTINE PIEN, LAUREEN M. MAZZOLA, ANTONIA DESTREE, LANA PARENT, and JULIAN ADAMS

Introduction

The pain, swelling, and tissue destruction that accompanies inflammatory disease result from a cascade of events that is initiated and propagated by the production of cytokines and chemokines, and the cell surface expression of cell adhesion molecules.[1] NF-κB is a key transcription factor in inflammation which is required for the expression of many proinflammatory

[1] P. J. Barnes and M. Karin, *New. Eng. J. Med.* **336,** 1066–1071 (1997).

0076-6879/99 $30.00

mediators. NF-κB is a member of the Rel family of proteins and is typically a heterodimer composed of a p50 and p65 (RelA) subunit.[2] In quiescent cells, NF-κB resides in the cytosol in latent form bound to an inhibitory protein κB, which masks the nuclear localization sequences of p50/65.[2] Stimulation of these cells with various cytokines, LPS, viruses, or oxidants triggers a series of signaling events that ultimately leads to the phosphorylation and proteolytic degradation of κB, and activation of the latent NF-κB.[2,3] It is thought these stimuli activate multiple signaling pathways that ultimately converge to enhance reactive oxygen metabolism within the cell. This intracellular oxidative stress has been proposed to activate, via several intermediate reactions, one or more redox-sensitive kinases which specifically phosphorylate IκB, resulting in IκB polyubiquitination and subsequent degradation. Once activated, the liberated NF-κB translocates into the nucleus and stimulates transcription by binding to cognate κB sites in the promoter regions of various target genes such as cytokines, chemokines, and cell adhesion molecules. These proteins are involved in the immune and inflammatory response by controlling leukocyte trafficking and activation, and many of them act in an autocrine or paracrine fashion to stimulate cells to further activate NF-κB.[1,2,4,5] If the activation of NF-κB is inhibited, the production of this array of proinflammatory molecules should be suppressed.[2,6–9] Thus, inhibitors of NF-κB activation should be antiinflammatory.

Studies in several laboratories have now established that the regulated proteolysis of IκB is mediated by the ubiquitin–proteasome pathway of protein degradation.[6,10–12] This is the principal pathway for intracellular protein turnover.[13,14] In addition, numerous examples of regulatory proteins

[2] A. S. Baldwin, *Annu. Rev. Immunol.* **14,** 649–681 (1996).

[3] D. Thanos and T. Maniatis, *Cell* **80,** 1–20 (1995).

[4] M. Grilli, J. J.-S. Chiu, and M. J. Lenardo, *Int. Rev. Cytol.* **143,** 1–62 (1993).

[5] P. A. Baeuerle and D. Baltimore, *Cell* **87,** 13–20 (1996).

[6] V. J. Palombella, O. J. Rando, A. L. Goldberg, and T. Maniatis, *Cell* **78,** 773–786 (1994).

[7] N. Auphan, J. DiDonato, C. Rossette, A. Helmbertg, and M. Karin, *Science* **270,** 286–290 (1995).

[8] E. Kopp and S. Ghosh, *Science* **265,** 956–959 (1994).

[9] R. Scheinman, P. Cogswell, A. Lofquist, and A. Baldwin, *Science* **270,** 283–286 (1995).

[10] Z. Chen, J. Hagler, V. J. Palombella, F. Melandri, D. Scherer, D. Ballard, and T. Maniatis, *Genes & Develop.* **9,** 1586–1597 (1995).

[11] D. C. Scherer, J. A. Brockman, Z. Chen, T. Maniatis, and D. W. Ballard, *Proc. Natl. Acad. Sci. USA* **92,** 11259–11263 (1995).

[12] J. DiDonato, F. Mercurio, C. Rosette, J. Wuli, H. Suyang, S. Ghosh, and M. Karin, *Mol. Cell. Biol.* **16,** 1295–1304 (1996).

[13] M. Hochstrasser, *Curr. Opin. Biol.* **7,** 215–223 (1995).

[14] A. Ciechanover, *Cell* **79,** 13–21 (1994).

which control cell cycle progression, transcriptional activity, and class I antigen presentation have also been found to undergo ubiquitin-dependent proteolysis. Some of these proteins include p53,[15–17] cyclins,[18] c-Fos,[19] and c-Jun.[20]

Protein substrates that enter the ubiquitin–proteasome pathway are first "marked" for degradation by their covalent ligation to a polyubiquitin chain. Ubiquitin (Ub) is a heat-stable polypeptide composed of 76 residues (molecular mass 8.6 kDa) that is evolutionarily conserved and is found in all eukaryotic cells.[13,14] In a series of enzyme-catalyzed reactions, the C-terminal Gly of Ub is ligated onto the ε-amine group of Lys residues of the targeted protein to form a covalent, isopeptide bond (reviewed in refs. 13, 14). Ubiquitin is activated by Ub activating enzyme, also referred to as E1, in the first step of the series of enzyme-catalyzed reactions. In an ATP-dependent reaction, the carboxylate of the C-terminal Gly of Ub forms a thioester with the active site thiol of E1. In the second step, the Ub is transferred from E1 to the active site thiol of a Ub carrier protein, or E2, to form yet another thioester bond. Finally, the Ub is transferred from E2 to the protein substrate. This reaction is often catalyzed by a Ub protein ligase, or E3 (reviewed in refs. 13, 14). Lysine-48 of this Ub molecule then donates its ε-amine for isopeptide bond formation with another molecule of Ub. In this way, subsequent rounds of ubiquitination ultimately build a long poly-Ub chain on the substrate.[21] Ubiquitinated proteins can either be deubiquitinated through the action of Ub C-terminal hydrolases or degraded to peptides through the action of the 26S proteasome.[13,14] Both of these reactions regenerate intact Ub.

The proteasome exists in multiple forms within eukaryotic cells, and at the heart of all these forms is the catalytic core known as the 20S proteasome. The structure of this large particle (~700 kDa) is best described as a stack of four oligomeric rings, each composed of seven subunits (Fig. 1; reviewed in refs. 22, 23). Substrates enter the central channel of the proteasome and undergo hydrolysis at active sites that are contained in the β

[15] M. Scheffner, J. M. Huibregste, R. D. Vierstra, and P. M. Howley, *Cell* **75,** 495–505 (1993).
[16] M. Rolfe, P. Beer-Romero, S. Glass, J. Eckstein, I. Berdo, A. Theodoras, M. Pagano, and G. Draetta, *Proc. Natl. Acad. Sci. USA* **92,** 3264–3268 (1995).
[17] D. Shkedy, H. Gonen, B. Bercovich, and A. Ciechanover, *FEBS Lett.* **348,** 126–130 (1994).
[18] A. Murray, *Cell* **81,** 149–152 (1995).
[19] I. Stancovski, H. Gonen, A. Orian, A. L. Schwartz, and A. Ciechanover, *Mol. Cell. Biol.* **15,** 7106–7116 (1995).
[20] M. Treler, L. M. Staszewski, and D. Bohmann, *Cell* **78,** 787–798 (1994).
[21] V. Chau, J. W. Tobias, A. Bachmair, D. Marriott, D. J. Ecker, D. K. Gonda, and A. Varshavsky, *Science* **243,** 1576–1583 (1989).
[22] J. Adams and R. Stein, *Ann. Reports Med. Chem.* **31,** 279–288 (1996).
[23] A. L. Goldberg, R. L. Stein, and J. Adams, *Chemistry & Biology* **2,** 503–508 (1995).

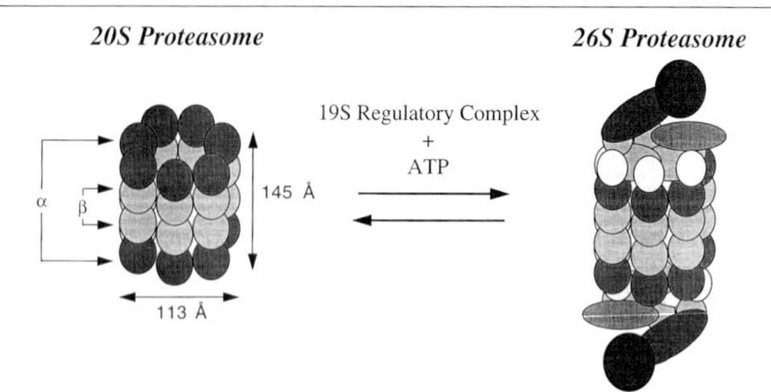

FIG. 1. Structures of the 20S and 26S proteasome. The proteasome exists in multiple forms and at the heart of all these forms is the catalytic core known as the 20S proteasome. The structure of this large particle (~700 kDa) is described as a stack of four oligomeric rings, each composed of seven subunits. Substrates enter the central channel of the proteasome and undergo hydrolysis at active sites that are contained in the β subunits. The 26S proteasome forms in an ATP-dependent manner from the combination of one copy of the 20S proteasome with two copies of the 19S regulatory complex, and this is the species that is responsible for degrading ubiquitinated proteins.

subunits. The natural substrates of proteasome are polyubiquitinated proteins, and the form of the proteasome that degrades these proteins is the 26S proteasome. This species forms in an ATP-dependent manner from the combination of one copy of the 20S proteasome with two copies of the 19S regulatory complex.[24] The identity of the catalytic nucleophile as the N-terminal threonine was determined by a combination of X-ray crystallographic and mutagenesis studies.[25,26]

Degradation of certain proteins is subject to a further level of regulation by phosphorylation, which targets the substrate protein for recognition by specific ubiquitin conjugating enzymes. For example, in the case of IκB-α, a specific kinase first phosphorylates Ser-32 and Ser-36 in response to various stimuli, such as cytokines and LPS.[10,12,27] Phosphorylation targets IκB-α for ubiquitination, and the ubiquitinated IκB-α is then selectively degraded by the 26S proteasome.[10] Taken together, it is clear that inhibitors of enzymes of the ubiquitin–proteasome pathway should suppress the activation of NF-κB by stabilizing the inhibitor IκB, thereby reducing levels of an array of proinflammatory proteins and, ultimately, providing therapeutic antiinflammatory effects.

[24] M. Chu-Ping, J. H. Vu, R. J. Proske, C. A. Slaughter, and G. N. DeMartino, *J. Biol. Chem.* **269,** 3539–3547 (1994).

[25] J. Lowe, D. Stock, B. Jap, P. Swickl, W. Baumeister, and R. Huber, *Science* **268,** 533–539 (1995).

[26] E. Seemuller, A. Lupas, D. Stock, J. Lowe, R. Huber, and W. Baumeister, *Science* **368,** 579–581 (1995).

[27] T. Maniatis, *Science* **278,** 818–819 (1997).

TABLE I

COMPARISON OF PEPTIDE ALDEHYDE AND BORONATE INHIBITORS OF PROTEASOME

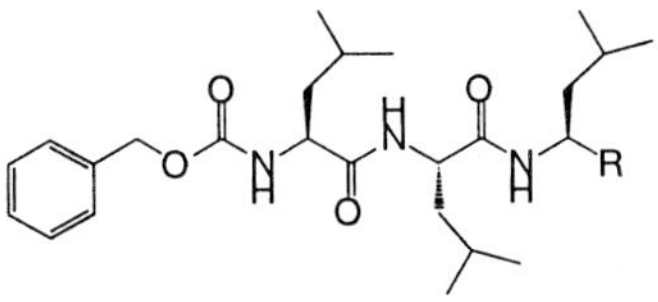

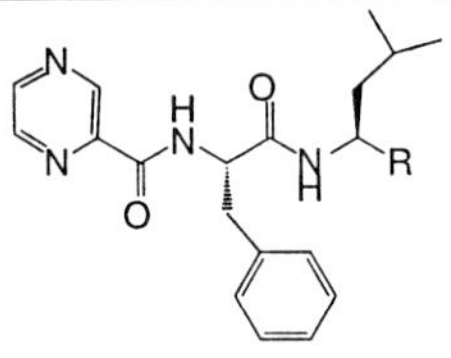

R	Compound	K_i (nM)	Compound	K_i (nM)
CHO	MG-132	4	PS-402	1600
$B(OH)_2$	PS-262	0.03	PS-341	0.6

Proteasome Inhibitors

Peptide Aldehydes

The best-characterized inhibitors of the proteasome are peptide aldehydes. Studies in a number of laboratories established the requirement for hydrophobic residues at P_2 and P_3 of the inhibitor.[28–30] A systematic study of P_1 in the series Z-Leu-Leu-Xaa-H[29] further established the requirement for a hydrophobic residue at this position with Leu or Phe giving the most potent inhibition. In a more recent mechanistic study, these compounds were found to have identical K_i values of 4 nM.[31] The peptide aldehyde inhibitor Z-Leu-Leu-Leu-H (MG 132) has been studied extensively by many investigators (Table I).

Lactacystin

Lactacystin is a natural product that inhibits cell cycle progression.[32,33] Mechanistic studies[34] have shown that lactacystin, and its synthetic precur-

[28] M. Orlowski, C. Cardozo, and C. Michaud, *Biochemistry* **32,** 1563–1572 (1993).

[29] S. Tsubi, H. Kawasaki, Y. Saito, N. Miyashita, M. Inomata, and S. Kawashima, *Biochem. Biophys. Res. Commun.* **196,** 1195–1201 (1993).

[30] A. Vinitsky, C. Michaud, J. C. Powers, and M. Orlowski, *Biochemistry* **31,** 9421–9428 (1992).

[31] L. Dick, A. Cruikshank, L. Grenier, F. Melandri, L. Plamondon, and R. Stein, unpublished observations, 1997.

[32] S. Omura, T. Fujimoto, K. Otoguro, K. Matsuzaki, R. Moriguchi, H. Tanaka, and Y. Sasaki, *J. Antibiot.* **44,** 113–116 (1991).

[33] S. Omura, K. Matsuzaki, T. Fujimoto, K. Kosuge, T. Furuya, S. Fujita, and A. Nakagawa, *J. Antibiot.* **44,** 117–118 (1991).

[34] G. Fenteany, R. F. Standaert, W. S. Lane, S. Choi, E. J. Corey, and S. L. Schreiber, *Science* **268,** 726–730 (1995).

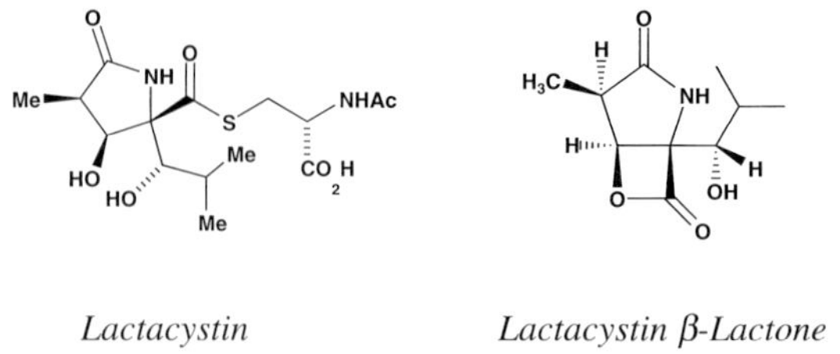

Lactacystin *Lactacystin β-Lactone*

FIG. 2. Structures of lactacystin and lactacystin β-lactone.

sor β-lactone, is a selective proteasome inhibitor (Fig. 2). It selectively and irreversibly binds to subunit X of the proteasome and the active site N-terminal threonine is acylated.

Boronic Acids

Boronic acid peptides have been used to inhibit serine proteases[35,36] and have been examined as inhibitors of the proteasome.[37] Boronate inhibitors of the proteasome are typically much more potent than their structurally analogous aldehydes. A direct comparison is shown in Table I for these two classes of proteasome inhibitors. Studies have led to potent dipeptide boronate proteasome inhibitors, and as a class, these compounds are active in cells and are active in animal models of disease.[37] The dipeptide boronates also demonstrate a high degree of enzyme selectivity and are inactive against many common proteases.[37]

Use of Proteasome Inhibitors *in Vitro*

Proteasome inhibitors of diverse structural class can penetrate cells and, in a dose-dependent manner, inhibit activation of NF-κB. Below we describe methods for blocking NF-κB activation by inhibiting proteasome-mediated degradation of IκBα in tissue culture cells. In addition, we have demonstrated that this inhibition blocks activation of NF-κB dependent gene expression, namely, cell adhesion molecule expression on human umbilical vein endothelial (HUVE) cells and interleukin-2 (IL-2) produc-

[35] A. S. Shenvi, *Biochemistry* **25,** 1286–1291 (1986).

[36] D. H. Kinder, C. A. Elstad, G. G. Meadows, and M. M. Ames, *Inv. Metastasis* **12,** 309–319 (1992).

[37] R. Stein, J. Adams, M. Behnke, S. Brand, Z. Chen, S. Chen, A. Cruickshank, E. Conner, A. Destree, L. Dick, L. Grenier, M. Grisham, J. Klunder, L. M. Mazzola, Y-T. Ma, L. F. Melandri, S. Nunes, L. Plamondon, L. Parent, S. Tagereud, K. Vaddi, and V. J. Palombella, manuscript submitted (1998).

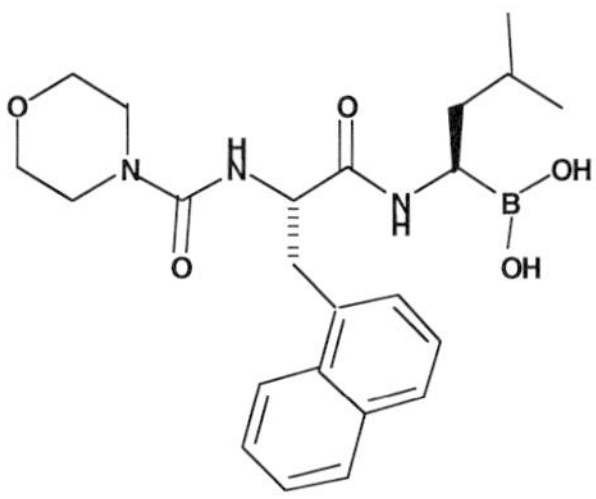

FIG. 3. Structure of PS-273.

tion in Jurkat T cells. The proteasome inhibitor used in these studies is exemplified by the dipeptide boronic acid PS-273 (Fig. 3; $K_i = 0.15$ nM).

NF-κB Activation

NF-κB is sequestered in the cytoplasm of cells bound to IκB. This form of NF-κB is unable to bind to DNA. However, IκB is rapidly degraded by the proteasome in response to various inducers such as TNF-α and lipopolysaccharide (LPS).[2] This leads to the release of NF-κB, which translocates to the nucleus, where it now binds to DNA. This is referred to as "NF-κB activation." The assay described here measures the amount of free NF-κB that can bind to DNA using an electrophoretic mobility shift assay (EMSA). Since the proteasome is required for the rapid degradation of IκB, this assay is used to measure the inhibition of NF-κB activation in cells treated with proteasome inhibitors and activators of NF-κB.

PS-273 can inhibit TNF-α-induced NF-κB activation in HeLa cells (Fig. 4). HeLa cells are pretreated for 1 hr with PS-273 (5 μM), MG-132 (5 μM), and as a control MG-102 (50 μM). MG-102 or calpain inhibitor II (Ac-Leu-Leu-Met-H) does not inhibit the proteasome. The cells are then treated for 0, 5, 15, and 30 min with TNF-α (10 ng/ml). Whole-cell extracts are

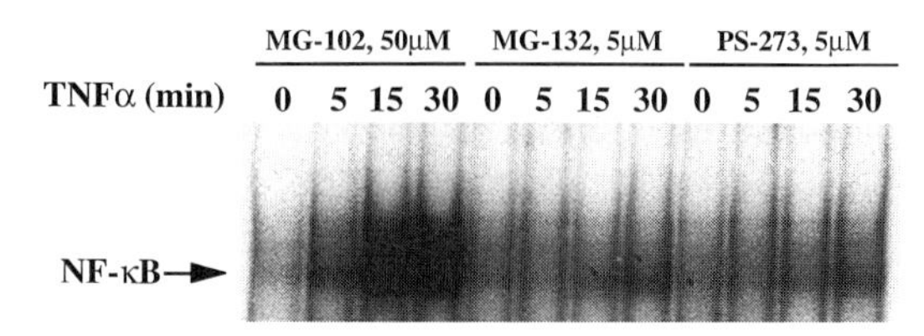

FIG. 4. PS-273 and MG-132 inhibit the activation of NF-κB in TNF-α treated HeLa cells. HeLa cells were pretreated for 1 hr with PS-273 (5 μM), MG-132 (5 μM), and as a control MG-102 (50 μM). The cells were then treated for 0, 5, 15, and 30 min with TNF-α (10 ng/ml) as indicated. Whole-cell extracts were prepared and NF-κB activation was monitored by EMSA as described in the text.

prepared and NF-κB activation is monitored by gel shift assay. Both PS-273 and MG-132 inhibited NF-κB activation at all times of TNF-α treatment. As expected, MG-102 did not inhibit NF-κB activation (Fig. 4). Similar results have been seen previously for MG-132 and MG-102.[6]

Protocol. This assay is a modification of the procedure by Palombella et al.[6] HeLa cells (ATCC, Rockville, MD) are seeded into 6-well plates and cultured overnight at 37° in 5% (v/v) CO_2 in Dulbecco's modified Eagle's medium (DMEM) (GIBCO-BRL Gaithersburg, MD) supplemented with 10% FBS (Sigma, St. Louis, MO). The next day the confluent cell monolayers are refed with fresh medium containing proteasome inhibitors [i.e., PS-273, 0–50 μM diluted in dimethyl sulfoxide (DMSO)]. The final concentration of DMSO in the culture medium is 0.1%. The cells are preincubated with the inhibitors for 1 hr. Next, the cells are treated with TNF-α (1000 IU/ml; 10 ng/ml) (Boehringer) for different periods of time (0–30 min). The cells are then washed 1× with ice-cold PBS and are scraped in 0.5 ml PBS. All manipulations are done on ice or at 4°. The cells are quickly centrifuged and the supernatants are removed. The cell pellets are resuspended in 20 μl of lysis buffer to make whole-cell extracts. Lysis buffer is 10% (v/v) glycerol, 0.4 *M* KCl, 20 m*M* Tris, pH 7.5, 20 μg/ml leupeptin (Sigma), 20 μg/ml pepstatin A (Sigma), 1 m*M* phenylmethylsulfonyl fluoride (PMSF) (Boehringer), 2 m*M* dithiothreitol (DTT), and 0.1% Nonidet P-40 (NP-40) (Sigma). The cells are frozen on dry ice and thawed on ice a total of three times. After the last thaw the lysates are incubated on ice for 10 min. The lysates are centrifuged in a microfuge for 10 min, 4°, to remove cell debris. All whole-cell extracts are stored at −80°.

Electrophoretic mobility shift assay (EMSA) is performed as described by Fried and Crothers[38] and Palombella *et al.*[6] Equal concentrations of whole-cell extract protein are mixed with 0.2–0.8 ng of end-labeled human IFNb PRDII DNA probe (NF-κB binding site; [(GTGGGAAATTCC)$_2$]) in the presence of binding buffer at 0° for 30 min. EMSA binding buffer is 5% glycerol, 10 m*M* Tris, pH 7.5, 0.25 m*M* EDTA, 0.5 m*M* DTT, 0.5 m*M* PMSF, 0.1% NP-40, and 2 μg of poly(dI-dC) · poly(dI-dC) (Pharmacia, Piscataway, NJ). The concentration of KCl in any given reaction varies from 20 to 100 m*M*. The PRDII oligo is end-labeled using [α-^{32}P]dCTP (NEN) and the Klenow fragment of DNA Pol I (New England Biolabs). The radiolabeled oligonucleotide is then purified using Sephadex G-50 (Pharmacia, Piscataway, NJ). The reaction is analyzed by electrophoresis in a nondenaturing 4.5% polyacrylamide gel at 4° in 0.5× TBE buffer (Sigma) run at 150 V. The gel is then fixed in 10% acetic acid and 10% methanol, dried, and exposed to X-ray film or visualized and quantitated using a phosphorimager.

[38] M. Fried and D. M. Crothers, *Nucleic Acids Res.* **9,** 6505–6525 (1981).

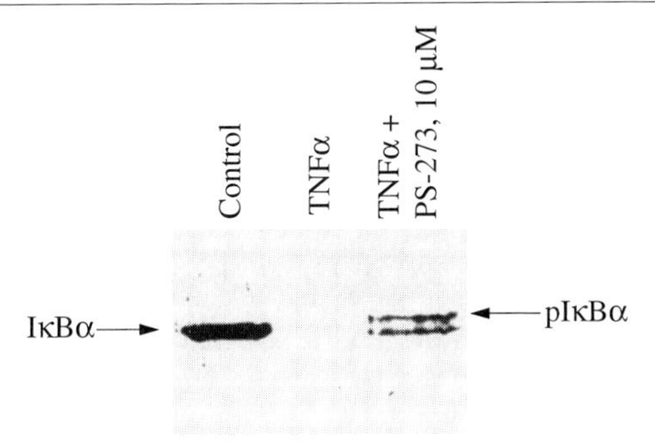

FIG. 5. PS-273 blocks TNF-α-induced IκB-α degradation in HeLa cells. HeLa cells were pretreated for 1 hr with PS-273 (10 μM). The cells were then treated for 30 min with TNF-α (10 ng/ml). Whole-cell extracts were prepared and IκB-α degradation was monitored by SDS–PAGE and Western blot analysis using a specific antibody for IκB-α as described in the text.

IκB-α Degradation

IκB-α is rapidly degraded by the ubiquitin–proteasome pathway in response to various inducers such as TNF-α and LPS. This degradation leads to the release of free NF-κB, which translocates to the nucleus where it binds to DNA. Since the proteasome is required for the rapid degradation of IκB-α, this assay is used to measure the stabilization of IκB-α in cells treated with proteasome inhibitors and activators of NF-κB.

PS-273 can block TNF-α-induced IκB-α degradation in HeLa cells (Fig. 5). HeLa cells are pretreated for 1 hr with PS-273 (10 μM). The cells are then treated for 30 min with TNF-α (10 ng/ml). Whole-cell extracts are prepared and IκB-α degradation is monitored by SDS–PAGE and Western blot analysis using a specific antibody for IκB-α. PS-273 inhibited the degradation of IκB-α and led to the accumulation of the phosphorylated form of IκB-α. Similar results have been seen previously for MG-132.[6]

Protocol. This assay is a modification of the Palombella *et al.* procedure.[6] HeLa cells (ATCC) are seeded into 6-well plates and cultured overnight at 37° in 5% CO_2 in DMEM (GIBCO-BRL) supplemented with 10% FBS (Sigma). The next day the confluent cell monolayers are refed with fresh medium containing proteasome inhibitors (i.e., PS-273, 0–50 μM diluted in DMSO). The final concentration of DMSO in the culture medium is 0.1%. The cells are preincubated with the inhibitors for 1 hr. Next, the cells are treated with TNF-α (1000 IU/ml; 10 ng/ml) (Boehringer) for 30 min. The cells are then washed 1× with ice-cold PBS and are scraped in 0.5 ml PBS. All manipulations are done on ice or at 4°. The cells are quickly centrifuged and the supernatants are removed. The cell pellets are resuspended in 20 μl of lysis buffer to make whole-cell extracts. Lysis buffer is

10% glycerol, 0.4 *M* KCl, 20 m*M* Tris, pH 7.5, 20 μg/ml leupeptin (Sigma), 20 μg/ml pepstatin A (Sigma), 1 m*M* PMSF (Boehringer), 2 m*M* DTT, and 0.1% NP-40 (Sigma). The cells are frozen on dry ice and thawed on ice a total of three times. After the last thaw the lysates are incubated on ice for 10 min. The lysates are centrifuged in a microfuge for 10 min, 4°, to remove cell debris. All whole-cell extracts are stored at −80°.

The whole-cell extracts are boiled in SDS sample buffer and separated by SDS–10% PAGE using a minigel apparatus (Bio-Rad, Richmond, CA). The proteins are then transferred to a nitrocellulose filter (Millipore) using a minitransblot apparatus (Bio-Rad) at 100 V for 2 hr, 4°. The filter is washed in Tris-buffered saline (TBS; 20 m*M* Tris, pH 7.5, 150 m*M* NaCl), then blocked for 30 min with 5% fat-free milk in TBS. The filter is washed 1× with TBS–0.5% Tween 20 (TTBS) and 1× with TBS (5 min each). Anti-C-terminal IκB-α antibody (Santa Cruz) is diluted 1 : 700 in 1% milk/TBS (4 ml total volume) and is incubated with the blot for 1 hr at room temperature. The blot is then washed 2× with TTBS (10 min each) and 1× with TBS (5 min). The secondary antibody, alkaline phosphatase conjugated anti-rabbit IgG (Promega, Madison, WI), is diluted 1 : 7000 in 1% milk/TBS (4 ml total volume) and added to the blot for 1 hr at room temperature. The blot is then washed 2× with TTBS (10 min each) and 1× with TBS (5 min). Finally, the blot is developed with the alkaline phosphatase color reagents. To develop the blot, 66 μl NBT and 33 μl BCIP (Promega) are added fresh to 10 ml of AP buffer. AP buffer is 100 m*M* Tris, pH 9.5, 100 m*M* NaCl, and 5 m*M* $MgCl_2$. This mixture is immediately added to the blot until the labeled bands appear. Color development is stopped by washing the blot with PBS and then with dH_2O.

Endothelial Cell Adhesion Molecule Expression

NF-κB is required for the transcription of a number of proinflammatory mediators, including cell surface adhesion molecules (CAMs) and cytokines, involved in leukocyte homing and activation. The ubiquitin–proteasome pathway is required for the activation of NF-κB by many inducers such as TNF-α. Therefore, proteasome inhibitors should block NF-κB induced expression of these inflammatory mediators. Since PS-273 inhibits NF-κB activation, it was examined for its ability to block NF-κB inducible expression of various cell surface adhesion molecules. The nonselective proteasome inhibitor MG-132 has been shown previously to inhibit the activation of NF-κB and the expression of cell adhesion molecules on HUVE cells.[39] Similarly, the more selective boronic acid proteasome

[39] M. A. Read, A. S. Neish, F. W. Luscinskas, V. J. Palombella, T. Maniatis, and T. Collins, *Immunity* **2,** 493–506 (1995).

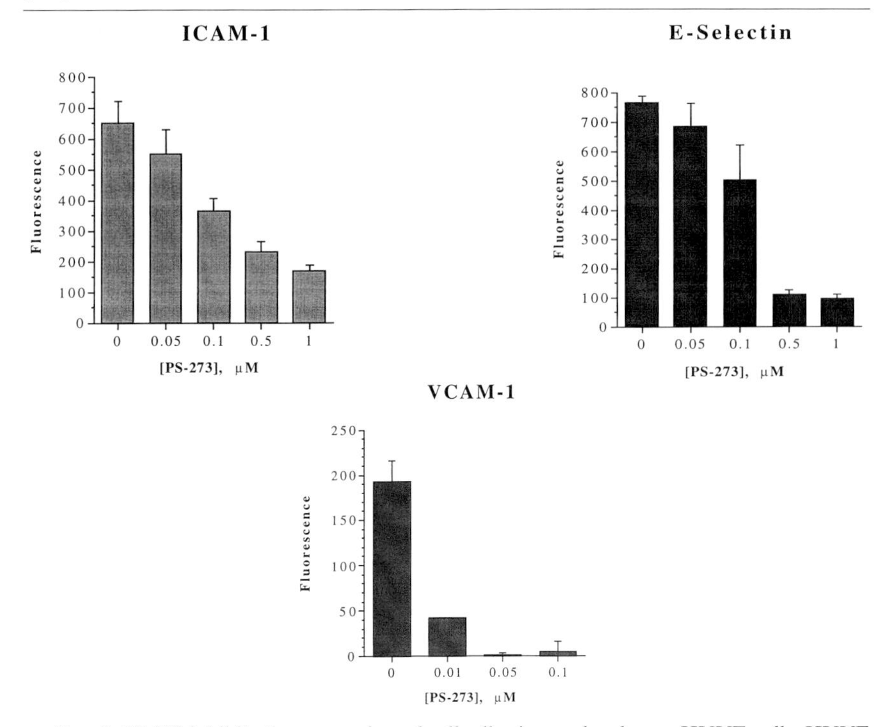

FIG. 6. PS-273 inhibits the expression of cell adhesion molecules on HUVE cells. HUVE cells were pretreated with different concentrations of PS-273 for 1 hr. TNF-α (1000 IU/ml; 10 ng/ml) was then added for 4 hr (E-selectin, ICAM-1) or 15 hr (VCAM-1). Cell surface expression of the cell adhesion molecules was measured by fluorescent immunoassay as described in the text.

inhibitor PS-273 inhibits TNF-α-induced expression of the cell surface adhesion molecules E-selectin, ICAM-1, and VCAM-1 on HUVE cells (Fig. 6). VCAM-1 is especially sensitive to PS-273 with an IC_{50} value that is less than 10 n*M*. Northern blot analysis demonstrates that this inhibition is at the level of transcription (data not shown). As a control, PS-273 has no effect on the expression of the NF-κB-independent, constitutively expressed, HUVE cell surface protein p96[39] (not shown).

Protocol. This assay is a modification of the Read *et al.* procedure.[39] HUVE cells (Cell Systems) are seeded into attachment factor (Cell Systems) coated 96-well plates and cultured overnight at 37° in 5% CO_2. The next day the confluent cell monolayers are refed with fresh HUVE cell medium (Cell Systems) containing proteasome inhibitors (0–50 μM diluted in DMSO). The final concentration of DMSO in the culture medium is 0.1%. The cells are preincubated with the inhibitors for 1 hr. Next, the cells are treated with TNF-α (Boehringer) (1000 IU/ml; 10 ng/ml) for either

4 hr (ICAM-1, E-selectin) or 15 hr (VCAM-1). The expression of p96 is also monitored since it is constitutively expressed and therefore serves as a control for both time points. At the end of the incubation period, the medium is decanted and the cells are washed with RPMI medium (JRH Biosciences) containing 1% FBS (Sigma). Primary anti-CAM monoclonal antibodies (Becton Dickinson) and anti-p96 antibody (gift from Dr. Tucker Collins, Harvard Medical School) are then added to the cells (diluted 1:500 in 1% FBS-RPMI; 150 μl/well). The cells are incubated on ice for 1 hr. A nonbinding antibody (mouse IgG_1 control, also diluted 1:500) is also included in the assay (Becton Dickinson). The cells are washed 3× with ice-cold 1% FBS-RPMI and then 150 μl of FITC $F(ab')^2$ goat anti-mouse IgG (Caltag, San Francisco, CA) (diluted 1:50 in Dulbecco's PBS) is added to the cells. The cells are again incubated for 1 hr on ice. The cells are washed 2× with Dulbecco's PBS (GIBCO-BRL, Grand Island, NY) containing 20% FBS and then 1× with Dulbecco's PBS without serum. The fluid is decanted and the cells are solubilized with 75 μl of lysis buffer (0.01% NaOH, 0.1% SDS). The plates are shaken and the fluorescence is measured with a microplate fluorometer. Fluorescence is measured using an excitation wavelength of 485 nm and an emission wavelength of 530 nm. The results are expressed as the amount of fluorescence with the CAM antibodies (minus the background fluorescence using the mouse IgG_1 control antibody) vs the concentration of the inhibitor. An IC_{50} for inhibiting CAM expression on TNF-α-treated HUVE cells is then determined from the graph of the plotted data.

Cytokine Expression

NF-κB is required for the gene transcription of a number of cytokines, including the T cell cytokine interleukin-2 (IL-2).[40] The proteasome inhibitor PS-273 was examined for its ability to block the production of IL-2 in human Jurkat T cells stimulated with phorbol 12-myristate 13-acetate (PMA) and ionomycin (Fig. 7). In this experiment, Jurkat cells were pretreated with different concentrations of PS-273 for 30 min. The cells were then stimulated with PMA (100 ng/ml) and Ionomycin (3.5 μg/ml) for 4 hr. Secreted IL-2 protein was measured by ELISA. Figure 5 demonstrates that PS-273 inhibited the production of IL-2 with an $IC_{50} \sim 50$–$100\ nM$.

Protocol. Briefly, cells are plated at a density of 5×10^6/ml and incubated at 37° for 30 minutes in RPMI 1640 supplemented with 10% FBS. Different concentrations of PS-273 are then added to the cells (in duplicate) and preincubated at 37° for 30 minutes. The cells are then stimulated with 100

[40] J. W. Rooney, Y.-L. Sun, L. H. Glimcher, and T. Hoey, *Mol. Cell. Biol.* **15,** 6299–6310 (1995).

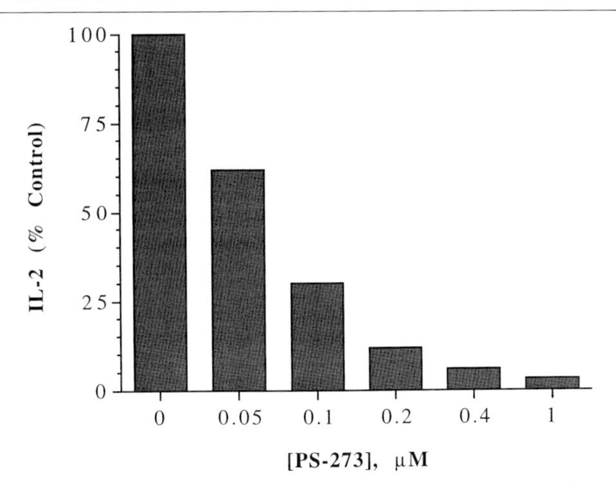

FIG. 7. PS-273 inhibits the production of IL-2 in Jurkat T cells. Jurkat cells were pretreated with different concentrations of PS-273 for 30 min. The cells were then stimulated with PMA (100 ng/ml) and Ionomycin (3.5 μg/ml) for 4 hr. Secreted IL-2 protein was measured by ELISA. The results are expressed as percent control or the relative amount of IL-2 production (pg/ml) from cells treated with different concentrations of PS-273 compared to IL-2 production from control cells treated with the inhibitor diluent DMSO.

ng/ml PMA and 3.5 μg/ml ionomycin for an additional 4 hr at 37°. After 4 hr, the cells are centrifuged in microcentrifuge tubes and the supernatants are collected. The culture supernatants are then tested for the presence of IL-2 using a human IL-2 immunoassay kit (R&D Systems, Minneapolis, MN) according to the manufacturers instructions.

Summary

The results observed with PS-273 have also been demonstrated with other boronic acid proteasome inhibitors, as well as the peptide aldehyde and lactacystin class of proteasome inhibitors. The relative potencies for inhibition in the above mentioned assays correlates generally with K_i values for proteasome inhibition. This is evidence that the intracellular mode of action of the inhibitors is indeed through proteasome inhibition.

Effect of Proteasome Inhibition on Bacterial Cell Wall-Induced Arthritis

Induction of Polyarthritis in Rats

Specific pathogen-free female Lewis rats (Harlan Sprague Dawley, Indianapolis, IN) weighing between 140 and 180 g receive a single injection (i.p.) of a sterile solution of peptidoglycan/polysaccharide (PG/PS; Lee Labs, Grayson, GA) derived from Group A streptococcal bacterial cell walls (25 μg rhamnose per gram body weight).[41–45] Control animals are injected with an equal volume of sterile saline. Animals are then randomized into three groups consisting of a saline-injected control group, a vehicle treated PG/PS group receiving 0.2 ml of 0.5% methylcellulose p.o. daily commencing 7 days after arthritis induction, and a drug-treated group receiving 0.3 mg kg^{-1} PS-341 p.o. daily commencing 7 days after the induction of arthritis. On the day of arthritis induction, animals are weighed, anesthetized via inhalation of isofluorane (Aerrane, Ohmeda PPD, Inc., Liberty Corner, NJ), and the appropriate volume of PG/PS injected intraperitoneally. Changes in rear paw volume are quantified daily by plethysmography using a Buxco Edema Table (Buxco Electronics, Sharon, CT). Joints are also scored daily for symptoms of arthritis using a standard scoring system modified from Dabbagh *et al.*[46] Each joint could receive a maximum score of 4, for a maximum score of 16 for each animal.

Leukocyte Isolation and Determination of 20S Proteasome Activity

Heparinized peripheral blood is obtained from anesthetized female Lewis rats (140–180 g) by cardiac puncture of the surgically exposed heart with an 18- to 21-gauge needle. Blood is then diluted 1:1 with normal isotonic saline and the sample is layered over Nycoprep (Grand Island, NY) separation medium (1 ml and 2 ml samples over 1 ml Nycoprep in a 14 × 75 mm polystyrene test tube; 4 ml and 6 ml samples over 4 ml Nycoprep in a 15 ml conical centrifuge tube). The entire sample is then centrifuged at 500*g* for 30 min at room temperature with no brake. After centrifugation, the top layer is aspirated off to within 2–3 mm of the cell band lying at the interface between the top and bottom layers. The cell band, which is composed of mononuclear cells (lymphocytes and monocytes) and platelets, is then pipetted off and transferred to a fresh tube, washed with ice-cold PBS, and centrifuged at 400*g* for 10 min at 4°. Neutrophils with pellet along with the erythrocyte fraction at the bottom of the tube. The

[41] W. J. Cromartie, J. C. Craddock, J. H. Schwab, S. K. Anderie, and C. H. Yang, *J. Exp. Med.* **146,** 1585–1602 (1977).

[42] R. L. Wilder, J. B. Allen, L. M. Wahl, G. B. Calandra, and S. M. Wahl, *Arth. Rheum.* **26,** 1442–1451 (1983).

[43] J. B. Allen and R. L. Wilder, *Arth. Rheum.* **28,** 1318–1319 (1985).

[44] J. W. Fuseler, E. M. Conner, J. M. Davis, R. E. Wolf, and M. B. Grisham, *Inflammation* **21,** 113–131 (1997).

[45] E. M. Conner, J. W. Fuseler, J. M. Davis, F. S. Laroux, V. J. Palombella, S. Brand, R. E. Wolf, and M. B. Grisham, *Arth. Rheum.* **40,** S322 (1997).

[46] A. J. Dabbagh, D. R. Blake, and C. J. Morris, *Ann. Rheum. Dis.* **51,** 516–521 (1992).

wash supernatant is then poured off and the cell pellet is resuspended with 1 ml of cold PBS. The cell suspension is transferred to a 1.5 ml Eppendorf microfuge tube and microfuged at 9000 rpm for 10 min at 4°. The wash supernatant is aspirated off as completely as possible and the cell pellet either can be processed further for the 20S assay or can be stored indefinitely at −80°.

To isolate platelet populations, a differential centrifugation step is performed in order to take advantage of the size and density differences between platelets and leukocytes. The cell pellet in the 15 ml conical centrifuge tube obtained after the first wash is resuspended again with cold PBS and then centrifuged at 60*g* for 5 min at 4° in order to pellet larger and denser cell types. The supernatant is then pipetted off and transferred to a fresh 15 ml conical tube and centrifuged at 400*g* for 5 min. The resulting cell pellet is composed primarily of platelets and can then be resuspended with 1 ml cold PBS and transferred to a microfuge tube. The cell pellet recovered after the platelet suspension was pipetted off is composed primarily of mononuclear cells and can be resuspended 1 ml cold PBS and also transferred to a microfuge tube. Both samples are then treated as described above.

Neutrophil populations can be isolated by diluting the erthrocyte pellets (obtained from the initial Nycroprep centrifugation) 1:1 with 3% dextran/PBS (molecular weight 500,000). The suspension is allowed to stand at room temperature for 30–40 min. The dextran will cause the red cells to aggregate and sediment faster than the neutrophil population. The top layer is pipetted off and transferred to a fresh 15 ml conical centrifuge tube and centrifuged at 400*g* for 5 min. The cell pellet is resuspended with 1 ml cold PBS and transferred to a 1.5 ml microfuge tube. The sample is then treated as previously described.

Determination of Residual 20S Proteasome Activity

Cell lysates are made from the cell pellets obtained as described above by the addition of a hypotonic lysing agent. Depending on its size, the cell pellet is resuspended with 50–200 μl of 5 m*M* EDTA. The samples are then placed on ice for 30–60 min. The samples are then microfuged at 10,000*g* for 10 min at 4°. The cell lysate is then pipetted off from the debris pellet and transferred to a fresh microfuge tube. Samples are held on ice until assayed.

20S activity is determined by measuring the rate of proteolytic hydrolysis of a fluorescent (AMC)-tagged peptide substrate by the cell lysate sample and normalizing the activity to the amount of cell specific protein present in the lysate. This is accomplished by adding 5 μl of cell lysate to a cuvette containing 2 ml of assay reaction buffer (20 m*M* HEPES, 0.5 *M* EDTA,

0.035% SDS, 20 m*M* Suc-LLVY-AMC) and a magnetic stir bar. The cuvette is placed in a fluorimeter and maintained at 37° while the amount of hydrolyzed AMC is measured by monitoring (λ_{em} 380 nm; λ_{ex} 440 nm) the increase is detectable fluorescence over a 5 min period. A linear regression fit of the reaction progress curve gives the rate of hydrolysis in fluorescent units per second (FU/sec). Protein and hemoglobin concentrations are determined using a modified Bradford assay (Pierce) and a hemoglobin-specific enzymatic-based assay (Sigma), respectively. In order to determine the 20S activity that is attributed only to the leukocyte portion of the cell lysate sample, the total amount of protein contributed by erythrocytes is estimated from the hemoglobin concentration, and this value is subtracted from the total protein concentration in order to give the amount of protein contributed solely by leukocytes. 20S specific activity is then determined by normalizing the rate of proteolytic activity with the amount of protein in the sample using Eq. (1) where C is conversion factor that equates the amount of fluorescence to the concentration of free AMC (*FU*/pmol AMC).

$$\text{Specific activity (pmol AMC/sec/mg protein)} = \frac{(FU/\text{sec})/(5 \times 10^{-6}\ \text{ml})(\text{protein}\ \mu\text{g/ml})}{C} \tag{1}$$

Results

Treatment with PS-341 significantly reduced 20S proteasome activity in circulating WBCs by approximately 60% when compared to animals receiving vehicle only. We found that oral administration of PS-341 beginning 7 days following PG/PS injection significantly attenuated the progression of chronic polyarthritis (Fig. 8[45]). Animals receiving PS-341 from day 7 to day 28 showed a significant inhibition in hind paw arthritis index which continued throughout the experimental period. These changes in arthritic indices were mirrored by a comparative inhibition of the PG/PS-induced increases in paw volume (Fig. 8[45]). The average increase in hind paw volume is approximately 0.5 ml during the acute flare of arthritis and increases further to approximately 1.3–1.4 ml during the chronic phase.[45] In contrast, joints from animals treated with PS-341 beginning on day 7 demonstrate a dramatic decrease in hind paw swelling within 48 hr which is then maintained throughout the course of the experimental period.

Histological inspection of the joints obtained from the drug-treated group revealed major differences in appearance when compared to joints from vehicle-treated arthritic animals. We found that PG/PS induces a massive synovial thickening with extensive inflammatory cell infiltrate, pannus formation along and overlying the articular cartilage, accompanied by

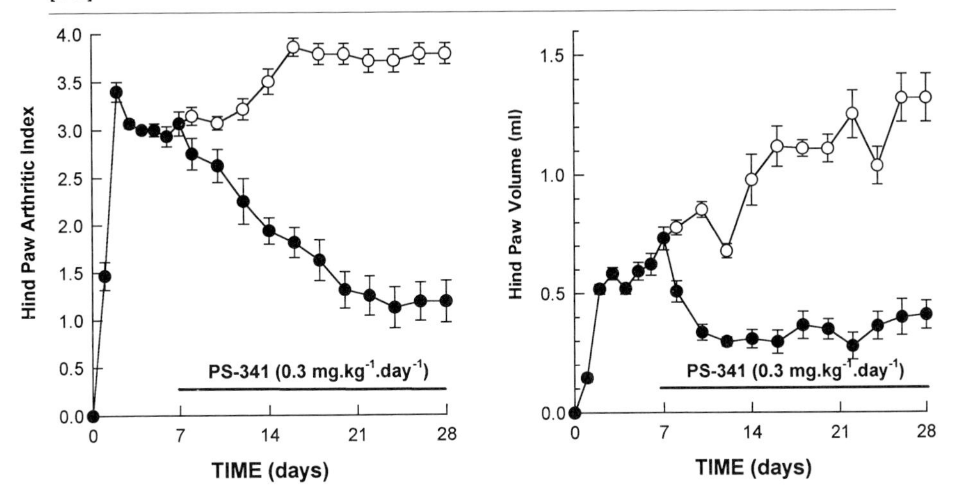

FIG. 8. Effects of PS-341 on polyarthritis induced by a single intraperitoneal injection of peptidoglycan/polysaccharide (PG/PS) into female Lewis rats. PS-341 was administered daily (p.o.) beginning 7 days following induction of arthritis and continuing through day 28. Hind joint inflammation was scored using a standardized scoring system.[46] Hind joint swelling was quantified by plethysmography.[45]

cartilage degradation and bone erosion. Treatment with PS-341 inhibited cellular infiltration and joint degradation induced by PG/PS. Pannus formation was still evident in these joints; however, the extensive cellular infiltrate observed in the vehicle-treated arthritic rats was largely attenuated, as was the cartilage degradation and bone erosion.

Effects of Proteasome Inhibition on Delayed-Type Hypersensitivity Reaction (DTH) *in Vivo*

Female BALB/c mice, 18–20 g, are obtained from Taconic Farms (Germantown, NY). Animals are housed in the animal facility at ProScript for at least 1 week before study initiation. Fluorescent lighting is controlled to automatically provide alternate light and dark cycles of approximately 12 hr each. Temperature and humidity are centrally controlled and recorded daily. Temperature readings range between 20° and 23°, and humidity readings ranged between 40 and 50% relative humidity. Pellets of standard rodent chow and water are available *ad libitum* throughout the observation and study periods.

Dose formulations of PS-341 are prepared daily during the course of the study. The stock solution of PS-341 is made up in 100% DMSO (Sigma). Dilutions of each stock solution is made in 0.9% sterile saline. Final dosing

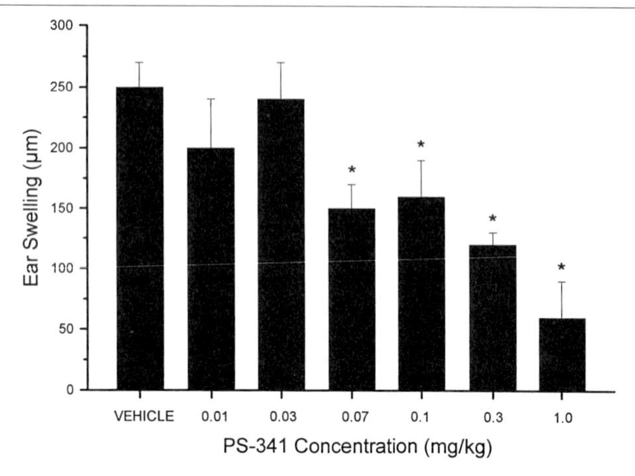

FIG. 9. Effect of PS-341 on DTH-induced ear swelling. PS-341 or vehicle was administered (IV) 30 min prior to DNFB challenge. Ear swelling was recorded 24 hr later. * represents $p < 0.05$ compared to vehicle-treated mice.

solutions of PS-341 contain a maximum of 1% (v/v) DMSO. Vehicle-treated animals are administered with a 1% (v/v) DMSO solution in 0.9% (v/v) saline. Dinitrofluorobenzene (DNFB) is made up in acetone/olive oil (4:1) and protected from light and given topically.

PS-341 is administered in vehicle as a single intravenous (IV) bolus injection. Control groups are administered with the appropriate vehicle. All mice are dosed volumes of 2.5 ml per kg of body weight.

Mice are sensitized to DNFB (0.5%) on days 0 and 1 by applying 20 μl of the solution to the hind foot pads. On day 5, DNFB (0.2%) is applied in a 10 μl volume to both the inside and outside of the right ear. The proteasome inhibitor and vehicle treatments are administered 30 min prior to the DNFB challenge on day 5. On day 6, the thickness of both ears is taken using a digital micrometer. The difference between the two ear measurements is taken to signify ear swelling in response to the DNFB challenge.[47]

Results

Doses of PS-341 from 0.01 to 1.0 mg/kg were administered IV to mice, 30 min before challenge with DNFB. One day later ear measurements were taken and results demonstrate that this proteasome inhibitor, PS-341, can significantly decrease ear swelling in a dose-dependent manner (Fig. 9).

[47] A. Scheynius, R. K. Camp, and E. Pure, *J. Immunol.* **150,** 655–663 (1993).

Summary and Conclusions

It is becoming increasingly apparent that NF-κB plays a critical role in regulating the inflammatory response. Data obtained from studies in our laboratories demonstrate that the proteasome plays an important role in the inflammatory cascade by regulating the activation of NF-κB. Indeed, the availability of selective and orally active proteasome inhibitors should prove useful in delineating the roles of the proteasome and NF-κB in other pathophysiological conditions such as cancer and heart disease.

[36] Nuclear Factor κB Activity in Response to Oxidants and Antioxidants

By YVONNE M. W. JANSSEN and CHANDAN K. SEN

Introduction

Oxidation–reduction (redox)-based regulation of gene expression appears to represent a fundamental regulatory mechanism in cell biology.[1–4] The transcription factor, nuclear factor κB (NF-κB), is of considerable interest to the field of free radical biology because it has been shown to be activated by oxidants and its induced activity is inhibited by antioxidants.[1–7] The exact oxidant sensitive sites, the localization of antioxidant action, and signaling events that are responsible for NF-κB activation remain elusive. The discoveries of the TRAF adaptor proteins, NF-κB inducing kinase (NIK), and IκB kinases (IκK) required for the activation of NF-κB by tumor necrosis factor (TNF) α may provide insight into these redox-sensitive sites.[8–10] Because free radicals are involved in numerous

[1] C. K. Sen and L. Packer, *FASEB J.* **10,** 709 (1996).
[2] J. M. Muller, M. R. A. Rupec, and P. A. Baeuerle, *Methods* **11,** 301 (1997).
[3] C. K. Sen, *Biochem. Pharmacol.* **55,** 1747 (1998).
[4] C. K. Sen, *Curr. Top. Cell Regul.* **36,** in press (1998).
[5] C. K. Sen, S. Khanna, A. Z. Reznick, S. Roy, and L. Packer, *Biochem. Biophys. Res. Commun.* **237,** 6645 (1997).
[6] R. Schreck, P. Rieber, and P. A. Baeuerle, *EMBO J.* **10,** 2247 (1991).
[7] R. Schreck, B. Meier, D. N. Mannel, W. Droege, and P. A. Baeuerle, *J. Exp. Med.* **175,** 1181 (1992).
[8] Z. Liu, H. Hsu, D. V. Goeddel, and M. Karin, *Cell* **87,** 565 (1996).
[9] N. L. Malinin, M. P. Boldin, A. V. Kovalenko, and D. Wallach, *Nature (London)* **385,** 540 (1997).
[10] T. Maniatis, *Science* **278,** 818 (1997).

0076-6879/99 $30.00

pathological conditions, the ability to measure NF-κB activation in these circumstances should provide important insight to elucidate the etiology of these disorders.

NF-κB activation leads to the transcriptional activation of genes, many of which are involved in the orchestration of an inflammatory response and immune modulatory events, as well as other processes.[11-13] These observations point to an important role of NF-κB in normal cellular function and disease.

In its latent form, NF-κB is complexed in the cytoplasm to IκB inhibitor proteins. Exposure to an NF-κB activating agent leads to dissociation and, in some cases, degradation of the inhibitory complex via a proteasome pathway.[13,14] This enables the NF-κB protein to translocate into the nucleus under the guidance of a nuclear localization signal, bind to κB sites in promoter regions of target genes, and activate transcription. Five members of the NF-κB/Rel family and seven members of IκB related proteins have been identified in mammalian cells, and new family members continue to be discovered.[15] It is thus evident that the regulation of NF-κB activation by various NF-κB/IκB proteins may be quite complex. Most of the research has focused on classical p65–p50 heterodimers (NF-κB) and its control via release of IκBα or IκBβ. Based on findings from these studies, a number of protocols have been established to measure the activity of p65–p50 and degradation of IκBα/β and are discussed in the following sections. Knowledge of the mechanisms involved in the activation of NF-κB and its target genes in response to different stress situations will be enhanced once the functions of other NF-κB/IκB family member are established.

To clarify the role of NF-κB in various aspects of cell function, several approaches to measure NF-κB activation in cell culture as well as intact tissue models have been used. For example, NF-κB activity can be monitored by electrophoretic mobility shift assays (EMSA), nuclear translocation, and Western blotting, as well as transactivation assays and the measurement of NF-κB regulated protein expression. The goal of this chapter is to review these approaches and discuss their strengths and weaknesses.

I. Electrophoretic Mobility Shift Assay

Translocation of NF-κB proteins from the cytosol to the nucleus leads to increased concentration of NF-κB proteins in the nucleus. Incubation

[11] P. A. Baeuerle and V. R. Baichwal, *Adv. Immunol.* **65,** 111 (1997).
[12] P. A. Baeuerle and T. Henkel, *Ann. Rev. Immunol.* **12,** 141 (1994).
[13] T. S. Finco and A. S. Baldwin, *Immunity* **3,** 263 (1995).
[14] V. Imbert, R. A. Rupec, A. Livolsi, H. L. Pahl, E. B. Traenckner, C. Mueller-Dieckmann, D. Farahifar, B. Rossi, P. Auberger, P. A. Baeuerle, and J.-F. Peyron, *Cell* **86,** 787 (1996).
[15] P. A. Baeuerle and D. Baltimore, *Cell* **87,** 13 (1996).

of a ^{32}P-labeled oligonucleotide containing an NF-κB consensus sequence with nuclear extracts enriched in NF-κB proteins results in enhanced binding of NF-κB proteins to the oligonucleotide. Such binding results in retardation ("shift") of the electrophoretic mobility of the oligonucleotide on a nondenaturing polyacrylamide gel. These "shifts" can be visualized by autoradiography and can be used as a marker of the presence of NF-κB proteins in the nucleus. Thus, EMSA is a measure of the translocation of activated NF-κB proteins to the nucleus, and this assay does not provide any information regarding the transcription regulatory effect of NF-κB.

A. Preparation of Nuclear Extracts

Cells ($2–10 \times 10^6$), adherent or grown in suspension, are harvested and centrifuged at 750*g* for 5 min at 4°. The cell pellet is gently resuspended in ice-cold phosphate-buffered saline (PBS, pH 7.4) and centrifuged at 750*g* for 5 min at 4°. The cell pellet thus obtained is resuspended in 400 μl of ice-cold hypotonic buffer containing 10 m*M* HEPES (pH 7.8), 10 m*M* KCl, 2 m*M* $MgCl_2$, 0.1 m*M* ethylenediaminetetraacetic acid (EDTA, sodium salt), 0.2 m*M* NaF, 0.2 m*M* Na_3VO_4, 0.4 m*M* phenylmethylsulfonyl fluoride (PMSF), 0.3 mg/ml leupeptin, and 1 m*M* dithiothreitol (DTT). When available, the use of molecular biology grade reagent is recommended for all procedures described in this work. Stock solutions of PMSF (10 m*M*) in isopropanol and leupeptin (10 mg/ml) in water may be stored as single-use aliquots at −20°. PMSF, leupeptin, and DTT should be added fresh. The cell suspension is kept on ice for 15 min, following which 25 μl of 10% Nonidet P-40 is added. The suspension is vortexed vigorously for 15 sec and then centrifuged at 16,000*g* for 1 min. This results in a pellet of nuclei.

Nuclei obtained as described above are resuspended in 50 μl of ice-cold high-salt extraction buffer [50 m*M* HEPES (pH 7.8), 50 m*M* KCl, 300 m*M* NaCl, 0.1 m*M* EDTA, 0.2 m*M* NaF, 0.2 m*M* Na_3VO_4, 0.4 m*M* PMSF, 0.3 mg/ml leupeptin, 1 m*M* DTT and 10% sterile glycerol]. PMSF, leupeptin, and DTT should be added fresh. The nuclear suspension is placed on a rocking platform for 30 min at 4° to facilitate lysis of the nuclei. The nuclear lysates are then centrifuged at 16,000*g* for 10–20 min at 4°. The supernatant is collected and stored as 15 μl aliquots at −80°. Nuclear extracts should not be thawed more than twice. The protein concentration in the supernatant is determined from a 5 μl aliquot using an assay method (Bradford protein assay reagent, Bio-Rad, Hercules, CA) that does not interfere with DTT present in the nuclear protein extract.

B. Labeling of NF-κB Consensus Oligonucleotide Sequence and EMSA

NF-κB oligonucleotide (8.75 pmol; Promega, Madison, WI) is incubated with 50 μCi [γ-^{32}P]ATP in the presence of 10 U T4 polynucleotide kinase and 1× kinase buffer supplied with the kinase (Boehringer, Indianapolis, IN) at 37° for 30 min. The reaction is stopped by addition of 1 volume of phenol : chloroform : isoamyl alcohol (25 : 24 : 1). After gentle mixing by pipette, the mixture is centrifuged at 16,000*g* for 2 min. The upper aqueous phase is collected and subjected to column filtration. Nonincorporated ^{32}P is removed by centrifugation through a Sephadex G-50 column (Boehringer) equilibrated with saline–Tris–EDTA buffer (10 m*M* Tris, 100 m*M* NaCl, and 10 m*M* EDTA, pH adjusted to 7.5 using HCl). Labeled oligonucleotide is precipitated at −20° until use.

C. DNA Binding Reaction and Electrophoresis

The nuclear protein extract (5 μg) obtained as described in Section A in a total of 4 μl of high-salt extraction buffer is incubated in the presence of 0.1 pmol ^{32}P-end-labeled NF-κB oligonucleotide, 40 m*M* HEPES (pH 7.8), 10% (v/v) glycerol (or 4% Ficoll 400), 1 m*M* $MgCl_2$, 0.1 m*M* DTT, and 200 ng poly(dI-dC) · poly(dI-dC) (Pharmacia, Piscataway, NJ) in a final volume of 20 μl for 20 min at room temperature. Following incubation, the sample is loaded on a 0.8 mm 6% nondenaturing polyacrylamide gel in 0.5× Tris/boric acid/EDTA (TBE) buffer [10× stock solution is available from Bio-Rad (Hercules, CA)]. The gel is prepared using a 40% acrylamide/bis acrylamide solution 29 : 1 (3.3% C) in 1× TBE. The mixture is polymerized using 0.1% (w/v) ammonium persulfate and 0.1% TEMED (Bio-Rad). Samples are electrophoresed in 0.5× TBE at 120 V for 2 hr. Gels are dried under heat and vacuum and band shifts examined by autoradiography using a ^{32}P-sensitive film (Hyperfilm, Amersham Life Science Inc., Arlington Heights, IL).

For the DNA binding reaction it is critical that the volume of high-salt extraction buffer present in each binding assay be identical. This can be achieved by diluting the more concentrated nuclear protein extracts in the same extraction buffer. The salts present in the extraction buffer determine the binding kinetics of the protein to the oligonucleotide and should therefore be strictly controlled for relative concentration in different samples.

D. Competition and Supershift Assays

To ascertain the specificity of NF-κB DNA binding, reactions are performed in the presence of a 10-fold molar excess of unlabeled NF-κB oligonucleotide, which will result in the disappearance of all NF-κB complexes. Incubation with an unrelated oligonucleotide or a mutated NF-κB oligonucleotide (one base pair altered, Santa Cruz Biotechnology Inc., Santa Cruz, CA) should not compete for binding (Fig. 1).

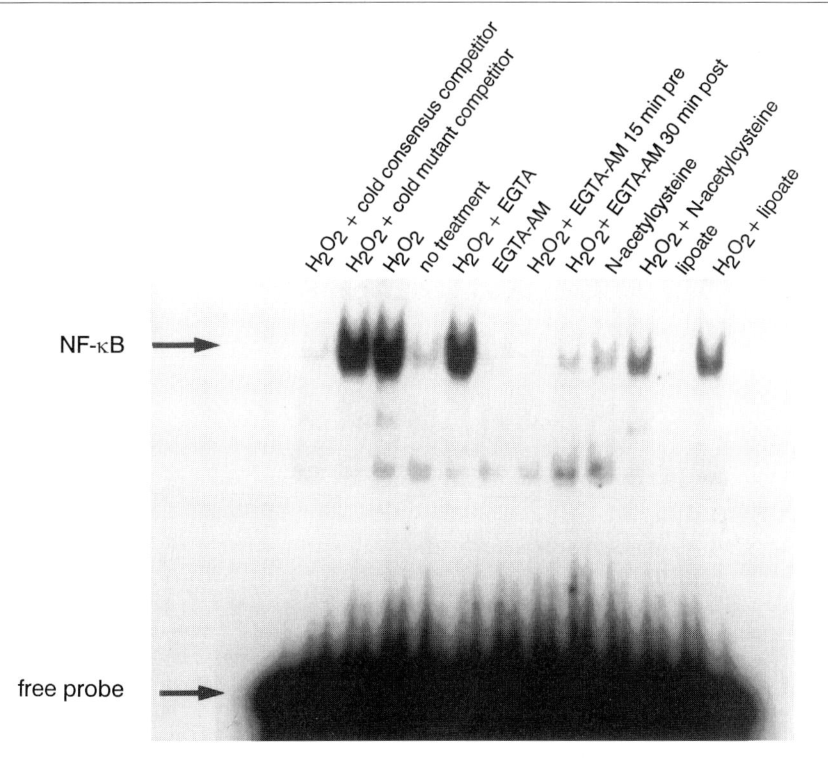

FIG. 1. Inhibition of H_2O_2-induced NF-κB activation in Wurzburg cells by intracellular calcium chelators and thiol antioxidants. To test the specificity of the NF-κB band, nuclear extracts from activated cells were treated with an excess of unlabeled consensus NF-κB oligonucleotide (lane 1; from left to right) or with an excess of cold mutant NF-κB oligonucleotide (lane 2) before incubation with ^{32}P-labeled consensus NF-κB probe. Lanes 3 and 4, cells were either treated or not treated with 0.25 m*M* H_2O_2, respectively; lane 5, cells treated with 1 m*M* EGTA (ethylene glycol tetraacetic acid) 10 min before H_2O_2 challenge; lane 6, cells treated with 0.5 m*M* lipophilic EGTA-AM (acetomethoxyl ester of EGTA); lane 7, cells treated with 0.5 m*M* EGTA-AM 15 min before H_2O_2 challenge; lane 8, cells treated with 0.5 m*M* EGTA-AM 30 min after H_2O_2 challenge; lane 9, cells treated with 20 m*M* *N*-acetyl-L-cysteine (NAC) for 6 hr; lane 10, cells pretreated with 20 m*M* NAC for 2 hr before H_2O_2 treatment for 4 hr; lane 11, cells treated with 1 m*M* α-lipoate for 22 hr; lane 12, cells pretreated with α-lipoate for 18 hr followed by H_2O_2 treatment for 4 hr. Reprinted from Sen *et al.* (1996),[16] with kind permission from Elsevier Science, Amsterdam, The Netherlands.

To assess the subunit composition of DNA binding proteins, specific antibodies are available to detect p65 and p50 proteins (Santa Cruz Biotechnology Inc., Santa Cruz, CA). DNA binding reactions (18 μl) are carried out as described, and are followed by the addition of 2 μl antibody for an additional 30 min prior to electrophoresis. In cases where antibody binding

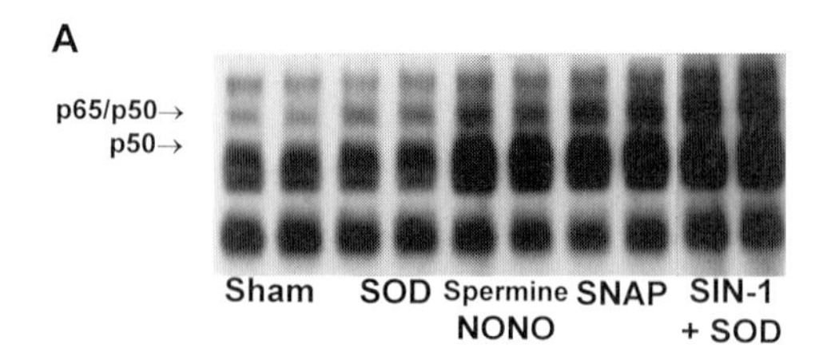

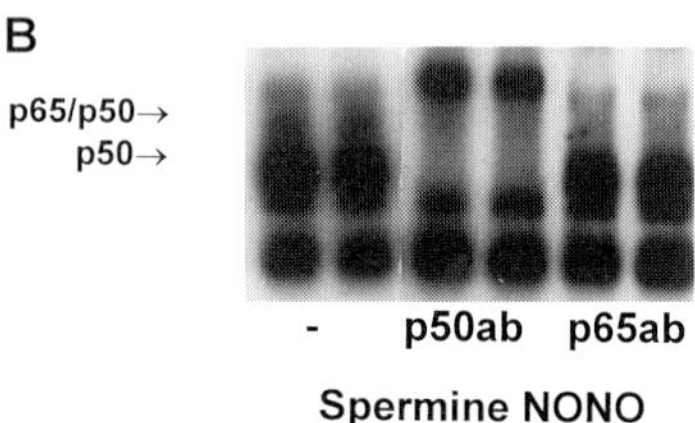

FIG. 2. Electrophoretic mobility shift assay. (A) Enhanced binding of nuclear protein complexes to the NF-κB consensus oligonucleotide in rat lung epithelial cells exposed to NO· donors. Following 6 hr of exposure to spermine NONOate (1 m*M*). SNAP (1 m*M*) plus 100 μ*M* L-cysteine, or SIN-1 in presence of SOD, nuclear extracts were prepared and evaluated. Note the increases compared to sham controls. (B) Supershift assays employing antibodies directed against p65 or p50 subunits of NF-κB. Incubation of DNA binding complexes with antibodies prior to electrophoresis abrogates two specific bands, verifying the presence of p65/p50 and p50 subunits. Spermine NONO, [*N*-2-(2-aminoethyl)-*N*-(2-hydroxy-2-nitrosohydrazino)-1,2-ethylenediamine]; SIN-1, 3-morpholinosydnonimine; SOD, superoxide dismutase; SNAP, (±)-*S*-nitroso-*N*-acetylpenicillamine.

results in an additional electrophoretic retardation, this is referred to as "supershift" (Fig. 2B). Examples of EMSA and supershifts are shown in Figs. 1 and 2. Results in Fig. 2A demonstrate increases in NF-κB DNA binding in response to exposure of rat lung epithelial (RLE) cells to the NO· generating compound spermine NONOate. Several antioxidants have been shown to inhibit inducible NF-κB activation. Figure 1 illustrates the inhibition of hydrogen peroxide induced NF-κB activation in Würzburg T cells by thiol antioxidants and calcium chelators.[16]

II. Nuclear Translocation of p65

Activation of NF-κB is characterized by translocation of NF-κB proteins into the nucleus. This can be detected by immunocytochemical or immuno-

[16] C. K. Sen, S. Roy, and L. Packer, *FEBS Lett.* **385,** 58 (1996).

fluorescence approaches or by performing a Western blot of nuclear extracts.

A. Evaluation of Nuclear p65 by Immunofluorescence

Cells are grown on glass coverslips and treated with NF-κB activating agents. After the appropriate incubation times, cells are washed twice in PBS (4°) and fixed in 100% methanol for 30 min at room temperature, followed by two washes with PBS. To permeabilize cells, sections are treated with 0.1% Triton X-100 in PBS for 20 min. To block nonspecific binding of antibody, sections are incubated in PBS containing 1% bovine serum albumin (BSA) twice for 30 min each at room temperature. Coverslips are then incubated with primary antibody against p65 (2.5 μg/ml, Santa Cruz Biotechnology Inc.) in PBS/BSA by inverting the coverslips onto 50 μl of antibody suspension. After 1 hr of incubation in primary antibody at room temperature, sections are washed in PBS/BSA (3 × 20 min each) and incubated with 10 μg/ml goat anti-rabbit antibody conjugated with the fluorophore Cy3 (Jackson Immunoresearch, West Grove, PA) for 45 min (10 μg/ml in PBS/BSA). To verify staining specificity, control sections are not incubated with primary antibody. Only secondary antibody is applied to these samples. Such treatment should result in a lack of immunostaining. Coverslips are then washed with PBS and incubated with the DNA stain YOYO. The use of a nuclear stain such as YOYO-1 iodide (500 n*M*, Molecular Probes, Eugene, OR) allows colocalization of p65 in the nuclear component.[17] An example of immunofluorescence of p65 in control cells, as well as cells treated with the NF-κB activator lipopolysaccharide (LPS), is shown in Fig. 3.

B. Detection of NF-κB Proteins in Nuclear Extracts by Western Blotting

An alternative approach is to perform Western blot analysis of the nuclear protein extract for the specific detection of NF-κB proteins. Increased presence of these proteins in the nucleus indicates NF-κB activation.

III. Transactivation of NF-κB Directed Gene

Activation of the transcription factor NF-κB should enhance the transcription of NF-κB regulated target genes. Translocation of p65 into the nucleus, evidenced by the techniques described above, does not provide any information regarding the transcriptional activity of NF-κB. The tran-

[17] Y. M. W. Janssen, K. E. Driscoll, B. Howard, T. R. Quinlan, M. Treadwell, A. Barchowsky, and B. T. Mossman, *Am. J. Pathol.* **151,** 389 (1997).

scription regulatory aspect of NF-κB activity may be studied in cells transiently transfected with NF-κB consensus sequence linked to a reporter gene, e.g., luciferase, β-galactosidase, human growth hormone, or green fluorescent protein, that is not expressed endogenously in mammalian cells. Increased reported gene expression in these transiently transfected cells reflects the transcription regulatory activity of NF-κB. Firefly luciferase is a convenient reporter enzyme that gives a linear response over a wide range of activity and has an enhanced sensitivity compared to β-galactosidase, which can occur endogenously in some mammalian cells.[18] A major drawback of this assay is that the transient transfection of NF-κB luciferase plasmid results in an artificially high amount of target sequence that is not physiologically relevant. Thus, this method could result in the overestimation of the transcription regulatory effect of NF-κB.

A. Transient Transfection

Cells are transiently transfected using optimal protocols, e.g., calcium phosphate coprecipitation, DEAE dextran, lipofectamine, or electroporation. Detailed descriptions of these basic laboratory protocols are available in the literature.[19] In our hands, cells difficult to transfect have responded well to electroporation (Bio-Rad, Genepulser electroporator) using complete medium with 10% serum and $0.8–1 \times 10^6$ cells at 240 V. Cells are transfected with an NF-κB–luciferase reporter vector in presence of the control vector PSV β-Gal (Promega), which serves as a vector control and allows to correct for variabilities in transfection efficiencies. Cells are transfected using 5–10 μg of DNA, plated in 35 mm dishes, incubated for 4–6 hr, and allowed to recover overnight in complete culture medium. After this, the cells are stimulated with appropriate agents and harvested at selected time points. In general, cells may be harvested after 48 hr of transfection or at time points during which expression of the transfected reported gene is optimal.

B. Reporter Activity Assay

To harvest cells for assessment of luciferase and β-galactosidase activities, cells in 35 mm dishes are washed twice in ice-cold PBS and lysed in 200 μl reporter lysis buffer (Promega). Dishes are scraped with a rubber

[18] J. R. de Wet, K. V. Wood, M. DeLuca, D. R. Helinski, and S. Subramani, *Mol. Cell. Biol.* **7,** 725 (1987).

[19] "Short Protocols in Molecular Biology: A Compendium of Methods from Current Protocols in Molecular Biology" (F. Ausubel, R. Brent, R. E. Kingston, *et al.*, Eds.), 3rd ed. Wiley, New York, 1995.

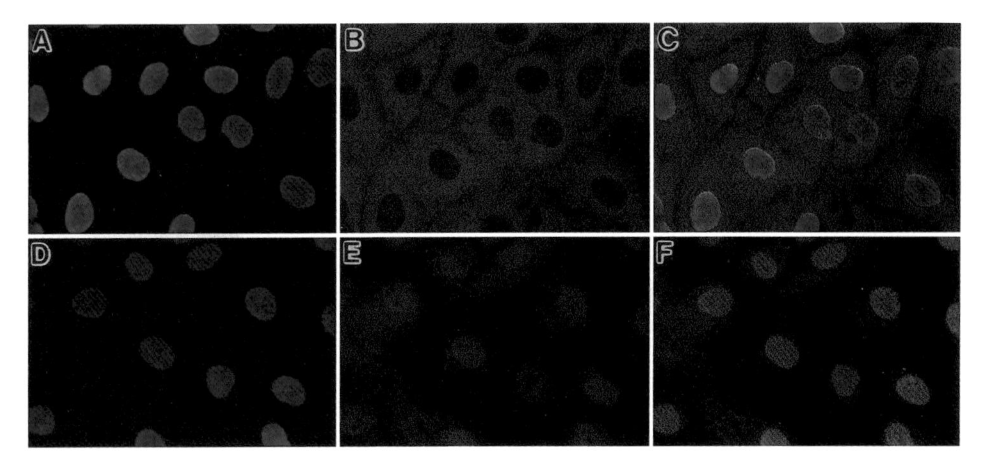

FIG. 3. Confocal microscopy illustrating the translocation of p65 into the nucleus of rat pleural mesothelial cells exposed to lipopolysaccharide (LPS) for 2 hr. Cells were incubated with p65 antibody, followed by incubation with a secondary antibody coupled to the fluorochrome Cy3™, resulting in a red staining to illustrate the subcellular localization of p65 (Panels B and E). Cells were simultaneously stained with YOYO, to stain the DNA green (Panels A and D). Overlap of green and red colors results in an orange/yellow color formation and verifies the presence of p65 in the nucleus (Panels C and F). Panels A–C are sham controls, panels D–F are LPS-exposed rat pleural mesothelial cells (magnification 1080×).

TABLE I
TIME COURSE OF ACTIVATION OF NF-κB IN RAT LUNG EPITHELIAL CELLS EXPOSED TO TNFα[a]

Step	Luc	β-Gal	Luc/β-Gal
Sham	15,863	43,296	366
	11,472	46,887	245
TNFα, 1 hr	13,214	44,553	297
	9,238	45,126	205
TNFα, 2 hr	41,872	29,004	1,444
TNFα, 8 hr	82,508	29,548	2,792
	116,372	28,407	4,097

[a] Luc, Luciferase activity; β-Gal, β-galactosidase activity; Luc/β-Gal, Luc activities normalized to β-Gal. The two rows of data for each treatment group indicate replicate values from two different experiments. Data are arbitrary luminescence values.

policeman and lysates are stored at $-80°$ until analysis. Lysates are centrifuged at 16,000g for 2 min at 4°. The supernatant is used for the measurement of reporter gene activities. The amount of extract required for the assay may be optimized. For this, 100 μl of luciferase assay reagent containing luciferin and ATP (Promega) is added to the extract and the generated luminescence is detected using a standard luminometer.

β-Galactosidase activity may be measured similarly using a Galactolight assay kit (Tropix, Bedford, MA) and a luminometer. Instrument settings should be optimized according to the detected activities in the sample. Typically, luciferase activities are measured in 10 sec, and β-galactosidase activities in 5 sec. Increased NF-κB activity is manifested as increased luciferase activity. β-Galactosidase is constitutively expressed in cells and its activity is used to normalize luciferase activity results. This controls for variations in transfection efficiencies that may occur in these transient assays. As a sham control, cells can be transfected with a luciferase vector not regulated by NF-κB or a vector where NF-κB sites are mutated. Luciferase activity detected from such samples should not respond to the treatment of cells with NF-κB activating agents. An example of NF-κB regulated luciferase activities and β-galactosidase activities in control RLE cells as well as after various times of exposure to TNF (tumor necrosis factor) α is illustrated in Table I. We have also demonstrated increases in NF-κB-luciferase activities in RLE cells exposed to a range of reactive oxygen or

TABLE II
ACTIVATION OF NF-κB IN RAT LUNG EPITHELIAL CELLS[a]

Step	Luciferase/β-Gal	
	I	II
Sham	488	648
300 U/ml SOD	631	
300 μM SIN-1	1337	1127
SIN-1 + SOD	1514	1020
300 μM ONOO-	1151	1895
300 μM H_2O_2	9410	10,741

[a] Exposed to reactive oxygen or reactive nitrogen species for 8 hr. β-Gal, β-galactosidase activity; luciferase/β-Gal, luciferase activities normalized for β-Gal. The two columns of data (I, II) for each treatment group indicate replicate values obtained from two different experiments. Data are arbitrary luminescence values.

nitrogen species for 8 hr as is shown in Table II. More striking increases in NF-κB luciferase activities are apparent after 16 hr of activation (unpublished data).

IV. NF-κB Regulated Protein Expression

The transient transfection of NF-κB luciferase plasmid results in an artificially high amount of target sequence that is not physiologically relevant. Thus, a more reasonable approach to assess NF-κB activation is to evaluate the expression of an NF-κB target gene that (1) contains NF-κB sequences in its promoter and (2) requires NF-κB activity for transcription. The latter can be confirmed by promoter deletion assays where loss of NF-κB sequences results in a loss of gene transcription. The power of this approach is that it can be performed in both *in vitro* and *in vivo* situations. The process requires minimal manipulation of the cell, and this is clearly an advantage compared to transfection requiring methods where cells must be subjected to considerable chemical or physical treatments for transfection (see Section III).

A number of target genes have been identified to be reasonable end points that reflect NF-κB activation in certain species. These genes and their products include macrophage inflammatory protein-2, interleukin-1, interleukin-2 receptor, interleukin-6, interleukin-8, intercellular adhesion molecule (ICAM)-1, and inducible nitric oxide synthase. In this context it should be noted that the expression of a gene is mostly, if not always, regulated by multiple transcription factors and that enhanced expression of a particular gene may be the result of cooperation between multiple transcription factors that include NF-κB. Additionally, the involvement of NF-κB in the activation of these genes appears to be species specific.

A. Determination of ICAM-1 Expression

Cells (1×10^6/ml) may be either pretreated or not with an antioxidant for an optimized duration necessary for the antioxidant to function best. Following the pretreatment, cells are either activated or not by appropriate agents, e.g., 100 μM H_2O_2 for 18 hr. To study ICAM-1 expression, cells (0.5×10^6/ml) are incubated with 20 μl of ICAM-1 monoclonal antibody coupled with the fluorochrome fluorescein isothiocyanate (FITC) (Immunotech, Coulter Corporation, Miami, FL) for 30 min at 4°. After the incubation, cells are washed twice in ice-cold Dulbecco's PBS, pH 7.4 (D-PBS), and finally resuspended in fresh D-PBS. Expression of ICAM-1 may be detected by Western blotting or using a flow cytometer as described below.

The fluorescence and light scattering properties (forward scatter, FS; side scatter, SS) of cells may be determined by using a standard EPICS-XL (Coulter, Miami, FL) flow cytometer. In each sample, at least 10,000 gated viable cells are examined. Cells stained with FITC-conjugated antibodies are studied at a rate of 300–500 events per second. FITC fluorescence is detected using a 488 nm argon ion laser, and the emission is recorded

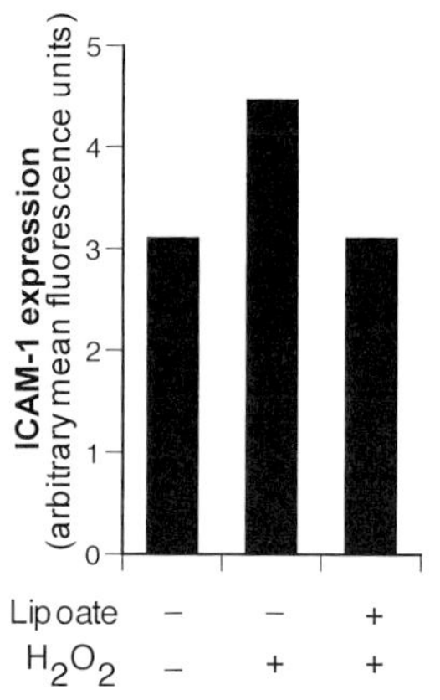

FIG. 4. Pretreatment of Wurzburg T cells with α-lipoate inhibits H_2O_2-induced expression of ICAM-1. α-Lipoate, 250 μM, 24 hr; H_2O_2, 100 μM, 18 hr. ICAM-1 expression was measured using a flow cytometer as described in the text.

at 525 nm. Autofluorescence of cells is recorded and subtracted from total mean fluorescence detected from FITC stained cells to obtain ICAM-1 specific fluorescence data. Pretreatment of cells with the antioxidant α-lipoate inhibited H_2O_2 induced expression of ICAM-1 (Fig. 4) in cells where H_2O_2 is known to result in NF-κB activation as shown by EMSA (Fig. 1).

V. Summary

The goal of this chapter was to review the current protocols that are available to measure the activation of NF-κB. The methods discussed all have their pitfalls when used in isolation. To obtain meaningful information. nuclear translocation and transcriptional activation should be studied in conjunction. Study of NF-κB regulated protein expression is the most physiologically relevant approach to monitoring the transcription regulatory effect of NF-κB. Because of the limitations of transcriptional analysis in primary cell cultures or tissues, incorporation of multiple approaches is recommended when the involvement of NF-κB in a disease process is evaluated.

[37] Assessing Induction of IκB by Nitric Oxide

By MARTIN SPIECKER and JAMES K. LIAO

Introduction

The activation of nuclear factor-κB (NF-κB) is required for the transcriptional induction of many proinflammatory mediators involved in vascular inflammation and atherogenesis such as cellular adhesion molecules, cytokines, and growth factors.[1,2] Factors which modulate the activity of NF-κB activation, therefore, could potentially regulate inflammatory processes and atherogenesis.

NF-κB is a family of homo- or heterodimeric cytosolic proteins, which can be further divided into two groups based upon their structure and function. The first group consists of p65 (Rel A), c-Rel, and Rel B, which contain transcriptional activation domains necessary for gene induction.[3] The second group consists of p105 and p100, which on proteolytic processing

[1] T. Collins, *Lab. Invest.* **68,** 499 (1993).

[2] H. B. Peng, T. B. Rajavashisth, P. Libby, and J. K. Liao, *J. Biol. Chem.* **270,** 17050 (1995).

[3] D. W. Ballard, E. P. Dixon, N. J. Peffer, H. Bogerd, S. Doerre, B. Stein, and W. C. Greene, *Proc. Natl. Acad. Sci. USA* **89,** 1875 (1992).

0076-6879/99 $30.00

give rise to p50 (NF-κB1) and p52 (NF-κB2), respectively.[4,5] On activation, NF-κB translocates into the nucleus. With the exception of Rel B, which cannot form homodimers, members of both groups can bind in a tissue-specific manner as homo- or heterodimers to enhancer elements of target genes.

The activation of NF-κB involves the degradation of its cytoplasmic inhibitor, IκB.[6] Several IκB proteins have been described.[7] They retain NF-κB in the cytoplasm by masking the nuclear localization sequence of NF-κB. Of the different IκB proteins, the best described is IκB-α.

Factors such as glucocorticoids, antioxidants, and salicylates which increase IκB-α expression or prevent IκB-α degradation inhibit NF-κB activation. We and others have shown that the endogenous mediator nitric oxide (NO) inhibits NF-κB and endothelial cell activation through a non-cGMP-dependent mechanism.[8–10] One of the mechanisms by which NO inhibits NF-κB activation is via the transcriptional induction of IκB-α.[11,12] The following sections, therefore, describe methods used to study the induction of IκB-α by NO.

Methods

General Considerations

Several methods are useful in assessing the induction of IκB-α by NO. On the protein level, immunoblotting with IκB-α-specific antibody can detect increases in IκB-α protein levels within 1 hr after treatment with NO donors. A more sensitive method for determining the induction of IκB-α by NO is via IκB-α mRNA analyses by Northern blotting. To determine the association of IκB-α with NF-κB subunits in subcellular compartments such as the nucleus or cytoplasm, immunoprecipitation of the subcellular fraction with agarose- or Sepharose-conjugated NF-κB antibodies followed by Western blotting with an IκB-α antibody is a useful technique.

[4] N. R. Rice, M. L. MacKichan, and A. Israël, *Cell* **71,** 243 (1992).

[5] R. I. Scheinman, A. A. Beg, and A. S. Baldwin, *Mol. Cell. Biol.* **13,** 6089 (1993).

[6] T. Henkel, T. Machleidt, I. Alkalay, M. Krönke, Y. Ben-Neriah, and P. A. Baeuerle, *Nature* **365,** 182 (1993).

[7] P. A. Baeuerle and D. Baltimore, *Cell* **87,** 13 (1996).

[8] R. De Caterina, P. Libby, H. B. Peng, V. J. Thannickal, T. B. Rajavashisth, M. A. Gimbrone, Jr., W. S. Shin, and J. K. Liao, *J. Clin Invest.* **96,** 60 (1995).

[9] B. V. Khan, D. G. Harrison, M. T. Olbrych, R. W. Alexander, and R. M. Medford, *Proc. Natl. Acad. Sci. USA* **93,** 9114 (1996).

[10] P. S. Tsao, R. Buitrago, J. R. Chan, and J. P. Cooke, *Circulation* **94,** 1682 (1996).

[11] H. B. Peng, P. Libby, and J. K. Liao, *J. Biol. Chem.* **270,** 14214 (1995).

[12] M. Spiecker, H. B. Peng, and J. K. Liao. *J. Biol. Chem.* **272,** in press (1997).

In addition, a general method for determining the effects of NO on NF-κB activation is electrophoretic mobility shift assay (EMSA). The effects of NO on IκB-α gene transcription can be determined by nuclear run-on assays. In addition, transient transfection of endothelial cells with various deletional/mutational IκB-α promoter constructs linked to a reporter gene will enable the localization of NO-responsive *cis*-acting element(s).

Detection of IκB-α Resynthesis by Western Blot

Principle of SDS–PAGE and Western Blotting for IκB-α. The temporal relationship between degradation and resynthesis of IκB-α protein can be accurately assessed by Western blotting. Since IκB-α protein is degraded within 15 min following TNF-α (tumor necrosis factor α) stimulation, it is important that cells be lysed quickly in order to minimize the effects attributed to further degradation or resynthesis of IκB-α. The resynthesis of IκB-α protein occurs as early as 1 hr after TNF-α stimulation via an inducible autoregulatory mechanism. In the absence of TNF-α stimulation, treatment with S-nitroso-glutathione (GSNO) (200 μM) alone also results in an increase in IκB-α protein levels after 1 hr.

Denatured and solubilized proteins from whole cell lysates or from subcellular fractions are separated by one-dimensional sodium dodecyl sulfate–polyacrylamide gel electrophoresis (SDS–PAGE) and transferred to a polyvinylidene fluoride (PVDF) membrane. Probing the membrane with a specific IκB-α antibody followed by a secondary antibody coupled to horseradish peroxidase will produce a protein band of 36 to 37 kDa corresponding to IκB-α by chemiluminescence.

Procedure for SDS–PAGE Electrophoresis and Western Blot. We use human saphenous vein endothelial cells (HSVEC) of less than 3 passages grown in Medium 199 (BioWhittaker, Walkersville, MD) containing 5% fetal calf serum (Hyclone), 50 μg/ml endothelial cell growth factor (Collaborative Research, Bedford, MA), 100 μg/ml heparin (Sigma, St. Louis, MO), and penicillin/streptomycin (BioWhittaker). Fresh medium should be added 6–12 hr before treatment. Degradation of IκB-α can be achieved in a concentration-dependent manner by stimulation with TNF-α (200–1000 U/ml). Although the amount of TNF-α is frequently indicated in ng/ml, we find that the activity of TNF-α per ng/ml can vary by more than 10-fold depending on the vendor. Thus, the amount of TNF-α used should preferably be indicated in U/ml.

As a source of NO, we use the NO donor *S*-nitrosoglutathione (GSNO), which has a relatively long half-life in terms of NO release. The molecular weight of GSNO is 373. We usually dissolve 23.25 mg of GSNO in 250 μl H_2O (final volume) just before treatment. Because the half-life of GSNO is greatly shortened by light, the preparation is shielded from light by foil. Following the addition of 2.5 μl of 10 *N* HCl, the GSNO preparation should form a clear red solution containing gaseous bubbles. These bubbles correspond to actual NO formation. The GSNO solution is sterilely filtered and 8 μl of the solution is added per 10 ml medium to achieve a final concentration of 200 μ*M*. We usually pretreat the cells with GSNO for 15–30 min prior to TNF-α stimulation. Again, exposure to light should be minimized. For confirmation, other NO donors should be used to verify the effects of NO on IκB-α.

Since IκB-α is rapidly degraded, preparation of the cell lysate is an important step in assaying IκB-α protein by Western blotting. In order to avoid degradation of the target protein, cell lysates must be prepared rapidly in the presence of various protease inhibitors. Thus, we suggest that the number of cell culture dishes be kept at a minimum in order to conserve time. At the appropriate time point, the culture medium is rapidly and completely removed and the endothelial monolayer is washed twice and ice-cold PBS prior to addition of the cellular lysis buffer. Rapid cell lysis is achieved with a strong lysis buffer (see below). Other methods for cellular lysis may also be adequate as long as the cellular lysis occurs rapidly. Add 300 μl of 2× lysis buffer (room temperature) for each 10 mm culture dish. Keep other culture dishes on ice while working with one dish. Scrape the lysate and transfer it to a microfuge tube on ice. Boil the lysates for 5 min and centrifuge for 2 min at 14,000 rpm in a microcentrifuge at 4°. It is important to keep the lysate on ice prior to boiling to keep the activity of intracellular proteases to a minimum. We determine the protein concentration with the BCA kit (Pierce, Rockford, IL). Before separation on SDS–PAGE, proteins are denatured and reduced by the addition of 5% (v/v) 2-mercaptoethanol to the cellular lysate. The cellular lysate can then be separated on SDS–PAGE or be stored at −80°.

Two different sizes of SDS–PAGE are used. Although IκB-α protein degradation and resynthesis can be detected using a minigel system, quantitation by densitometry and recognition of the slower migrating phosphorylated IκB-α are improved by using larger size gels (i.e., 18 × 16 cm glass plates). With a 12% acrylamide gel, the IκB-α protein band migrates approximately halfway along the length of the gel. Pouring two separating gels with 1.5 mm spacers requires 60 ml solution with 19.8 ml H_2O, 24 ml 30% acrylamide/bisacrylamide, 15 ml Tris (1.5 *M*, pH 8.8), 600 μl 10% SDS, 600 μl 10% ammonium persulfate, and 24 μl TEMED. After the separating gels solidify (i.e., 30–60 min), the stacking gel solution (5% acrylamide) is added to the top of the separating gel. For two gels, prepare 15 ml with 10.2 ml H_2O, 2.53 ml 30% acrylamide/bisacrylamide, 1.88 ml Tris (1 *M*, pH 6.8), 150 μl 10% SDS, 150 μl 10% ammonium persulfate, and 15 μl TEMED.

We use 1 × Towbin buffer as the electrophoresis buffer (for recipe, see below). Approximately 100 μg protein is loaded for a standard gel and 40 μg protein for a minigel. If the samples were frozen, they are boiled again for 2 min. No additional loading buffer is required if the cells were previously lysed in the buffer described. Bromphenol blue is added to the samples to facilitate with the visual loading into the lanes. A low range SDS–PAGE protein standard (i.e., Bio-Rad, Richmond, CA, low range) should be loaded on each gel in a separate lane.

Standard size gels are electrophoresed at 150 V (cooling system or cold room required) for 6–8 hr or overnight at a lower voltage (24–40 V). Minigels can be run at 150 V for 1 hr without cooling. Separated proteins are then transferred to a membrane using a semidry blotting system (Bio-Rad or Hoefer). The type of transfer membrane can strongly affect the final quality of the blots. With the horseradish peroxidase chemiluminescence, better results are achieved with PVDF membranes (Millipore Bedford, MA Immobilon P, pore size 0.45 μm). Soak the membrane for 5 sec in methanol, wash briefly in H_2O. Then soak the gel, the transfer membrane and filter paper 5 min in transfer buffer (for recipe, see below).

For semidry transfer, mount the following layers in order from bottom (anode of transfer apparatus) to top (cathode of transfer apparatus): one buffer-soaked thick filter paper, the transfer membrane, the gel, and one buffer-soaked thick filter paper. Air bubbles between these layers should be avoided and removed by gently rolling a glass pipette over the transfer membrane. Negatively charged proteins will move downward (from the gel into the membrane). With a Bio-Rad semidry transfer apparatus, all of the IκB-α proteins are transferred at 25 V for 30 min. With longer transfer times, some proteins may migrate through the PVDF membrane into the bottom filter paper. Blocking of unspecific binding sites is achieved by incubating the membrane in PBS with 0.1% (v/v) Tween and 5% (w/v) dry milk for 1 hr at room temperature or overnight at 4°.

To detect IκB-α protein, we use an affinity-purified rabbit polyclonal antibody directed against amino acids 6–20 of the human IκB-α (Santa Cruz, IκB-α, C15). Dilutions of 1:250 in PBST and 5% dry milk are satisfactory when the membrane is incubated for 1 hr at 37°. Membranes are washed twice for 5 min and twice for 10 min with PBST at room temperature and then incubated with a secondary antibody (anti-rabbit IgG, coupled to horseradish peroxidase, Amersham) for 30 min. The secondary antibody is diluted 1:4000 in PBST and 5% dry milk. If another IκB-α antibody is used, appropriate dilutions will first have to be determined. Visualization is achieved using the ECL (enhanced chemiluminescence) reagent (Amersham) and autoradiography.

Reagents

GSNO. Add 5 mmol glutathione and 5 mmol sodium nitrite to 8 ml ice-cold H_2O. Stir for 5 min, then add 2.5 ml 2 *N* HCl and stir at 4° for another 30 min. The solution should turn bright pink. Then add 10 ml acetone and stir for another 10 min. GSNO forms a red precipitate. Filter the precipitate and wash it with ice-cold water (5 × 1 ml), acetone (3 × 10 ml), and finally ether (3 × 10 ml). The final yield of GSNO should be 3.8–4 mmol. Store GSNO at room temperature away from light and moisture.

Lysis buffer

100 m*M* Tris, pH 6.8
4% SDS
20% glycerol
30 μl/ml aprotinin (Sigma)
1 m*M* Na_3VO_4
1 m*M* PMSF
1 m*M* NaF

The last four substances should be added immediately before use.

Towbin buffer

10 × (use dilution 1×):
250 m*M* Tris
1.92 *M* glycine
1% SDS, pH 8.3

Western blot transfer buffer, 1×

25 m*M* Tris, pH 8.3
190 m*M* glycine
10% methanol

PBST, pH 7.4

0.1% polyoxyethylene sorbitan
monolaureate (Tween 20) dissolved in phosphate-buffered saline (PBS)

Detection of IκB-α Bound to NF-κB by Immunoprecipitation

To assess whether resynthesized IκB-α binds to NF-κB subunits, cellular extracts are immunoprecipitated with agarose-conjugated Rel A (p65) antibody. The immunoprecipitate is then boiled, denatured, and separated by

SDS–PAGE. Immunoblotting of IκB-α is performed as described above for Western blotting. This technique is particularly useful in assessing the functional relevance of nuclear translocated IκB-α (i.e., ability to bind NF-κB) if nuclear extracts are used. Indeed, nuclear IκB-α has been shown to displace NF-κB from its cognate DNA.[13]

Preparation of Nuclear Extracts for Immunoprecipitation

Wash the endothelial cell monolayer twice with ice-cold PBS and collect cells by scraping in ice-cold PBS (10 ml for a 150 cm^2 flask). Centrifuge for 5 min at 650*g* (4°). Aspirate the supernatant carefully and resuspend cells in a microfuge tube with 360 μl ice-cold Buffer A (for recipe, see below). Incubate 15 min on ice, than add 50 μl of Buffer A containing 6% Nonidet P-40 (NP-40) or Ipegal (Sigma). Spin immediately in a microfuge for 10 min at 4° (16,000*g*). Save supernatant as cytosolic extract and resuspend pellet in 50 μl of Buffer B (for recipe, see below). Incubate on ice for 30 min and centrifuge for 15 min at 4° (16,000*g*). Save supernatant as nuclear extract and determine its protein concentration by the BCA method. Freeze 100 μg aliquots at 80° to avoid repetitive freezing and thawing.

Reagents

Buffer A

10 m*M* HEPES (pH 7.8)
10 m*M* KCl
0.1 m*M* EDTA
0.1 m*M* EGTA
1m*M* DTT
1 m*M* PMSF
Store at 4°.

Buffer B

20 m*M* HEPES (pH 7.9)
0.4 *M* NaCl
1 m*M* EDTA
1 m*M* EGTA
Prepare 10× solution of buffer B and store at −20°.

Immunoprecipitation

Dilute 200–500 μg of nuclear extracts with immunoprecipitation buffer (for recipe, see below) in a total volume of 500 μl. For immunoprecipitation,

[13] U. Zabel, T. Henkel, M. S. Silva and P. A. Baeuerle, *EMBO J.* **12,** 201 (1993).

we use an agarose-conjugated Rel A antibody (Santa Cruz, NF-κB p65 AC). Approximately 5 μl of the agarose-conjugated Rel A antibody is added to the nuclear extract and the mixture is incubated at 4° overnight with gentle rocking. We spin the mixture for 2 min at 14,000 rpm in a microcentrifuge at 4°. The supernatant is saved as a control. The immunoprecipitates (i.e., pellets) are washed 4 times with 500 μl immunoprecipitation buffer. Avoid aspirating the buffer completely; otherwise, a certain amount of the agarose-conjugate pellet may be lost during each washing step. After the final washing step, add 40 μl gel loading buffer, and boil for 5 min to release the antigen–antibody complexes from the agarose. The samples are loaded on an SDS–PAGE gel as previously described for Western blotting for IκB-α.

Reagents

Immunoprecipitation (IP) buffer

PBS
1% Ipegal (Sigma)
2 m*M* EDTA

Add inhibitors just before use: 10 μg/ml leupeptin and aprotinin, 1 m*M* PMSF, 1 m*M* NaF, 1 m*M* Na_3VO_4.

Gel loading buffer, 2×

100 m*M* Tris, pH 6.8
4% (w/v) SDS
20% (v/v) glycerol
0.2% (v/v) bromphenol blue
200 m*M* DTT

Steady-State mRNA Expression of IκB-α

IκB-α is barely detectable by Northern blotting in resting HSVEC. However, on TNF-α stimulation, the IκB-α mRNA level is significantly induced since the IκB-α promoter possesses several functional κB *cis*-acting elements. Both glucocorticoids and NO induce IκB-α gene transcription, but the putative *cis*-acting element(s) have not yet been identified. To determine IκB-α mRNA level, we use the guanidine isothiocyanate method to isolate total RNA from cultured endothelial cells. All solutions should be prepared using DEPC (diethyl pyrocarbonate)-treated H_2O to minimize ribonuclease activity. After washing the cell monolayers with PBS at room temperature, 2.5 ml guanidine isothiocyanate solution (for recipe, see below) is added directly to the 10 cm culture dish. After 2–3 min, the lysate is scraped and transferred into polyallomer centrifuge tubes prefilled with an equal volume of CsCl (2.5 ml, for recipe see below). To balance the filling of the centrifuge tubes, add CsCl solution. Ultracentrifuge at 20° with an SW 55 rotor for 18 hr at 35,000 rpm. RNA forms a small pellet at the bottom of the tube. Discard supernatant completely and clean the inside of the tube with sterile cotton tips to dry out remaining liquid. Dry the pellet until it turns white. Dissolve the pellet in 50–70 μl DEPC-treated H_2O. RNA samples should be stored at −70°. Before the RNA gel is run, double-stranded RNA is denatured by the RNA loading buffer (≥1/3 of RNA volume) and heated at 55° for 15 min. The samples are placed on ice for 2 min and spun briefly in a centrifuge before loading. We usually load 20 μg of total RNA in each lane of a 1% formaldehyde-agarose gel (for 100 ml: 10 ml 10× MOPS, 1 g agarose, 75 ml DEPC-treated H_2O, and 15 ml 12.3 *M* formaldehyde) and electrophoresed at 70 V until the loading dye has migrated one-half the length of the gel (~12 hr).

RNA is transferred overnight to a positively charged nylon membrane by capillary action and baked for 2 hr at 80° prior to prehybridization for at least 4 hr in hybridization solution (recipe see below). Radiolabeling of IκB-α and human β-actin cDNA probe should be performed using random hexamer priming with [α-^{32}P]CTP and Klenow fragment of DNA polymerase I (Pharmacia). Incubate membranes overnight at 45° in hybridization solution. Wash several times (0.2× SSC and 0.1% SDS) at 65° before exposing to an autoradiography film at −80° for 24–72 hr.

Reagents

DEPC–H_2O. Add 1 ml DEPC to 1000 ml H_2O, stir overnight, and autoclave.

MOPS–buffer, 10×. 0.2 *M* 3-morpholinopropanesulfonic acid (MOPS), pH 7.0, 0.05 *M* sodium acetate, 0.01 *M* EDTA in DEPC-H_2O.

Guanidine isothiocyanate solution

4 *M* guanidine isothiocyanate
25 m*M* sodium acetate, pH 6.0
0.1 m*M* DTT
DEPC–water

Sterile filter the solution and store at −80°.

CsCl solution:

5.7 *M* CsCl
25 m*M* sodium acetate, pH 6
DEPC–water

Sterile filter the solution and store at −20°.

Hybridization buffer

50% formamide
5× SCC
2.5× Denhardt's solution
25 m*M* sodium phosphate buffer (pH 6.5)
0.1% SDS
250 μg/ml salmon sperm DNA

Electrophoretic Mobility Shift Assay

To investigate the functional effect of IκB-α induction on NF-κB activity, the activation of NF-κB is assessed by electrophoretic mobility shift assay (EMSA). This method has been described in detail for NF-κB previously.[14] We use nuclear extracts from 10^7 cells prepared according to the method described above for immunoprecipitation. The NF-κB consensus sequence 5′-AGTTGAGGGGACTTTCCCAGG-3′ is end-labeled with [γ-^{32}P]ATP and T_4 polynucleotide kinase (New England Biolabs). Although other oligonucleotides may require gel purification after labeling, the NF-κB consensus oligonucleotide yields good results with low background when purified by Sephadex G-50 columns (Pharmacia Biotech). To label the blunt end oligonucleotide, prepare the following reaction (total volume 10 μl): 50 ng oligonucleotide (1–3 μl), 0.1 *M* DTT (1 μl), 10× T_4 kinase buffer (1 μl), [γ-^{32}P]ATP (3000 Ci/mmol, 4 μl) and T_4 kinase (1 μl). Incubate for 30 min at 37°, then terminate the reaction with 40 μl 0.1 *M* EDTA. After purification, dilute the final probe to 4000 cpm/μl and store the labeled probe at −20°. Incubate 5–10 μg nuclear extracts with ^{32}P-labeled NF-κB oligonucleotide (~20,000 cpm) in a buffer containing 2 μg poly(dI · dC), 10 μg bovine serum albumin, and 2 μl 10× EMSA buffer (for recipe, see below) in a total volume of 20 μl for 20 min at room temperature followed by 10 min on ice. To determine the specificity of the NF-κB band, mutated and unlabeled oligonucleotide are added (20- to 50-fold excess) to the binding reaction prior to the addition of labeled probe. The unlabeled oligonucleotide should compete for and attenuate the NF-κB band while that of the mutated oligonucleotide would have no effect.

Additionally, specificity can be determined by supershifting the specific NF-κB band(s) with antibodies directed against Rel A or p50. For the supershift assay, 150 ng/ml specific antibody is added to the binding reaction for 10 min prior to the addition of the labeled probe (at room temperature). We routinely use nuclear extracts from TNF-α stimulated cells in the su-

[14] R. Schreck and P. A. Baeuerle, *Methods Enzymol.* **234,** 151 (1994).

pershift assay. The antibody–NF-κB complex migrates slower and is "supershifted" compared to that of NF-κB alone.

To resolve DNA–NF-κB complexes, load the binding reactions on a vertical 4% nondenaturating polyacrylamide gel. A tracking dye (bromphenol blue/xylene cyanol) should be loaded on each gel in a separate lane as a marker of oligonucleotide position. The gel is electrophoresed in a low ionic strength buffer (0.5× TBE; for recipe, see below) at 200 V for 2–3 hr (until bromphenol blue is 2.5 cm from the bottom of the gel). Gel separation of the binding reaction should be performed at 4° or with a buffer cooling system. The gel is transferred onto filter paper (Whatman), covered with Saran wrap, and vacuum dried at 80° for approximately 90 min. The vacuum-dried gel is subjected to autoradiography at −80° for 18–36 hr.

Interpretation of EMSAs. Nuclear extracts from HSVEC should not have NF-κB/DNA binding activity under basal conditions. Cell lysates should always be kept on ice during preparation of nuclear extracts. Otherwise, artifactual NF-κB activation during the extraction procedure may occur. The activation of NF-κB activation is detectable as early as 15 min following TNF-α stimulation. The NF-κB band migrates relatively slowly during gel separation and is commonly found in the upper portion of the gel. However, it is always important to prove the specificity of the band, preferably by excess unlabeled oligonucleotide and by antibody supershifting. Nonspecific bands should not be affected by unlabeled oligonucleotides or NF-κB antibodies.

Reagents

EMSA buffer, 10×

100 m*M* Tris, pH 7.5
500 m*M* NaCl
10 m*M* DTT
10 m*M* EDTA
50% glycerol
Store at −20°.

TBE(Tris–borate) buffer, 5×

0.45 *M* Tris–borate (per liter: 54 g Tris base, 27.5 g boric acid)
10 m*M* EDTA
H_2O

Methods for Assessing IκB-α Gene Transcription

On TNF-α stimulation, IκB-α protein is completely degraded within 15 min. Treatment with NO donors does not affect this rapid degradation process. Treatment with NO donors in the presence and absence of TNF-α, however, induces IκB-α after 60–120 min. Thus, acquiring an accurate assessment of IκB-α gene transcription by NO is necessary. We use two methods for assessing IκB-α gene transcription. The first method is the "gold standard" or nuclear run-on assay which measures *in vitro* transcriptional activity of nascent IκB-α mRNA from NO-treated endothelial cells. The second method involves transiently transfecting various deletional/mutational IκB-α promoter constructs linked to a reporter gene (i.e., firefly luciferase) into bovine endothelial cells and determining the reporter gene activity as an index of IκB-α promoter activity. The advantage of the second method is that potential NO-responsive *cis*-acting element(s) can be localized.

Procedure for Nuclear Run-on Assay. This *in vitro* transcription assay requires nuclei from 10^8 endothelial cells for each condition. A careful extraction of nuclei is a crucial step.

Preparation of nuclear extracts. We collect cells after washing them 3× in ice-cold PBS. Spin down cells at 1500 rpm in 50 ml conical tubes for 5 min at 4°. The samples should be kept on ice for all following steps. Discard the supernatant and add 12 ml of Buffer 1 (for recipe see below). Slightly resuspend the pellet and transfer it to an autoclaved Dounce homogenizer (Wheaton 15 ml, type A). The number of strokes necessary to disrupt cytoplasmic membranes differs for cell types. For HSVEC, approximately 40 strokes are adequate. Check 10 μl of nuclei under the microscope every 20 strokes. Removal of all cytoplasmatic membranes is not necessary. Transfer buffer with nuclei into a 15 ml sterile conical tube and centrifuge again for 5 min at 4° (1500 rpm). Wash the pellet 2× with 5 ml Buffer 1 and spin down. Finally, resuspend nuclear pellet with 100 μl of Buffer 2 (storage buffer; for recipe, see below) and freeze with dry ice/ethanol. Store at −80°.

Preparation of DNA blot. We use a Schleicher & Schuell (Keene, NH) vacuum transfer slot-blot apparatus. Soak the apparatus with 0.2 *N* NaOH for 1 hr at room temperature. Rinse once with distilled water and dry. Then mount a presoaked cut nylon membrane on Schleicher & Schuell gel blot paper. Denature IκB-α cDNA and, as a control, human β-actin cDNA and linearized pGEM containing plasmids (for each condition 1 μg DNA in 50 μl 0.2 *N* NaOH for 30 min at room temperature; use autoclaved microfuge tubes). Mix denatured DNA with 500 μl 6× SCC and transfer the solution into the specific slots of the blot apparatus. Absorb the DNA with mild vacuum and add another 1 ml of 6× SCC into each slot. Then air dry the removed membrane and cut it for each condition. UV cross-link the membrane and bake at 80° for 2 hr.

In vitro transcription. Thaw the nuclei on ice and mix with 100 μl of Buffer 3 (transcription buffer). The transcription is carried out in a 30°

shaking water bath for 30 min. Terminate the reaction by incubation with 40 U of DNase I (RNase free) at 30° for 20 min in a shaking bath. Degrade the proteins with 200 μl Buffer 4 with proteinase K. Extract radiolabeled RNA transcripts 1× with 400 μl phenol/chloroform, and 1× with chloroform. Add 2 *M* ammonium acetate and 2.5 volumes ethanol to the transcripts and keep in dry ice for 20 min prior to centrifugation at 12,000*g* for 15 min. Air-dry the pellet.

Hybridization. Prehybridize membranes in hybridization bags at 42° for more than 2 hr (for recipe prehybridization buffer see Northern blot reagents). Resuspend pellets in 1 ml prehybridization buffer and add equal amounts of radioactivity ($1–5 \times 10^7$ cpm) to all blots. Hybridization of radiolabeled transcripts to the nylon membranes is carried out at 42° for 48 hr. Before exposing the membranes to an autoradiography film (at −80°), wash with 2× SSC + 0.1% SDS at 42° for 20 min and then with 0.2× SSC + 0.1% SDS at 65° for 20 min.

Reagents

Buffer 1 (lysis buffer). 10 m*M* Tris-HcL (pH 7.4), 10 m*M* NaCl, 3 m*M* $MgCl_2$, DEPC–H_2O. Autoclave and add 0.5% NP-40 (or Ipegal).

Buffer 2 (storage buffer). 20 m*M* Tris-HCl (pH 8.1), 75 m*M* NaCl, 0.5 m*M* EDTA, DEPC–H_2O. Autoclave and add 1 m*M* DTT and 50% glycerol freshly before use.

Buffer 3 (transcription buffer). 10 m*M* Tris-HCl (pH 8.0), 5 m*M* $MgCl_2$, 300 m*M* KCl, 50 μ*M* EDTA, DEPC–H_2O, autoclave. Use 88 μl of Buffer 3 for each sample of nuclei and add 1 m*M* DTT, 250 U/ml RNAsin, 0.5 m*M* CTP, ATP, GTP, 250 μCi [α-^{32}P]UTP.

Buffer 4 (with proteinase K). 0.4% SDS, 40 m*M* Tris-HCl (pH 7.4), 10 m*M* EDTA, DEPC–H_2O, autoclave. Use 198 μl of Buffer 4 for each sample and add 400 μg/ml proteinase K (2 μl of 10 mg/250 μl stock solution).

Procedure for Transient Transfection. Transfection of endothelial cells with the IκB-α promotor linked to a reporter gene is another useful method for quantitating IκB-α gene transcription. Efficiency and reproducibility of transient transfection in human endothelial cells, however, is very low (<2% by β-galactosidase staining). We find that efficiency is significantly improved in bovine aortic endothelial cells (~15%). The preparation of IκB-α promoter constructs has been previously described.[15,16] For transfection studies, bovine aortic endothelial cells of less than 3 passages are seeded in 100 mm or 6-well dishes (uncoated). We use DME medium with 10% heat-

[15] O. Le Bail, R. Schmidt-Ullrich, and A. Israel, *EMBO J.* **12,** 5043 (1993).

[16] C. Y. Ito, N. Adey, V. L. Bautch, and A. S. Baldwin, Jr., *Genomics* **29,** 490 (1995).

inactivated fetal calf serum to grow the cells to 60–70% confluency. The culture medium should be changed 4 hr before transfection to medium containing 5% (v/v) fetal calf serum (5 ml for 100 mm dishes and 1.5 ml for 6-well dishes). All transfection reagents should be passed through 0.22 μm pore filters. We use the calcium phosphate coprecipitation method to introduce plasmids into bovine aortic endothelial cells. Another plasmid vector (i.e., pRSV.β-gal) is cotransfected in order to serve as an internal control for transfection efficiency.

The principle of the calcium phosphate coprecipitation method is the formation of DNA/calcium phosphate precipitates which adhere to the endothelial cell surface.[17] To allow DNA precipitation, add plasmid DNA (plasmid with IκB-α promoter and plasmid with reporter gene for internal control) gently on top of dH_2O (final volume 500 μl). The amount of plasmid DNA sufficient for transfection depends on the purity. With both CsCl gradient centrifugation and Qiagen column purification, 10 μg of each plasmid DNA per 10 cm dish is sufficient. Place 31 μl of 2 *M* $CaCl_2$ on top of the DNA. Finally, add 250 μl of 2× HBSP buffer from the bottom of the sterile reaction tube. Mix gently by bubbling air with the pipette tip from the bottom of the tube (1× 250 μl). The precipitates form slowly (within 30–60 min) at room temperature. The solution should turn slightly cloudy. Drop the DNA/$CaPO_4$ mixture gently into the medium and incubate for 4 hr at 37°. Then add another 5 ml or 1.5 ml of culture medium containing 10% fetal calf serum (for 100 cm or 6 well dishes, respectively) and incubate overnight. By then, a fine precipitate should be visible on top of the cells on microscopic inspection. We typically incubate endothelial cells for 48 hr after DNA transfection prior to treatment with TNF-α or GSNO.

Reagents

Sterile distilled H_2O
2× HBSP buffer:
- 1.5 m*M* Na_2HPO_4
- 10 m*M* KCl
- 280 m*M* NaCl
- 12 m*M* glucose
- 50 m*M* HEPES, pH 7.0

2 *M* $CaCl_2$

Principle of Reporter Gene Assay. The assessment of promoter activity requires a reporter gene, preferably absent in the cell type of interest. The IκB-α promoter constructs inserted upstream to a reporter gene without

[17] F. L. Graham and A. J. van der Eb, *Virology* **52,** 456 (1973).

intrinsic promoter activity allows indirect assessment of transcriptional activity. We use two different reporter genes. The reporter gene encoding for the prokaryotic chloramphenicol acetyltransferase (CAT) is more cumbersome to use, but adequate in sensitivity for determining IκB-α promoter activity. In the firefly luciferase assay, the promoter of interest is linked to the luciferase gene. Luciferin and other components are added to the cellular lysate. The production of light is monitored by a luminometer. We use a plasmid containing the firefly luciferase gene (pGL$_2$, Promega). Advantages of the luciferase system are rapid nonradioactive assaying, high sensitivity, and minimal endogenous activity.[18] The emitted light signal is measured by a luminometer (Berthold Luminat Wallac Inc., Gaithersburg, MD). Cotransfection with another reporter gene (i.e., β-galactosidase) may require a different assay system. We recommend using a β-galactosidase reporter gene as internal control. The Dual-Light kit from Tropix (Bedford, MA), detects luciferase and β-galactosidase in the same cell extract.[19] Other dual reporter gene kits are available. Using the same cell extract for the internal control saves time and reduces methodical variability.

Procedure for Luciferase Assay. The light signal produced by the luciferin/luciferase reaction decays with a half-life of approximately 1 min; the light signal from the galacton-plus/β-galactosidase reaction has a half-life of up to 180 min. Therefore, the luciferase signal should be measured immediately after adding the substrate and β-galactosidase signal. For preparation of the cell lysate and incubation with substrates we refer to the instructions of the kit used. With the Tropix kit we use only 100 μl lysis buffer per 10 cm dish. The time intervals between measuring the first (luciferase) and second (β-galactosidase) light signal should be constant for individual samples.

Acknowledgments

This work was supported by a grant from the National Institutes of Health (HL-52233). Martin Spiecker is a recipient of the Feodor Lynen Fellowship (Alexander von Humboldt-Stiftung). James K. Liao is an Established Investigator of the American Heart Association.

[18] S. J. Gould and S. Subramani, *Anal. Biochem.* **7,** 5 (1988).
[19] C. S. Martin, P. A. Wight, A. Dobretsova, and I. Bronstein, *BioTechniques* **21,** 520 (1996).

[38] Nitrosative Stress

By ALFRED HAUSLADEN and JONATHAN S. STAMLER

Introduction

Nitric oxide (NO) and its aerobic reaction products have long been recognized as cytotoxic, mutagenic, and carcinogenic. The discovery of endogenous NO biosynthesis and a multitude of NO signaling functions has greatly increased interest in identifying the targets involved and the nature of the interactions. Nitros(yl)ation of transition metals or of basic amino acids, in particular cysteines, form the basis of NO-related activity[1]: The best-characterized mechanism of signal transduction is the activation of guanylate cyclase by nitrosyl-heme formation[1]; other mechanisms include the activation of the bacterial transcription factors soxR[2] and oxyR[3] by nitrosylation of an iron-sulfur cluster and a cysteine, respectively, and the activation of G proteins by nitrosylation of a conserved cysteine conforming to a consensus motif.[4] However, nitros(yl)ation—either when excessive or when happening in less than regulated fashion—can lead to a loss of cellular function. The molecular correlates of nitrosative injury include enzyme inactivation and damage to DNA.

We have termed such impairment of metabolism "nitrosative stress,"[3] and the causal species nitrosants, emphasizing the relationship to—and analogy with—oxidative stress and oxidants. Here we focus on the differences, the parallels, and the synergism between nitrosative stress and oxidative stress with emphasis on mechanism of cytotoxicity and the protective response, in particular, the activation of antinitrosative defenses that lead to the acquisition of resistance. An approach to assessment of nitrosative stress is also provided.

Interactions between Oxidants and Nitrosants

Nitric oxide does not react with most cellular constituents and is usually quite selective in the choice of cysteine or heme-containing target protein, with which it reacts in the presence of molecular oxygen. However, in combination with reactive oxygen species or with increases in the NO level,

[1] J. S. Stamler, *Cell* **78,** 931 (1994).
[2] T. Nunoshiba, T. DeRojas-Walker, J. S. Wishnok, S. R. Tannenbaum, and B. Demple, *Proc. Natl. Acad. Sci. USA* **90,** 9993 (1993).
[3] A. Hausladen, C. T. Privalle, T. Keng, J. DeAngelo, and J. S. Stamler, *Cell* **86,** 719 (1996).
[4] J. S. Stamler, E. J. Toone, S. A. Lipton, and N. J. Sucher, *Neuron* **18,** 691 (1997).

METHODS IN ENZYMOLOGY, VOL. 300
0076-6879/99 $30.00

nitrosation and oxidation are more readily observed. This may extend the signaling repertoire of reactive nitrogen species (e.g., beyond activation of guanylate cyclase or inhibition of caspases) on the one hand,[5,6] but predisposes to toxic and mutagenic effects on the other hand. Subtle changes in the molecular mechanism of nitrosation may account for such a loss of selectivity toward targets during stress. For instance, NO may be directly responsible for S-nitrosylation in certain regulated functions, in which case the reaction is first-order in NO and critically dependent on thiol pK,[7] or S-nitrosylation may be enzymatically controlled—as exemplified in hemoglobin catalysis—and thus highly selective for substrate.[8] On the other hand, N_2O_3[9–11] probably acts as the principal nitrosant when the NO level is $>10\ \mu M$.[12] Nitrosation and deamination of DNA bases and mutagenicity are much more likely in this case because N_2O_3 donates reactive NO^+.

Other examples of powerful nitrosating and oxidizing species are the NO-superoxide product peroxynitrite,[13,14] the poorly characterized derivatives that form when NO and O_2^- are generated in unequal amounts, and the species with NO^+ and hydroxyl radical character that form in transition metal-catalyzed reactions between hydrogen peroxide and NO.[15,16] Indeed, the connection between oxidative and nitrosative stress is well exemplified by, but not limited to, reactions of NO/O_2^-, which can nitrate,[17–22] nitrosate,

[5] J. Haendeler, U. Weiland, A. M. Zeiher, and S. Dimmeler, *Nitric Oxide* **1,** 282 (1997).
[6] J. B. Mannick, X. Q. Miao, and J. S. Stamler, *J. Biol. Chem.* **272,** 24125 (1997).
[7] A. J. Gow, D. G. Buerk, and H. Ischiropoulos, *J. Biol. Chem.* **272,** 2841 (1997).
[8] A. J. Gow and J. S. Stamler, *Nature* **391,** 169 (1998).
[9] V. G. Kharitonov, A. R. Sundquist, and V. S. Sharma, *J. Biol. Chem.* **270,** 28158 (1995).
[10] D. A. Wink, J. A. Cook, S. Y. Kim, Y. Vodovotz, R. Pacelli, M. C. Krishna, A. Russo, J. B. Mitchell, D. Jourd'heuil, A. M. Miles, and M. B. Grisham, *J. Biol. Chem.* **272,** 11147 (1997).
[11] N. Hogg, R. J. Singh, and B. Kalyanaraman, *FEBS Lett.* **382,** 223 (1996).
[12] D. A. Wink, K. S. Kasprazak, C. M. Maragos, R. K. Elespuru, M. Misra, T. M. Dunams, T. A. Cebula, W. H. Koch, A. W. Andrews, J. S. Allen, and L. K. Keefer, *Science* **254,** 1001 (1991).
[13] J. S. Beckman, *Chem. Res. Toxicol.* **9,** 836 (1996).
[14] W. H. Koppenol, J. J. Moreno, W. A. Pryor, H. Ischiropoulos, and J. S. Beckman, *Chem. Res. Toxicol.* **5,** 834 (1992).
[15] R. Pacelli, D. A. Wink, J. A. Cook, M. C. Krishna, W. DeGraff, N. Friedman, M. Tsokos, A. Samuni, and J. B. Mitchell, *J. Exp. Med.* **182,** 1469 (1995).
[16] R. Farias-Eisner, G. Chaudhuri, E. Aeberhard, and J. M. Fukuto, *J. Biol. Chem.* **271,** 6144 (1996).
[17] A. Gow, D. Duran, S. R. Thom, and H. Ischiropoulos, *Arch. Biochem. Biophys.* **333,** 42 (1996).
[18] H. Ischiropoulos, M. F. Beers, S. T. Ohnishi, D. Fisher, S. E. Garner, and S. R. Thom, *J. Clin. Invest.* **97,** 2260 (1996).
[19] A. J. Gow, D. Duran, S. Malcolm, and H. Ischiropoulos, *FEBS Lett.* **385,** 63 (1996).
[20] I. Y. Haddad, S. Zhu, H. Ischiropoulos, and S. Matalon, *Am. J. Physiol.* **270,** L281 (1996).
[21] J. P. Crow and H. Ischiropoulos *Methods Enzymol.* **269,** 185 (1996).

and oxidize[23–26] biological molecules. Thus, NO by itself has little inhibitory effect on bacteria or cultured cells, while N_2O_3, NO/O_2^-, SNOs, and peroxynitrite can be highly cytotoxic.[15,16,27,28]

Manifestations of Nitrosative Stress

The targets of nitrosative stress—protein, DNA, lipids—are in many cases the same as those damaged by oxidative stress, and exposure to high levels of nitrosants can result in thiol oxidation[11,29,30] or induce an oxidative stress through the depletion of antioxidants. However, while a change in redox state is the *sine qua non* of an oxidative stress, it is not a prerequisite for a nitrosative stress. This state is alternatively characterized by (excessive) covalent attachments (e.g., NO group attachment) or additions to target nucleophiles. At the molecular level, for example, proteins experience nitrosative modifications that impair function.[31] Such nitros(yl)ation can occur in the absence of oxidation. The following examples are cases in point:

$$RSH + NO_2^- + H^+ \rightarrow RSNO + H_2O \quad (1)$$

$$RSH + N_2O_3 \rightarrow RSNO + NO_2^- + H^+ \quad (2)$$

Acidified nitrite is equivalent, through dehydration, to NO^+, and N_2O_3 to $[NO^+ \, NO_2^-]$, that is, the reactions involve substitution of H^+ for NO^+ rather than a change in oxidation state of either N or S. Inasmuch as the assignment of electrons to S ($RS^- \, NO^+$) depends on the nature of the R group, oxidation may occur in some instances. However, many RSNOs transfer NO^+, preserving the reduced thiol, which suggests that the electron assignment is appropriate as written.

[22] J. S. Beckman, H. Ischiropoulos, L. Zhu, M. van der Woerd, C. Smith, J. Chen, J. Harrison, J. C. Martin, and M. Tsai, *Arch. Biochem. Biophys.* **298,** 438 (1992).

[23] M. G. Salgo and W. A. Pryor, *Arch. Biochem. Biophys.* **333,** 482 (1996).

[24] R. M. Uppu, G. L. Squadrito, and W. A. Pryor, *Arch. Biochem. Biophys.* **327,** 335 (1996).

[25] M. G. Salgo, E. Bermudez, G. L. Squadrito, and W. A. Pryor, *Arch. Biochem. Biophys.* **322,** 500 (1995).

[26] R. Radi, J. S. Beckman, K. M. Bush, and B. A. Freeman, *J. Biol. Chem.* **266,** 4244 (1991).

[27] I. Ioannidis, M. Bätz, M. Kirsch, H. G. Korth, R. Sustmann, and H. De Groot, *Biochem. J.* **329,** 425 (1998).

[28] M. A. De Groote, D. Granger, Y. Xu, G. Campbell, R. Prince, and F. C. Fang, *Proc. Natl. Acad. Sci. USA* **92,** 6399 (1995).

[29] E. G. DeMaster, B. J. Quast, B. Redfern, and H. T. Nagasawa, *Biochemistry* **34,** 11494 (1995).

[30] S. P. Singh, J. S. Wishnok, M. Keshive, W. M. Deen, and S. R. Tannenbaum, *Proc. Natl. Acad. Sci. USA* **93,** 14428 (1996).

[31] D. I. Simon, M. E. Mullins, L. Jia, B. Gaston, D. J. Singel, and J. S. Stamler, *Proc. Natl. Acad. Sci. USA* **93,** 4736 (1996).

Nitrosylation can also have effects distinct from oxidation. This is best exemplified in regulation of the cardiac calcium release channel (ryanodine receptor). Poly-S-nitrosylation reversibly activates the calcium channel, whereas thiol oxidation leads to irreversible channel activation and a loss of control.[32] In other words, channel regulation is not the result of a general change in redox state. Rather, the channel can sense and distinguish between the nitrosative and oxidative modifications. Likewise, the activation of the bacterial transcription factor oxyR by S-nitrosylation seems to be more readily reversible than activation by oxidation[3]. Taken together, the data suggest the possibility of differential control by nitrosative and oxidative signals. Moreover, SNO modifications may protect thiols from irreversible oxidation.

Resistance to Nitrosative Stress

The first line of defense against production of nitrosants is glutathione (GSH). The high intracellular concentrations make it a primary target for oxidants and nitrosants alike. *S*-Nitrosoglutathione, however, does not accumulate to any significant degree, suggesting the presence of SNO metabolizing activity.[3] Such "SNOase" candidates include the enzymes thioredoxin and thioredoxin reductase, which cleave *S*-nitrosoglutathione more readily than glutathione disulfide. Ubiquitous reductants like ascorbate, moreover, reduce SNO to NO.[4,33–35] Protein thiol may also serve to buffer nitrosants; proteins readily modified without loss of function are cases in point.[31,32,36] But ultimately, SNO accumulation in proteins will be associated with untoward effects[3] and glutathione depletion is likely to be a contributing factor. Indeed, several enzymes involved in maintaining the GSH pool, including γ-glutamylcysteine synthetase (glutamate–cysteine ligase), glutathione reductase, and glutathione peroxidase[37–39] are themselves inhibited by S-nitrosylation or NO-mediated oxidation. Additional regulatory pro-

[32] L. Xu, J. P. Eu, G. Meissner, and J. S. Stamler, *Science* **279,** 234 (1998).

[33] D. Nikitovic and A. Holmgren, *J. Biol. Chem.* **271,** 19180 (1996).

[34] G. Scorza, D. Pietraforte, and M. Minetti, *Free Radic. Biol. Med.* **22,** 633 (1997).

[35] M. Kashiba-Iwatsuki, M. Yamaguchi, and M. Inoue, *FEBS Lett.* **389,** 149 (1996).

[36] J. S. Stamler, O. Jaraki, J. Osborne, D. I. Simon, J. Keaney, J. Vita, D. Singel, C. R. Valeri, and J. Loscalzo, *Proc. Natl. Acad. Sci. USA* **89,** 7674 (1992).

[37] M. A. Keese, M. Bose, A. Mulsch, R. H. Schirmer, and K. Becker, *Biochem. Pharmacol.* **54,** 1307 (1997).

[38] M. Asahi, J. Fujii, T. Takao, T. Kuzuya, M. Hori, Y. Shimonishi, and N. Taniguchi, *J. Biol. Chem.* **272,** 19152 (1997).

[39] J. Han, J. S. Stamler, H.-L. Li, and O. Grifith, *in* "Biology of Nitric Oxide (IV)" (J. S. Stamler, S. Gross, S. Moncada, and A. Higgs, Eds.), Portland Press, London (1995).

teins, lipids, and DNA[40,41] are subject to nitrosative damage as well. Nitrosative stress might increase mutations, inhibit growth, promote cell death,[3,28,40–46] or contribute to the pathophysiological processes of atherosclerosis, chronic inflammation, and sepsis.[47]

Accumulating evidence suggests that cells possess inducible mechanisms to counter the toxic effects of nitrosants in much the same way as oxidative stress is controlled by an elaborate system of inducible antioxidative defenses. Demple and co-workers first showed that the soxRS regulon in *Escherichia coli,* which controls the expression of antioxidative genes in response to increased fluxes of superoxide, is also induced by NO and provides resistance to its untoward effects.[2,48] We have shown that the hydrogen peroxide responsive trancription factor oxyR, which controls the expression of peroxide scavenging genes, also provides resistance to the cytostatic effects of SNOs.[3] Molecular recognition of NO by soxRS and SNO by OxR is performed by an iron-sulfur cluster and cysteine, respectively. Thus, metals and thiols in proteins are not only the sites most prone to the damaging effects of NO/SNO exposure, but also serve to alert the cell to their presence. None of the genes controlled by soxRS or oxyR are known to detoxify nitrosants, so the protective effect has been explained by virtue of (elimination of) harmful interactions with oxidants. There is, however, evidence that *E. coli* metabolize nitrosants and recovery from SNO-induced growth inhibition correlates well with SNO decomposition.[3] Resistance to SNOs has also been shown in *Salmonella typhimurium,* where mutation of a dipeptide transporter or homocysteine-producing gene provide protection.[28,44] Rat hepatocytes are protected from SNO toxicity by pretreatment with a low dose of SNO, indicating an inducible mechanism of SNO resistance,[49] and macrophages became resistant to NO induced-growth inhibition by continuous stimulation of NO biosynthesis.[50,51] NO

[40] T. Nguyen, D. Brunson, C. L. Crespi, B. W. Penman, J. S. Wishnok, and S. R. Tannenbaum, *Proc. Natl. Acad. Sci. USA* **89,** 3030 (1992).

[41] A. Gal and G. N. Wogan, *Proc. Natl. Acad. Sci. USA* **93,** 15102 (1996).

[42] M. N. Routledge, D. A. Wink, L. K. Keefer, and A. Dipple, *Carcinogenesis* **14,** 1251 (1993).

[43] S. Tamir, S. Burney, and S. R. Tannenbaum, *Chem. Res. Toxicol.* **9,** 821 (1996).

[44] M. A. De Groote, T. Testerman, Y. Xu, G. Stauffer, and F. C. Fang, *Science* **272,** 414 (1996).

[45] J. C. Zhuang and G. N. Wogan, *Proc. Natl. Acad. Sci. USA* **94,** 11875 (1997).

[46] F. C. Fang, *J. Clin. Invest.* **99,** 2818 (1997).

[47] J. F. Kerwin, Jr., J. R. Lancaster, Jr., and P. L. Feldman, *J. Med. Chem.* **38,** 4343 (1995).

[48] T. Nunoshiba, T. DeRojas-Walker, S. R. Tannenbaum, and B. Demple, *Infect. Immun.* **63,** 794 (1995).

[49] Y. M. Kim, H. Bergonia, and J. R. Lancaster, Jr., *FEBS Lett.* **374,** 228 (1995).

[50] B. Brüne, C. Götz, U. K. Mebmer, K. Sandau, M.-H. Hirvonen, and E. G. Lapatina, *J. Biol. Chem.* **272,** 7253 (1997).

[51] M. R. Hirvonen, B. Brune, and E. G. Lapetina, *Biochem. J.* **315,** 845 (1996).

TABLE I
DETECTION OF NITROSANTS AND PRODUCTS OF NITROSATIVE STRESS

Nitrosant or product	Method	Ref.
NO_2^-	Capillary electrophoresis	57
	HPLC	57
	Griess reaction	58
SNO	Photolysis/chemiluminescence	59
	Saville assay	59
	HPLC/postcolumn reactor	60
	HPLC/mass spectrometry	61
	Capillary electrophoresis	62
Nitrosamines	GC/chemiluminescence	63, 64
	Fluorescence	57
Metal nitrosyls	Electron paramagnetic resonance	65, 66
Exhaled NO	GC/mass spectrometry	64
Deaminated DNA	Mutagenesis	67, 68
$NO\cdot + O_2^{\cdot -}/NO^+ + H_2O_2$	Tyrosine nitration, thiol nitrosation	21, 59–62

scavenging by superoxide,[50] metallothionein,[52] or heat shock protein[53] has been implicated as a resistance mechanism. In addition, a number of enzymes have been implicated in NO or SNO resistance, but their physiological significance remains to be established.[33,54–56]

Assessment of Nitrosative Stress

The assessment of a nitrosative stress is an important objective. One way is to measure products of NO metabolism that can serve as nitrosants. These include nitrite (which has the potential to serve as an NO^+ equivalent), SNOs, nitrosamines (or derivatives thereof) and metal nitrosyls, particularly low mass species. The formation of a fluorescent triazole following nitrosation of 2,3-diaminonaphlalene at physiological pH[57] or an increasing ratio of nitrite to nitrate are useful measures of nitrosative stress. A second approach is to measure the biological end products of nitrosation, in particu-

[52] M. A. Schwarz, J. S. Lazo, J. C. Yalowich, W. P. Allen, M. Whitmore, H. A. Bergonia, E. Tzeng, T. R. Billiar, P. D. Robbins, J. R. Lancaster, Jr., and B. R. Pitt, *Proc. Natl. Acad. Sci. USA* **92,** 4452 (1995).

[53] K. Bellmann, M. Jaattela, D. Wissing, V. Burkart, and H. Kolb, *FEBS Lett.* **391,** 185 (1996).

[54] M. P. Gordge, J. S. Hothersall, G. H. Neild, and A. A. Dutra, *Br. J. Pharmacol.* **119,** 533 (1996).

[55] N. Hogg, R. J. Singh, E. Konorev, J. Joseph, and B. Kalyanaraman, *Biochem. J.* **323,** 477 (1997).

[56] V. Borutaité and G. C. Brown, *Biochem. J.* **315,** 295 (1996).

[57] P. J. Andrew, M. Auer, I. J. Lindley, H. F. Kauffmann, and A. J. Kungl, *FEBS Lett* **408,** 319 (1997).

lar, nitrosylated proteins and deaminated DNA bases. Table I provides a list of nitrosants and nitrosated products as well as references to some chemical, electrophoretic, spectroscopic, and chemiluminescent methods used for their detection.

[57] A. M. Leone and M. Kelm, *in* "Methods in Nitric Oxide Research" (M. Feelisch and J. S. Stamler, Eds.), p. 499, Wiley, Chichester, UK (1996).
[58] H. H. H. W. Schmidt and M. Kelm, *in* "Methods in Nitric Oxide Research" (M. Feelisch and J. S. Stamler, Eds.), p. 491. Wiley, Chichester, UK (1996).
[59] J. S. Stamler and M. Feelisch, *in* "Methods in Nitric Oxide Research" (M. Feelisch and J. S. Stamler, Eds.), p. 521. Wiley, Chichester, UK (1996).
[60] T. Akaike, K. Inoue, T. Okamoto, H. Nishino, M. Otagiri, S. Fujii, and H. Maeda, *J. Biochem. (Tokyo)* **122,** 459 (1997).
[61] I. Kluge, U. Gutteck-Amsler, M. Zollinger, and K. Q. Do, *J. Neurochem.* **69,** 2599 (1997).
[62] J. S. Stamler and J. Loscalzo, *Anal. Chem.* **64,** 779 (1992).
[63] L. K. Keefer and D. L. H. Williams, *in* "Methods in Nitric Oxide Research" (M. Feelisch and J. S. Stamler, Eds.), p. 509. Wiley, Chichester, UK (1996).
[64] A. M. Leone, L. E. Gustafsson, P. L. Francis, M. G. Persson, N. P. Wiklund, and S. Moncada, *Biochem. Biophys. Res. Commun.* **201,** 883 (1994).
[65] D. J. Singel and J. R. Lancaster, *in* "Methods in Nitric Oxide Research" (M. Feelisch and J. S. Stamler, Eds.), p. 341. Wiley, Chichester, UK (1996).
[66] Y. A. Henry and D. J. Singel, *in* "Methods in Nitric Oxide Research" (M. Feelisch and J. S. Stamler, Eds.), p. 357. Wiley, Chichester, UK (1996).
[67] S. Christen, P. Gee, and B. N. Ames, *Methods Enzymol.* **269,** 267 (1996).
[68] S. Tamir, T. deRojas-Walker, J. S. Wishnok, and S. R. Tannenbaum, *Methods Enzymol.* **269,** 230 (1996).

[39] Determination of Cell–Cell Adhesion in Response to Oxidants and Antioxidants

By SASHWATI ROY, CHANDAN K. SEN, and LESTER PACKER

Introduction

Adhesion of leukocytes to endothelial cells is the earliest step in immune recognition process and is mediated by cell adhesion molecules (CAM).[1] In the early phases of cell adhesion, leukocytes transiently adhere to the vessel wall in a process termed "rolling." Rolling of leukocytes is mediated by a family of adhesive molecules called selectins, expressed both on the leukocyte and endothelial cell surface.[2] After "rolling," leukocytes firmly

[1] M. P. Bevilacqua, *Annu. Rev. Immunol.* **11,** 767 (1993).
[2] C. W. Smith, *Semin. Hematol.* **30,** 45 (1993).

0076-6879/99 $30.00

adhere to endothelial cells. This process is mediated by the binding of lymphocyte function-associated antigen-1 (LFA-1, CD11a/CD18) and very late antigens-4 (VLA-4) expressed on leukocytes to intercellular adhesion molecule-1 (ICAM-1, CD54) and vascular cell adhesion molecule-1 (VCAM-1, CD 106) present on the endothelial cell surface, respectively.[3] Platelet endothelial cell adhesion molecule-1 (PECAM-1) is one of the CAM that is involved in transendothelial migration of leukocytes through the vessel wall into inflamed foci.[4] CAM expression and adhesive properties of cells are greatly modified in several conditions such as cancer, atherosclerosis, diabetes, chronic inflammation, and ischemia–reperfusion injury.[5–10] Redox imbalances have been suggested to play a critical role in the etiology of the above-mentioned diseases.

The expression of CAM are known to be induced in response to several stimuli such as cytokines (tumor necrosis factor-α, TNF-α; interleukin-1α and -1β), phorbol 12-myristate 13-acetate (PMA), and lipopolysaccharide.[3] Oxidants and antioxidants have been shown to directly or indirectly influence the expression of CAM and other cell–cell adhesion processes.[11–14]

Significance and Overview of Cell–Cell Adhesion Assays

Cell–cell adhesion assay is a powerful tool for the study of interaction between leukocytes and endothelial cells during inflammation, immunological disorders, and other disorders where altered cell adhesion processes

[3] S. M. Albelda, C. W. Smith, and P. A. Ward, *FASEB J.* **8,** 504 (1994).

[4] S. M. Albelda, W. A. Muller, C. A. Muller, C. A. Buck, and P. J. Newman, *J. Biol. Chem.* **114,** 1059 (1991).

[5] J. Thiery, D. Teupser, A. K. Walli, B. Ivandic, K. Nebendahl, O. Stein, Y. Stein, and D. Seidel, *Atherosclerosis* **121,** 63 (1996).

[6] J. M. Munro, *Eur. Heart J.* **14** (Suppl. K), 72 (1993).

[7] K. Pantel, G. Schlimok, M. Angstwurm, B. Passlick, J. R. Izbicki, J. P. Johnson, and G. Riethmuller, *Ciba Found. Symp.* **189,** 157 (1995).

[8] C. Dosquet, D. Weill, and J. L. Wautier, *Nouv. Rev. Fr. Hematol.* **34** (Suppl.), S55 (1992).

[9] M. Cutolo, A. Sulli, A. Barone, B. Seriolo, and S. Accardo, *Clin. Exp. Rheumatol.* **11,** 331 (1993).

[10] G. L. Kukielka, K. A. Youker, H. K. Hawkins, J. L. Perrard, L. H. Michael, C. M. Ballantyne, C. W. Smith, and M. L. Entman, *Ann. NY Acad. Sci.* **723,** 258 (1994).

[11] H. Sellak, E. Franzini, J. Hakim, and C. Pasquier, *Blood* **83,** 2669 (1994).

[12] V. Lakshminarayanan, D. W. A. Beno, R. H. Costa, and K. A. Roebuck, *J. Biol. Chem.* **272,** 32910 (1997).

[13] T. Aoki, Y. Suzuki, K. Suzuki, A. Miyata, Y. Oyamada, T. Takasugi, M. Mori, H. Fujita, and K. Yamaguchi, *Am. J. Respir. Cell. Mol. Biol.* **15,** 319 (1996).

[14] N. Marui, M. K. Offermann, R. Swerlick, C. Kunsch, C. A. Rosen, M. Ahmad, R. W. Alexander, and R. M. Medford, *J. Clin. Invest.* **92,** 1866 (1993).

are implicated in the etiology of disease.[15] Cell–cell adhesion is expected to be mediated by a large number of CAM present on leukocytes and target cells; only some such molecules are characterized at present. Thus, although study of the expression of individual adhesion molecules is informative, prediction of actual cell–cell adhesion based on such data may not always be correct. A number of different methods have been used for quantifying cell–cell adhesion. Microscopic investigation of adhered cells and radioactive (^{51}Cr or ^{3}H) labeling of leukocytes in cell adhesion assays are some of the widely used methods.[16,17] Such assays are time consuming and require individual handling of samples. Several spectrophotometric assays such as biotin labeling of leukocytes have also been reported to study cell–cell interactions.[18] Most of these techniques include several steps, and therefore chances of error are high. Since the fluorescence microtiter plate reader became available, fluorescence based techniques have proven to be useful to study cell–cell interactions.[15,19] Here we describe a simple and highly sensitive fluorescent method using calcein to estimate cell–cell adhesion in response to oxidants and antioxidants. Among the various methods available for the determination of cell–cell adhesion, fluorescence labeling of adherent cells using calcein has several advantages, particularly in studies where the roles of oxidants and antioxidants are evaluated in cell adhesion processes. Several fluorescent probes are sensitive to changes in intracellular pH, Ca^{2+}, or redox state and thus may not be effectively used to study the effect of oxidants or antioxidants. Also, loading of some of these probes to live cells may interfere with several aspects of cell function. Calcein has proven to be safe in these respects.[19]

Cells and Culture Conditions

Human Jurkat T cells clone E6-1 (American Type Culture Collection, ATCC, Rockville, MD) are grown in RPMI 1640 medium (GIBCO-BRL, Life Technologies Inc., Gaithersburg, MD) supplemented with 10% (v/v) fetal calf serum, 100 U/ml penicillin and 100 μg/ml streptomycin, 110 mg/liter sodium pyruvate, and 2 m*M* L-glutamine (University of California, San Francisco, CA). ECV304 (ECV), a spontaneously transformed immortal endothelial cell line established from the vein of an apparently normal

[15] F. Braut-Bocher, J. Pichon, P. Rat, M. Adolphe, M. Aubery, and J. Font, *J. Immunol. Methods* **178,** 41 (1995).
[16] N. Oppenheimer-Marks, L. S. Davis, and P. E. Lipsky, *J. Immunol.* **145,** 140 (1990).
[17] M. H. Thornhill, U. Kyan-Aung, T. H. Lee, and D. O. Haskard, *Immunology* **69,** 287 (1990).
[18] R. Pearce-Pratt, D. M. Phillips, and A. S. Bourinbaiar, *J. Immunol. Methods* **140,** 159 (1991).
[19] L. S. De Clerck, C. H. Bridts, A. M. Mertens, M. M. Moens, and W. J. Stevens, *J. Immunol. Methods* **172,** 115 (1994).

human umbilical cord, are obtained from ATCC (Bethesda, MD). ECV cells are grown in medium 199 (GIBCO-BRL, Life Technologies Inc., Gaithersburg, MD) supplemented with 10% (v/v) fetal calf serum, 100 U/ml penicillin, and 100 μg/ml streptomycin. Primary cultures of human umbilical vein endothelial cells (HUVEC), are obtained from Cascade Biologics Inc. (Portland, OR). HUVEC cells are grown in medium 200 (Cascade Biologics Inc., Portland, OR) supplemented with low serum growth supplement (LSGS, Cascade Biologics Inc., Portland, OR). Cells are maintained in a standard culture incubator with humidified air containing 5% (v/v) CO_2 at 37°.

α-Lipoate (racemate mixture, ASTA Meica, Frankfurt, Germany) stock solution for ECV pretreatment is prepared fresh in basic phosphate-buffered saline (PBS) and pH is adjusted to 7.4. Hydrogen peroxide (H_2O_2, Sigma, St. Louis, MO) stock solution is prepared fresh in endotoxin-free type sterile distilled water. PMA (Sigma, St. Louis, MO) stock solution is prepared in sterile dimethyl sulfoxide (DMSO).

Calcein Labeling of Cells

Calcein acetoxymethyl ester (calcein-AM, Molecular Probes, Eugene, OR) is used to fluorescently label Jurkat T cells. Nonfluorescent calcein-AM is lipophilic and is cleaved by intracellular esterases to yield highly charged fluorescent calcein that is retained by viable cells. The fluorescence labeling of Jurkat T cells is achieved by incubating cells (1×10^7 cells/ml) with 5 μM calcein-AM in RPMI 1640 for 30 min at 37°. Calcein-AM is prepared as a 1 mM stock in DMSO and stored in aliquots at $-20°$. After loading of calcein-AM, cells are washed three times with phosphate-buffered saline (pH 7.4, PBS) to remove excess dye. Cell viability, as detected by trypan blue exclusion, is $>95\%$ up to 3 hr after loading. The cells are finally resuspended in phenol red free RPMI 1640 containing 10% fetal calf serum at a density of 2×10^6 cells/ml.

Cell–Cell Adhesion Assay

Monolayers of HUVEC or ECV are seeded at a density of 10^4 cells/well in 96 well tissue culture plates (Falcon 3072, Becton Dickinson, Franklin Lakes, NJ). After 24 hr of seeding, the cells are pretreated with α-lipoate for 48 hr and then activated with H_2O_2 (100 μM, 1 hr), TNF-α (50 ng/ml, 6 hr), or PMA (100 nM, 24 hr). The relative contributions of ICAM-1 and VCAM-1 to the adherence of Jurkat T cells to ECV may be investigated by treating ECV with anti-human ICAM-1 (10 μg/ml) or anti-human VCAM-1 (10 μg/ml) antibodies for 30 min before coculture of ECV with

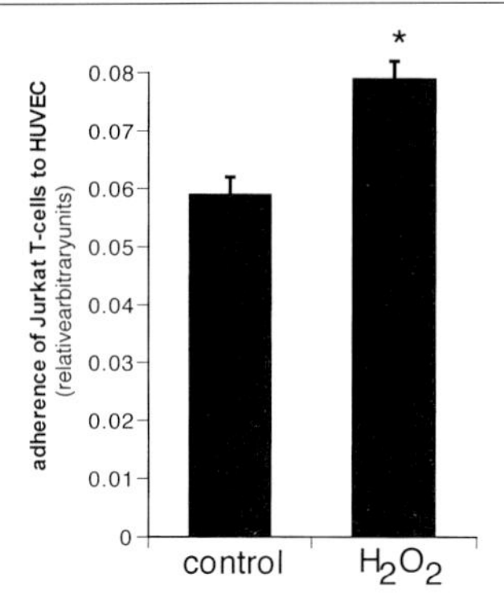

FIG. 1. Hydrogen peroxide (H_2O_2) increases adhesion of human Jurkat T cells to HUVEC. HUVEC were activated with 0.1 m*M* H_2O_2 for 1 hr. Following 1 hr of activation, cells were washed three times with PBS and then cocultured with calcein-AM labeled Jurkat T cells for 1 hr. Data are mean ±SD of at least five experiments. *, $p < 0.01$ (Student's *t*-test) when compared with nonactivated HUVEC.

Jurkat T cells. Confluent cultures of HUVEC/ECV should be used to avoid any nonspecific attachment of T cells to the plastic bed of the 96-well plate. Before the assay of cell–cell adhesion, the HUVEC or ECV monolayers are washed three times with PBS. Calcein-AM labeled Jurkat T cells (2×10^5 cells/well) are cocultured with HUVEC or ECV monolayer for 1 hr in a culture incubator with humidified air containing 5% CO_2 at 37°. Blank wells with HUVEC or ECV monolayer alone are maintained in final Jurkat T-cell suspension medium. After the coculture period, the nonadherent Jurkat T cells are removed from monolayers by washing each well four times with D (Dulbecco's)-PBS using a multichannel pipette with wide open tips.

Fluorescence Microtiter Plate Reader

The fluorescence intensity of each well is measured using a fluorescence plate reader (CS-9301, Shimadzu, Corporation, Columbia, MD). The excitation and emission wavelengths for the calcein molecule are 480 and 530 nm, respectively. Data from the plate reader are collected and processed using a CS-9301PC software (Shimadzu Corporation, Columbia, MD).

Oxidant-Induced Adherence of Human Jurkat T Cells to HUVEC

We observed a significant ~25% ($p < 0.01$) increase in the adherence of Jurkat T cells to HUVEC following activation of HUVEC with 0.1 m*M* H_2O_2 for 1 hr (Fig. 1). Reactive oxygen species (ROS) generated by hypo-

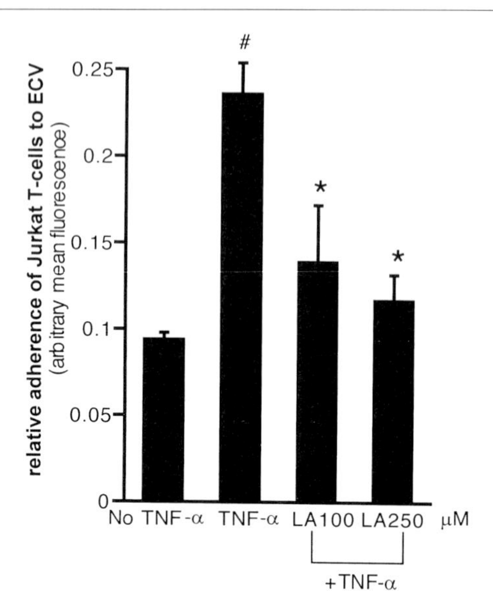

FIG. 2. TNF-α induced adhesion of human Jurkat T cell to ECV cell is inhibited by α-lipoate pretreatment. ECV cells were pretreated with α-lipoate (100 or 250 μM) for 72 hr and then activated with 50 ng/ml TNF-α for 6 hr. Cells were washed three times with PBS and then co-cultured with calcein-AM labeled Jurkat T cells for 1 hr. Data are mean $\pm$ SD of at least five experiments. #, $p < 0.01$ (Student's t-test) when compared with control (no TNF-α) cells. *, $p < 0.01$ (Student's t-test) when compared with LA nontreated and TNF-α-treated cells.

xanthine (HX)-xanthine oxidase (XO) have been previously reported to increase the adherence of polymorphonuclear neutrophil adherence to HX-XO treated HUVEC both immediately and after 2 hours.[11] We and others[11] have observed that such increased binding of leukocytes to oxidant-exposed HUVEC involves ICAM-1, but is independent of its upregulation.

Antioxidant Regulation of TNF-α Induced Adhesion of Human Jurkat T Cell

To investigate the role of antioxidants on cytokine-induced lymphocyte–endothelial cell adhesion, ECV cells are pretreated with thiol antioxidant α-lipoate and then activated with TNF-α (50 ng/ml) for 6 hr. Treatment of ECV cells markedly increased (1.5-fold) Jurkat T cell adhesion to ECV compared to cells that were not treated with TNF-α. Pretreatment of ECV cells with α-lipoate (100 or 250 μM) for 72 hr effectively downregulated TNF-α induced ICAM-1 expression compared to cells that were not pretreated with α-lipoate (Fig. 2).

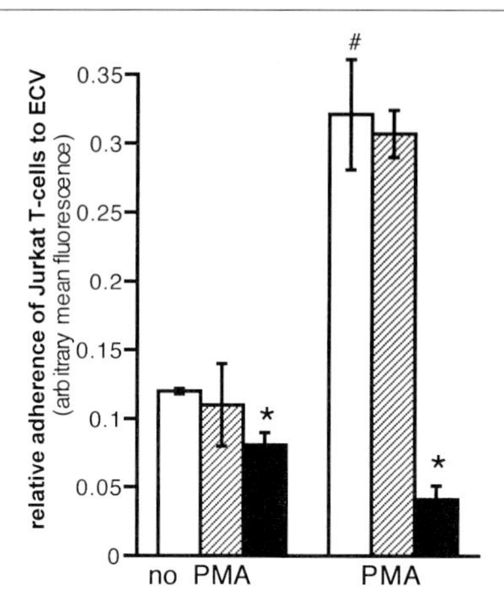

FIG. 3. PMA-induced adhesion of Jurkat T-cells to ECV cells and relative contributions of ICAM-1 and VCAM-1 molecules in such adhesion. ECV cells were treated with or without 100 nM PMA (open bars) for 24 hr. Following activation the cells were treated with anti-human VCAM-1 (10 μg/ml, hatched bars) or anti-human ICAM-1 (10 μg/ml, solid bars) antibodies for 30 min prior to coculture with Jurkat T cells. Adhesion assay was performed with calcein-AM labeled Jurkat T cells for 1 hr as described in the text. Data are mean ± SD of at least five experiments. #, $p < 0.01$ (Student's t-test) when compared with control (no PMA) cells. *, $p < 0.01$ (Student's t-test) when compared PMA-treated cells.

Effect of Anti-ICAM-1 and Anti-VCAM-1 on PMA-Induced Cell–Cell Adhesion

The relative contributions of ICAM-1 and VCAM-1 in Jurkat T cells' adherence to ECV cells was characterized following treatment of PMA-activated or nonactivated ECV cells with anti-human ICAM-1 and anti-human VCAM-1 antibody. Treatment of ECV cells with anti-ICAM-1 markedly (85–90%) blocked PMA-induced Jurkat T-cell adhesion to ECV cells. Anti-human VCAM-1 antibody treatment of ECV cells had no significant effect on such cell–cell adhesion, suggesting that ICAM-1, but not VCAM-1, plays a major role in agonist induced adhesion of Jurkat T cells to ECV cells (Fig. 3). ICAM-1 antibody also decreased (~35%) the adherence of Jurkat T cells to nonactivated ECV. Again, anti-human VCAM-1 antibody treatment of ECV cells had no significant effect on such cell–cell adhesion.

[40] Antioxidant Regulation of Gene Expression: Analysis of Differentially Expressed mRNAs

By KISHORCHANDRA GOHIL, SASHWATI ROY, LESTER PACKER, and CHANDAN K. SEN

Introduction

To develop a better understanding of the exact mechanisms that underlie reactive oxygen species-dependent disorders in biological systems, studies have been directed to investigate the regulation of gene expression by oxidants, antioxidants, and other determinants of the intracellular reduction–oxidation (redox) state. [1–5] Progress in genome sciences has led to the development of sensitive and rapid techniques for DNA sequencing.[6] These techniques, combined with improved application of the techniques for total RNA and mRNA isolations, amplification of DNA fragments by polymerase chain reaction (PCR), and the techniques for comprehensive analysis of transcription products, such as serial analysis of gene expression (SAGE)[7] and differential display of mRNA,[8] are useful to examine changes in gene expression profile in response to treatment of cells by different antioxidants.

Differential display of mRNA enables a simultaneous comparison of mRNA transcripts of control and antioxidant treated cells (see Fig. 1A and 1B). Unlike substraction–hybridization, the differential display method enables the detection of both upregulated and downregulated mRNAs in the same experiment. After selection of the appropriate biological model for study and its exposure to an antioxidant or a complex mixture of antioxidants such as in extracts (EGb761) of *Ginko biloba,* the technique involves seven major steps that lead to the definition of specific gene(s) affected by the treatment. These steps are (1) extraction of total and mRNA from the biological sample, (2) cDNA synthesis with the three distinct one base-anchored oligo(dT) primers, (3) amplification of cDNAs with the

[1] C. K. Sen and L. Packer, *FASEB J.* **10,** 709 (1996).
[2] J. M. Muller, M. R. A. Rupec, and P. A. Baeuerle, *Methods* **11,** 301 (1997).
[3] C. K. Sen, *Biochem. Pharmacol.* **55,** 1747 (1998).
[4] C. K. Sen, *Curr. Top. Cell Regul.* **36,** in press (1998).
[5] K. Polyak, Y. Xia, J. L. Zweier, K. W. Kinzler, and B. Vogelstein, *Nature* **389,** 300–305 (1997).
[6] M. D. Adams, C. Fields, and J. C. Venter (Eds.), "Automated DNA Sequencing and Analysis," Academic Press, London, 1994.
[7] E. V. Velculescu, L. Zhang, B. Vogelstein, and K. W. Kinzler, *Science* **270,** 484 (1995).
[8] P. Liang and A. B. Pardee, *Science* **257,** 967 (1992).

METHODS IN ENZYMOLOGY, VOL. 300
0076-6879/99 $30.00

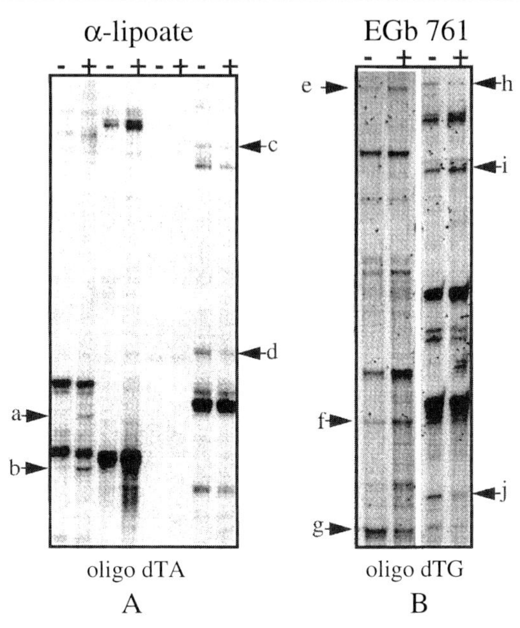

FIG. 1. α-Lipoate and EGb761 sensitive mRNAs. Selected autoradiographic images of cDNA fragments resolved on 6% polyacrylamide gel (Step 4). Jurkat T lymphocytes and human endothelial cells were treated with α-lipoate (100 μM, 72 hr) and EGb 761 (100 μg/ml, 72 hr), respectively. Total RNA was extracted as described in step 1 and cDNA synthesized, amplified, and resolved on polyacrylamide gel as described in steps 2, 3, and 4, respectively. (A) The arrows indicate cDNA fragments that are clearly modulated by the antioxidant treatment; a and b show upregulated, whereas c and d indicate downregulated cDNAs by exposure to α-lipoate. (B) Treatment with EGb 761 extract resulted in upregulation of three cDNAs, e, f, and i, and downregulation of cDNAs indicated as g, h, and j.

combination of anchored and random primers in the presence of a radiolabeled deoxynucleotide triphosphate, (4) separation of the amplified radiolabeled cDNA fragments on polyacrylamide gel followed by exposure of the dried gel to an autoradiographic film, (5) lane-by-lane comparison of the cDNA bands displayed on the film followed by identification and selection of differentially displayed bands, (6) excision of the selected bands from the gel followed by extraction of cDNA, amplification, and purification, and (7) cloning and sequencing of the gel-purified cDNAs and verification of the specific mRNA change by Northern analysis.

Differential Display of mRNA

This technique for the identification of mRNA transcripts affected by a defined treatment can be divided into seven distinct analytical steps,

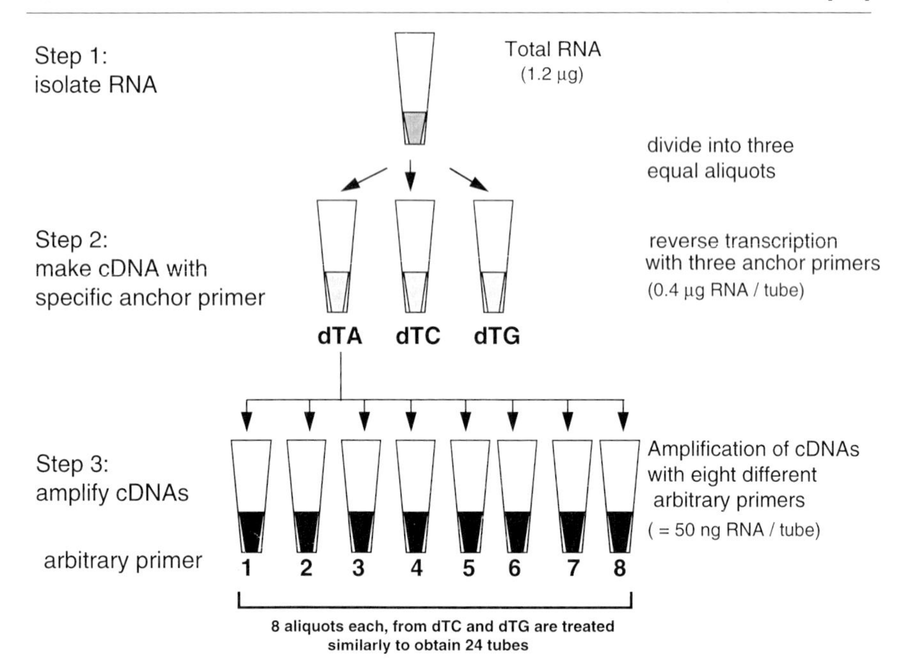

FIG. 2. Schematic representation of the first three steps of differential display of mRNA by polymerase chain reaction (PCR) to illustrate that a single RNA sample is divided into 24 equal aliquots, each amplified with a different pair of primers to screen a maximum number of mRNA transcripts (see text, step 3, for details). Aliquots of PCR products are further analyzed as described in steps 4–7 to identify differentially expressed mRNAs.

starting with the isolation of total RNA and ending with the verification of differentially expressed transcripts by Northern analysis. The first three steps are illustrated in Fig. 2.

Step 1: Extraction of Total RNA and mRNA

The ubiquitous presence of RNase activity requires special precautions to ensure that RNA under investigation is preserved in its native state. We routinely use sterile, disposable, polypropylene labware and hand gloves whenever possible. These precautions have proved sufficient to maintain integrity of isolated RNA as measured by the appearance of 18S and 28S ribosomal RNA on 1% agarose gel (see below, step 7).

Total RNA is extracted from cell pellets ($\sim 10^8$ cells) using TRIzol Reagent (GIBCO-BRL, Life Technologies, Gaithersburg, MD), according to the manufacturer's protocol. The yield of total RNA is about 0.7–1.0 mg/10^8 cells. An aliquot of total RNA (~ 100 μg) is frozen immediately and stored at −80° (for subsequent steps such as DNase I treatment, cDNA synthesis, and RNA blots, see below). The remaining total RNA is used for mRNA preparation by poly(A)$^+$ selection using MessageMaker Reagent Assembly (GIBCO-BRL). The yield of mRNA from total RNA thus obtained is 15–20 μg. This amount is sufficient for a number of subsequent steps, including Northern analysis. We have found that mRNA prepared from Jurkat cells by this method can be stored as a concentrated solution (1–2 μg/μl) at −80° for at least 1 week without degradation and for at least 2 months as pellet in ethanol (80–100%). In addition to obtaining total RNA and mRNA, this protocol also enables the isolation of total DNA and protein, both of which may serve as useful markers for normalization of data from one experiment to another.

Step 2: cDNA Synthesis

Total RNA obtained in step 1 is used as a template to prepare cDNA with oligo(dT) one-base anchor primers. To remove contaminating DNA which frequently copurifies with RNA the total RNA is treated with DNase I. This step is essential to minimize the occurrence of differentially displayed bands of non-mRNA origin. The most essential requirement for this step is to use DNase with no detectable RNase activity. We routinely use MessageClean Kit (GenHunter Corporation, Nashville, TN). DNase-treated total RNA can be stored at −80° as a concentrated solution (1–2 μg/μl) or as a pellet in ethanol (80–100%). It is very important that DNase treated RNA from control and treated cells is accurately requantitated by spectral analysis (40 μg/ml of RNA give A_{260nm} of 1 unit; 1 cm light path in a quartz cuvette) which also enables the determination of the ratio at 260 nm/280 nm, a useful indicator of the purity of RNA; a ratio of 1.8–2.0 is obtained for most preparations.

DNase-treated total RNA is reverse transcribed with three different oligo(dT)$_{11}$N primers, where N can be either dATP or dCTP or dGTP. Total RNA (0.2 μg/μl) from Jurkat T cells or from human endothelial cells is reverse transcribed with recombinant or purified reverse transcriptase of Moloney murine leukemia virus. We routinely use the RNAimage Kit (GenHunter Corporation, Nashville, TN) for this purpose and all the subsequent steps up to step 6. Superscript II reverse transcriptase (GIBCO-BRL) can be substituted for the GenHunter reverse transcriptase. This step requires 1.2 μg of total RNA from each of the control and antioxidant treated groups, and all of it is used in the following amplification step with eight different primers (see Fig. 2). The cDNA obtained following reverse transcription can be stored at −80°; however, we perform the amplification step (step 3) on the same day.

Step 3: Amplification of cDNA

Each aliquot of cDNA from Step 2 is subdivided into eight equal aliquots to which each of the eight different random primers (supplied with the RNAimage Kit) are added. Since there are three different anchored primers a total of 24 different tubes are generated from each sample of total RNA. Hence, to compare the possible effects of antioxidant treatment with cells that were untreated a grand total of 48 tubes is generated. The commonly used thermal cycler (Perkin-Elmer, Norwalk, CT) is equipped with a 48-tube sample block and therefore can accommodate all the tubes generated in this step. The set of eight random primers in combination with the three different oligo(dT)-anchored primers allows the assessment of ~30% of the mRNAs. It is necessary to use 80 different random primers (all available from GenHunter Corporation either individually or as part of RNAimage Kits) to screen most (>95%) of the ~15,000 different mRNAs in a cell.[9,10]

We have routinely used the set of random primers supplied in the RNAimage Kit (GenHunter Corporation) at concentrations recommended in the manufacturer's protocol. It is critical that the cDNA concentration and the reagents used for the preparation of the eight different amplification tubes for the control and antioxidant-treated cells be identical to ensure valid comparison of the changes in autoradiographic band intensities of amplified transcripts. The amplification conditions specified in the manufacturer's protocol are applied without modification. If [α-^{35}S]dATP is used to label the transcripts, then the thermal cycler must be operated in an appropriately certified fume hood because at high temperatures volatile, ^{35}S-labeled products are formed.[11,12]

Step 4: Electrophoretic Separation of cDNA Fragments

The radiolabeled cDNA transcripts from step 3 are resolved on a 6% polyacrylamide gel by high voltage electrophoresis (power supply, 2000 V; Sigma, St. Louis, MO) using Tris–borate–EDTA (100 m*M* Tris, 90 m*M* boric acid, 1 m*M* EDTA, pH 8.4) as the running buffer. To resolve a maximum number of transcripts it is necessary to analyze these samples on a large gel such as that used for the determination of DNA sequence. The complete electrophoresis apparatus can be purchased from a number of suppliers. We routinely use the apparatus (GenHunter Corporation) that enables a simultaneous analysis of 48 samples; this is the number of

[9] P. Liang and A. B. Pardee (Eds.), "Methods in Molecular Biology: Differential Display Methods and Protocols," Vol. **85,** Humana Press, Totoya, NJ, 1997.

[10] P. Liang, L. Averboukh, and A. B. Pardee, *Meth. Molec. Genet.* **5,** (1994).

[11] S. F. Trentmann, E. van der Knapp, and H. Kende, *Science* **267,** 1186 (1995).

[12] P. Liang and A. B. Pardee, *Science* **267,** 1186 (1995).

samples generated for screening ~30% of the mRNAs in a single experiment (see step 2 or Fig. 2 for more details). The use of a 48-square-well comb rather than a shark-tooth comb for sample wells is strongly recommended because the former generates well separated lanes and eliminates sample cross-contamination during loading and during the excision of differentially expressed bands.

In order to obtain sharply resolved bands it is also essential to prerun the gel for at least 30 min and flush the sample wells with the running buffer to remove urea before loading the samples. An aliquot (3.5 μl) from each tube from Step 3 is mixed with 2 μl of the loading dye (RNAimage Kit). The samples are incubated at 80° for 2 min and loaded into wells thoroughly flushed free of urea. The gel is developed for 3.5 to 4 hr at about 60 W constant power (as recommended in the RNAimage Kit protocol). The gel is then transferred to 3M Chr paper (Whatman, Clifton, NJ) and dried for about 1 hr under vacuum at 80°. An autoradiographic film (BioMax MR, Kodak Scientific, Rochester, NY) with enhancer screens (VWR Scientific) is exposed to the dried gel at −70°. The film must be stapled to the dried gel so that permanent marks are retained on the gel and the film for exact alignment of the two for excision of the differentially displayed bands. At least two such autoradiographic films are obtained so that one can be filed for permanent record and the other can be used for identification and excision of differentially expressed transcripts from the dried gel. Alternatively, if an imaging system is available a permanent record of a digital image of the autoradiograph can be stored, and one can proceed to step 5 without the necessity for a second exposure.

Step 5: Identification of Differentially Displayed cDNA Fragments

The autoradiographic image of the gel must show sharp, well-defined bands with minimal or no regular ladders (see Figs. 1A and 1B). Ideally, the lanes should have low background between bands. High background usually indicates poor or multiple hybridization of the templates with the primers generating multiple amplification products that are poorly resolved during electrophoresis and appear as streaks on the autoradiographs. The top 70% of the autoradiographic film, representing about 500–100 bp fragments, is then carefully examined lane by lane, for each of the anchor primers, namely, $(dT)_{11}A$ or $(dT)_{11}C$ or $(dT)_{11}G$. Bands that show clear difference between lanes in each pair of lanes are noted (for example, see the first two lanes in Fig. 1A). Frequently such differences are likely to be real, especially if the bands above and below the identified bands are of very similar or identical intensities. Such differentially displayed bands are good candidates for further investigations as described in steps 6 and 7 below.

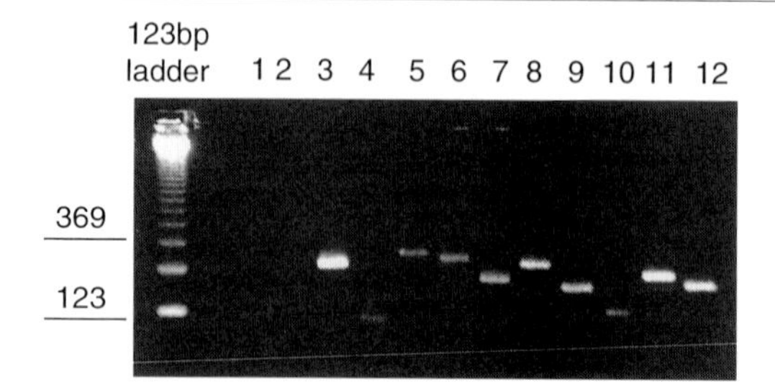

FIG. 3. Differentially displayed cDNA fragments excised from dried polyacrylamide gel and amplified with the appropriate pair of primers as described in step 6. Ethidium bromide-stained DNA fragments (123–400 bp) are electrophoretically resolved on 2% agarose gel. Nine of the 12 fragments excised from polyacrylamide gel could be reamplified in the first round of amplification.

Step 6: Excision and Amplification of cDNA Fragments

After identification of differentially displayed bands on the autoradiographic film, further characterization of corresponding cDNA fragments requires the excision of the radiolabeled cDNA fragments from the dried polyacrylamide gel. The film is carefully superimposed on the dried gel using the alignment markers punched by the staples as described in step 4. A thin slice of gel containing the radiolabeled cDNA fragment is excised and transferred to a 1.5 ml tube by cutting through the film, into the gel and the 3M blot paper. The DNA fragment is then eluted from the gel by soaking and boiling in 100 μl of water [diethyl pyrocarbonate-treated, autoclaved] as described in the protocol of the RNAimage Kit. Eluted DNA is pelleted and redissolved in 10 μl of water (DEPC-treated, autoclaved) as described in the protocol of the RNAimage Kit. We use 8 μl of the eluted and concentrated, excised DNA fragment for amplification with the primer set that generated the radiolabeled cDNA on the large gel. The PCR products (30 μl) of excised cDNA fragments are analyzed on low melt agarose (Apex, McFrugal's Lab Depot) gel [2% (w/v) in Tris(40 m*M*)-HCl, pH 8.0, EDTA (1 m*M*)]. Most (60–70%) of such excised fragments could be reamplified (see Fig. 3) by the first round of amplification using the protocol recommended by GenHunter Corporation in the RNAimage Kit. We have had a limited success in obtaining good sequence information directly from fragments purified from agarose gel. Therefore, we routinely clone the PCR fragments into a plasmid using the pCR-TRAP cloning system as described in step 7.

Step 7: Cloning of Amplified cDNA Fragments and Confirmation of Differential Display of mRNA

In the final step of the differential display technique for identifying the genes modulated by antioxidant treatment, the amplified cDNA fragments

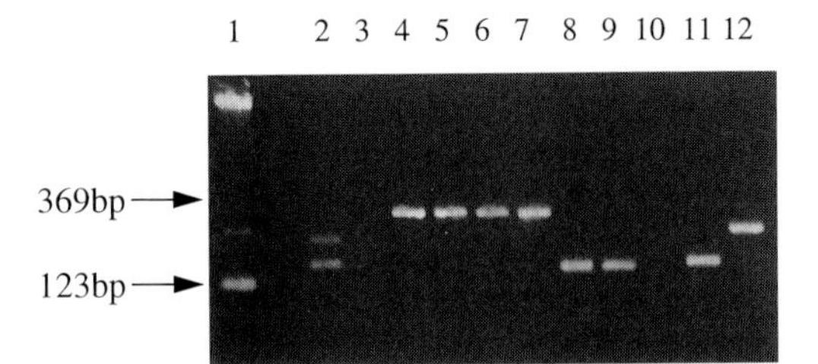

FIG. 4. cDNA inserts obtained by PCR of lysed samples of *Escherichia coli* colonies transformed with the pCR-TRAP vector (pCR-TRAP Cloning System) ligated with an aliquot of PCR solution (containing the amplified cDNA fragment) from step 6. Colonies of transformed *E. coli* were picked from three different Luria–Bertani plates, each plated with *E. coli* transformed with a different cDNA fragment containing the pCR-TRAP plasmid from step 6. Lane 1 represents a 123 bp DNA ladder. Lanes 3–7 and 8–12 are samples of five different colonies from two different plates. Lane 2 shows an insert in a randomly sampled colony from another plate.

obtained in the previous step are cloned into an appropriate plasmid vector such as pCR-Script vector (Stratagene, La Jolla, CA). We use the pCR-TRAP Cloning System (GenHunter Corporation). The transformation efficiency may be increased by doubling the amount suggested in the kit protocol of PCR product used for ligation. The presence of a DNA insert is evaluated by PCR amplification of the insert in lysed *Escherichia coli* colonies using the manufacturer's protocol (see Fig. 4). The colonies with the insert were streaked as described in the protocol and samples of single colonies were seeded in Luria–Bertani medium (5 ml) containing tetracycline (20 μg/ml). DNA is isolated from pelleted cells using QIAprep Spin Miniprep Kit (Qiagen Inc., Santa Clara, CA). An aliquot of DNA (0.5–1 μg) is sufficient for obtaining a good nucleotide sequence. Alternatively, sequences can also be obtained from DNA (200–500 bp) of insert amplified by PCR such as those shown in Fig. 4.

It is essential to confirm that the change in a candidate cDNA fragment identified by differential display is also displayed in total RNA or mRNAs obtained from the same cell type subjected to identical treatment with the antioxidant. For this analysis it is necessary to probe blots of mRNAs (5–10 μg/lane, for low abundance mRNAs) or total RNAs ($\sim$20 μg/lane, for high abundance mRNAs) separated electrophoretically on an agarose gel (Ambion Inc., TX, NorthernMax). Such blots are probed with radiolabeled DNA obtained by the random prime method (Amersham, Arlington Heights, IL) using purified cDNA fragments as templates (Fig. 5). Alternatively, radiolabeled oligodeoxyonucleotide probes (20–30 bases) of sequence derived from the DNA sequence of the differentially expressed fragments can be used. The DNA sequence data can then be used to search the DNA database (NCBI, Blastin 2.0.3.) to identify the sequence of full-

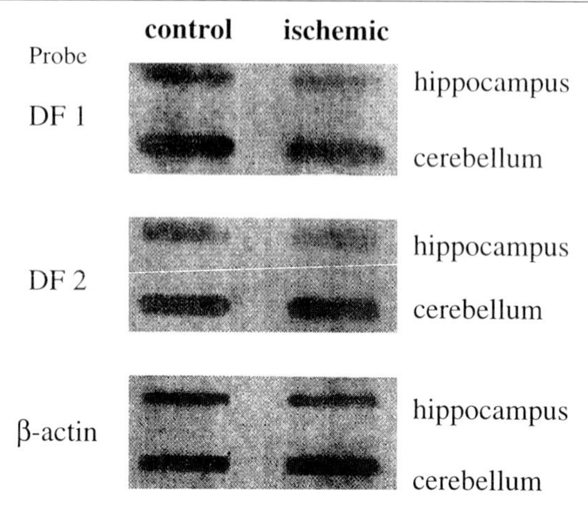

FIG. 5. RNA slot-blot analysis to confirm the differentially displayed mRNA transcripts. Two downregulated DNA fragments (DF 1 and DF 2) were identified from hippocampi of four adult rats (Sprague-Dawley, male) subjected to 15 min of global ischemia followed by 6 hr reperfusion. Total RNA (20 μg/slot) from hippocampi and cerebellum of control and ischemic brains was blotted (Bio-Rad, Richmond, CA, Bio-Dot SF Apparatus) and probed with [α-^{32}P]dCTP-labeled probes (obtained by random prime method, see step 7) using either DF1 or DF2 or β-actin as templates. The figure shows that both DF 1 and DF 2 are downregulated in ischemic hippocampus but not in cerebellum, which is known to be less vulnerable to ischemic damage compared to hippocampus. β-Actin showed no change with the ischemic insult.

length mRNA if there is a match. However, if such searches reveal a novel sequence it is necessary to probe the full-length cDNA library of the tissue of origin to identify the new gene using conventional techniques of molecular biology.

Section IV

Noninvasive Methods

Articles 41 through 46

[41] Noninvasive Measurement of α-Tocopherol Gradients in Human Stratum Corneum by High-Performance Liquid Chromatography Analysis of Sequential Tape Strippings

By JENS J. THIELE and LESTER PACKER

Introduction

As the outermost organ of the body, the skin is frequently and directly exposed to environmental sources of oxidative stress, including air pollutants, ultraviolet radiation, and chemicals.[1] To counteract oxidative insults, human skin is naturally equipped with a variety of enzymatic and nonenzymatic antioxidants. Based on studies demonstrating protective effects of topically applied α-tocopherol against UVB-induced tumorigenesis and immunosuppression,[2] ultraviolet (UV)B-induced photooxidation of α-tocopherol has become a subject of interest.[3] While epidermal and dermal α-tocopherol have been investigated in animal and human skin,[4–6] little is known about the presence and distribution of this major chain-breaking antioxidant in the outermost skin layer and site of the penetration barrier, the stratum corneum (SC, Fig. 1). The SC, in most human skin sites only 5–15 μm thin, comprises a unique two-compartment system of structural, enucleated cells (corneocytes) embedded in a lipid-enriched intercellular matrix, forming stacks of bilayers that are rich in ceramides, cholesterol, and free fatty acids.[7] The SC lipid composition and structure plays a key role in determining barrier integrity, which is essential for skin moisturization, normal desquamation, and healthy skin condition.[8]

We describe here a sensitive tape stripping-based technique to noninvasively obtain SC samples from human subjects, and an extraction and HPLC analysis procedure to measure α-tocopherol and cholesterol in SC samples. This technique allows measurement of these compounds in consecutively

[1] J. Fuchs, "Oxidative Injury in Dermatopathology." Springer, Berlin, 1992.
[2] H. L. Gensler and M. Magdaleno, *Nutr. Cancer* **15,** 97 (1991).
[3] K. A. Kramer and D. C. Liebler, *Chem. Rev. Toxicol.* **10,** 219 (1997).
[4] Y. Shindo, E. Witt, D. Han, W. Epstein, and L. Packer, *J. Invest. Dermatol.* **102,** 122 (1994).
[5] Y. Shindo, E. Witt, and L. Packer, *J. Invest. Dermatol.* **100,** 260 (1993).
[6] M. Lopez-Torres, Y. Shindo, and L. Packer, *J. Invest. Dermatol.* **102,** 476 (1994).
[7] P. M. Elias, *J. Invest. Dermatol.* **80,** 44 (1983).
[8] A. V. Rawlings, I. R. Scott, C. R. Harding, and P. A. Bowser, *J. Invest. Dermatol.* **103,** 731 (1994).

0076-6879/99 $30.00

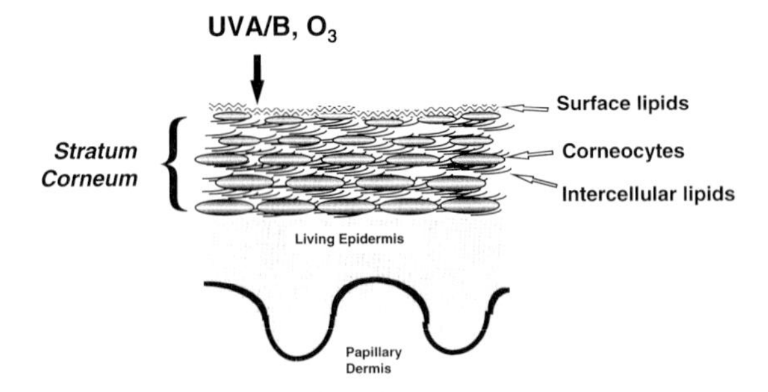

FIG. 1. The stratum corneum: a structural barrier model. Note that the stratum corneum is emphasized in this scheme; in skin, its width is normally 5–10 μm and accounts for less than 5–10% of the total epidermis.

obtained SC layers in human skin *in vivo,* and thus determination of specific α-tocopherol gradients. We have demonstrated in hairless mice that SC α-tocopherol is almost 10-fold more susceptible to depletion by ozone,[9,10] and that it is depleted by suberythemogenic doses (i.e., doses that do not cause skin redness 24 hr after UVA/B exposure) of solar simulated ultraviolet radiation.[11] The methodology described herein allows establishment of physiological SC gradients of α-tocopherol in various sites of human skin, and may, moreover, serve to investigate the role of α-tocopherol in pathophysiological states of the skin barrier.

Tape Stripping of Human Stratum Corneum

"Tape stripping" is a frequently used method in the fields of skin pharmacology and stratum corneum (SC) research. By applying and removing adhesive tapes onto the skin it is possible to subsequently remove the SC layer by layer from the underlying epidermis. For analytical methods to determine selected compounds using electrochemical or UV detection, the use of the right tape material is critical; most commercially available adhesive book tapes (we tested 20 different products) interfere strongly with the electrochemical detection of α-tocopherol. For tape stripping of human SC, the use of D-Squame adhesive disks (22 mm diameter, CuDerm Corporation, Dallas, TX) is recommended, since these are designed exclusively

[9] J. J. Thiele, M. G. Traber, K. G. Tsang, C. E. Cross, and L. Packer, *Free Rad. Biol. Med.* **23,** 385 (1997).

[10] J. J. Thiele, M. G. Traber, T. G. Polefka, C. E. Cross, and L. P. Packer, *J. Invest. Dermatol.* **108,** 753 (1997).

[11] J. J. Thiele, M. G. Traber, and L. Packer, *J. Invest. Dermatol.* **110,** 756 (1998).

for SC removal and therefore improve reproducibility and, as opposed to adhesive book tapes, do not have to be cut or otherwise prepared prior to use. Tape stripping must be performed using a standardized protocol: Each disk is smoothly adhered onto the skin, equally flattened three times, and gently removed using moderate and even traction. Using this technique, it takes usually between 20 and 30 tape strippings to remove the entire SC. Note, however, that this number varies intraindividually with the skin site, and that interindividual variations in the SC thickness of identical sites are a common phenomenon. The amount of SC is determined by the difference in weight before and after application and immediate removal from the skin. Since the SC amount obtained per disk is very small, a number of disks should be pooled and subsequently weighed out, extracted, and analyzed together. The experimental data described herein were obtained by pooling tape strippings #1–6, #7–12, #13–18, and #19–24 from upper arm SC (n = 12 human volunteers). It should be noted that a better reproducibility of the weighed SC amounts is achieved when the adhesive material is removed from its cover 3–6 hr prior to use and stored in a clean container; otherwise, evaporation of organic solvents after tape stripping can lead to weighing artifacts. Furthermore, only a very sensitive scale that is able to measure 0.5–1.5 mg amounts can be used. Since static charge can interfere with the weighing process, it is recommended to weigh out the disks/tapes on aluminum foil. The human SC is covered with surface lipids (Fig. 1), which contain sebum and environmental contaminants. Therefore, either the chosen skin site should be cleaned carefully, or the first two tape strippings, which contain the highest amount of surface lipids, should be discarded. When organic solvents are used for cleaning purposes, it must be considered that thorough use can lead to lipid extraction of even lower SC layers.

Lipid Extraction

After weighing, six consecutive tape strippings (i.e., six disks) are transferred to a 50 ml polypropylene centrifuge tube (Corning Costar Corporation, Cambridge, MA) containing 2 ml phosphate-buffered saline with 1 mM EDTA, 50 μl butylated hydroxytoluene (BHT: 1 mg/ml), 1 ml 0.1 M sodium dodecyl sulfate (SDS), and 4 ml ethanol. This is vortexed vigorously for 1 min at room temperature; trapping of the disks at the bottom of the tube should be avoided. Then, 4 ml hexane are added and mixed vigorously for 1 min. The whole content of the 50 ml tubes, except for the disks, is transferred to a 10 ml conical glass tube and centrifuged for 3 min. Following the centrifugation, 3 ml of the top layer (hexane layer) is transferred to another 10 ml conical glass tube. The hexane is taken to dryness under nitrogen; note that when about three-fourths of the hexane has dried down, a film of glue will form. The latter can prevent the rest of the hexane from evaporating. At this point the glue film has to be punctured using the drying needle or a clean pipette tip. Evaporator needles must be washed and cleaned thoroughly with ethanol before reapplying. Resuspend the residue in 500 μl of ethanol : methanol (1 : 1) by vortexing for 15 sec, transfer into an Eppendorf tube, and centrifuge for 1 min at maximum speed. A known amount of the clear supernatant should be analyzed by HPLC as soon as possible, since oxidation may occur during storage, even in the presence of added antioxidants.

HPLC Separation and Detection

HPLC separation and detection of α-tocopherol was modified after Lang *et al.*[12,13] Samples and standards are injected onto the HPLC system by an SIL-10A autoinjector with a sample cooler which is operated by a SCL-10A system controller (both from Shimadzu, Kyoto, Japan). The mobile phase consists of 20 m*M* lithium perchlorate in methanol : ethanol (1 : 9, v/v) and must be filtered and degassed by bubbling with nitrogen or sonicating prior to use. It can be stored at 4°C for up to 1 week, but should be filtered and degassed again before use. An isocratic delivery system is used with a flow of 1.2 ml/min performed by a LC-10AD Shimadzu pump. Chromatographic separation is achieved by an Ultrasphere ODS C_{18}, 4.6 mm i.d., 25 cm, 5 μm particle size column (Beckman, Fullerton, CA) with an All-Guard precolumn system (Alltech, Deerfield, IL). The column should be allowed to equilibrate for at least 2 hr. α-Tocopherol is detected by a Bioanalytical Systems LC-4B amperometric electrochemical detector with a glassy carbon electrode (Bioanalytical Systems, West Lafayette, IN). Cholesterol is detected by a Hewlett-Packard 1050 diode array detector (DAD) which is set up in line with the electrochemical detector. The tubing distance between the detectors should be as short as possible with an eluent delay of less than 2 sec. The diode array detector is set to 208 nm for detection of cholesterol: the electrochemical detector is in the oxidizing mode, potential 500 mV, full recorder scale at 50 nA. Data is collected with a Perkin-Elmer Interface and Turbochrom software (P.E. Nelson, Cupertino, CA).

Identification and Quantitation of α-Tocopherol

A stock solution of α-tocopherol is prepared by dissolving the pure compound (obtained from Henkel, La Grange) in reagent alcohol to yield

[12] J. K. Lang and L. Packer, *J. Chromatogr.* **385,** 109 (1987).

[13] J. K. Lang, K. Gohil, and L. Packer, *Anal. Biochem.* **157,** 106 (1986).

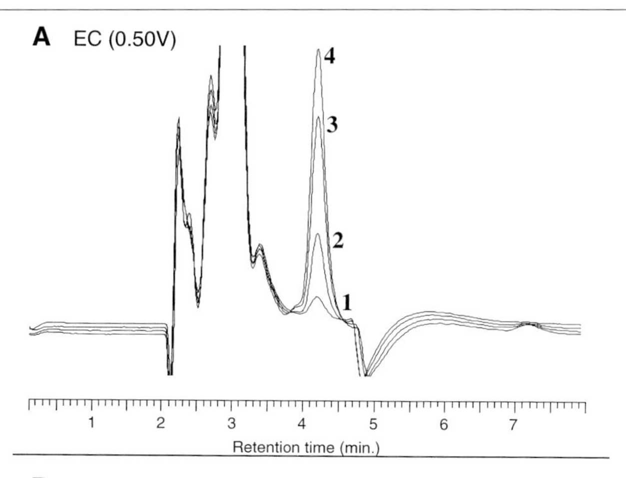

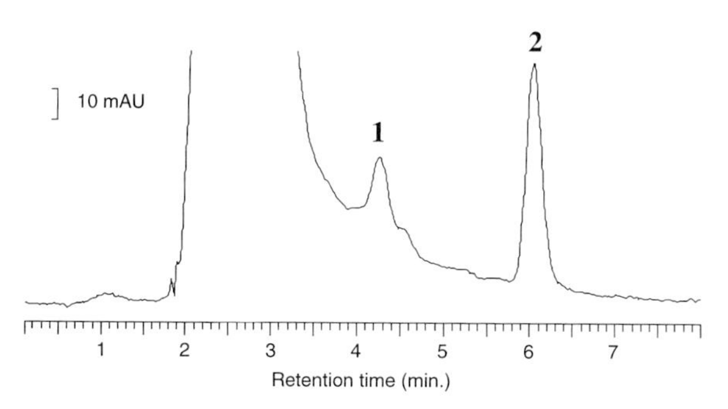

FIG. 2. HPLC chromatograms of a lipophilic human stratum corneum extract. (A) α-Tocopherol concentrations in consecutive layers of human SC obtained from a single suject: 1, upper SC (tape strippings #1–6); 2, upper intermediate SC (#7–12); 3, lower intermediate SC (#13–18); 4, lower SC (#19–24). (B) Example of an in-line UV detection of coextracted cholesterol (1) and the surface lipid squalene (2) in upper human SC.

a final concentration of approximately 50–100 μM. The accurate concentration is determined spectrophotometrically using the molar extinction coefficient ($\varepsilon = 3236\ M^{-1}\ \text{cm}^{-1}$ at 292 nm).[14] Chromatography grade cholesterol standards are obtained from Sigma (St. Louis, MO). Linear dose–response curves are obtained for α-tocopherol in the range of 0.2 to 50 pmol, and 0.2 to 20 nmol for cholesterol, respectively. Well-defined α-tocopherol and cholesterol peaks are detected both in standards and human SC samples (Fig. 2). The identity of the peaks is confirmed by coelution with authentic

[14] M. Kofler, P. F. Sommer, H. R. Bolliger, B. Schmidli, and M. Vecchi, *Vitam. Horm.* **20,** 407 (1962).

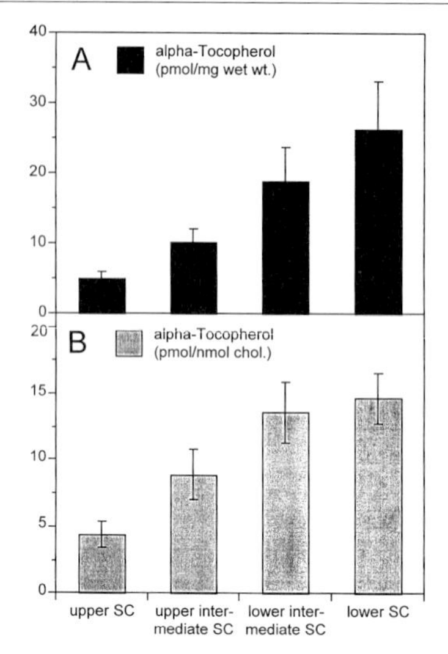

FIG. 3. α-Tocopherol gradient in human SC. α-Tocopherol concentrations in consecutive layers of human upper arm SC obtained from 12 donors: "upper SC," tape strippings #1–6; "upper intermed. SC," #7–12; "lower intermed. SC," #13–18; "lower SC," #19–24. (A) Expressed per mg wet weight of SC; (B) expressed per amount of coextracted cholesterol as determined by simultaneous in-line UV detection. Shown are means $\pm$ SEM.

tocopherol and cholesterol standards and comparison of in-line DAD-detected absorption spectra with those of α-tocopherol and cholesterol standards. Under the conditions described the recovery of α-tocopherol ranges between 70 and 75%. Expressing the SC α-tocopherol concentrations per coextracted and codetected cholesterol can be useful to normalize variations occurring during the multistep extraction procedure. Alternatively, an exogenously added vitamin E homolog such as γ-tocotrienol can be used as internal standard for α-tocopherol.

Results and Comments

α-Tocopherol Gradient in Human Skin

Figure 2A shows four superimposed chromatograms from four consecutively tape-stripped stratum corneum layers. A physiological gradient of α-tocopherol is found in the human SC of human upper arm skin, with concentrations ranging from 5 $\pm$ 0.8 to 26.2 $\pm$ 6.9 pmol/mg (Fig. 3A), and 4.4 $\pm$ 0.9 to 14.7 $\pm$ 1.9 pmol/nmol cholesterol, respectively (Fig. 3B).

As the outermost layer of skin, the SC is most directly and frequently exposed to environmental oxidative stressors. Although it does not contain nucleated cells, oxidation products of the lipid- and protein-rich SC may reveal signaling effects on adjacent keratinocytes. Previously, we have demonstrated that SC α-tocopherol depletion is a very early and sensitive marker of ozone- and UVA/UVB-induced photoxidation in skin, and that it can be noninvasively measured in human skin *in vivo*.[10,11] In addition to being the outermost skin layer, the SC is the result of a differentiation process initiated in lower epidermal layers. Therefore, measurement of α-tocopherol gradients in the SC may be of interest for studying not only oxidative stress initiated in the SC, but also oxidative insults that were set during the differentiation period of keratinocytes. Furthermore, differences in the regulation of α-tocopherol in different anatomical sites may be studied less invasively by the tape stripping technique than by regular skin biopsies.

In conclusion, the presented method may be a useful tool for *in vivo* studies of oxidative stress in humans, which are needed to validate the relevance of *in vitro* and animal experiments.

Acknowledgment

This work was supported by a postdoctoral fellowship from the Deutsche Forschungsgemeinschaft (Th 620/1-1) to J. Thiele.

[42] Ultraweak Photon Emission of Human Skin *in Vivo:* Influence of Topically Applied Antioxidants on Human Skin

By GERHARD SAUERMANN, WEI PING MEI, UDO HOPPE, and FRANZ STÄB

Introduction

Since ancient times humans have observed the emissions from fireflies, fungi, and marine organisms. In Bologna in the year 1603, Carcariolo described a stone phosphorescent after exposure to sunlight. In Hamburg in 1669, Brandt observed the light emission of newly discovered white phosphorus, caused by autoxidation and subsequent chemiluminescence. After the development of the theory of light, the quantum theory and

0076-6879/99 $30.00

quantum mechanics are the basis for the explanation of luminescence processes.[1]

Luminescence is the emission of electromagnetic radiation in the UV/VIS/IR spectral range from atoms or molecules that parallel relaxation of electronically exited states to a lower state of energy,[2,3] preferably the ground state. If the excitation of the emitting molecule is caused by chemical reactions, the observed emission is classified as chemiluminescence. If absorbed radiation is triggering the luminescence, the process may be called photoluminescence, which is related to fluorescence and phosphorescence. If the origin of excitation is obscure one may use the term ultraweak photon emission (UPE) or low level chemiluminescence.[4]

Low intensity photon emission can be detected with a photon counting device such as a photomultiplier.[4,5] The monitoring of UPE directly on the skin has the advantages of being noninvasive and providing continuous and convenient monitoring. Especially for the detection of peroxidative processes and the effectiveness of antioxidants for human skin *in vivo,* this method provides a unique technique for routine applications in the laboratory.[6–9] Three generations of UPE detectors have been used in our laboratory. The first investigation was carried out in 1985 using Showa Denko's Shonic Chemiluminescence Counter (Tokyo). After removal of the drawer for measuring samples *in vitro,* the emission from the back of the hands was determined *in vivo,* allowing comparison of two different treatments. The Shonic allowed insertion of cutoff filters to record stepwise emission spectra. An advanced system developed by SI Spectral Instrument (Munich) had comparable sensitivity but allowed recording UPE from other skin areas as well as hands. This was extended to whole body measurements using a newly developed system in cooperation between International Institute of Biophysics (Kaiserslautern) and the Skin Research Centre (Hamburg).

[1] A. K. Campbell, "Chemiluminescence, Principles and Applications in Biology and Medicine." Ellis Horwood, Chichester, UK, 1988.

[2] G. Cilento, *in* "Chemical and Biological Generation of Excited States" (W. Adam and G. Cliento, Eds.), p. 221. Academic Press, New York, 1982.

[3] I. V. Baskakov and V. L. Voeikov, *Biochemistry (Moscow)* **61,** 1169 (1996).

[4] W. P. Mei, Ph.D. Thesis, Hanover University, Germany (1991).

[5] M. E. Murphy and H. Sies, *Methods Enzymol.* **186,** 121 (1990).

[6] G. Sauermann, U. Hoppe, and F. Stäb, *Photochem. Photobiol.* **59,** 41s (1993).

[7] W. Mei, *in* "Biological Effects of Light 1995, Proceeding of a Symposium, Atlanta, USA" (F. Holick and E. Jung, Eds.), p. 93. Walter de Gruyter, Berlin, 1996.

[8] F. Stäb, U. Hoppe, R. Keyhani, and G. Sauermann, *Allergologie* **16,** 163 (1993).

[9] I. Hadshiew, F. Stäb, S. Untiedt, K. Bohnsack, F. Rippke, and E. Hölzle, *Dermatology,* in press (1997).

Most exergonic oxidation reactions produce heat; only a small percentage (<1%) of the enthalpy of the reaction is liberated as light. Some spectacular biological luminescence phenomena optimized by evolution produce "cold" light (i.e., at ambient temperatures) with a very high quantum efficiency (up to 95%). Triplet states of carbonyl are mainly considered a light emitting species in UPE.[1,4,5,10,11] A classical scheme was proposed by Sies[12] illustrating the peroxidative cleavage of olefinic bonds into aldehydes and ketones, producing triplet carbonyl which normally relaxes with low quantum yield (10^{-14}–10^{-10}) preferably in the visible spectral range. This low emission rate of 10–10^3 photons sec^{-1} cm^{-2} is much greater than the blackbody radiation at 37°C of approximately 10^{-9} photons sec^{-1} cm^{-2}.

Method to Detect UPE of Human Skin *in Vivo*

The detector head (with a photomultiplier) is located in a dark room and is shielded from any surrounding light and phosphorescent substances. In order to localize the detector head on human skin, it can be controlled through three stepper motors by computer or manually moved. The dark room should be free from any phosphorescent and synthetic color substance, and during the investigation the same temperature and humidity conditions should be maintained. For light, especially UV light-induced UPE, it is important to avoid any excitation of the control area and other body sites and clothing from stimulation. The volunteer should be kept in a dark environment for at least 5 min before starting the experiment.

The main part of the system is the detector head. A single-photon counting device, the photomultiplier (PM), which has a background noise of about 15 cps (counts/sec), is cooled using both Peltier elements and water cooling. The working temperature of the PM is about −30°. To avoid precipitation of moisture on the quartz window which has direct contact with the surface of the skin, double-vacuum quartz windows are used. To protect against the strong scattering of light during light stimulation (especially when using UVA light), two magnetic shutters are used. Two filter wheels are also employed for the purpose of spectral investigations. A short description of the technical setup of the unit is given in Fig. 1, which contain the following units:

1. Control unit. *Computer:* In this computer (486 DX) there are two special cards, one for I/O of experimental arrangement control (prototype

[10] E. Cadenas, *NATO ASI Ser. Ser. A* (*React. Oxygen Species Chem. Bio. Med.*) **146,** 117 (1988).

[11] N. Suzuki and H. Inaba, *Photomed. Photobiol.* **11,** 15 (1989).

[12] H. Sies, Ed., "Oxidative Stress: Oxidants and Antioxidants." Academic Press, New York, 1991.

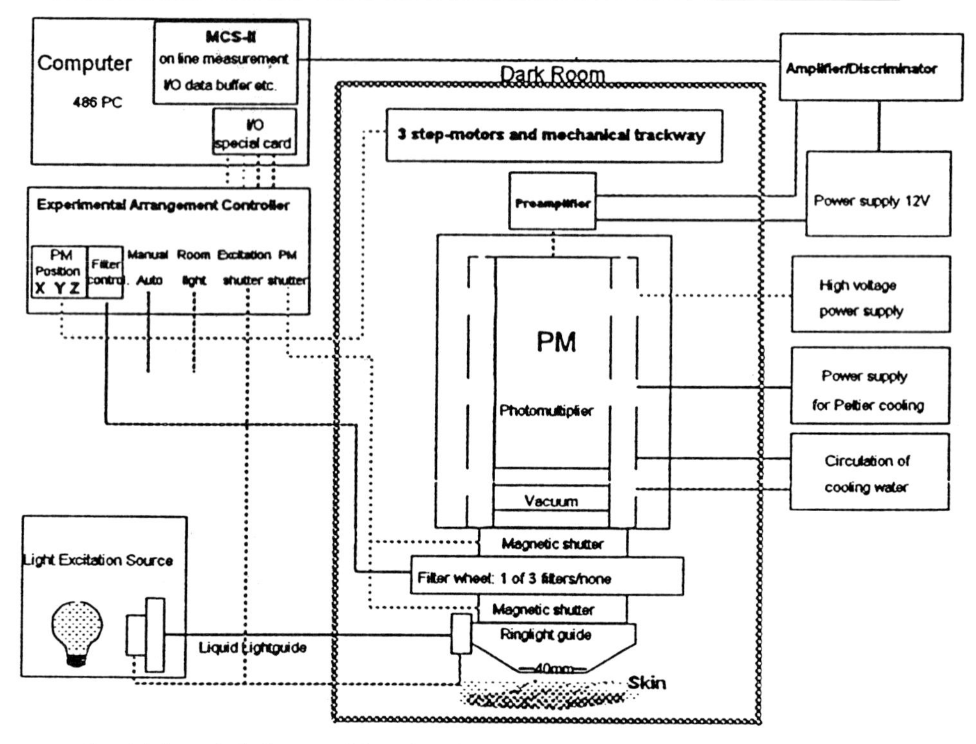

FIG. 1. The technical setup of detecting ultraweak photon emission (UPE) of human skin *in vivo*.

construction) and another for *I/O* of data acquisition (MCS, Oxford Instrument Inc.) *Experimental arrangement control unit:* In this unit (prototype construction) all the mechanical movement (via three stepper motors for movement of the detector head, photomultiplier and light excitation shutters, the stepper motor for the filter wheel, the room light, and the light sensor) can be controlled both manually and automatically.

2. Detector head. *Photomultiplier (PM):* A Thorn-EMI photomultiplier type 9558QA with a 48 mm diameter cathode is used. In order to reduce the noise, the PM should be cooled down to $-30°$ by setting it in a cooling chamber which is cooled by both Peltier elements and cooling water. *Filter wheel:* This is controlled through the computer operating a stepper motor and has four positions with three filters (Schott, Mainz). *Two magnetic shutters:* These parallel shutters (Pronto, E/64) are used to keep the PM from strong light, especially during the light excitation. They can only be opened when the door from the dark chamber is closed; there is no light in the measuring room and the two shutters of the excitation pathway are

closed. *Photosensor:* This is used to protect the PM from being damaged by strong light. This sensor is weakly light sensitive. The computer will only be allowed to open the PM when this photosensor does not sense any light.

3. Light excitation unit: *UV light source:* A 200 W mercury lamp (Lumatec, Superlite, S-UV 201AV-F) is used. *Two magnetic shutters:* These are used for light excitation. They exclude the background light from the surroundings, light source and light guide, and also ensure that there is no phosphorescence going into the PM. *Liquid light guide:* This connects the light source and detector head. Because we work in the UV region, a special liquid light guide (Lumatec, Series 2000) for the UV is used. *Ring light guide:* To meet the various geometric constraints, a ring light guide is used to illuminate the skin homogeneously within the area of interest. This ring light guide is specially constructed to suit UV light stimulation. *Filter wheel:* This is controlled manually and has six positions with four filters (Lumatec).

4. Data processing unit: *Preamplifier* (Otec 9301): This is directly and closely connected to the PM. *Amplifier and discriminator* (Phillips 6930): The signal from the preamplifier is amplified and converted into an electrical pulse chain through discriminating windows. The pulse chain is input to the MCS card. *Multichannel scaler (MCS):* This has two 24-bit 200 MHz counters, on-board data memory, a 48-bit internal dwell timer, and a 48-bit pass counter. This card with standard software transforms the personal computer into a versatile multichannel scaler. A negative TTL input is the only external signal necessary for MCS operation.

Treatment-Induced Photon Emission

As already mentioned, cell[4,13–16] and skin[6,7] show a spontaneous photon emission of 10–100 photons sec^{-1} cm^{-2}. Treatment with UV (especially UVA)[6,7,13,15,17] enhances induced photon emission, which indicates increased oxidative processes in the skin. A typical measurement on forearm *in vivo* (Fig. 2) using UVA-induced emission (12.5 J/25 cm^2) shows the running setup of the experiment (10 volunteers).

[13] W. Mei, *J. Biol. Syst.* **2,** 25 (1994).

[14] R. Van Wijk, J. M. van Aken, W. P. Mei, and F. A. Popp, *J. Photochem. Photobiol. B: Biol.* **18,** 489 (1993).

[15] H. J. Niggli, *J. Photochem. Photobiol. B: Biol.* **18,** 281 (1993).

[16] W. P. Mei, *in* "Biologic Effects of Light 1993, Proceeding of a Symposium, Basel, Switzerland" (F. Holick and E. Jung, Eds.), p. 458. Walter de Gruyter, Berlin, 1994.

[17] W. Mei and F. Popp, "Proceedings of European Symposium on Biomedical Optics. Lille, France," **2331** (Medical Sensors II and Fiber Optics Sensors), 203 (1995).

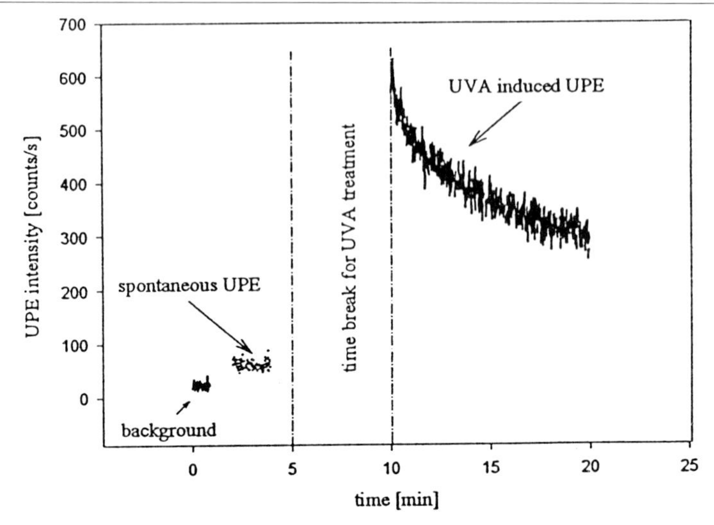

FIG. 2. A typical measurement on forearm of volunteer *in vivo* before and after UVA illumination (UVA doses: 12.5 J/25 cm^2).

Other possible treatments inducing UPE include ultrasound, topical application of peroxides (benzoyl peroxide), iron salts especially in combination with ascorbate, and rubbing contact with other surfaces (textiles).

Emission Spectrum of UVA-Induced UPE. The emission spectrum of spontaneous UPE from human skin tends to be strongest at the red end of the spectrum and weakest at the blue.[17] UVA-induced UPE, however, shows that the maximum is spectrally located between 400 and 580 nm, supporting the assumption that triplet states of newly synthesized carbonyl groups are a main source of UPE and that peroxidative processes induced by mechanisms listed above are basic sources for excited triplets.

Influence of Photosensitizers. Topical application of Rose Bengal or hypericine (both singlet-oxygen producers) and subsequent UVA irradiation cause a dramatic rise in UPE, i.e., induces prooxidative events in skin producing triplet states. Aminolevulinic acid, a precursor of photosensitizing porphyrins, reduces UPE significantly 1 hr after application of a 1% solution in ethanol.

Influence of UV Filters. Topical application of UVA filter Parsol 1789 increases protective power against UVA, with linear dose response, and reduces UVA-induced UPE, but not in a steady dose response. It soon approaches a saturation level with rising concentrations. TiO_2 shows qualitative differences with UVA-induced UPE: uncoated material very strongly triggers the UPE, while coated material reduces this. Uncoated TiO_2 liber-

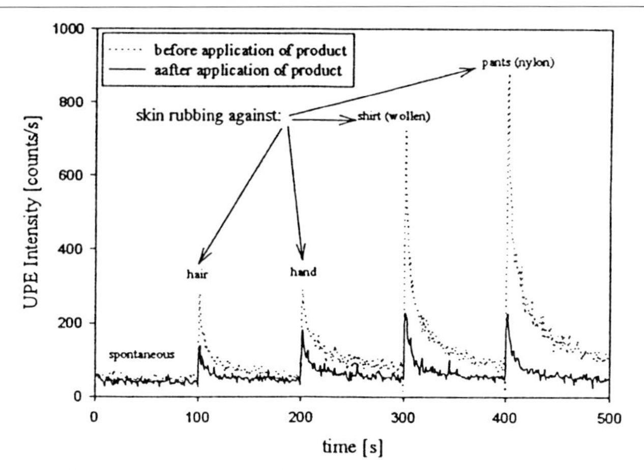

FIG. 3. Rubbing-induced UPE (triboluminescence) of human skin *in vivo* and its dependence on the surface material of rubbing substance and influence of product application.

ates electrons in the surrounding medium under bombardment of UVA and produces O_2^-, another inducer of UPE-stimulating processes.

Influence of Rubbing. Depending on the intensity of contact and movement, of the rubbing material and its surface, more or less UPE will be induced. Materials in contact will have different potentials; separation may cause electron transfer resulting in electric charges (see classical experiments in electrostatics) and subsequent molecular rearrangements, superficial chemical reactions. Therefore, UPE (triboluminescence) seems to represent the sum of oxidative reactions caused by contact, friction, and separation. There are two main triboluminescence influencing factors in skin: surface moisture content and superficial sebum content. Figure 3 shows a typical development of such investigation.

Influence of Horny Layer Moisture Content. Increasing corneal moisture content decreased UVA-induced UPE. Therefore, measurements of UPE should be performed under standardized environmental conditions (relative humidity and temperature). Horny layer proteins are more prone to changes by water than lipids and these changes subsequently affect the solubilities of oxygen species. The oxidation of horny layer proteins is accompanied by UPE and horny layer proteins are oxidized by UVA, peroxides, and ferrous/ferric ions.[18]

Oxidative Status of Young and Elderly Human Skin. The degree of oxidation of stratum corneum proteins increases gradually with increasing

[18] S. Richert, Ph.D. Thesis, Hamburg University, Germany, 1996.

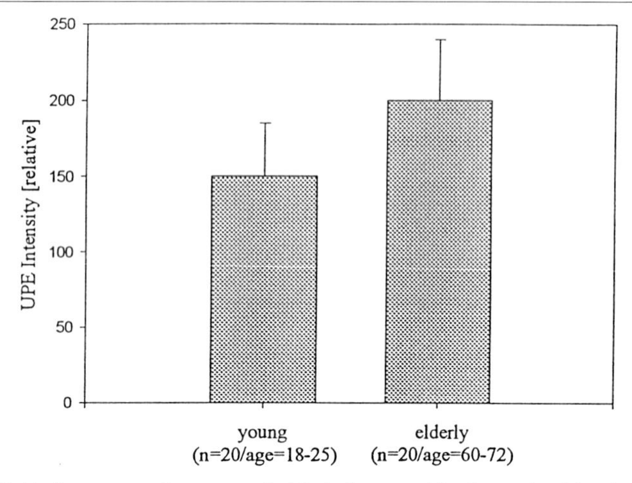

FIG. 4. Oxidative status of young and elderly human skin determined by detecting UPE *in vivo*.

age of the donor and parallels the rising spontaneous and UV-induced UPE. Humidification of the horny layer reduces the tendency of those proteins to oxidize. The investigation of two age groups (Fig. 4) shows a clear correlation of UPE intensity and age. The same results have been reported in several other publications.

Efficacy of Topically Applied Antioxidants on Human Skin: The usefulness of UPE method (Fig. 5) is shown by the efficacy of topically applied antioxidants to reduce the UVA-induced UPE. Figure 5 shows the difference between effects of the test product and placebo after topical application of α-glucosylrutin[9] for 7 days.

Concluding Remarks

Very little work has been done so far on UPE from human subjects. In one report[19] a mouse model was used to evaluate the occurrence of oxidative stress in skin exposed to UVA radiation. The effect of the topical application of α-tocopherol and β-carotene was also determined in this study. It is well known that radiation with UVA decreases the levels of antioxidants, inactivates the antioxidant enzymes, and increases lipid perox-

[19] P. Evelson, C. P. Ordonez, S. Llesuy and A. Boveris, *J. Photochem. Photobiol. B: Biol.* **38,** 215 (1997).

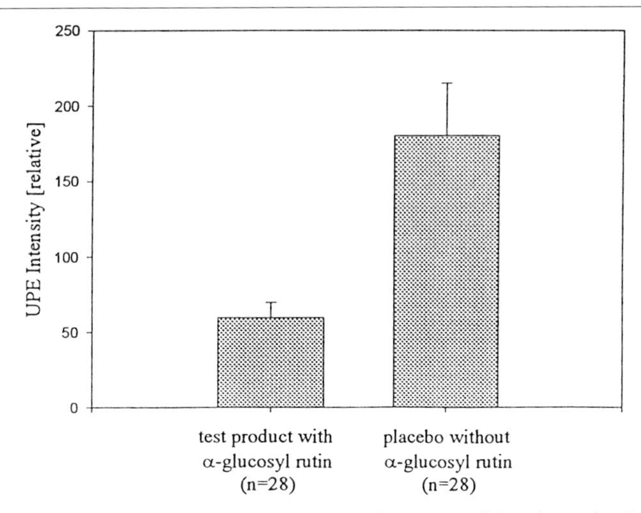

FIG. 5. Efficacy of topically applied antioxidants (the flavonoid α-glucosylrutin) on human skin *in vivo*.

idation in skin.[20-23] The UPE detection method provides a useful technique *in vivo* to determine peroxidative events and efficacy of topically applied antioxidants on human skin.[6-9] In order to record the emissions from the skin of human volunteers *in vivo*, the instruments have to be adapted to special applications. It is necessary to replace small sample compartments and to keep the distance between photocathode and skin surface as short as possible. Avoidance of light from external sources is also necessary. The entire detector head has to be installed in rooms without phosphorescent walls, surfaces, and lamps, and it should be freely movable.

Irrespective of theoretical considerations as to whether some kind of physical or biochemical phenomena may be occurring, skin as an organ designed for protection against noxious materials in the environment (among which numerous UPE triggering influences can be identified) may be useful as an indicator of free-radical metabolism. Because damage in-

[20] S. Ahmad, Ed., "Oxidative Stress and Antioxidant Defenses in Biology." Chapman & Hall, New York, 1995.

[21] L. Packer and J. Fuchs, Eds., "Vitamin E in Health and Disease." Marcel Dekker, New York, 1993.

[22] Y. Shindo, E. Witt, D. Han, W. Epstein, and L. Packer, *J. Invest. Dermatol.* **102,** 122 (1994).

[23] E. Cadenas and L. Packer, Eds., "Handbook of Antioxidants." Marcel Dekker, New York, 1996.

duced by these radicals is thought to be the cause of many of the phenomena of aging, light could serve as an indicator of the rate of aging and of the success of measures taken to slow this process, for example, by the use of the flavonoid α-glucosylrutin.[9]

[43] Noninvasive in Vivo Evaluation of Skin Antioxidant Activity and Oxidation Status

By RON KOHEN, DAVID FANBERSTEIN, ABRAHAM ZELKOWICZ, OREN TIROSH, and SHARON FARFOURI

Introduction

Most of the existing methods for evaluation of skin antioxidant capacity and its oxidation status are invasive and require removal of the skin, its separation into its various layers, homogenization, and analysis of its antioxidants or its oxidation products. It is obvious that such measurements cannot be done on a large scale in humans. Therefore, only a few comprehensive reviews can be found in the scientific literature which describe the oxidant/antioxidant status of skin. Shindo *et al.*[1] demonstrated that skin possesses antioxidant activity from both antioxidant enzymes and low molecular weight antioxidants. Of the various layers of the skin, the epidermis contains the highest levels of antioxidant molecules.[1] Evaluation of the antioxidant status by invasive means revealed that the total antioxidant activity of young skin is much higher than that of old skin.[2] Since a major contribution to the antioxidant activity is derived from low molecular weight antioxidants (LMWA) which can easily penetrate via cell membranes and provide local protection, we developed a procedure for evaluating antioxidant activity contributed by LMWA which are present on the surface of the skin (stratum corneum) and can be secreted from deeper layers to the outer environment. The method described here is suitable for both indirect and direct evaluation of skin antioxidant activity. It is also suitable for evaluating lipid hydroperoxide levels on the surface of the skin as an indicator of oxidation status of the skin.[3]

[1] Y. Shindo, E. Witt, D. Han, W. Epstein, and L. Packer, *J. Invest. Dermatol.* **102,** 122 (1994).

[2] R. Kohen, D. Fanberstein, and O. Tirosh, *Arch. Gerontol. Geriatr.* **24,** 103 (1997).

[3] R. Kohen, D. Fanberstein, and O. Tirosh, Israeli Patent Application 107240 (1994).

0076-6879/99 $30.00

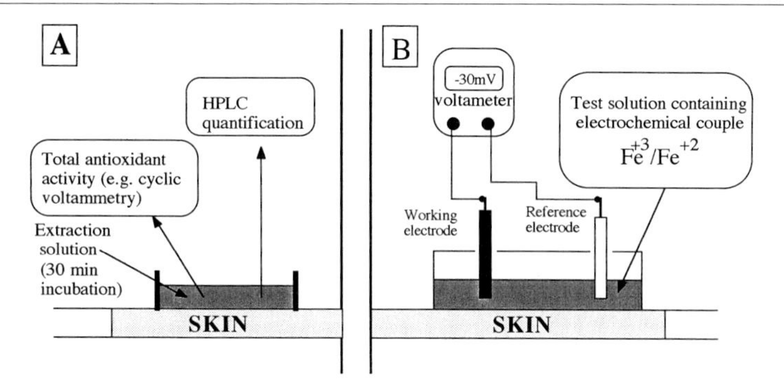

FIG. 1. Schematic representation of the well containing the extraction solution placed on the surface of the skin. (A) A well designed to measure the antioxidant activity by an indirect approach. Following incubation the analysis is carried out using methods for evaluating total antioxidant activity or specific antioxidant molecules. (B) A well designed to measure antioxidant activity by a direct approach. The well contains cytochrome *c* solution, including both the oxidized and reduced forms of cytochrome *c*. Two electrodes (working electrode: platinum wire or glassy carbon; reference electrode: Ag|AgCl) are placed in the solution and are connected to a voltmeter. The changes in the potential of the electrochemical couple of cytochrome *c* are measured and recorded.

General Principles of Noninvasive Skin Evaluation

The epidermis provides the first line of defense against oxidative stress and reactive oxygen species. It has been found that skin releases LMWA which act directly with the reactive oxygen species (ROS) (scavengers) from its outer layer (stratum corneum) into an outside solution. We have designed a well which is placed on the surface of the skin (Fig. 1).[3] An extraction solution is introduced into the well for a 30-min period. Following incubation the extraction solution is analyzed (Fig. 1A). Alternatively, antioxidant activity can be analyzed directly as shown in Fig. 1B. In this approach the extraction solution contains chelates of iron in its oxidized form. Reducing equivalents which are released from the outer surface of the skin reduce ferric ions to their ferrous form. The electrochemical couple created produces a potential which can be measured using two electrodes connected to a voltmeter. The changes recorded in the potential reveal the levels of the reducing equivalents secreted as indicated by the Nernst–Peters equation (Fig. 2). A decrease in the potential indicates a decrease in the oxidized form of the iron ions and an increase in the levels of the reduced form of the iron ions. The potential change is proportional to the level of the reducing LMWA secreted which reduces ferric ions to ferrous. A modification of this method allows the quantification of organic hydroper-

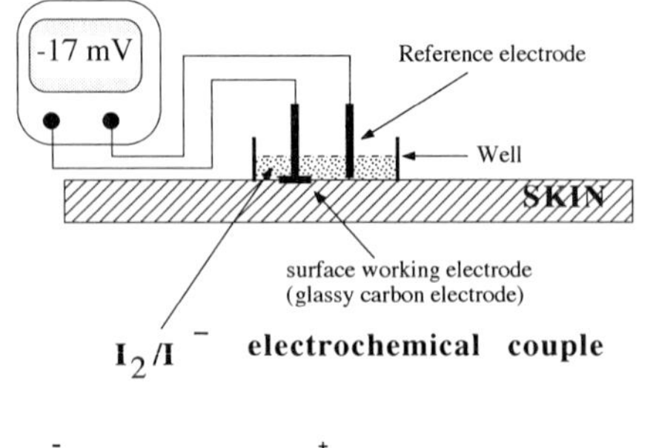

$$2I^- + ROOH + 2H^+ \longrightarrow I_2 + ROH + H_2O$$

Nernst-Peters equation $E_h = E_0 + RT/nF \ln [I_2]/[I^-]$

FIG. 2. Schematic representation of the well designed for measuring lipid hydroperoxide on the skin surface. The well placed on the skin contains 1 ml of a solution of iodide/iodine (300 mg of KI and 100 mg of iodine). A glassy carbon electrode (disk electrode, 3.3 mm in diameter) and Ag|AgCl as the reference electrode are introduced into the well. The glassy carbon electrode should be in contact with the skin surface. The electrodes are connected to a voltmeter to measure the potential of the electrochemical couple.

oxides on the skin surface by using a different test solution as described below (Fig. 2).[4]

Skin Antioxidant Determination

Procedure

A well (made of Plexiglas) 1 cm in diameter is placed on the surface of the skin. In the rat animal model the hair from the back of the animal is shaved 10 hr before the measurement, and the well is attached to the back of the animal using silicone rubber. In humans the well is placed on the inner side of the wrist and attached to the arm using an adhesive pad (Parafilm). Extraction solution of phosphate-buffered saline (PBS) at pH 7.2 (1–1.5 mL) (or another buffer at the desirable pH) is introduced into the well for a 30-min incubation period. During this time the well is covered using an adhesive pad (Parafilm). In order to prevent oxidation of the secreted reducing equivalents, oxygen is removed from the extraction buffers prior to its placement on the surface of the skin by bubbling pure nitrogen (99.999%) for 20 min in the buffer stock solution and flushing

[4] R. Ezra, S. Benita, I. Ginsburg, and R. Kohen, *Eur. J. Pharm. Biopharm.* **42,** 291 (1996).

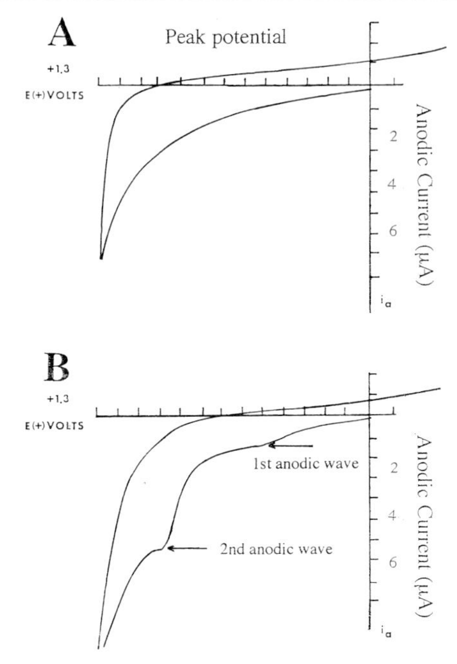

FIG. 3. Cyclic voltammograms of the extraction solution which was placed on the backs of rats for a 30-min incubation period. (A) Phosphate-buffered saline (PBS) prior to its placement on the skin. (B) Phosphate-buffered saline following 30 min of incubation.

above the extraction solution in the well during the measurement. The extraction solution is analyzed following the incubation period to determine the levels of the reducing LMWA.

Results

Secretion of Reducing LMWA from Skin

An extraction solution (PBS) is placed on 3-month-old rat skin for 30 min. Following incubation the solution is collected from the well using a 1 ml pipette and is analyzed in the cyclic voltammeter to determine total LMWA.[5] Figure 3 shows that the PBS itself does not exhibit any anodic waves (Fig. 3A) prior to its placement on the skin. Following a 30-min incubation period, two anodic waves are detected, indicating two groups

[5] R. Kohen, E. Beit-Yannai, E. M. Berry, and O. Tirosh, *Methods Enzymol.* **300,** [30], (1998) (this volume).

of reducing LMWA which have been secreted from the skin into the well (Fig. 3B).

Antioxidant Activity of Secreted Equivalents

In order to determine the antioxidant activity of the extracted solution containing the reducing equivalents, we measure the ability of the solution to scavenge reactive oxygen species induced by activated neutrophils. Neutrophils (PMNs) are obtained from healthy donors following separation on a Ficoll–Hypaque gradient and dextran sedimentation.[6] PMNs are suspended in Hanks' balanced salt solution (HBSS) buffered with 0.3 M HEPES, pH 7.4. The neutrophils (10^6/ml) are incubated with luminol (0.9 μM) and with washed group A streptococci (10 μl of a saline suspension of OD 2.0 at 550 nm) opsonized with poly(L-histidine) (25 μg/ml). The mixture is agitated briefly over a mechanical vortex and immediately transferred to a LUMAC/3M Biocounter, Netherlands. The light-induced, luminol-dependent chemiluminescence (LDCL) is monitored for 4 min at 37° or until peak LDCL is reached. Increasing volumes of the extraction solution that was placed on the skin for 30 min are added to the reaction mixture at the peak, and the LDCL is recorded.[7] The significant decrease in the light induced indicates the scavenging potential of the LMWA present in the extraction solution (Fig. 4).

Quantification of Compounds Secreted

In order to identify the various compounds composing the wave, aliquots from the extraction solution are injected into a high-performance liquid chromatography (HPLC) system equipped with an electrochemical detector.[8] Uric acid and ascorbic acid are the major LMWA secreted from the skin and make up the first anodic wave.

Effect of Aging on Reducing Equivalents Released from Skin

Determination of the reducing LMWA secreted from rat skin in young rats (2 months old) and old rats (24 months old) reveals that there is a significant change in the levels of reducing LMWA. Figure 5 shows that the levels of uric acid secreted from the skin, as measured by HPLC–ECD (electrochemical detection), into the well are significantly lower in old animals as compared to young rats. Similar results have been obtained for the overall concentration of reducing LMWA secreted.

[6] I. Ginsburg, R. Borinski, and D. Malamud, *Inflammation* **9,** 245 (1985).
[7] R. Kohen, R. Misgav, and I. Ginsburg, *Inflammopharmacology* **2,** 15 (1993).
[8] P. A. Motchnik, B. Frei, and B. N. Ames, *Methods Enzymol.* **234,** 269 (1994).

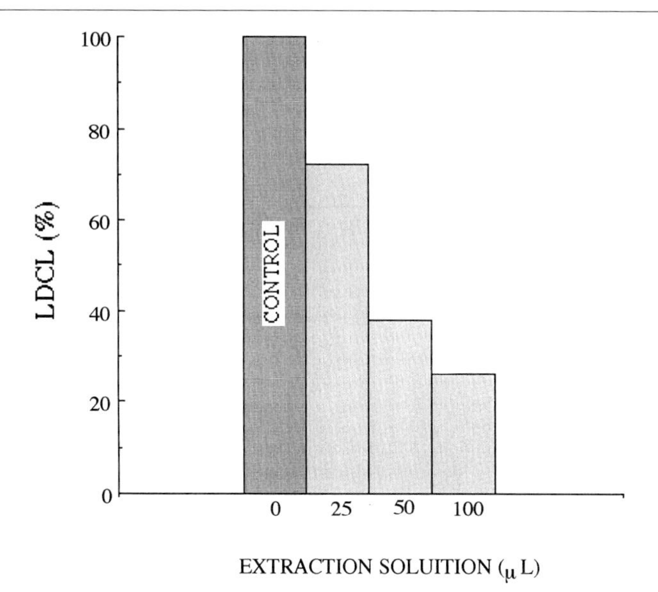

FIG. 4. Effect of the extraction solution on LDCL. PMNs (10^6/ml) in HBSS were stimulated by polyhistidine-opsonized streptococci in luminol.

Direct Measurement of Skin Antioxidants

Procedure

As shown in Fig. 2, two electrodes (platinum or glassy carbon) are introduced into the extraction mixture in the well [PBS, containing cyto-

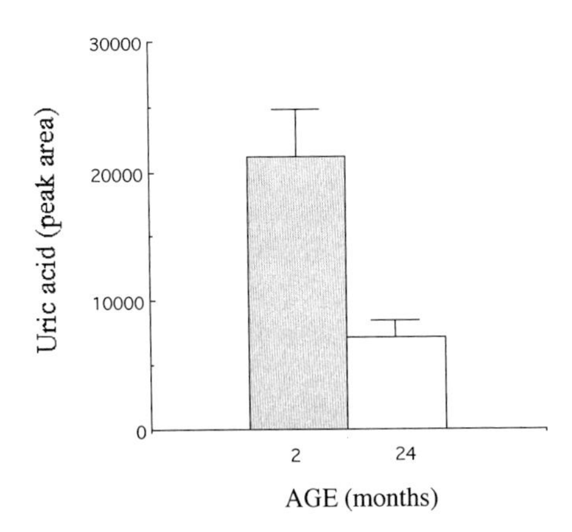

FIG. 5. Effect of age on the levels of uric acid secreted by rat skin.

chrome c (50 μM)]. The potential created by the oxidized and reduced species of the cytochrome c solution is recorded using a voltmeter. The changes in the potential are monitored for 30 min. It has been found that the potential decreases in a linear mode and stabilizes after 25 min. The decrease in the potential is due to an increase in the reduced form of cytochrome c [cyt c (Fe^{2+})] and a decrease in the oxidized form of cytochrome c [cyt c (Fe^{3+})]. Reducing equivalents secreted from the skin into the extraction solution are responsible for reducing the iron in the cytochrome and for changing the potential according to the Nernst–Peters equation (Fig. 2). In order to determine the exact concentrations of the two cytochrome c species, the sample is analyzed prior to its placement on the skin and following an incubation period by a spectrophotometer at 550 nm where the reduced form of the cytochrome c is absorbed.

Skin Oxidation Status Determination

The lipid peroxidation process is characterized by three major steps: initiation, propagation, and termination. During the peroxidation process (which may occur in proteins and amino acids as well as in lipids) there is a continuous production and accumulation of hydroperoxide. The level of hydroperoxide in biological tissue may serve as an indicator of oxidative damage. Although there are numerous methods for evaluating peroxides, Gebicki *et al.* have already suggested the beneficial use of the iodometric hydroperoxide assay over the other methods.[9] This method is not specific to certain molecules containing hydroperoxides. In fact, hydroperoxide in a wide range of molecules can react with iodide quantitatively and stoichiometrically as shown in the reaction in Fig. 2. In the presence of iodide the iodine produced is converted to I_3^-. For this reaction to indeed be stoichiometric it should be conducted under specific conditions. The reaction has been found to accelerate at acidic pH. However, this reaction also proceeds at physiological pH, although at a slow rate. We have developed a method for determining the iodine (I_3^-) produced on the surface of the skin from the interaction between hydroperoxide in the outer layer and iodide in the solution present in the well. Monitoring of the reaction is done using an electrochemical approach which determines the potential produced by the electrochemical couple in the well.[2] The procedure described next has been found to be suitable for assessing oxidation status of the skin in physiological situations and following exposure to oxidative stress of various kinds.[2]

[9] W. Jessup, R. T. Dean, and J. M. Gebicky, *Methods Enzymol.* **233,** 289 (1994).

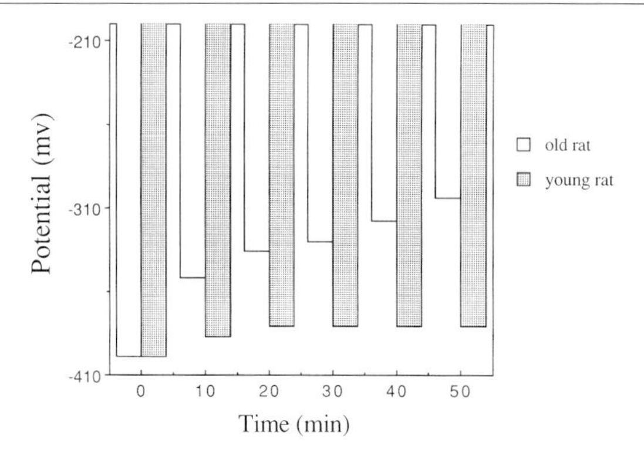

FIG. 6. Noninvasive measurement of the oxidation status of young and old rat skin. The reaction between organic peroxides and iodide in the wells on the backs of the rats was monitored using the changes in the potential as derived from the Nernst–Peters equation.

Procedure

A well 1 cm in diameter is placed on the surface of the skin using silicone glue (in animals) or a Parafilm pad (in humans). The test solution is prepared as follows: 300 mg of KI and 100 mg of iodine solution are dissolved in 10 ml of double-distilled buffer or PBS. The redox potential of the solution is analyzed by introducing two electrodes. A glassy carbon disk (3.3 mm) is used as the working electrode, and as a reference electrode an Ag|AgCl electrode is used. The glassy carbon electrode should be in close contact with the surface of the skin, as shown in Fig. 2. The close contact of the electrode with the skin surface allows the reaction to occur at the interface between the disk electrode and the skin outer layer where lipid hydroperoxide interacts with the iodide ions. The close proximity prevents possible interference between the reducing equivalents released into the solution and the lipid hydroperoxide and the iodine. The initial redox potential of this solution has been found to be −360 mV. The electrodes are attached to a digital voltmeter, allowing the changing potential to be monitored.[3]

Results

Determination of the Oxidation Status of Young and Old Rat Skin

Figure 6 shows the changes in the potential of the electrochemical couple iodine|iodide measured on the backs of young (2 months old) and old (24 months old) rats. The results are the mean potential obtained from 10 animals in each group. The results show that the potential does not change

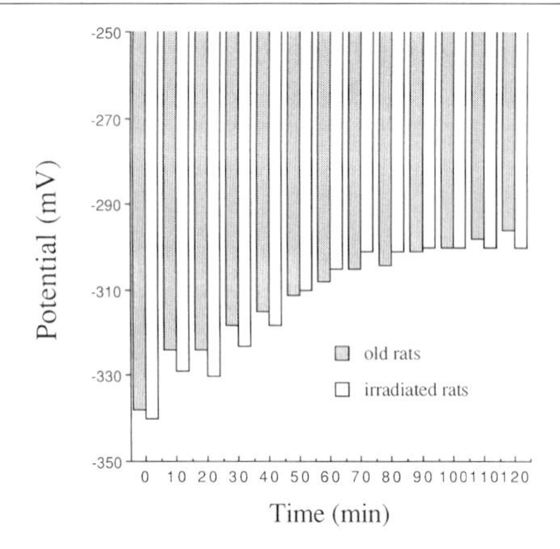

FIG. 7. Noninvasive measurement of the oxidation status of irradiated young rat skin. The reaction between organic peroxides and iodide in the wells on the backs of the rats was monitored using the changes in the potential as derived from the Nernst–Peters equation. The rats were irradiated by γ-irradiation (950 rad, total body irradiation). Irradiation was carried out using a Co source.

during the time the solution is placed on young rat skin. However, significant changes occur in the potential recorded during the incubation time of the iodine/iodide solution on the backs of old rats. These results indicate that old rat skin contains lipid hydroperoxides in its outer layer which is capable of interacting with the iodide ions to produce iodine. The change in the ratio of these two species leads to the change in the potential as derived from the Nernst–Peters equation (Fig. 2).[2]

Determination of Rat Skin Hydroperoxide in γ-Irradiated Animals

Young rats (2 months old) ($n = 10$) are exposed to γ-irradiation (950 rad, total body irradiation). Three hours following irradiation lipid hydroperoxides on the surface of the skin are analyzed and compared with the control animals (same age, no irradiation). The results presented in Fig. 7 show that while there is no change in the potential in skin evaluation of the control rats (Fig. 6), a significant change occurs in the solution that was placed for 30 min on the backs of the young irradiated rats similar to changes observed in old rats. These results suggest an accumulation of lipid hydroperoxide on the surface of the skin following irradiation.[2]

Summary and Conclusions

The method described here allows noninvasive quantification of reducing LMWA or the lipid hydroperoxide present on the surface of the skin.

Quantification of reducing antioxidants can be achieved because they are secreted from the skin surface into a well containing an extraction solution. Analysis of the reducing equivalents released indicates the presence of uric acid and ascorbic acid. Other LMWA released from the skin are as yet unidentified. The secretion of the LMWA reaches a plateau following 20–30 min of incubation. Therefore, a 30-min incubation period was chosen as the optimal time for the extraction solution to be present in the well and in contact with the skin. This extraction procedure can be repeated after 24 hr. This period of time is needed for regeneration of the LMWA to their initial levels. Direct measurement allows continuous determination of the release of LMWA and their interaction with the iron chelate. The reaction is completed after 25–35 min, at which time the final potential can be recorded. When organic peroxides on the surface of the skin are determined, it is important that the glassy carbon electrode be in close contact with the skin, since the reaction occurs on the surface of the electrode and the bound peroxide on the outer layer of the skin. Furthermore, close contact is needed to avoid interference of reducing equivalents secreted from the skin into the well.

Acknowledgments

This work was supported in part by the Foundation for Research and Development, by the Bergman Foundation, and by the David R. Bloom Center of Pharmacy.

[44] Antioxidative Homeostasis: Characterization by Means of Chemiluminescent Technique

By IGOR POPOV and GUDRUN LEWIN

In contrast to the dead, the living incubate themselves at 37°, they are moist, they are exposed to light, and they are in perpetual if not always purposeful motion. What stops them going rancid? (T. L. Dormandy, 1978.)

Antioxidative System

The main feature of living organisms is their capability to protect themselves against uncontrolled oxidation. Although all organisms are subject to the permanent influence of oxygen and other oxidatively active causes (UV sun irradiation, atmospheric noxae, natural and artificial ionizing radia-

0076-6879/99 $30.00

tion, etc.), they maintain their integrity due to the effect of a special antioxidative system which developed in the course of phylogenesis.[1,2] This is in contrast to avital compounds in which oxidizability depends only on their chemical composition. Following general biological principles, the antioxidative protection mechanisms can be classified into four categories: compartmentation, detoxification, repair, and utilization. Compartmentation means both spatial separation of potentially harmful but essential substances (e.g., storage of iron in ferritin) and cell and tissue-specific distribution of antioxidative compounds and serves the prevention of uncontrolled oxidation. Detoxification of prooxidatively active molecules (radicals, peroxides) is ensured by enzymatic and nonenzymatic substances. With the exception of the so-called extracellular superoxide dismutase (SOD), which plays a not yet resolved role, the detoxificating enzymes (SOD, catalase, and glutathione peroxidase) are intracellularly present and protect the cells from the destructive side effects of the physiological metabolism by prevention of initiation and branching of free-radical chain reactions. The outstanding role of the antioxidative enzymes is the detoxification from prooxidative compounds which are formed in certain cell compartments both under physiological and pathological conditions before they attack cell-proprietary structures. This is in contrast to repair enzymes which act specifically as a "second level of defense," reverting the originated changes when they are still reversible. Utilization in the meaning of further degradation of denatured and peroxidized potentially toxic proteins and lipids by proteolytic enzymes and phospholipase A_2 is also considered as a secondary antioxidative protection mechanism.

Parameters of Antioxidative Homeostasis

Despite the obvious significance of all protection mechanisms, the main focus of practical attention is detoxification, because this is most probably subject to a therapeutic approach. Nonenzymatic antioxidants, which in contrast to enzymes are present in all compartments of an organism, have an essential meaning for detoxifying protection from oxidative damage. This more or less uniform distribution can be phylogenetically explained by the fact that everywhere in the organism an oxidative breakout can occur which is due to exogenous stimulation such as UV or ionizing irradiation, injuries, poisoning, or inflammations. Therefore, easily oxidizable structures

[1] G. Lewin and I. Popov, "Entwicklung und experimentell-klinische Nutzung der Photochemolumineszenz-Methode für die Beurteilung der antioxidativen Homöostase im Organismus." Humboldt-Universität, Berlin (1990).

[2] G. Lewin and I. Popov, *Med. Hypotheses* **42,** 269 (1994).

in the organism are protected by a permanent influx of antioxidants. The antioxidative system with all its components and mechanisms[1,2] ensures the maintenance of antioxidative homeostasis *in vivo*. In *ex vivo* examinations, these mechanisms are no longer present.

Nevertheless, conclusions about the state of antioxidative homeostasis can be made based on the measurement of antioxidative reserves (capacity) in a biological substrate and the extent of its oxidative damage.

Antioxidative Capacity

Under experimental conditions, a number of nonenzymatic compounds such as tocopherols, carotinoids, vitamins C and D, steroids, ubiquinones, thiols, uric acid, bilirubin, inosine, taurine, pyruvate, C-reactive protein (CRP), and many more, demonstrate qualitatively antioxidative properties. Nevertheless, the practical quantitative relevance of the majority of these findings is not yet clear.

The selective determination of the physiologically most relevant antioxidants in blood plasma, vitamins C and E, carotinoids, and the endogenously synthesized compounds, uric acid and bilirubin, has been published several times. For at least two reasons this determination does not seem relevant to the assessment of the antioxidative state of the organism: (1) Since investigations of biological antioxidants and particularly of the antioxidative effect of vitamins exist, several authors have demonstrated the very complicated interactions of antioxidants with each other as well. An extensive terminology has been developed to describe these phenomena, including terms such as synergism, antagonism, competition, potentiation, mimicry, saving, rebound effect, and pseudoactivity. (2) The results of individual determinations of antioxidants can only be interpreted with difficulty. This is due to both their different biological and biochemical activity and (particularly under extreme or pathological conditions) the fact that uncontrollable causes of nonoxidative nature can influence endogenous synthesis, absorption, and excretion of the individual antioxidative compounds.

The measurement of the summarized overall antioxidative effect in a specialized test system can serve as a physiologically relevant alternative to the selective determination of individual compounds in biological substrates, such as blood plasma. Normally, such a test system is composed of two components: (1) a generator of free radicals and (2) a detection system for free radicals which enables their quantification and which changes the measured signal when the investigated compounds are present.

Different generators can be used, i.e., physical (radiolysis, photolysis, electrolysis, etc.), physicochemical (thermal decomposition of nitrogen compounds, photosensitized generation), chemical (Fe^{2+}/H_2O_2 system, KO_2

decomposition, autoxidation of several compounds), and biochemical systems of different complexity from individual enzymes (e.g., xanthine oxidase) to subcellular fractions (NADPH-consuming microsomes), and tissue homogenates (e.g., brain homogenates).

The effect of the antioxidants can be detected by the measurement of O_2 consumption, light absorption, electrical conductivity, fluorescence, and chemiluminescence. Despite the diversity of detection methods it is not possible to select the "right" method based on objective criteria because the reactivity of the ingredients with free radicals depends on their composition[3] and the nature of the free radicals is affected by "external conditions" such as hydrophobicity and viscosity of the medium, oxygen partial pressure, and pH value.[4,5] Therefore, in the examination of biological problems, more information is obtained with methods which not only measure the concentration of individual compounds, but also record their integrative reaction against free radicals including the total hierarchy of antioxidants.

The type of free radical selected for the generator system is important in expressing measurements. This means that in an adequate measuring system the biological target substrate should react with radicals in which size, lipophilicity, and reactivity are relevant. Because the superoxide radical is a precursor of the other reactive oxygen species and interacts with blood plasma components under both physiological and pathological conditions, systems that deal with its generation play a special role. Nevertheless, the reactivity of the superoxide radical in the initiation of lipid peroxidation is very limited and only the toxicity of its protonized form is comparable to that of lipid peroxyl radicals.[6]

Systems which generate hydroxyl radicals are of little relevance because their extreme reactivity oxidizes all types of molecules. Based on these systems, practically every compound can be classified as "antioxidant" which cannot be verified in *in vivo* experiments. A definition of practical relevance for antioxidants was proposed by Halliwell and Gutteridge[7]: "any substance that, when present at low concentrations compared to those of an oxidisable substrate, significantly delays or prevents oxidation of that substrate." This means that there are no antioxidants *per se,* and compounds can be considered as antioxidants which are active in a defined radical-

[3] P. Palozza, C. Luberto, and G. M. Bartoli, *Free Rad. Biol. Med.* **18,** 943 (1995).

[4] M. C. Hanlon and D. W. Seybert, *Free Rad. Biol. Med.* **23,** 712 (1997).

[5] L. Landi, D. Fiorentini, M. Lucarini, E. Marchesi, and G. F. Pedulli, *in* "Book of Abstracts; SFRR Europe Summer Meeting, June 26–28 1997, Abano Terme," p. 265.

[6] E. Niki, *in* "Atmospheric Oxidation and Antioxidants" (G. Scott, Ed.), Vol. 3, pp. 1–32. Elsevier, Amsterdam, 1993.

[7] B. Halliwell and J. M. C. Gutteridge, *Arch. Biochem. Biophys.* **280,** 1 (1990).

generating system under defined physicochemical conditions. Therefore, biologically relevant free radicals must be generated and examined under biologically relevant conditions for the determination of biological antioxidants. None of these conditions can be met with absolute precision because only the nature of radicals initiating oxidation processes is known. Radicals of lipids, proteins, fatty and amino acids, vitamins, hormones, and other compounds may be formed after reactions of primary radicals with biological substrates. The reactivity of these radicals with antioxidants depends on the physicochemical parameters of the medium.

None of the model systems discussed above can generate results which are sufficiently substantiated. In addition to sensitivity and manageability of the measuring system, the correlation with laboratory chemical, immunological, and clinical parameters is a crucial criterion. Impairment of antioxidative homeostasis as in an antioxidant deficiency exists as the cause or a concomitant phenomenon of diseases and pathological conditions. As the determination of antioxidant consumption *in vivo* is only possible in exceptional cases (such as arterial–venous difference in the umbilical cord blood), this information is not sufficient for a well-founded antioxidative therapy. A valuable approach is the combination of antioxidant determination and determination of the extent of oxidative damage. For the first task, a method of photochemical generation of free radicals with their chemiluminometric assay was developed. For the second task, hydrogen peroxide-stimulated chemiluminescence was used as the method.

Photoinduced Chemiluminescence. Photochemiluminescence (PCL) assay is based on the approximately 1000-fold acceleration of the oxidative reactions *in vitro* compared to normal conditions. This effect is achieved by optical excitation of a suitable photosensitizer S which exclusively results in the generation of the superoxide radical $O_2^{\cdot -}$ (and does not generate the singlet oxygen $^1O_2$):

$$S + h\nu + O_2 \rightarrow [S^*O_2] \rightarrow S^{\cdot +} + O_2^{\cdot -}$$

The free radicals are visualized with a chemiluminescent detection reagent. The diagram of the apparatus for the PCL measurements is shown in Fig. 1. Table I lists the major components for the determination of integral antioxidative capacity[8,9] of water- (ACW) and lipid-soluble (ACL) antioxidants, SOD,[10] and ascorbic acid.[11] Luminol plays a double role as photosensitizer and also as oxygen radical detection reagent.

[8] I. Popov and G. Lewin, *Free Rad. Biol. Med.* **17,** 267 (1994).

[9] I. Popov and G. Lewin, *J. Biochem. Biophys. Methods* **31,** 1 (1996).

[10] I. Popov, J. Hörnig, and R. v. Baehr, *Z. med. Lab. diagn.* **26,** 417 (1985).

[11] G. Lewin and I. Popov, *J. Biochem. Biophys. Methods* **28,** 277 (1994).

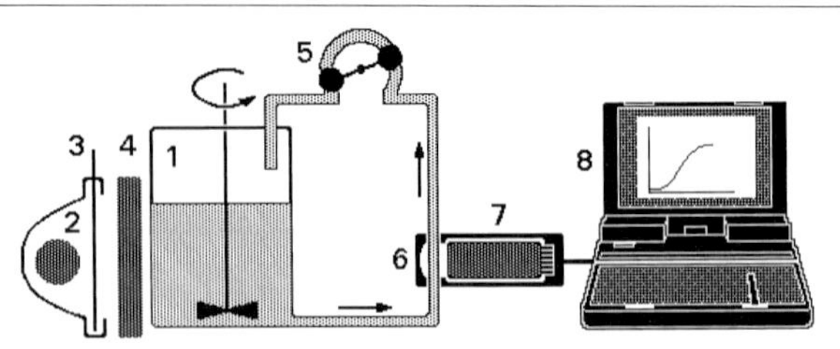

FIG. 1. Diagram of the apparatus for the measurement of photoinduced chemiluminescence. 1, irradiation cell; 2, low-pressure mercury lamp; 3, shutter; 4, UV filter; 5, peristaltic minipump; 6, measuring cell; 7, photomultiplier; 8, computer. (Reprinted from[11] with kind permission of Elsevier Science - NL.)

During measurement, the irradiated solution is transferred into the measuring cell where chemiluminescence is determined. Immediately after the start of irradiation, the intensity of the photoinduced chemiluminescence increases, after approximately 1 min it reaches a maximum and then decreases slowly (Fig. 2). The change of the signal in the presence of antioxidants allows their quantification.

The evaluation parameter for the determination of water-soluble antioxidants is the lag phase L in seconds:

$$L = L_0 - L_1$$

where L_0 and L_1 are the respective parameters of blank and sample.

Figure 3 represents a typical example of the measurement of lipid-soluble antioxidants in blood plasma. The evaluation parameter for ACL

TABLE I
COMPARISON OF THE PCL ASSAY KITS[a]

Assay KIT	Buffer	Water	Methanol	Luminol	Plasma	Extract	Ery lysate[b]	Eluate
ACW	1000	1500	0	25	2	0	0	0
ACL	200	0	2200	25	0	100	0	0
SOD	1000	1500	0	25	0	0	3	0
ASC	200	400	1800	25	0	0	0	100

[a] For integral measurement of water- (ACW) and lipid-soluble (ACL) antioxidants, selective determination of vitamin C (ASC) in human blood plasma and the SOD activity in erythrocytes. Principal kit components: carbonate buffer, 0.1 M, pH 10.5; luminol 1 mM. All volumes are in microliters. Lipid extraction was carried out as follows: 200 μl of plasma were mixed with 200 μl H_2O and 400 μl ethanol. After adding 800 μl hexane, the sample was agitated for 1 min and centrifuged 5 min at 1000g and 10°; 200 μl of the hexane phase were dried under N_2 and stored in a refrigerator. For measurements, the extracts were dissolved in methanol.

[b] 1:10 diluted whole blood.

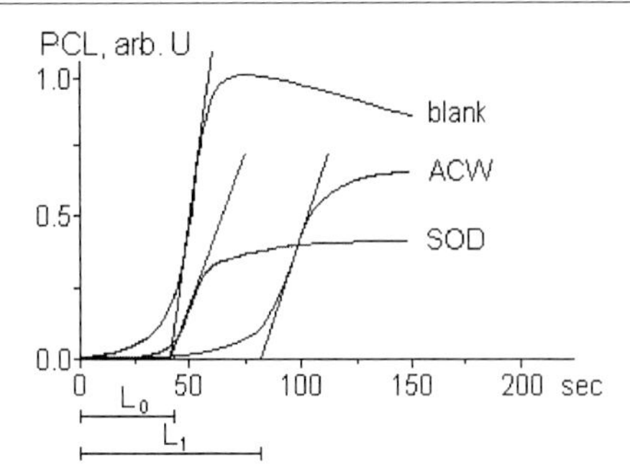

FIG. 2. Time course of the photoinduced chemiluminescence of luminol registered with the device PHOTOCHEM: blank, without additions; ACW, effect of ascorbic acid (1 nmol); SOD, effect of superoxide dismutase (100 ng).

and SOD measurements is the degree of PCL inhibition I, which is calculated as follows:

$$I = 1 - S/S_0$$

where S_0 is the integral under the blank curve and S is the integral under the sample curve.

In the measurement of both water-soluble and lipid-soluble antioxidants, the measuring range is between 0 and 2 nmol of the respective calibration reagents, ascorbic acid and α-tocopherol. The calibration curve is linear for ACW, and for ACL and SOD it can be linearized by representation in a reciprocal coordinate system $1/I$ and $1/m$, where m is the amount of reagent used.

The intraassay variability of ACW measurement is below 2%. For ACL and SOD, comparable results can be obtained for measurements in the

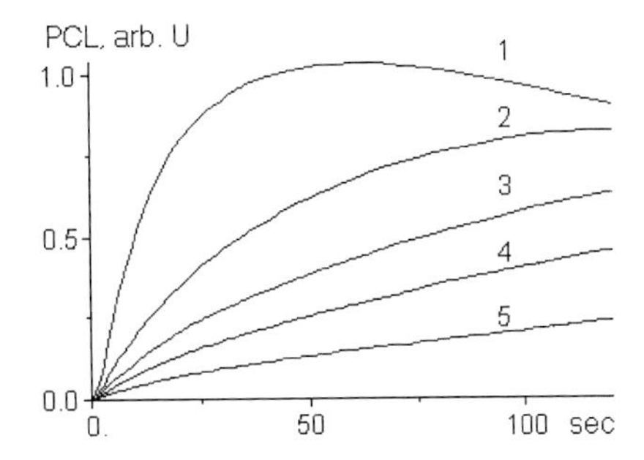

FIG. 3. Time course of the photoinduced chemiluminescence of luminol registered with the device PHOTOCHEM: 1, blank; 2–5, effect of the addition of lipid extracts from 20, 40, 60, and 80 μl blood plasma of a healthy blood donor. (Reprinted from[9] with kind permission of Elsevier Science - NL.)

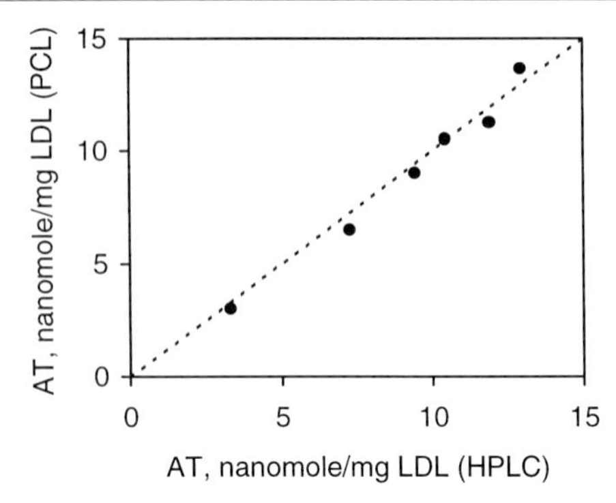

FIG. 4. Photochemiluminescent (PCL) and HPLC determination of the α-tocopherol content (AT) in *in vitro* supplemented LDL. (Reprinted from[9] with kind permission of Elsevier Science - NL.)

optimum inhibition range between 50 and 70%.[12] Besides speed and precision, the major benefits of the PCL method are the possibility of examining water- and lipid-soluble compounds in the same system and of changing the type of generated oxygen species without any problems due to the selection of the respective sensitizer. In contrast to other known procedures, this method is not restricted to a certain pH value or temperature range.

PCL Compared to Other Methods. In Fig. 4, the results of the ACL determinations in LDL samples supplemented with α-tocopherol *in vitro* are compared to HPLC assays.[9]

The PCL determination of ascorbic acid (ASC KIT) is compared with the results of the photometric KIT of Boehringer Mannheim (Fig. 5).[11] Both methods are in good agreement ($r = 0.994$). The results of the enzymatic ASC determination are not linear; this could be explained by incomplete oxidation of ascorbic acid at higher concentrations due to product inhibition.

The application of the PCL method for ACW determination in various animal species demonstrated that this parameter is age-dependent[8] and animal species-specific[8] and that under normal stress-free conditions the parameter is very stable for several days[1] (see Tables II, III and Fig. 6). The high ACW in spontaneous hypertensive (SH) rats and in vitamin C synthesis-incompetent guinea pigs can be compared to the situation in humans.

Under normal conditions, the individual coefficient of variation was smaller than 6% for the ACW of all 7 subjects. On the other hand, each

[12] I. Popov, G. Lewin, and R. v. Baehr, *Z. med. Lab. diagn.* **28,** 320 (1987).

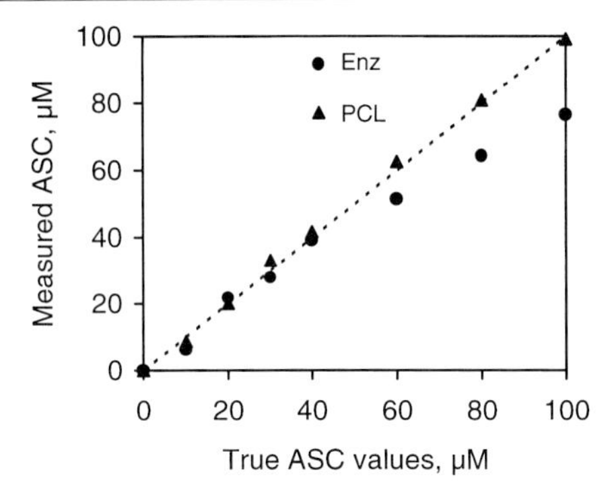

FIG. 5. Photochemiluminescent (PCL) and photometric enzymatic (Enz) determination of ascorbic acid in blood samples supplemented *in vitro*, in which its initial content was not detectable. (Reprinted from[11] with kind permission of Elsevier Science - NL.)

TABLE II
ACW OF BLOOD PLASMA[a] OF SOME ANIMAL SPECIES[b]

Species	n	m	$\pm m_e$
Rat (Wistar)	25	176	5.7
Rat (SH)	10	244	8.5
Rat (Lewis)	9	131	4.8
Mini-pig	24	39.8	2.8
Guinea pig	10	242	17
Mouse (strain 17)	10	108	14.2

[a] In micromolar ascorbic acid.
[b] m, Mean; m_e, mean error.

TABLE III
ACW OF BLOOD PLASMA[a] FROM HEALTHY DONORS OF DIFFERENT AGE[b]

Parameter	Newborns	Children	Youth	Elderly
Age (years)	0	7.3 ± 3.0	28.7 ± 6.5	46.0 ± 9.4
n	7	10	22	16
m (μM)	592	282	344	438
± SD	116	74	77	148
CV (%)	19.6	26.2	22.4	33.8
m_e	44	23	16	37

[a] In μM of ascorbic acid.
[b] m, mean; SD, standard deviation; CV, coefficient of variation; m_e, mean error.

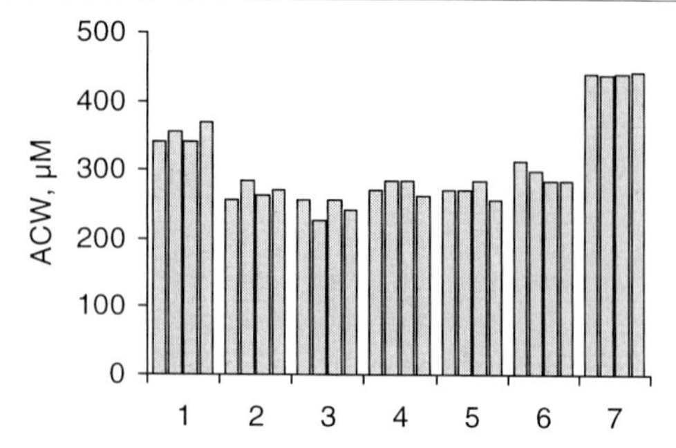

FIG. 6. ACW stability in seven healthy subjects. The period between all four blood samplings was 1 week each.

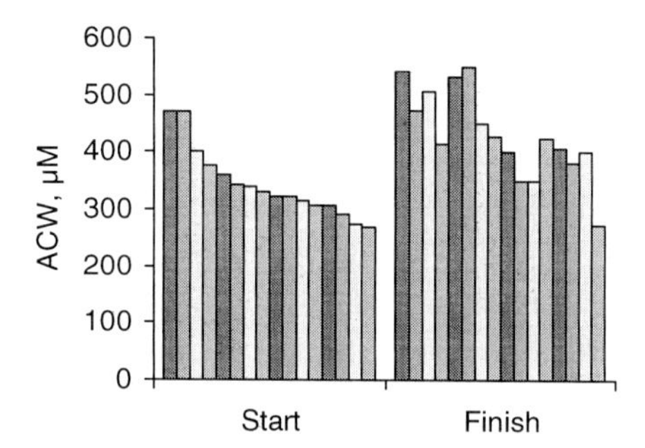

FIG. 7. Antioxidant capacity of the blood plasma of marathon runners at the start and at the finish, in the same order of subjects.

stress factor, such as physical exercise,[13,14] inflammation,[14,15] or noise,[14] results in a rapid increase of ACW and their physiologically most important component, ASC (see Figs. 7–11). This was confirmed by a direct check of the involvement of the stress axis in the regulation of antioxidative homeostasis in an animal study on Wistar rats. The β-receptor-active compounds adrenaline and isoprenaline (isoproterenol) lead to a dose-dependent ACW increase[14] (Figs. 12 and 13).

An increase of the antioxidative capacity of blood plasma could also be achieved by metabolic stimulation of the antioxidative system with pro-

[13] J. Henning, "Untersuchung zur Antioxidativen Kapazität bei Marathonläufern." Diploma Thesis, Free University, Berlin (1993).

[14] G. Lewin and I. Popov, "Praxisleitfaden der Antioxidativen Therapie." Karl F. Haug Verlag, Heidelberg, 1999.

[15] I. Popov and R. v. Baehr, *in* "Modulation, Mediation and Inhibition of Inflammation," Wiss. Beiträge MLU Halle-Wittenberg, p. 101. Halle (Saale) (1987).

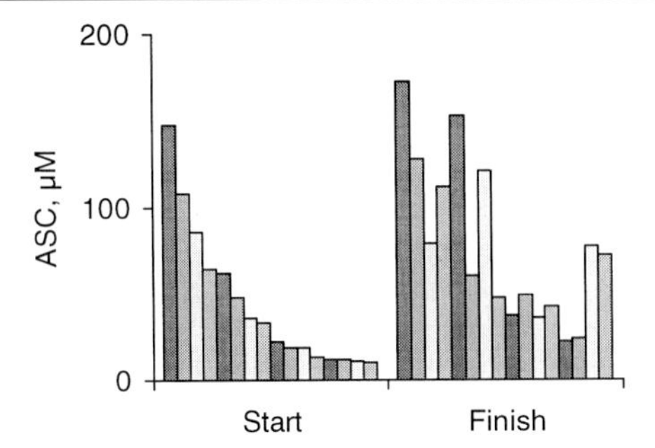

FIG. 8. Ascorbic acid in the blood plasma of marathon runners at start and at the finish, in the same order of subjects.

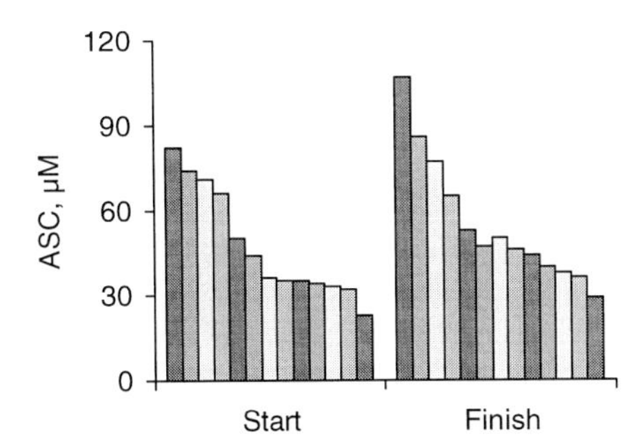

FIG. 9. Ascorbic acid in the blood plasma of athletes before and after veloergometric strain, in the same order of subjects.

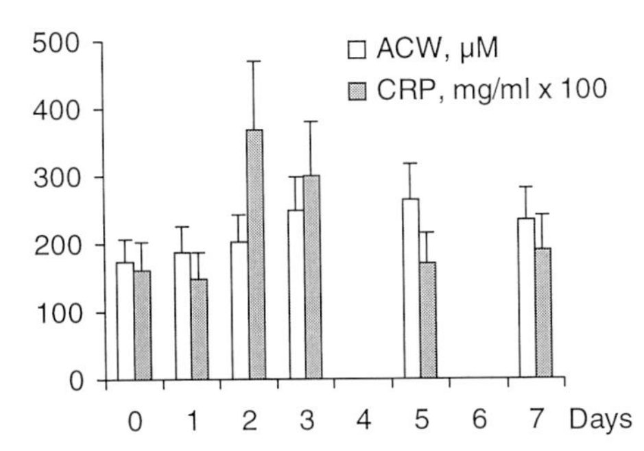

FIG. 10. Time course of ACW and CRP in the blood plasma of Wistar rats treated with turpentine oil (1 ml subcutaneous, as a model of sterile inflammation). $p < 0.01$ compared to the baseline value: Day 3–5 for ACW, day 2 and 3 for CRP.

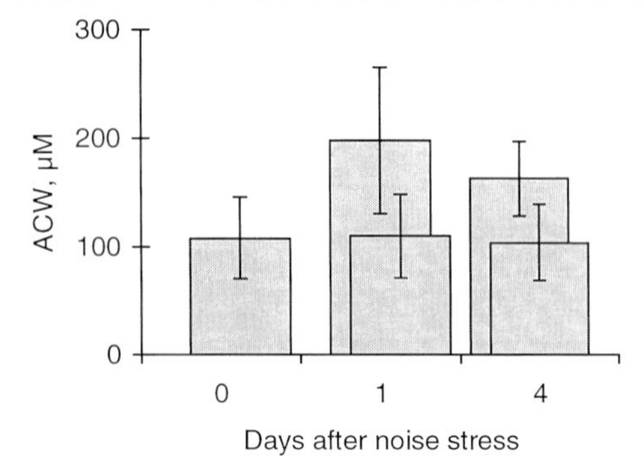

FIG. 11. Influence of an acute noise stress (1 hr, 90 dB) on the ACW in the blood plasma of mice (strain 17). Front bars: control values. For the stressed animals $p < 0.01$ was calculated for day 1 and $p < 0.05$ for day 4.

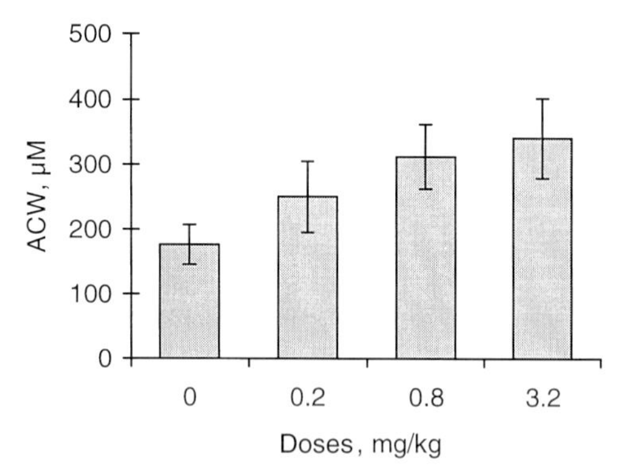

FIG. 12. Effect of adrenaline on the blood plasma ACW of Wistar rats. Measurements were taken 24 hr after subcutaneous application in the listed dosages (mg/kg body weight). $n = 10$ for each dosage, $p < 0.05$ for 0.2 mg/kg, $p < 0.01$ for 0.8 and 3.2 mg/kg.

oxidative stimuli such as ultraviolet irradiation of the blood[1,16] (Fig. 14). The increase in ACW demonstrated a positive correlation with the clinical effect of the blood UV irradiation therapy for peripheral arterial sclerosing disease—the prolongation of the claudication distance, CD (pain-free walking distance).

An impairment in antioxidative homeostasis may occur due to the insufficient supply of antioxidants during normal consumption (e.g., in vitamin deficiencies), or the incapability of the antioxidative system to react adequately to the requirements may be the cause. A pronounced decrease in

[16] I. Popov, G. Lewin, H.-P. Scherf, and H. Meffert, *Z. Klin. Med.* **44,** 1857 (1989).

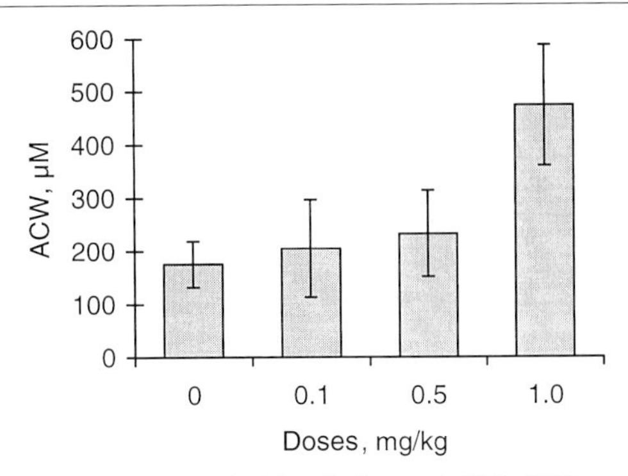

FIG. 13. Effect of isoprenaline on the blood plasma ACW of Wistar rats. Measurements were taken 24 hr after subcutaneous application in the listed dosages (mg/kg body weight). $n = 10$ for each dosage, $p < 0.05$ for 0.5 mg/kg, $p < 0.01$ for 1 mg/kg.

antioxidant levels occurs only in rare cases (Figs. 15 and 16). A slight inflammation such as a common cold does not cause drastic changes in the antioxidant level, for this reason both diagnostics and therapy control may be impeded. The cause of this phenomenon is the inherent property of the antioxidative system to react with suboptimal rather than maximal intensity to the local prooxidative stimuli due to dilution effects in blood circulation.[2] Therefore, an antioxidant determination cannot completely describe the state of antioxidative homeostasis. Additional parameters are necessary which can indicate whether the actual antioxidant level corresponds to its needs. In this case conclusions on recovery would be possible during therapy even when the level remains unchanged.

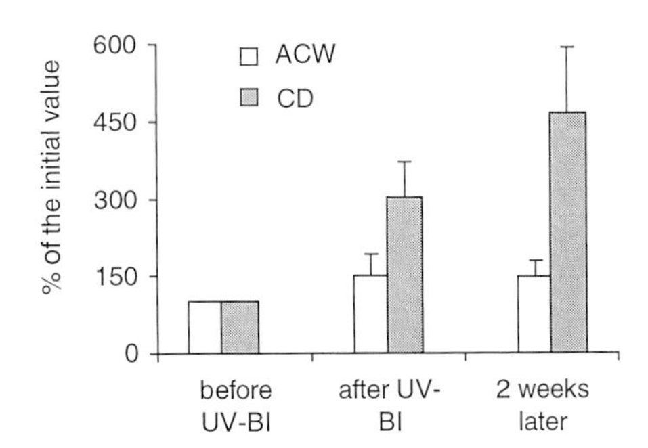

FIG. 14. Results of ACW and CD (claudication distance) measurements in patients with peripheral arterial sclerosing disease under therapy with ultraviolet autohemoirradiation. For all parameters after therapy, $p < 0.01$.

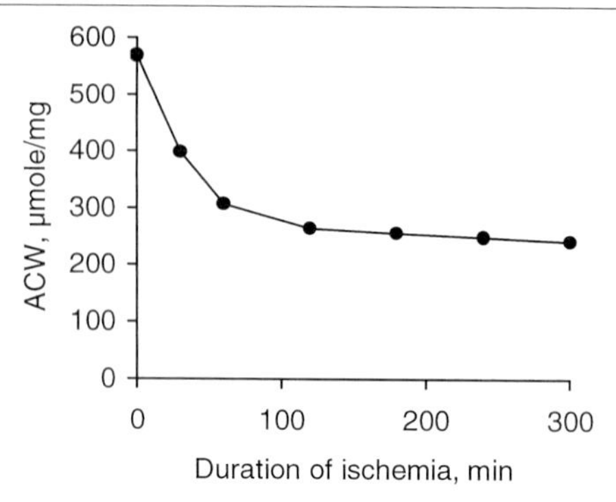

FIG. 15. Antioxidative capacity of rat liver (Wistar line) *in vitro* in μmol of a standard substance ascorbic acid per mg wet weight during warm ischemia of the organ at 37°.

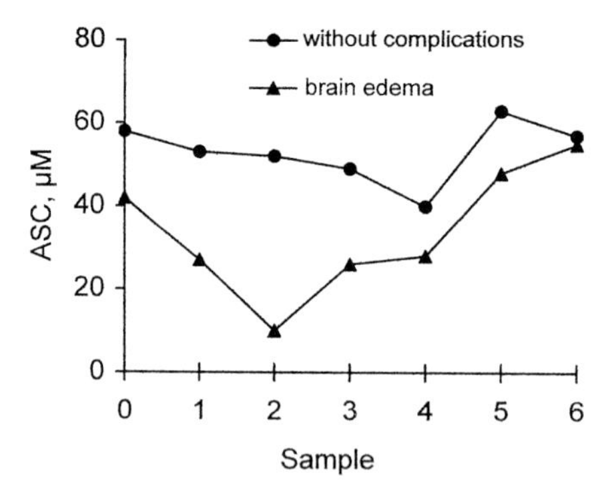

FIG. 16. Dynamics of ascorbic acid in the blood plasma of neurosurgical patients. Conditions of blood sampling referring to the sample number: (0) 1 day before operation; (1) before the first cut; (2) and (3) during operation, in 30- to 40-min intervals; (4) at the end of the operation; (5) the next day; (6) at discharge from the hospital. Mean values from $n = 14$ in the group without complications and $n = 6$ in the group with brain edema. $p < 0.05$ for (1) and (3) samples, and $p < 0.01$ for (2) sample.

Hydrogen peroxide-stimulated chemiluminescence seems to be an adequate parameter to determine the degree of oxidative damage to tissues.[17,18]

Hydrogen Peroxide-Stimulated Chemiluminescence and Oxidative Damage

Hydrogen peroxide-stimulated chemiluminescence was first discussed in the 1970s.[19] It was demonstrated that denaturation or oxidative modifica-

[17] I. Popov, G. Lewin, W. Gäbel, and R. v. Baehr, *Biomed. Biochim. Acta* **49,** 297 (1990).
[18] I. N. Popov and G. I. Lewin, *Phys. Chem. Biol. Med.* **1,** 75 (1994).
[19] A. E. Zakharyan and Yu. V. Babok, *Trans. Moscow Naturalists Soc.* **50,** 83 (1974).

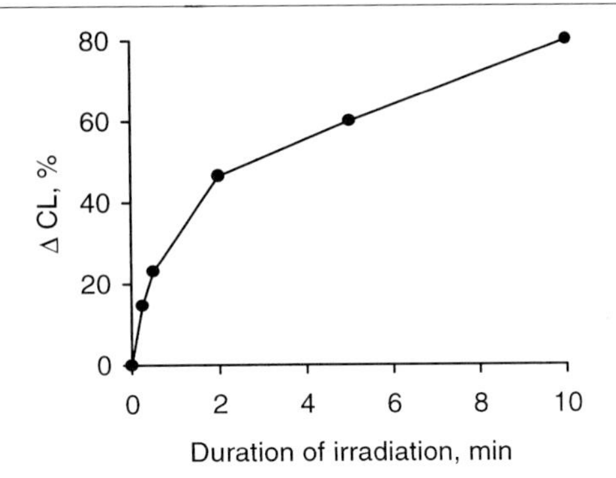

FIG. 17. Effect of UV irradiation on blood plasma *in vitro* on its H_2O_2-stimulated chemiluminescence presented as percent change compared to the baseline value.

tion of proteins and amino acids is accompanied by an increase in H_2O_2 chemiluminescence (CL) intensity. The validation of these findings in a model of spontaneous or UV-induced oxidation of blood plasma[20] and human serum albumin (HSA)[14] confirmed these results.

Routine measurements of the H_2O_2 CL of erythrocytes and blood plasma (referred to as CLE and CLP) are carried out with the universal instrument UNILUM which allows two measuring modes: PCL, which was described above, and H_2O_2 CL. In H_2O_2 CL, 0.5 ml of blood plasma or an erythrocyte suspension in saline is injected into the measuring cell which has been prefilled with 2 ml 3% (v/v) H_2O_2, and the time course of CL is recorded for 3 min. The parameter to be analyzed is the light production within this interval, calculated as an integral of the CL curve.

Before measurement, low-molecular weight compounds can be separated with a desalting column (Econo-pac, Bio-Rad, Richmond, CA). One ml plasma is applied and fractionated with phosphate buffer in 1 ml steps. CLP is measured in the fourth fraction of the eluate. Erythrocyte samples have a concentration of 10^6/ml in the measuring cell.

The dependence of CLP on the duration of the UV irradiation is presented in Fig. 17. Similar examinations on irradiated[21] and ischemic damaged[17] tissues confirm that this type of measurement yields information to assess the degree of the oxidative damage, i.e., that an increase in oxidative damage results in a similar increase of H_2O_2 CL. Obviously, the inclusion of the parameter CLP enables a better characterization of the state of antioxidative homeostasis.

[20] G. Lewin, B. Ramsauer, and I. Popov, *Wiener Klin. Wschr.,* submitted (1997).

[21] G. Lewin, G. A. Schreiber, K. W. Bögl, and I. N. Popov, *Bundesgesundhbl.* **2,** 61 (1993).

TABLE IV
CORRELATIONS BETWEEN MEASURED PARAMETERS UNDER NORMAL CONDITIONS[a]

Parameter	ACU	ASC	VE	AOL	UA
Age	n.s.	n.s.	0.2298*	−0.2565*	0.2069*
ACW	n.s.	n.s.	n.a.	n.s.	0.9970**
ASC	0.5968**	—	−0.2120*	n.s.	n.s.
AOW	0.5821**	−0.3049**	n.s.	n.s.	n.s.
ACL	n.s.	n.s.	0.7255**	0.2595**	n.s.
VE	−0.2317*	−0.2120*	—	−0.4764**	n.s.
AOL	0.2297*	n.s.	−0.4764**	—	n.s.
CLP	n.s.	−0.3947**	n.s.	n.s.	n.s.

[a] ACW, integral antioxidant capacity of water-soluble substances; ACU, urate-independent component of ACW; ASC, ascorbic acid; UA, uric acid (UA = ACW − ACU); AOW, sum of minor unidentified water-soluble antioxidants (AOW = ACU − ASC); ACL, integral antioxidant capacity of lipid-soluble substances; VE, vitamin E as a sum of α- and γ-tocopherols; AOL, sum of unidentified lipid-soluble antioxidants (AOL = ACL − VE); CLP, H_2O_2-induced chemiluminescence of the fourth fraction, containing the HSA and lipoproteins, after gel filtration of blood plasma on a desalting column. $^*p < 0.01$, $^{**}p < 0.001$, $n = 138$.

When the assumption of homeostatic regulation of the antioxidative capacity or its vitamin-dependent component is correct, it should be possible to detect certain deviations and correlations between antioxidative parameters and parameters of oxidative damage already affected by only a weak prooxidative cause. A check of the correlation of the relevant parameters was carried out on healthy volunteers without any influence and after a single UV total body exposure (Tables IV and V).

Table IV demonstrates clearly that under normal conditions the fine regulation of antioxidative homeostasis is ensured by the parameters ACU = ASC + AOW and ACL = VE + AOL. The unknown components

TABLE V
CORRELATIONS BETWEEN PARAMETERS OF ANTIOXIDATIVE HOMEOSTASIS[a]

Parameter	ASC	ACU	ACL	CLP	AOW	UA
ACW	−0.387**	n.s.	n.s.	0.307*	n.s.	0.993**
ASC	—	0.331*	−0.309*	−0.596**	−0.421**	−0.419**
ACU	0.331*	—	n.s.	−0.401**	0.717**	n.s.
CLP	−0.596**	−0.401**	n.s.	—	n.s.	0.349**

[a] Measured in a group of 12 healthy volunteers after a single solarium treatment with blood drawn before and after 15 and 30 min, 1, 6, 24, and 48 hr. $^*p < 0.01$, $^{**}p < 0.001$, $n = 84$. VE was not measured.

TABLE VI
URATE-INDEPENDENT ANTIOXIDATIVE CAPACITY[a]

Group	n	ACU	±SD	CLP	±SD
Healthy women	24	75.06	15.07	24.04	7.78
Benign tumors	40	47.63*	17.71	43.3*	16.18
Carcinoma stage I	14	44.71**	11.07	44.4**	9.97
Carcinoma stage II	13	42.98**	15.44	53.22**	18.36
Carcinoma stage III	9	22.36**	5.18	60.8**	16.13
Carcinoma stage IV	7	19.28**	5.24	80.0**	11.0

[a] ACU (in μM of the ascorbic acid equivalent) and hydrogen peroxide-stimulated chemiluminescence (CLP, arbitrary U) of blood plasma from patients with breast tumors. $*p < 0.05$, $**p < 0.01$ in comparison to control group.

AOW and AOL play a substituting role for the most biologically important antioxidants, vitamins C and E[7] (see highly significant correlations between ASC and AOW and between VE and AOL). Also, the negative correlation between ASC and CLP is remarkable. This is another indication of the protective antioxidative role of ascorbic acid with regard to proteins and lipoproteins. A comparison of the parameters within the subgroups of smokers and nonsmokers demonstrated that some parameters differ significantly from each other: for ACW and ACU $p < 0.05$, and for CLP $p < 0.001$. As expected, the ACW and ACU values are smaller in smokers, and CLP is higher in smokers than in nonsmokers[20] (data not shown).

After prooxidative stress, such as exposure to UV irradiation, antioxidative homeostasis becomes unbalanced for a short time. This enables one to see the connections between the parameters even better (Table V). Despite a lower n, the correlations ASC–AOW and VE–AOL discussed above are amplified, and new correlations become significant between ASC and UA. The last finding confirms the thesis which has been discussed in the literature several times concerning urate oxidase: in contrast to other mammals urate oxidase was lost in primates during evolution as a response to the loss of gulonolactone oxidase.[22] Finally, two examples of an impairment of antioxidative homeostasis in severely[23] and critically[24] ill patients are presented (Table VI and Fig. 18).

The diagram in Fig. 19 summarizes the processes of tissue damage due

[22] P. Proctor, *Nature (London)* **228,** 868 (1970).

[23] H. Völker, "Die Antioxidative Homöostase bei Patientinnen mit Mammatumoren. "Doctoral Thesis, Free University, Berlin (1996).

[24] M. Glatz, "Zum Verhalten der Antioxidativen Kapazität des Blutplasmas unter intensivmedizinischen Bedingungen." Doctoral Thesis, Humboldt University, Berlin (1997).

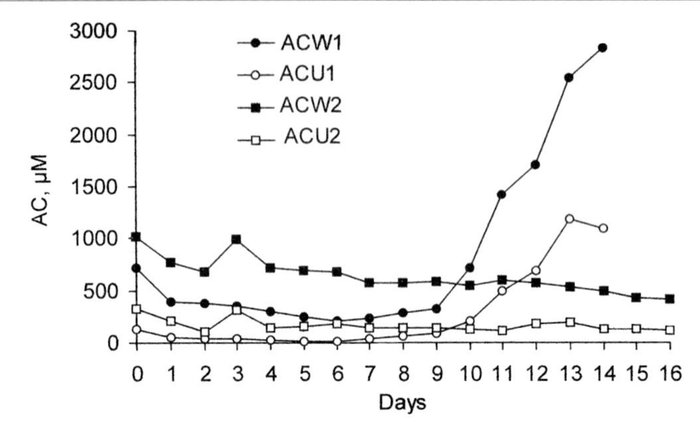

FIG. 18. Time course of the integral antioxidative capacity of blood plasma from two intensive care patients with peritonitis. Lethal outcome in patient 1.

to free radicals, their consequences, and detection methods.[14,17] It can be concluded that antioxidative capacity with all its components and hydrogen peroxide-stimulated chemiluminescence reflect the damage processes and the reactions of the antioxidative system.

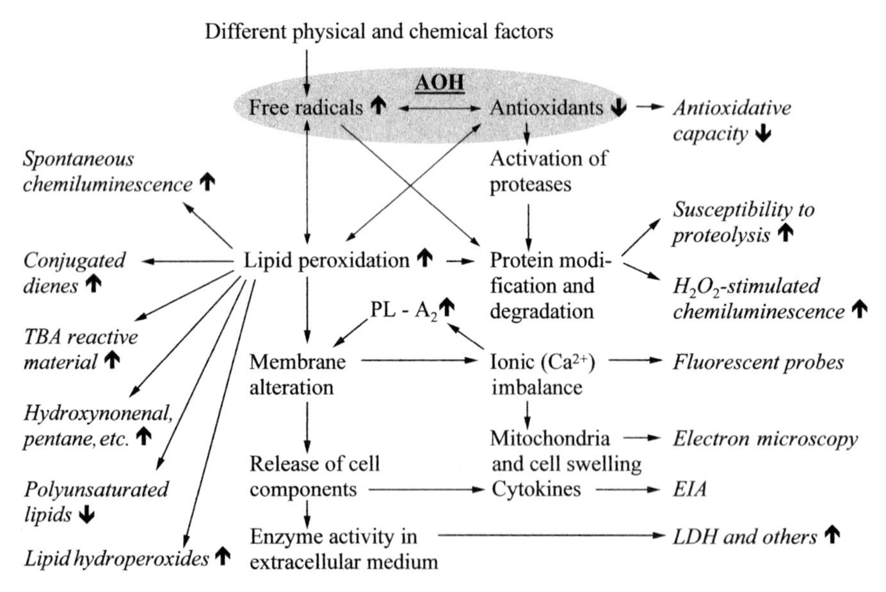

FIG. 19. Main constituents of free radical-mediated tissue damage and its detection. AOH, antioxidative homeostasis; PL, phospholipase; TBA, thiobarbituric acid; EIA, enzyme immunoassay; LDH, lactate dehydrogenase.

Summary and Perspectives

Until a few decades ago, natural radiation was relatively stable compared to the lifetime of living organisms and resulted in approximately two potentially damaging events per cell and day. During the past two centuries, because of new physical (e.g., artificial sources of UV and ionizing radiation, increase in the proportion of sun UV-B radiation on the earth surface caused by the destruction of the ozone layer) and chemical (industrial products and by-products, exhaust gases, etc.) changes, environmental pollution has increased more and more. Currently, the number of damaging elementary events is as high as 3×10^4 per cell and day.[25]

There is a connection between these phenomena and an increased incidence of diseases and pathological conditions such as cancer, atherosclerosis, and allergies, because the antioxidative system of the human organism is not capable of adapting to this rapid development of environmental prooxidative factors.

For oxidative damage to occur, a local decrease in antioxidative factors under a critical limit is the cause for subsequent pathological events. The centrally determined ACW reflects the reaction of the organism toward these local processes.

Isolated ACW measurements yield information on the actual antioxidant concentration but do not demonstrate whether their concentration is sufficient under the given circumstances or if it is suboptimal for an interruption of the local pathological oxidation process. Only an analysis of changes in ACW and its components during the course of disease and a comparison with the clinical pattern and the biochemical parameters of cellular, cell membrane, protein, and lipid damage make it possible to make some prognostic statements and to substantiate the necessity and type of antioxidative therapy.

In recent years, artificial, human, animal, plant, and gene technological antioxidatively acting compounds are being increasingly investigated in clinical trials. So far, however, real practical approaches to antioxidative therapy have not been developed.

An approach based on correcting the concentration of antioxidative vitamins that deviate from normal levels appears somewhat promising, as does the application of individual artificial antioxidants or so-called cocktails.

Clinically relevant approaches involve quality, dosage, and timing of an adequate effect on antioxidative homeostasis. These can be determined by systematic investigation of the antioxidative capacity of blood plasma with

[25] C. L. Greenstock, *Adv. Radiat. Biol.* **11,** 269 (1984).

particular emphasis on vitamin C as one of the biologically most relevant components of the water-soluble portion of ACW or vitamin E as the respective lipid-soluble antioxidant, together with hydrogen peroxide-stimulated chemiluminescence. In addition, some importance should be assigned to the measurement of a new parameter, the uric and ascorbic acid-independent proportion of the integral antioxidative capacity in blood plasma.

[45] Clinical Use of Photoionization Gas Chromatography for Detection of Lipid Peroxidation

By EMILE R. MOHLER III and DAVID R. HATHAWAY

Peroxidation of the lipid component of low density lipoproteins is key to the molecular pathophysiology of atherosclerosis.[1,2] The biochemical chain reaction of lipid peroxidation begins with oxidation of polyunsaturated fatty acids forming lipid peroxides which ultimately decompose to monohydroperoxides, aldehydes such as malondialdehyde, ketones, and volatile hydrocarbons. Some of the latter by-products, such as malondialdehyde, may react with lysine residues in apolipoprotein B-100 forming Schiff bases that change the configuration of the protein moiety of LDL enabling recognition by scavenger receptors. The modified LDL is more avidly incorporated into macrophages than native LDL and incorporation induces the formation of foam cells in atherosclerosis. This understanding has led investigators to develop clinically useful measurements of lipid peroxidation.

Several methods have been developed to measure *in vitro* products of lipid peroxidation.[3,4] However, *in vitro* techniques may not always reflect specific LDL oxidation and most require lengthy analytical procedures. The current interest in the role of lipid peroxidation in atherogenesis has stimulated the search for simple markers of *in vivo* lipid peroxidation. The volatile hydrocarbons ethane and pentane were first reported in the 1970s

[1] D. Steinberg, S. Parthasarathy, T. E. Carew, J. C. Khoo, and J. L. Witztum, *N. Engl. J. Med.* **320,** 915 (1989).
[2] U. P. Steinbrecher, *Clin. Cardiol.* **14,** 865 (1991).
[3] K. Yagi, *Biochem. Med.* **15,** 212 (1976).
[4] M. Tsuchida, T. Miura, K. Mizutani, and K. Aibara, *Biochem. Biophys. Acta* **834,** 196 (1985).

0076-6879/99 $30.00

to be measurable products of *in vivo* lipid peroxidation.[5–7] The hydrocarbon ethane is formed from peroxidation of omega-3 fatty acids, such as linolenic acids, while pentane is produced from peroxidation of omega-6 fatty acids, such as linoleic and arachidonic acids. These two volatile hydrocarbons are detected in the breath and have been demonstrated to reflect *in vivo* lipid peroxidation.[8,9]

Clinical experiments utilizing gas chromatography as a measure of lipid peroxidation are now numerous.[10] Lipid peroxidation accompanies the aging process and Zarling *et al.* found that exhaled pentane significantly increased with age.[11] Antioxidant vitamin supplementation has been shown to reduce pentane levels.[5,12] Also, pentane levels are thought to be elevated with acute myocardial infarction.[13] Interestingly, propranolol decreased pentane levels after psychological stress, although the mechanism for this finding is unknown.[14] Diseases such as rheumatoid arthritis[15] and Alzheimer's disease[10] are also associated with increased breath hydrocarbons.

Hydrocarbons have thermal stability and chemical inertness and thus can usually be adequately captured for analysis. A major problem encountered in sampling the breath is the large number of hydrocarbons present. Methane is the most common hydrocarbon exhaled and is present in large excess. Several techniques for capture of the breath have been reported in the literature. Most involve the use of either a bag which is impermeable to gas or a long gas-impermeable tube. The type of collection device used depends on the amount of sample needed to allow for adequate quantification of the desired measured gas. The sensitivity of the apparatus used for the gas detection ultimately determines whether the sample should be first concentrated on a precolumn. One difficulty with using precolumns is that sample may be lost in the transfer.

Two general techniques, flame ionization and photoionization, are used to measure breath hydrocarbons. Flame ionization for use in detection of

[5] C. J. Dillard, E. E. Dumelin, and A. L. Tappel, *Lipids* **12,** 109 (1977).

[6] E. E. Dumelin, C. J. Dillard, and A. L. Tappel, *Environ. Res.* **15,** 38 (1978).

[7] C. A. Riely, G. Cohen, and M. Lieberman, *Science* **183,** 208 (1974).

[8] S. R. B. Allerheiligen, T. M. Ludden, and R. F. Burk, *Drug Metab. Dispos.* **15,** 794 (1987).

[9] M. Phillips, J. Greenberg, and M. Sabas, *Free Radic. Res.* **20,** 333 (1994).

[10] C. M. F. Kneepkens, C. Ferreira, G. Lepage, and C. C. Roy, *Clin. Invest. Med.* **15,** 163 (1993).

[11] E. J. Zarling, S. Mobarhan, P. Bowen, and S. Kamath, *Mech. Aging Dev.* **67,** 141 (1993).

[12] E. Hoshino, R. Shariff, A. Van Gossum, J. P. Allard, C. Pichard, R. Kurian, and K. N. Jeejeebhoy, *J. Paren. Enter. Nutr.* **14,** 300 (1990).

[13] Z. W. Weitz, A. J. Birnbaum, P. A. Sobotka, E. J. Zarling, and J. L. Skosey, *Lancet* **337,** 933 (1991).

[14] J. Pincemail, G. Camus, A. Roesgen, E. Dreezen, Y. Bertrand, M. Lismonde, and G. Deby-Dupont, *Eur. J. Appl. Physiol.* **61,** 319 (1990).

[15] S. Humad, E. Zarling, M. Clapper, and J. L. Skosey, *Free Rad. Res. Commun.* **5,** 101 (1988).

breath hydrocarbons separated by chromatography has been widely reported in the literature.[16–18] Seabra *et al.* described such a system using a flame ionization detector (FID) with preconcentration of the breath on a solid adsorbent at 0° and desorption at 250°.[19] Guilbaud *et al.* also described a method to evaluate lipid peroxidation by measuring exhaled pentane with a flame ionization detector.[18] However, flame ionization techniques require larger samples to achieve adequate sensitivity, thereby necessitating a precolumn to concentrate the sample prior to chromatography. This lengthens the time of analysis. In addition, flame ionization detectors are typically bench type and not portable.

Photoionization chromatography utilizes a photoionization detector (PID) that emits photons from an ultraviolet lamp that ionize selected gas molecules in the carrier gas stream. Permanent gases (N_2, O_2, CO_2), including water vapor, are not ionized. An electrical signal generated from the ionized gas is proportional to the quantity of gas present and diluted samples of the purified gas of interest can be used to produce a calibration curve. The results are usually expressed in parts per million (ppm) or parts per billion (ppb). The chromatographic analyses are usually performed at a temperature of 30°C using an isothermal oven.

The photoionization detector is unable to distinguish between the thousands of samples that it can ionize. Thus, gas chromatography is necessary to separate each component of a mixed sample. A stream of pure air, usually ultra zero compressed air, is used as a carrier gas that moves a sample through a gas chromatographic column prior to analysis by the PID. The time required for a particular gas to move through the column (the retention time) depends on the chemical nature of the gas, the adsorbent in the column, the length of the column, and the temperature. Advantages of photoionization gas chromatographs are their relatively small size and portability. Also, samples can be automatically and repeatedly measured over a prolonged time period. A precolumn to concentrate the gas sample prior to analysis is usually not necessary.

The correct choice of chromatographic column packing material is essential for optimum results. Columns are packed with an adsorptive powder or lined with an adhesive coating. Solid adsorbents have been predominantly used to separate mixtures of permanent gases and hydrocarbons. Sorbents with polar surfaces, such as silica and alumina, are useful in selecting out alkenes and alkynes.

[16] M. Lemoyne, A. Van Gossum, R. Kurian, M. Ostro, J. Axler, and K. N. Jeejeebhoy, *Am. J. Clin. Nutr.* **46,** 267 (1987).

[17] E. J. Zarling and M. Clapper, *Clin. Chem.* **33,** 140 (1987).

[18] R. Guilbaud, A. C. Ricard, C. Daniel, S. Boileau, H. Van Tra, and G. Chevalier, *Toxicol. Meth.* **4,** 1 (1994).

[19] L. Seabra, J. M. Braganza, and M. F. Jones, *J. Pharm. Biomed. Anal.* **9,** 693 (1991).

The various packings described above are used in columns a few meters in length that are generally about 2–4 mm in internal diameter. The retention times of the gases can obviously vary, and most equipment is equipped with a computer system that can output the data on a digital display. Halasz and Heine were the first to describe the use of packed capillary columns for the rapid separation of hydrocarbon gases.[20] McTaggart *et al.* reported the use of alumina-packed glass capillary columns for quantitative analysis of C_1–C_5 hydrocarbons.[21] They found that the reproducibility obtained with conventional packed columns was similar, but there was a significant reduction of approximately 50% in analysis time. The columns they described are also stable enough for routine use.

Human breath is composed of several light hydrocarbons, including butane, propane, isopentane, *n*-pentane, and isoprene, that may be difficult to separate depending on the column selected. The importance of choosing the correct column was the subject of an investigation by Kohlmuller *et al.*, who demonstrated that many gas chromatographic column materials frequently used to uniformly separate pentane from other gases fail to resolve pentane from isoprene.[22] This is especially important since the amount of isoprene found in human breath and cigarette smoke may be very large compared to pentane. In addition, an investigation by Springfield and Levitt described several artifacts that may explain why values for breath pentane in healthy subjects as an indicator of lipid peroxidation can vary in literature reports by up to 1000-fold.[23] They found that a major component of human breath, probably isoprene, coeluted with pentane on columns used by some investigators, resulting in erroneously high determinations. Some commonly used gas column adsorbants that are not thought to adequately separate isoprene from pentane include Porasil C, Tenax GC, activated alumina, and GSQ/Poraplot Q, uncoded Carbopack B and Chromosorb 102.[22] Also, isoprene, to our knowledge, is not an accurate marker for lipid peroxidation and may bias results. Hence, precise analysis requires selecting and designing the best column for the separation.

Mohler *et al.* investigated the feasibility of a photoionization gas chromatographic system to analyze breath pentane that addresses many of the technical problems described above and that would provide a simple *in vivo* method for assessing lipid peroxidation.[24] *n*-Pentane and isopentane standards were easily separated from isoprene with an encapsulated Al_2O_3/KCl packed capillary column contained in a portable gas chromatograph

[20] I. Halasz and E. Heine, *Nature* **194,** 971 (1962).

[21] N. G. McTaggart, C. A. Miller, and B. Pearce, *J. Inst. Petrol.* **54,** 265 (1968).

[22] D. Kohlmuller and W. Kochen, *Anal. Biochem.* **210,** 268 (1993).

[23] J. R. Springfield and M. D. Levitt, *J. Lipid. Res.* **35,** 1497 (1994).

[24] E. R. Mohler, P. Reaven, J. E. Stegner, N. S. Fineberg, and D. R. Hathaway, *J. Chromatogr. B. Biomed. Appl.* **685,** 201 (1996).

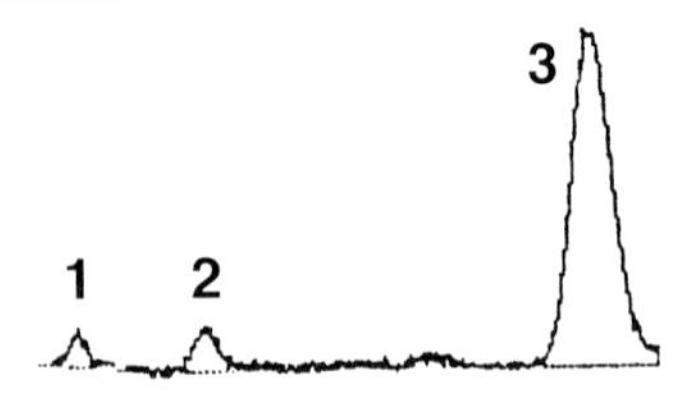

FIG. 1. GC/PID chromatogram showing isopentane (peak 1), *n*-pentane (peak 2), and isoprene (peak 3) using an Al_2O_3/KCl column.

equipped with a photoionization detector. The analysis of repeated measures showed a low coefficient of variation for measurements of *n*-pentane (10%) and isopentane (9%) (Fig. 1). Other investigators have also found that there is adequate separation of these and other breath gases using this alumina-type capillary column.[25]

Several potentially confounding variables in breath pentane measurement have been studied to determine their influence on breath pentane levels. One such variable is whether pentane produced by colonic flora influences breath pentane levels. Lemoyne *et al.* found that pentane output did not differ in the fed and fasted state and after feeding fat and carbohydrate.[16] Similarly, Regat *et al.* found no difference in breath ethane secretion before and after breakfast in healthy children.[26] Zarling *et al.* gave healthy subjects standardized liquid diet after an overnight fast and measured breath ethane and pentane.[27] There were no statistically significant changes in either alkane relative to the fasting state. Thus, it is unlikely that diet significantly influences measurement of breath pentane.

The production rate of pentane at steady state is the sum of pulmonary excretion and elimination from other organs such as the liver.[8,9,28] Allerheiligen *et al.* noted that inhibition of cytochrome P450 in the liver of rats caused a dramatic decrease in pentane clearance.[8] Thus, drugs that block hepatic P450 monooxygenase can influence pentane levels. Alternatively, Burk *et al.* noted that liver injury by thioacetamide markedly decreased pentane levels.[28] The administration of ethanol also decreased pentane clearance, suggesting that alcohol dehydrogenase may be involved in pen-

[25] K. F. Heim, U. M. Makila, R. Leveson, G. S. Ledley, G. Thomas, C. E. Rackley, and P. W. Ramwell, *in* "Lipid Mediators in the Immunology of Shock" (M. Paubert-Braquet, Ed.), p. 103. Plenum, New York, 1987.

[26] M. Refat, T. J. Moore, M. Kazui, T. H. Risby, J. A. Perman, and K. B. Schwarz, *Pediatr. Res.* **30,** 396 (1991).

[27] E. J. Zarling, S. Mobarhan, P. Bowen, and S. Sugerman, *J. Am. Coll. Nutr.* **11,** 349 (1992).

[28] R. F. Burk, T. M. Ludden, and J. M. Lane, *Gastroenterology* **84,** 138 (1983).

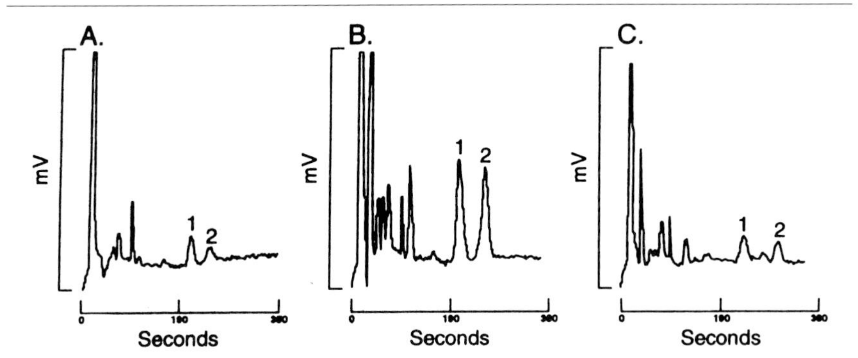

FIG. 2. Representative chromatogram of a breath sample after an overnight abstinence from smoking (A), and 10 min (B) and 1 hr (C) after smoking one cigarette, using the PID gas chromatograph. Isopentane is indicated as peak 1 and *n*-pentane as peak 2.

tane metabolism. These findings indicate that drugs affecting liver function should be considered when attempting to measure breath pentane.

Early use of a photoionization detector for identification of hydrocarbons was described by Driscoll *et al.*, who analyzed the effluent from a gas chromatograph with both photoionization and flame ionization detectors.[29] They found that the PID has a 35-fold greater sensitivity for aromatics and a 7-fold greater sensitivity for alkanes such as *n*-hexane than FID. Heim *et al.* used a photoionization detector and gas chromatography to evaluate lipid peroxidation in patients who underwent renal transplantation.[25] They found that 3 of 4 patients with clinical signs of transplant rejection also had an accompanying elevation in breath pentane. The PID system was described as simple, sensitive, and rapid compared to standard flame ionization techniques.[25]

Cigarette smoke contains numerous chemical components, including free radicals and hydrocarbons. The heightened oxidant stress from free radicals in cigarettes is thought to result in increased lipid peroxidation in the lung and in circulating lipoproteins.[30–32] In several studies, both ethane and pentane were found to be increased in the breath of smokers.[12,24,33,34] Mohler *et al.* used the above photoionization detector method to measure breath pentane in 27 subjects (15 smokers, 12 nonsmokers)[24]. There was no significant difference between the baseline and 4 week interval

[29] J. N. Driscoll and P. Warneck, *J. Air. Pollut. Control. Assoc.,* in press.

[30] B. Frei, T. M. Forte, B. N. Ames, and C. E. Cross, *Biochem. J.* **277,** 133 (1991).

[31] J. D. Morrow, B. Frei, A. W. Longmire, *et al. N. Engl. J. Med.* **332,** 1198 (1995).

[32] M. Reilly, N. Delanty, J. A. Lawson, and G. A. FitzGerald, *Circulation* **94,** 19 (1996).

[33] C. R. Wade and A. M. Van Riji, *Proc. Univ. Otago Med. Sch.* **64,** 75 (1986).

[34] B. K. Do, H. S. Garewal, N. C. Clements, Jr., Y. M. Peng, and M. P. Habib, *Chest* **110,** 159 (1996).

measurements of *n*-pentane for smokers both before and after cigarette smoking. The within-subject variability indicated that the assay is highly reproducible for both low and high pentane levels in smokers. Smokers were found to have higher levels of both *n*-pentane and isopentane than nonsmokers ($p < 0.001$). In addition, smokers had further significant elevation of pentane levels 10 min after smoking ($p < 0.001$), which returned to baseline by 1 hr (Fig. 2). These studies demonstrate that measurement of breath pentane using a gas chromatograph with a photoionization detector is simple and reproducible. Additionally, these results suggest that pentane elevation associated with smoking is secondary to the oxidant effects of cigarette smoke and an important temporal relationship exists between cigarette smoking and breath sample analysis. The sensitivity and clinical utility of measuring breath pentane using PID in nonsmokers remains to be determined.

In summary, the measurement of breath hydrocarbons provides for *in vivo* measurement of lipid peroxidation. Two types of detection techniques, flame ionization and photoionization, are generally employed in measuring breath pentane and ethane. Photoionization is more sensitive than flame ionization and affords a more rapid analysis time. Any methodology used in measuring breath hydrocarbons requires careful validation of the gas chromatography column adsorbent to ensure adequate separation of isoprene from pentane and ethane. The routine clinical utility of measuring breath hydrocarbons is currently undetermined; however, it is hoped that future investigations will better define its role in the clinical practice of medicine.

[46] Intravital Fluorescence Microscopy for the Study of Leukocyte Interaction with Platelets and Endothelial Cells

By HANS-ANTON LEHR, BRIGITTE VOLLMAR, PETER VAJKOCZY, and MICHAEL D. MENGER

Introduction

Reactive oxygen species (ROS) exert far reaching effects on the microcirculation with enormous clinical relevance in diverse pathophysiological conditions (reviewed extensively in Halliwell and Gutteridge[1]), ranging

[1] B. Halliwell and J. M. C. Gutteridge, "Free Radicals in Biology and Medicine," Clarendon Press, Oxford, 1989.

0076-6879/99 $30.00

from emphysema formation in smokers[2] to diabetic microangiopathy,[3] from myocardial infarction[4] to the rejection of transplanted organs.[5] Despite this wide array of diverse clinical and experimental conditions, the microcirculatory manifestations of ROS injury are limited to only a few phenomena, which affect the endothelium and/or the circulating blood elements. Well-characterized ROS-mediated endothelial dysfunction includes the upregulation of adhesion molecules for leukocytes and platelets,[6] endothelial cell swelling with subsequent compromise of capillary lumina,[7] and the breakdown of endothelial integrity, resulting in the leakage of fluid and macromolecules into the interstitium.[8] On the other hand, ROS affect corpuscular blood elements, i.e., via the activation, endothelial adhesion, and transendothelial emigration of leukocytes, leukocyte aggregation with platelets, and diverse metabolic functions, including the respiratory burst of leukocytes, which results in further generation and release of ROS at the site of microvascular damage (reviewed in McIntyre *et al.*[9] and Heppiheimer and Granger[10]). Most of these phenomena can be observed and studied at the light microscopic level, *in vitro* using cell culture methods, *ex vivo* using isolated perfused organ preparations, or *in vivo* using animal models. The present paper introduces the techniques of intravital microscopy as a tool to study the microcirculation in intact animals under conditions of oxidative stress.

Model 1: Dorsal Skinfold Chamber Model in Hamsters and Mice

History

In 1924, Sandison was the first to implant observation chambers into animals for intravital microscopy of living tissues.[11] Since then, various

[2] M. M. Dooley and W. A. Pryor, *Biochem. Biophys. Res. Commun.* **106,** 981 (1982).
[3] J. L. Wautier, C. Zoukourian, O. Chappey, M. P. Wautier, P. J. Guillausseau, R. Cao, O. Hori, D. Stern, and A. M. Schmidt, *J. Clin. Invest.* **97,** 238 (1996).
[4] R. A. Axford-Gately and G. J. Wilson, *Can. J. Cardiol.* **9,** 94 (1993).
[5] P. Vajkoczy, H. A. Lehr, C. Hübner, K. E. Arfors, and M. D. Menger, *Am. J. Pathol.* **150,** 1487 (1997).
[6] K. D. Patel, G. A. Zimmerman, S. M. Prescott, R. P. McEver, and T. M. McIntyre, *J. Cell. Biol.* **112,** 749 (1991).
[7] C. Willy, J. Thiery, M. D. Menger, K. Messmer, K. E. Arfors, and H. A. Lehr, *Free Rad. Biol. Med.* **19,** 919 (1995).
[8] M. D. Menger, S. Pelikan, D. Steiner, and K. Messmer, *Am. J. Physiol.* **263,** H1901 (1992).
[9] T. M. McIntyre, K. D. Patel, G. A. Zimmerman, and S. M. Prescott, *in* "Physiology and Pathophysiology of Leukocyte Adhesion" (N. D. Granger and G. Schmid-Schoenbein, Eds.), p. 261. Oxford University Press, New York, 1994.
[10] M. J. Eppiheimer and D. N. Granger, *Shock* **8,** 16 (1997).
[11] J. C. Sandison, *Am. J. Anat.* **41,** 447 (1928).

chambers have been developed and implanted with the aim to investigate the microcirculation in mice,[12] hamsters,[13] rats,[14] rabbits,[15,16] and even in human subjects.[17] The tissue under microscopic observation was either newly formed granulation tissue[12,14,16,18] or preformed tissue, mostly striated skin muscle and subcutaneous tissue.[19–21] To facilitate the access to the chamber and to allow the microscopic investigation in restrained but nonanesthesized animals, chambers were implanted into the dorsal skinfold in mice,[19,22] rats,[21] and hamsters[20] (Fig. 1). These models have been applied to the study of the native microvasculature, but also to the growth and microvascularization of transplanted neoplastic[21,23] and non-neoplastic tissues.[24–26] The adaption of the chamber technique to mice[22] has made it possible not only to make experimental use of the wide array of functionally blocking antibodies available in this species (i.e., directed toward adhesion molecules[27]), but also to use immunoincompetent animals (i.e., nude or SCID mice), genetically aberrant animals (i.e., MDX mice as a model of Duchenne muscular dystrophy or NOD mice for the study of diabetic microangiopathy), or genetically engineered animals (i.e., knockout mice or transgenic mice).

Surgical Technique

The chambers are specifically manufactured and can be obtained through the Institute for Surgical Research, University of Munich (Prof. Dr. K. Messmer) or the Department of Experimental Surgery, University of Heidelberg (Prof. Dr. M. Gebhard). The frames are made of titanium

[12] G. H. Algire, *J. Natl. Cancer Inst.* **4,** 1 (1943).
[13] M. Greenblatt and P. Shubik, *Cancer Bull.* **19,** 65 (1967).
[14] J. B. Hobbs, S. Chusilp, A. Hua, P. Kincaid-Smith, and M. A. McIver, *Clin. Sci. Mol. Med.* **3,** 73S (1976).
[15] K. E. Arfors, J. A. Jonsson, and F. N. McKenzie, *Microvasc. Res.* **2,** 516 (1970).
[16] E. R. Clark, H. T. Kirby-Smith, R. O. Rex, and R. G. Williams. *Anat. Rec.* **47,** 187 (1930).
[17] P. L. Branemark, K. Aspegren, and U. Breine, *Angiology* **15,** 329 (1964).
[18] S. Z. Cardon, C. F. Ostermeyer, and E. H. Bloch, *Microvasc. Res.* **2,** 67 (1970).
[19] K. H. Falkvoll, E. K. Rofstad, T. Brustad, and P. Marton, *Exp. Cell Biol.* **52,** 260 (1984).
[20] B. Endrich, K. Asaishi, A. Goetz, and K. Messmer, *Res. Exp. Med.* **177,** 125 (1980).
[21] H. D. Papenfuss, J. F. Gross, M. Intaglietta, and F. A. Treese, *Microvasc. Res.* **18,** 311 (1979).
[22] H. A. Lehr, M. Leunig, M. D. Menger, D. Nolte, and K. Messmer, *Am. J. Pathol.* **143,** 1055 (1993).
[23] T. Oda, A. Lehmann, and B. Endrich, *Biorheology* **21,** 509 (1984).
[24] K. Messmer, W. Funk, B. Endrich, and H. Zeintl, *Prog. Appl. Microcirc.* **6,** 77 (1984).
[25] W. Funk, B. Endrich, and K. Messmer, *Res. Exp. Med.* **186,** 259 (1986).
[26] M. D. Menger, S. Jäger, P. Walter, F. Hammersen, and K. Messmer, *Int. J. Microcirc. Clin. Exp.* **9,** 103 (1990).
[27] H. A. Lehr, M. Kröber, C. Hübner, P. Vajkoczy, M. D. Menger, D. Nolte, A. Kohlschütter, and K. Messmer, *Lab. Invest.* **68,** 388 (1993).

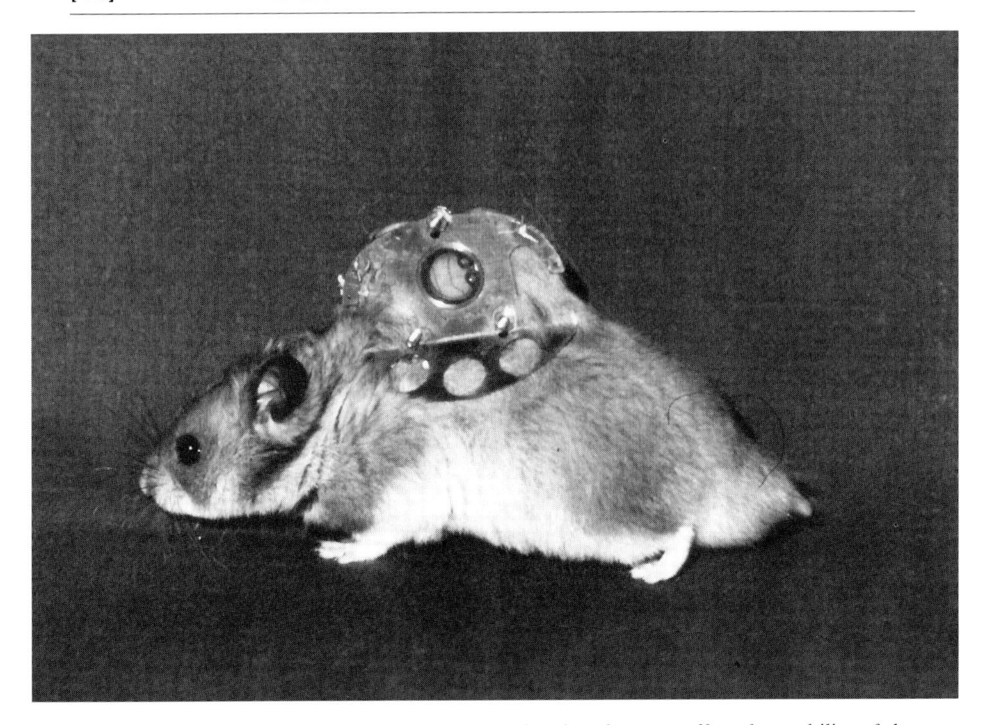

FIG. 1. Hamster with skinfold chamber. The chamber does not affect the mobility of the animal, nor its feeding, sleeping or cleaning habits. The microvessels and larger feeding arterioles/draining venules are seen traversing the observation chamber.

(Fig. 2). This material is characterized by a low weight (less than 3.5 g per frame) and a low temperature conductance coefficient and thus prevents the loss of body heat when chronically implanted into the animals (Fig. 1). The implantation of the chambers is performed in anesthesized, 5- to 7-week-old hamsters weighing a minimum of 45 g and a maximum of 70 g or in mice weighing a minimum of 25 g (smaller chambers, see Fig. 2). Implantation is performed under anesthesia (hamsters: 60 mg pentobarbital/kg bodyweight, i.p.; mice: 3.75 ml ketamine, s.c.). The fur at the back of the animals is carefully shaven and depilated using commercially available depilation creams. This is done very gently in order not to irritate the skin and the subcutaneous tissues. The depilation cream is removed with a soft sponge and lukewarm water. Heating pads or heating lamps are used to prevent heat loss during anesthesia and the postoperative period.

In a first step, the dorsal skin is pulled up and the vascular network visualized through the translucent skin in front of a regular light source. The skin flap is positioned so as to match the vessels of both sides. This

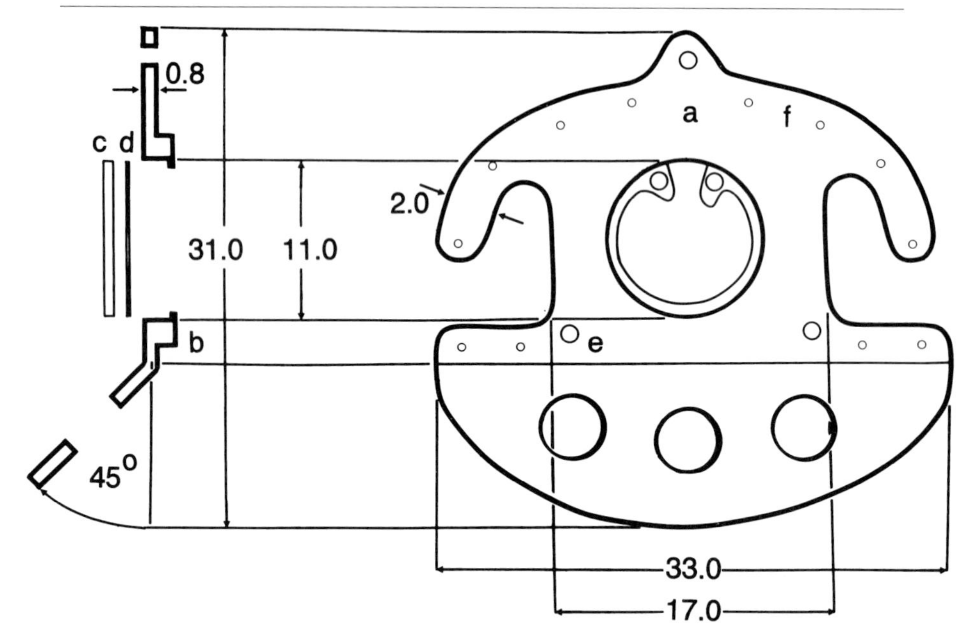

FIG. 2. Schematic representation of the titanium chamber frames used for implantation in mice. The three large holes in the chamber frames serve to minimize the weight of the chambers, which is only 3.5 grams. *a*, Titanium frame, frontal view; *b*, titanium frame, side view; *c*, metal ring; *d*, glass coverslip; *e*, holes for screws; *f*, holes for sutures. The numbers indicate dimensions in mm.

position is then fixed by suspending the skin flap using two sutures placed through the dorsal midline of the skin. The main feeding and draining vessel bundles show a striking degree of consistency between hamsters and mice from breeding stocks in different parts of the world and over extended periods of time (early 1980s till present). The site of the chamber implantation is immediately caudal to the main vessel bundle. This is the only area in the skinfold in which the retractor muscle and the cutaneous muscle have no communicating perforance vessels and can thus be dissected microsurgically without major hemorrhage. After selection of the implantation site, one frame of the chamber is sutured to the skin and small incisions are made to accommodate the screws. The frame is secured using hemostats attached to these screws. Using a waterproof marker, the area of the observation window is outlined and the most superficial layer of epidermis, subcutaneous tissue, and skin muscle removed. At every step of the dissection, exquisite efforts are made to prevent hemorrhage into the surgical field. Subsequently, the animal is repositioned and the surgical field placed horizontally on a small pile of surgical gauze. From that point on, the

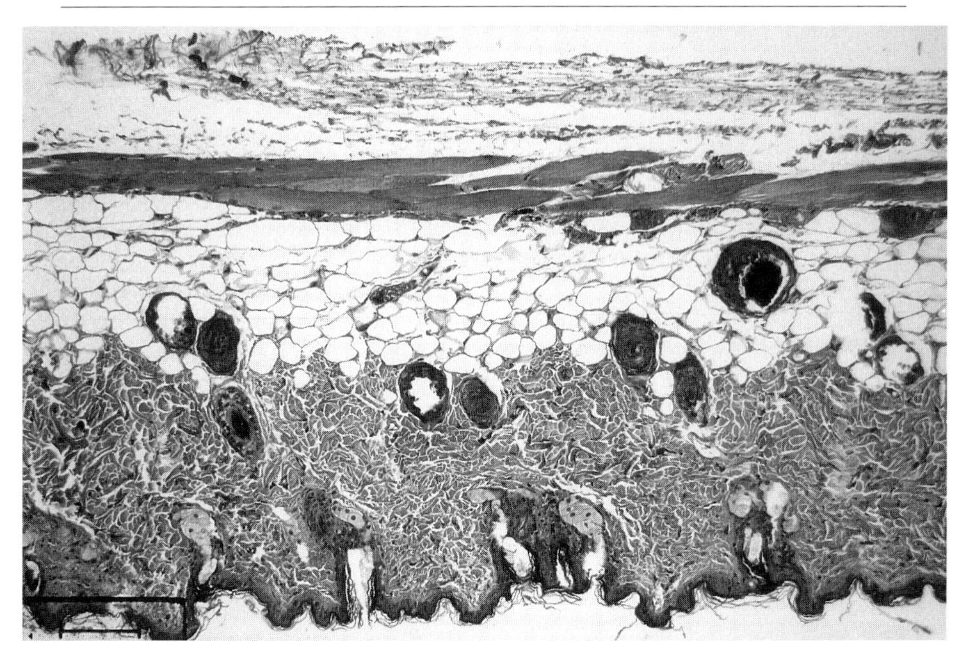

FIG. 3. Histological section through the skin at the dorsal aspect of a nude mouse. The fine skin muscle (upper third of picture) can be distinguished immediately adjacent to the subcutaneous adipose tissue (middle third of picture), dermal layer (lower third of picture), and epidermis (bottom of picture). The class coverslip must be imagined immediately on top of the skin muscle, separated only by a few micrometers of fat and/or matrix material (top of picture). The bar in the left lower corner represents 100 μm.

microdissection is performed using an operation microscope and microsurgical instruments. The tissue is covered with a few drops of normal saline to prevent desiccation. Next, the two layers of the retractor muscle are gently removed, leaving the remaining "contralateral" skin muscle and subcutaneous tissue for microcirculatory studies within the observation window (Fig. 3). At this point, the other frame (which contains the glass coverslip) is gently lowered onto the tissue and the surgical field is allowed to layer onto this frame by adhesion forces. Then, the two frames are approximated and secured using surgical sutures.

For the implantation of indwelling catheters, a mandrin and trocar is inserted immediately next to the cranial aspect of the chamber and led subcutaneously to the ventral aspect of the neck. The trocar is removed and a fine polypropylene catheter is inserted through the mandrin. The catheters are filled with normal saline and attached to a 1 ml syringe. The animal is then placed in a supine position. In order to accommodate the chamber in the dorsal skinfold, two plastic blocks are placed at a distance of 0.5 inches and covered by a surgical cloth. The fur may or may not be shaved in the anterior aspect of the neck. Where the mandrin pierces the skin, the skin is incised along the longitudinal axis of the animal over a length of 1 cm. The jugular vein is visualized by blunt dissection through the muscles of the neck and a silk suture placed around the vein at the caudal edge of the surgical field. The vein is then incised with fine microscissors; the catheter is inserted and gently pushed forward until it reaches the right atrium. For practical purposes, small pencil marks on the catheter help to accurately position the catheter within the jugular vein. The catheter is then fixed to the vein and surrounding tissues by prepositioned suture loops. A small loop of catheter is left in the surgical field to allow for sufficient mobility. The skin is closed at the neck. The catheter is sutured to the chamber through the most cranial suture hole (Fig. 2) and attached to the chamber using gauze tape.

The animals are then allowed to wake up and recover from surgery. The quality of the chamber preparation is visualized daily by transillumination at low magnification to identify evidence of inflammation (tortuous dilated vessels, edema, hemorrhage, pus, etc.). Only those chambers are used for intravital microscopic experiments which show no such features of inflammation (Fig. 4).

Techniques

Minor details often decide the quality of the chamber preparation and the outcome of the surgical endeavor. A major factor is the speed of the implantation. Ideally, the entire surgical implantation time for the skinfold chamber should not exceed 10–15 min. Meticulous care should be given not to irritate the skin prior to implantation. For instance, the skin in the area of the later observation field should not be touched, however gently, and the shaving and depilation should be done in as careful and nontraumatic a manner as possible. Clean instruments and chambers should be used (i.e., ultrasound cleaner), but they need not be sterilized. Failure of chamber implantation results from suboptimal surgical technique, not from contaminated instruments or chambers.

During implantation of the indwelling catheters, two aspects should be carefully observed: (1) Catheters should be gently pushed forward all the way into the right atrium. This is best accomplished by cutting the catheter tip at a fairly blunt angle (approximately 30°). Sometimes it helps to manipulate the ipsilateral forelimb of the animal during insertion of the catheter in order to obtain a better transition of the catheter from the jugular vein into the superior vena cava. (2) Meticulous attention should be paid to

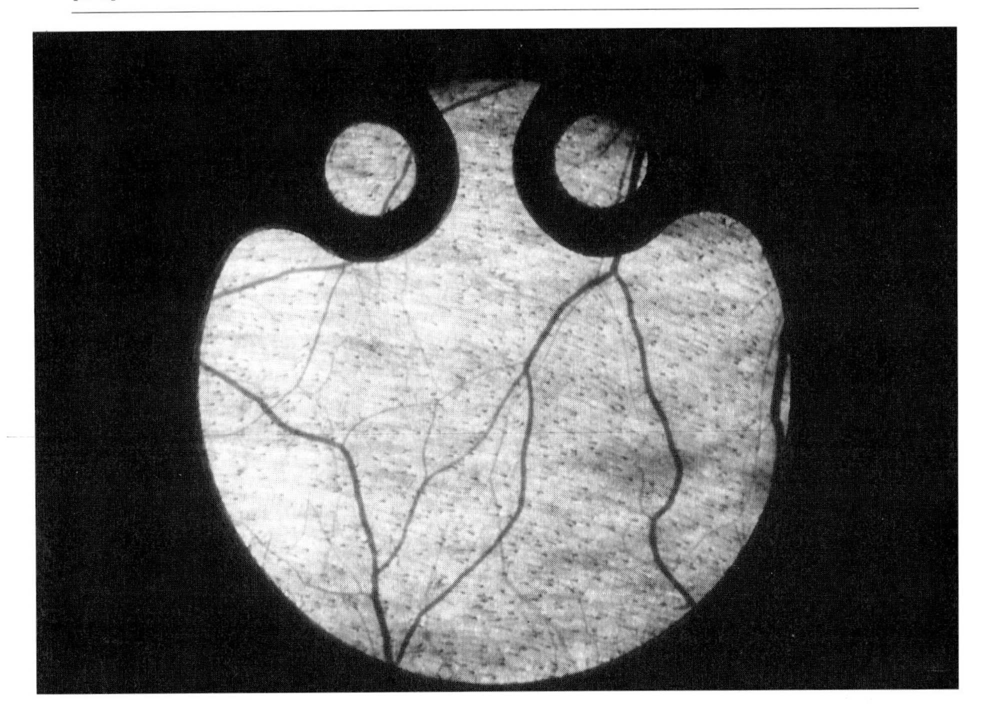

FIG. 4. Magnified view of the observation chamber within the skinfold chamber. The diameter of the chamber window is 11 mm. Note the fine, straight feeding arterioles and the broader, more irregular draining venules. Smaller vessels cannot be visualized at this magnification and image resolution. The small black dots represent hair follicles. The two circular areas at the top of the image represent the ends of the metal ring that holds the glass coverslip in place (item *c* in Fig. 2).

the firm attachment of the catheter to the chamber. This is best done by repetitively pulling the suture around the catheter and securing it with repeated series of surgical knots. Nothing is more frustrating than having a chamber well implanted and realizing during the experiment that the catheter is no longer in place because the animal had other plans for it.

Microscopic Observations

The intravital microscopic observations are performed in awake animals. This is best done during the early hours of the day, when the circadian rhythm of the animals has them mostly sleepy and inactive. The animals are gently positioned in a Plexiglas tube of an inner diameter that roughly corresponds to their circumference. The chamber protrudes out through a slit in this Plexiglas tube and can thus be examined under the microscope.

The tube is placed on a Plexiglas platform and secured using sutures and gauze tape. The platform is then placed under the microscope. Intravital microscopy can be performed either using transillumination with regular light sources or using epiillumination, which is best for fluorescence microscopy. For fluorescence microscopy, fluorescent markers [i.e., acridine orange, rhodamine 6G, fluorescein isothiocyanate (FITC) dextran] are injected i.v. via indwelling catheters (Fig. 5). Via these catheters drugs or other agents (i.e., inflammatory mediators) can be injected or infused using micropumps. The same vessel sites within an observation chamber can be examined over and over again, either by manually repositioning the platform or by using computer-assisted stepping motor-driven platforms. Depending on the time of day and the noise level in the laboratory, the animals usually fall asleep in the tubes after a few minutes of observation.

Role of ROS in Microcirculatory Dysfunction

Based on previous *in vitro* and *in vivo* findings, which had demonstrated a role of ROS in the induction of leukocyte rolling and adhesion,[28,9,10] we have stimulated leukocyte rolling and adhesion by ischemia and reperfusion[8,29,30] by injection of oxidized lipoproteins,[31,32] and by exposure of the animals to the smoke of research cigarettes.[33-35] By pretreatment of the animals with superoxide dismutase (SOD) or antioxidant vitamins, we could demonstrate a significant inhibition of leukocyte adhesion under the above pathophysiological conditions, supporting the role of ROS in these models of disease. Furthermore, we could observe that antioxidant treatment also prevented the formation in the bloodstream of leukocyte/platelet aggregates after injection of oxidized LDL[36] or after cigarette smoke exposure.[34] These leukocyte/platelet aggregates were blocked completely by pretreat-

[28] M. Suzuki, W. Inauen, P. R. Kvietys, M. B. Grisham, C. Meininger, M. E. Schelling, H. J. Granger, and D. N. Granger, *Am. J. Physiol.* **257,** H1740 (1989).

[29] H. A. Lehr, A. Guhlmann, D. Nolte, D. Keppler, and K. Messmer, *J. Clin. Invest.* **87,** 2036 (1991).

[30] M. Becker, M. D. Menger, and H. A. Lehr, *Am. J. Physiol.* **36,** H925 (1994).

[31] H. A. Lehr, C. Hübner, B. Finckh, S. Angermüller, D. Noltel, U. Beisiegel, A. Kohlschütter, and K. Messmer, *J. Clin. Invest.* **88,** 4 (1991).

[32] H. A. Lehr, M. Becker, S. L. Marklund, C. Hübner, K. E. Arfors, A. Kohlschütter, and K. Messmer, *Arterioscler. Thromb.* **12,** 824 (1992).

[33] H. A. Lehr, E. Kress, M. D. Menger, H. P. Friedl, C. Hübner, K. E. Arfors, and K. Messmer, *Free Rad. Biol. Med.* **14,** 573 (1993).

[34] H. A. Lehr, B. Frei, and K. E. Arfors, *Proc. Natl. Acad. Sci USA* **91,** 7688 (1994).

[35] H. A. Lehr, A. S. Weyrich, R. K. Saetzler, A. Jurek, K. E. Arfors, G. A. Zimmerman, S. M. Prescott, and T. M. McIntyre, *J. Clin. Invest.* **99,** 2358 (1997).

[36] H. A. Lehr, B. Frei, T. E. Carew, A. M. Olofsson, and K. E. Arfors, *Circulation* **91,** 1525 (1995).

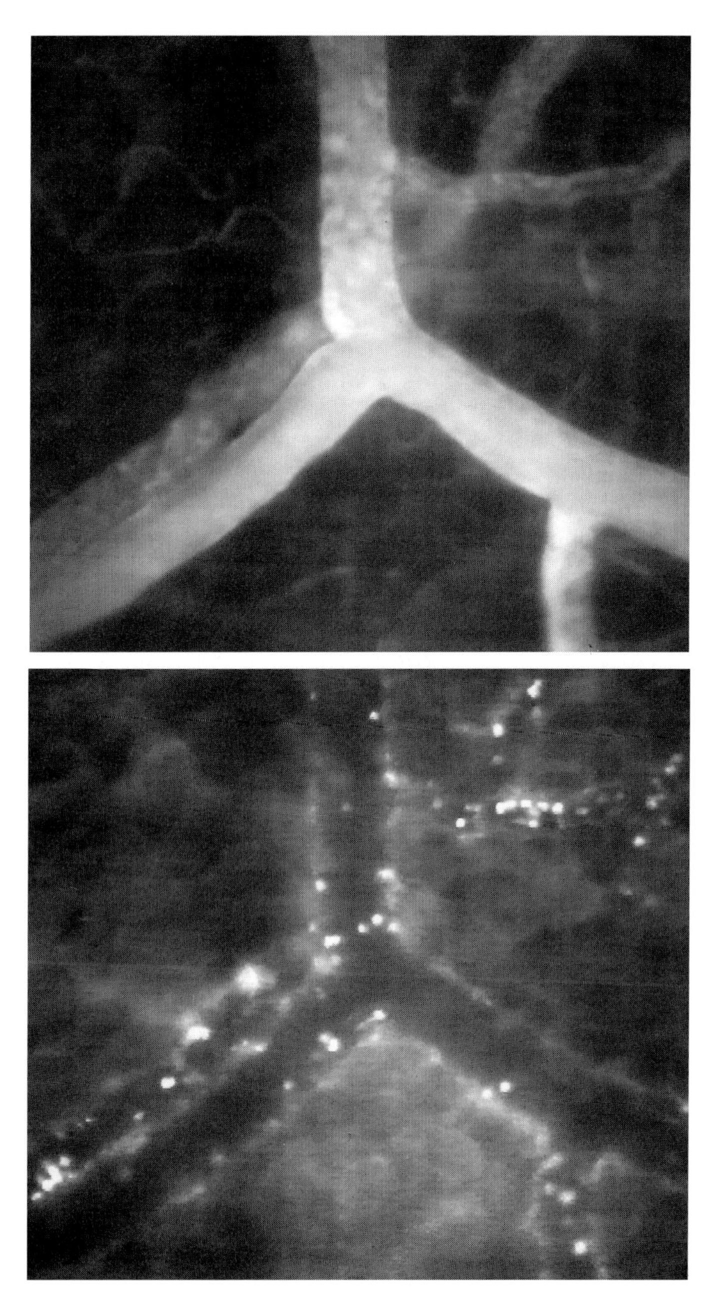

FIG. 5. Microvasculature of the striated skin muscle as visualized by *in vivo* epiillumination fluorescence microscopy after contrast enhancement with FITC-dextran (M_r 150,000, upper panel), or rhodamine 6G (lower panel). The white lines in the upper panel represent the inner diameter of the microvessels, the fluorescent marker being retained by the intact endothelial barrier found under physiological conditions. In the lower panel, the white dots represent leukoctes rolling or firmly adhering to the microvascular endothelium (nuclear staining with rhodamine 6G). Note the preferential adhesion of leukocytes in small postcapillary venules as seen in the upper right-hand corner (width of image is 0.45 mm).

ment of the animals with functionally blocking antibodies to P-selectin[37] or with platelet-activating factor receptor antagonists,[35,38] suggesting a role of P-selectin and platelet activating factor or related lipid mediators in their formation. Based on observations made in previous experiments, we propose that the formation of these leukocyte/platelet aggregates during oxidative stress is a prerequisite for the interaction of leukocytes with the arteriolar wall, where the high dispersal forces normally prevent the adhesion of individual leukocytes to endothelial cells.

Of particular interest for researchers interested in animal models of ROS-mediated tissue damage is a study in which we demonstrated the impact of vitamin E supplements in animal diets on leukocyte adhesion and functional capillary density after ischemia and reperfusion. In this study,[7] we found that supplementation of animal diets with as little as 60 mg/kg diet significantly attenuates postischemic leukocyte adhesion and that no further inhibition could be obtained by further raising the vitamin E supplement even into megadose ranges. From this and other analogous studies,[5] we conclude that the vitamin E supplement in animal diets (which range from 20 to 200 mg/kg in different countries) may jeopardize the outcome in animal models of ROS-mediated tissue injury.[39]

Model 2: Intravital Microscopy of Liver Surface

History

The study of hepatic microcirculation by intravital microscopy was first described in the frog by Knisely in 1936.[40] Further studies were performed in frogs,[41–42] mice,[42,43] rats,[42,44,45] hamsters,[45] guinea pigs,[46] rabbits,[42] dogs, and monkeys.[41] The hepatic microcirculation can be visualized by intravital microscopy using trans- or epiillumination techniques. Most of our current knowledge about the microvascular network of the liver is based on early investigations using transillumination. These early studies were directed

[37] H. A. Lehr, A. M. Olofsson, T. E. Carew, P. Vajkoczy, U. von Andrian, C. Hübner, M. C. Berndt, D. Steinberg, K. Messmer, and K. E. Arfors, *Lab. Invest.* **71,** 380 (1994).

[38] H. A. Lehr, J. Seemüller, C. Hübner, M. D. Menger, and K. Messmer, *Arterioscler. Thromb.* **13,** 1013 (1993).

[39] H. A. Lehr, P. Vajkoczy, M. D. Menger, and K. E. Arfors, *Free Radic. Biol. Med.* in press (1998).

[40] M. H. Knisely, *Anat. Rec.* **64,** 499 (1936).

[41] E. H. Bloch, *Angiology* **6,** 340 (1955).

[42] R. S. McCuskey, *Am. J. Anat.* **119,** 3455 (1966).

[43] A. M. Rappaport, *Bibl. Anat.* **16,** 116 (1977).

[44] A. Koo and I. Y. S. Liang, *J. Physiol.* **295,** 191 (1979).

[45] M. D. Menger, I. Marzi, and K. Messmer, *Eur. Surg. Res.* **23,** 158 (1991).

[46] J. W. Irwin and J. MacDonald, *Anat. Rec.* **117,** 1 (1953).

toward the microvascular anatomy[41,46] and in particular toward hepatic arterioles and sphincters.[41,42] Yet, even in rodents, in which the liver capsule is thin, the use of transillumination techniques is restricted to the most peripheral edges of liver lobes. As a consequence, the number of microvascular segments accessible for intravital microscopic studies is strictly limited. Moreover, in contrast to central aspects of the liver surface, peripherally located acini are known for their marked degree of perfusion heterogeneity—even under physiological conditions[47]—and cannot be considered representative of the liver microcirculation as a whole.

The introduction of fluorescent dyes and refined epiillumination techniques has significantly advanced the possibilities of intravital microscopic studies of the liver microcirculation, including quantitative analyses of both circulatory parameters and cellular mechanisms.[48] Using appropriate dyes, intravital microscopy allows the study of (i) microvascular (sinusoidal) perfusion, (ii) leukocyte–endothelial cell interaction, (iii) Kupffer cell and Ito cell function, (iv) hepatocellular membrane damage and mitochondrial redox state, (v) vascular–hepatocellular–canalicular bile transport, as well as (vi) microlymphatic drainage.[49–51]

Surgical Techniques

The epiillumination technique can be applied to the liver surface of virtually any small laboratory animal, including rats, mice, hamsters, and guinea pigs. In larger animals, such as rabbits and pigs, the thickness of the liver capsule, which cannot be penetrated by high magnification optics, limits the application of the epiillumination technique.

For *in vivo* investigations, animals are anesthesized with pentobarbital (50 mg/kg body weight) or chloral hydrate (30 mg/kg body weight) intraperitoneally, and placed in supine position on a heating pad. If necessary, mechanical ventilation can be applied via a tracheal tube after tracheotomy or oral intubation.[52,53] Even when no mechanical ventilation is intended, tracheotomy should be performed in order to facilitate spontaneous breathing.[51] One carotid artery and one jugular vein are cannulated with fine

[47] H. Eguchi, P. A. McCiskey, and R. S. McCuskey, *Hepatology* **13,** 751 (1991).

[48] M. D. Menger and H. A. Lehr, *Immunol. Today* **14,** 519 (1993).

[49] M. D. Menger, B. Vollmar, J. Glasz, P. Post, and K. Messmer, *Prog. Appl. Microcirc.* **19,** 106 (1993).

[50] B. Vollmar, M. Burkhardt, T. Minor, H. Klauke, and M. D. Menger, *Microvasc. Res.* **54,** 164 (1997).

[51] B. Vollmar, B. Wolf, S. Siegmund, A. Katsen, and M. D. Menger, *Am. J. Pathol.* **151,** 169 (1997).

[52] B. Vollmar, G. Lang, M. D. Menger, and K. Messmer, *Am. J. Physiol.* **266,** H1927 (1994).

[53] B. Vollmar, G. Preissler, and M. D. Menger, *Am. J. Physiol.* **267,** H1936 (1994).

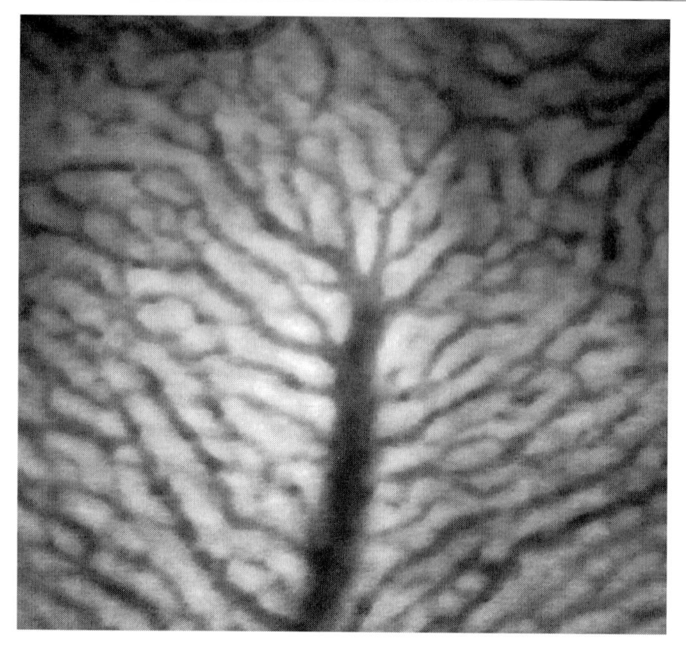

FIG. 6. Hepatic microvasculature as visualized by *in vivo* epiillumination fluorescence microscopy after contrast enhancement with intravenously administered sodium fluorescein. Low magnification (upper panel) demonstrates the hexagonal arrangement of the liver acini. Higher magnification allows differentiation between periportal, midzonal, and pericentral parts of the hepatic sinusoids, draining into postsinusoidal venules (middle panel). Microcirculatory parameters are analyzed quantitatively in high magnification (lower panel), which allows estimation of the individual perfusion pattern as well as distinct changes of perfusion under pathological conditions. Under physiological conditions (as presented in these images), all of the sinusoids are perfused (width of image is 3.6 mm in upper panel, 0.9 mm in middle panel, and 0.45 mm in lower panel).

polyethylene tubes (PE-50; inner diameter 0.58 mm), allowing for macrohemodynamic monitoring (mean arterial blood pressure, heart rate, and central venous pressure) and providing a route for administration of fluorescent dyes and pharmacological compounds. After transverse laparotomy, the animal is positioned on its left side, and the liver is prepared for intravital microscopy by placing the left lobe on a plastic disk held by an adjustable stage attached to the heating pad.[54] This way, the lower surface of the liver is situated horizontal to the microscope, yielding an adequately homogeneous focus level for the microscopic observation of the area of liver surface under investigation. The micromanipulator-controlled adjustment of the

[54] B. Vollmar, J. Glasz, R. Leiderer, S. Post, and M. D. Menger, *Am. J. Pathol.* **145,** 1421 (1994).

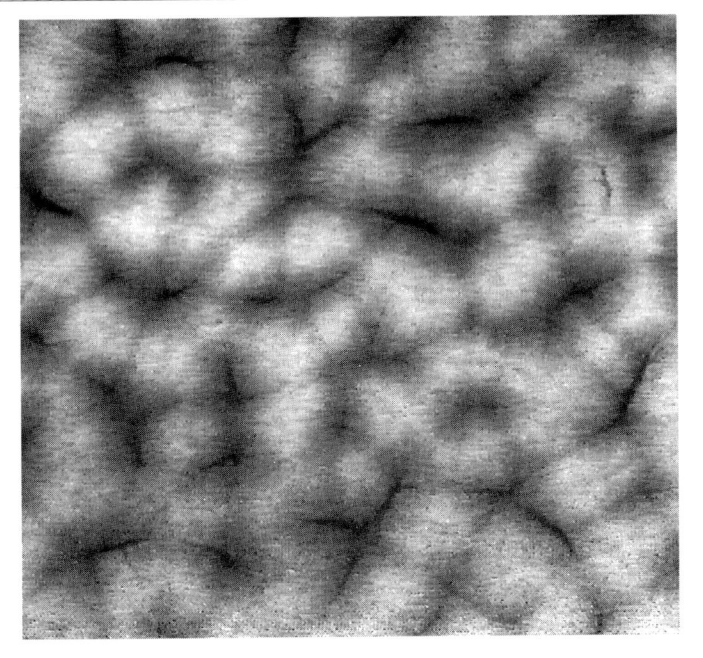

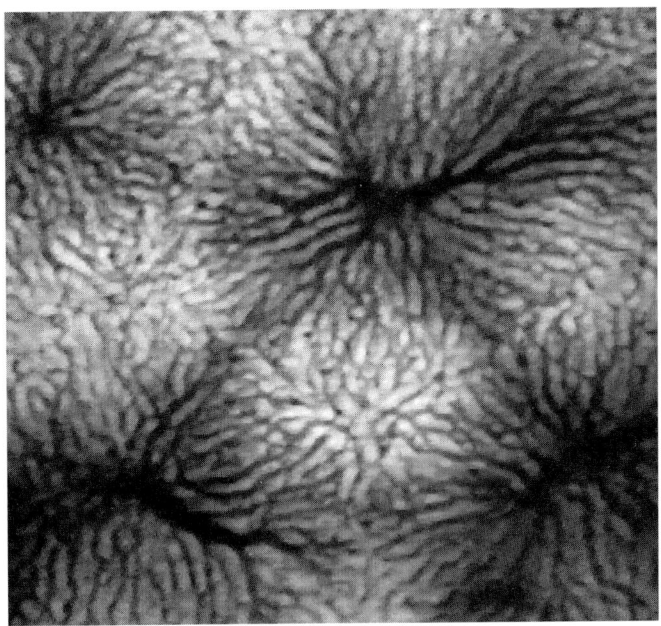

FIG. 6 (*continued*)

plastic disk in three dimensions allows to avoid mechanical obstruction of the feeding and draining macrovessels and to virtually exclude respiratory movements of the lobe, which would periodically move the tissue under investigation out of the focus level. The exposed area of the left liver lobe is immediately covered with a glass slide to prevent the influence of ambient air and desiccation of the tissue.[55] Alternatively, the exposed area of the liver may be superfused, as practised routinely in other tissue preparations for intravital microscopy, such as the mesentery or the cremaster muscle.[56,57]

Intravital Microscopy

Intravital fluorescence microscopy can be performed with almost all types of epiillumination microscopes available. Care must be given to use a microscope setup in which the distance between the lenses and the microscope desk is not restricted to a few centimeters (in commercially available systems usually 4 cm), but can be chosen freely up to distances of 20 cm. This distance is required in order to allow for adjustment of the stage—on which the animal is positioned—in three dimensions in order to obtain the best position of the liver lobe under the microscope lens.[58] Under ideal conditions, the entire surface area of the liver lobe can be studied in this way.

The quantitative analysis of the hepatic microcirculation includes the perfusion of sinusoids and of postsinusoidal venules as visualized after intravenous administration of sodium fluorescein. In this way, unstained red blood cells displace the plasma marker within the microvascular space and thus provide a distinct negative contrast to the hepatocellularly transported sodium fluorescein[54] (Fig. 6). The intravenous application of the fluorescent markers acridine orange or rhodamine 6G allows for *in vivo* staining of platelets and leukocytes, and thus for the quantitative analysis of the interaction of these cells with each other and/or with the endothelial lining of the microvasculature[59] (Fig. 7). Application of the nuclear marker bisbenzamid allows the visualization of hepatocyte distribution *in vivo*[60] (Fig. 8), and the hepatocellular transport may be estimated by intracellular brightness after application of sodium fluorescein or FITC-labeled bile

[55] B. Vollmar, S. Richter, and M. D. Menger, *Am. J. Physiol.* **270,** G798 (1996).

[56] M. B. Lawrence, G. S. Kansas, E. J. Kunkel, and K. Ley, *J. Cell Biol.* **136,** 717 (1997).

[57] A. R. Pries, D. Schönfeld, P. Gaehtgens, M. F. Kiani, and G. R. Cokelet, *Am. J. Physiol.* **272,** H2716 (1997).

[58] S. Post, P. Palma, M. Rentsch, A. P. Gonzalez, and M. D. Menger, *Prog. Appl. Microcirc.* **19,** 152 (1993).

[59] B. Vollmar, M. D. Menger, J. Glasz, R. Leiderer, and K. Messmer, *Am. J. Physiol.* **267,** G786 (1994).

[60] B. Vollmar, M. Rücker, and M. D. Menger, *Microvasc. Res.* **51,** 250 (1996).

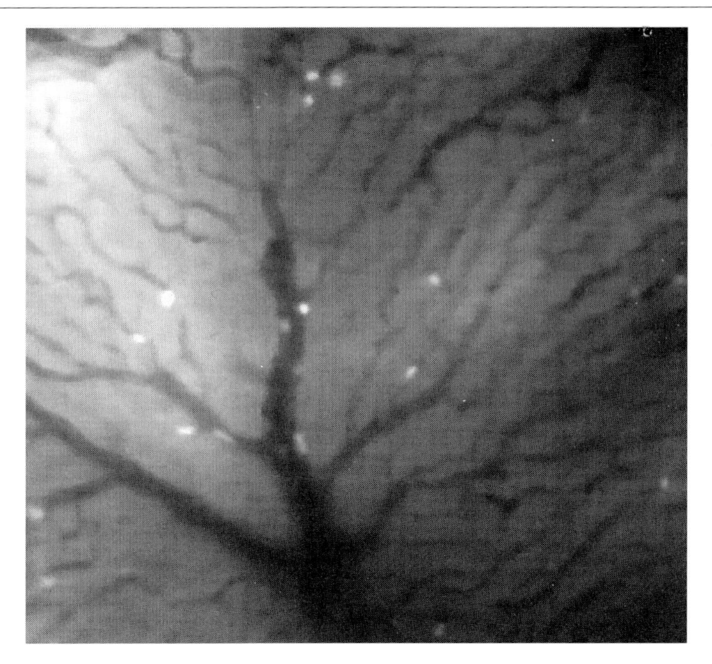

FIG. 7. Hepatic microvasculature after ischemia/reperfusion as visualized by *in vivo* epiillumination fluorescence microscopy after contrast enhancement with intravenously administered acridine orange. Note circumscribed sinusoidal perfusion deficits and microvascular accumulation of leukocytes (white spots due to nuclear staining with acridine orange), which are rolling along and firmly adhering to the endothelial lining of the microvessels (width of image is 0.45 mm).

salts. This way, intravital microscopy allows to study not only hepatocyte morphology, but also hepatocyte function.[49] Moreover, hepatocellular excretion function can be assessed by imaging the accumulation of the fluorescent dyes within the fine bile canalicular network[49] and the quantification of NADH (which exhibits blue autofluorescence) allows to assess hepatocellular mitochondrial function, and thus the redox state of the cells by the intravital fluorescence microscopic approach.[50] Intravenous injection of fluorescently labeled latex particles allows the estimation of phagocytic function of Kupffer cells[61] (Fig. 9), and the function of Ito cells can be visualized because of the autofluorescence of vitamin A by excitation with ultraviolet light[60] (Fig. 10). Of particular interest for the purpose of this review, the generation of ROS in microvessels and different cell populations may be visualized *in vivo* on the basis of oxygen radical-dependent photo-

[61] B. Vollmar, J. Glasz, S. Post, and M. D. Menger, *J. Hepatol.* **20,** 301 (1994).

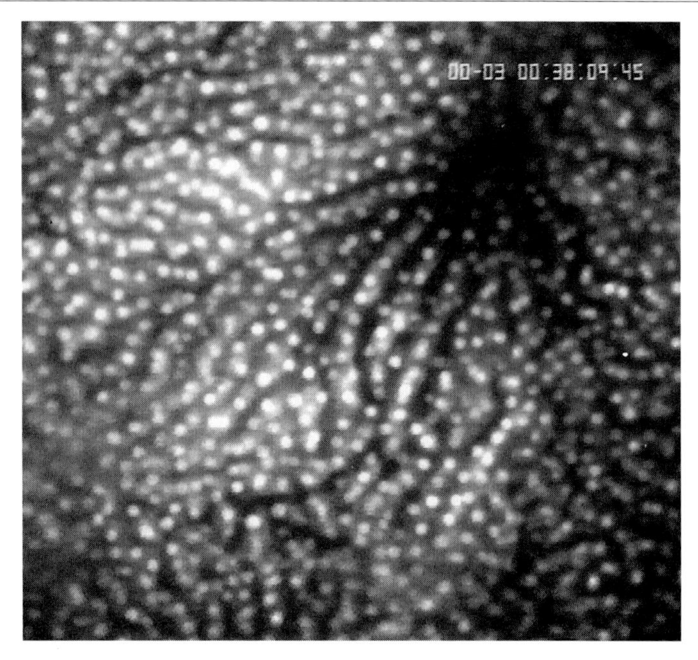

FIG. 8. Hepatic microvasculature as visualized by *in vivo* epiillumination fluorescence microscopy after intravenous administration of bisbenzamide. Under physiological conditions, this dye stains exclusively hepatocellular nuclei. Note the homogeneous hepatocyte distribution within the intersinusoidal spaces (width of image is 0.45 mm).

emission.[62,63] Finally, intravenous injection of propidium iodide or ethidium bromide allow *in vivo* detection of hepatocellular necrosis, indicated by the nuclear accumulation of the dyes in injured cells.[49]

Intrahepatic lymphatic transport, which is minimal in normal livers but markedly enhanced in liver disease such as cirrhosis, can be studied by fluorescently labeled macromolecules, including FITC-dextrans.[51] Overall, this broad spectrum of circulatory and cellular parameters, which can be analyzed simultaneously *in vivo,* makes the epiillumination fluorescence microscopic approach an interesting tool to study the pathophysiological mechanisms underlying hepatic disease.

Role of ROS in Hepatic Microcirculatory Dysfunction

A considerable number of studies have demonstrated that in ischemia/reperfusion ROS play a pivotal role to trigger manifestation of injury

[62] M. Suematsu, G. W. Schmid-Schonbein, R. H. Chavez-Chavez, T. T. Yee, T. Tamatani, M. Miyasaka, F. A. Delano, and B. W. Zweifach, *Am. J. Physiol.* **264,** H881 (1993).

[63] M. Tsuchiya, M. Suematsu, and H. Suzuki, *Methods Enzymol.* **233,** 128 (1994).

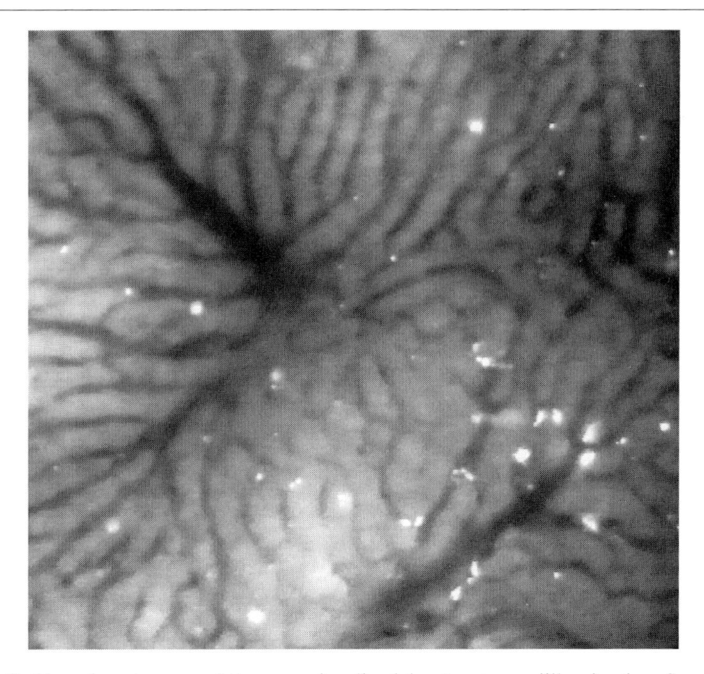

FIG. 9. Hepatic microvasculature as visualized by *in vivo* epiillumination fluorescence microscopy after intravenous administration of fluorescently labeled latex particles, which allows quantitative assessment of Kupffer cell phagocytic activity. Note that under normal conditions particles are primarily phagocytized by periportally located Kupffer cells (right lower aspect of the figure), while only few phagocytized particles are observed in the pericentral areas of the acinus (left upper aspect) (width of image is 0.45 mm).

(reviewed in Eppiheimer and Granger[10]). In the liver this is of clinical relevance during major hepatic surgery and, in particular, during hepatic transplantation. ROS are thought to mediate leukocytic activation and their infiltration into tissue, and application of radical scavengers successfully prevents the leukocytic infiltrative response.[64] Indeed, hepatic ischemia/reperfusion is associated with pronounced microvascular leukocyte recruitment and leukocyte–endothelial cell interaction in sinusoids and postsinusoidal venules.[54,55] These studies suggest that in liver tissue also, the microvasculature represents the primary target for the ischemia/reperfusion-induced inflammatory injury. The action of the ROS may include the activation of adhesion molecules, as suggested from *in vitro* studies. For instance, we could demonstrate using intravital microscopic studies

[64] H. Komatsu, A. Koo, E. Ghadishah, H. Zeng, J. F. Kuhlenkamp, M. Inoue, P. H. Guth, and N. Kaplowitz, *Am. J. Physiol.* **262,** G669 (1992).

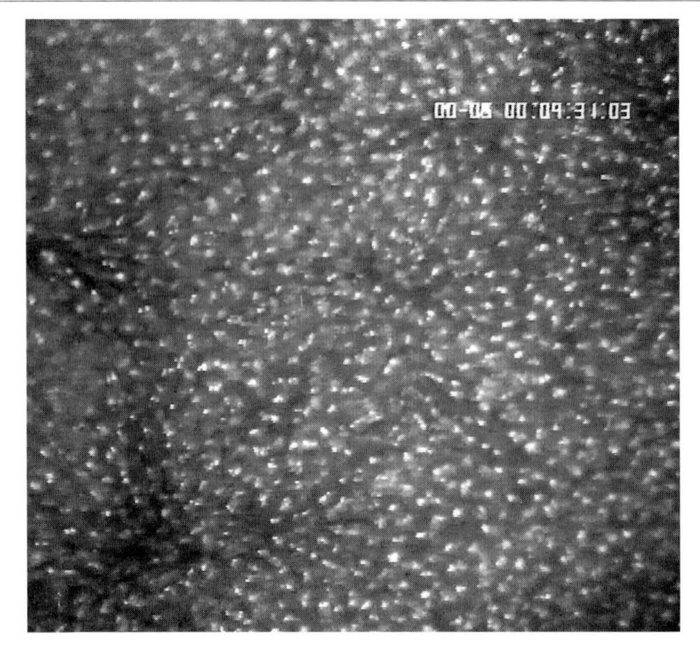

FIG. 10. Hepatic microvasculature as visualized by *in vivo* UV light epiillumination fluorescence microscopy. Autofluorescence of vitamin A indicates appropriate storage function of Ito cells, which—under normal conditions—are homogeneously distributed over the three zones of the hepatic acini (width of image is 0.9 mm.)

following experimental hepatic ischemia/reperfusion that application of functionally blocking monoclonal antibodies against adhesion molecules such as the intercellular adhesion molecule 1 (ICAM-1) effectively prevents leukocyte adherence and consequently the manifestation of hepatic tissue injury.[65]

Apart from the leukocytic response (reflow-paradox[8]), ischemia/reperfusion is also associated with nutritive perfusion failure (no-reflow[66]), resulting in perfusion deficits of individual sinusoids.[54] Because studies have demonstrated that these perfusion deficits are not necessarily linked to the infiltration and/or action of leukocytes, which lodge within the sinusoidal lumen,[55] the two phenomena have to be considered as two disparate entities of microvascular reperfusion injury. Despite their different nature, both types of injury seem to be linked to the formation of ROS during the postischemic reperfusion period. This view is based on intravital micro-

[65] B. Vollmar, J. Glasz, M. D. Menger, and K. Messmer, *Surgery* **117,** 195 (1995).
[66] M. D. Menger, D. Steiner, and K. Messmer, *Am. J. Physiol.* **263,** H1892 (1992).

both sinusoidal perfusion failure and leukocyte adherence to the hepatic microvascular endothelium after either warm or cold hepatic ischemia/reperfusion.[67–69] These results have an immediate clinical impact since they advocate the administration of ROS scavengers during reperfusion of transplanted livers as a promising approach for the prevention of posttransplant inflammatory (=ischemia/reperfusion) injury and primary nonfunction of the organ grafts. Indeed, intravital microscopic studies have shown that addition of antioxidative substances to preservation solutions (i.e., University of Wisconsin solution) and to rinse solutions (i.e., Caroline rinse) affords significant protective effects against the postischemic manifestation of experimental hepatic ischemia/reperfusion injury.[70,71]

Furthermore, ROS may not only activate circulating leukocytes but also Kupffer cells, the sessile macrophages of the liver. Although *in vivo* studies after warm ischemia/reperfusion demonstrated reduced phagocytic activity of Kupffer cells,[61] there is major evidence that Kupffer cells contribute to the systemic inflammatory response.[72] Correspondingly, experiments with cold hepatic ischemia/reperfusion showed marked activation of Kupffer cells as indicated by a significantly increased phagocytic activity.[73] The role of ROS in this scenario was suggested from studies in which rinsing the liver grafts with antioxidative compounds before the onset of reperfusion prevented excessive Kupffer cell activation and which was associated with a significant attenuation of postischemic liver injury.[71]

[67] I. Marzi, J. Knee, V. Bühren, M. Menger, and O. Trentz, *Surgery* **111,** 90 (1992).
[68] A. Koo, H. Komatsu, G. Tao, M. Inoue, P. H. Guth, and N. Kaplowitz, *Hepatology* **15,** 507 (1992).
[69] M. J. Müller, B. Vollmar, H. P. Friedl, and M. D. Menger, *Free Rad. Biol. Med.* **21,** 189 (1996).
[70] S. Post, M. Rentsch, A. P. Gonzalez, P. Palma, G. Otto, and M. D. Menger, *Transplantation* **55,** 972 (1993).
[71] S. Post, P. Palma, M. Rentsch, A. P. Gonzalez, and M. D. Menger, *Hepatology* **18,** 1490 (1993).
[72] G. Wanner, W. Ertel, P. Müller, Y. Höfer, R. Leiderer, M. D. Menger, and K. Messmer, *Shock* **5,** 34 (1996).
[73] S. Post, A. P. Gonzalez, P. Palma, M. Rentsch, A. Stiehl, and M. D. Menger, *Hepatology* **16,** 803 (1992).

Author Index

Numbers in parentheses are footnote reference numbers and indicate that an author's work is referred to although the name is not cited in the text.

A

B

C

D

E

F

G

H

I

L

M

O

P

S

T

W

X

Y

Z

Subject Index

A

B

C

D

E

H

I

J

L

M

N

O

P

R

S

T

U

V

W

ISBN 0-12-182201-X